ANNUAL REVIEW OF BIOPHYSICS AND BIOPHYSICAL CHEMISTRY

ANNUAL REVIEW OF BIOPHYSICS AND BIOPHYSICAL CHEMISTRY

VOLUME 15, 1986

DONALD M. ENGELMAN, *Editor*
Yale University

CHARLES R. CANTOR, *Associate Editor*
Columbia University

THOMAS D. POLLARD, *Associate Editor*
The Johns Hopkins University School of Medicine

ANNUAL REVIEWS INC.   4139 EL CAMINO WAY   PALO ALTO, CALIFORNIA 94306 USA

144655/.

R̸ ANNUAL REVIEWS INC.
Palo Alto, California, USA

International Standard Serial Number: 0883-9182
International Standard Book Number: 0-8243-1815-3
Library of Congress Catalog Card Number: 79-188446

TYPESET BY AUP TYPESETTERS (GLASGOW) LTD., SCOTLAND
PRINTED AND BOUND IN THE UNITED STATES OF AMERICA

PREFACE

The subject of biophysics and biophysical chemistry is biology. We take the editorial position that biology is not to be used as an excuse for studying the properties of physical methods or for exploring issues in physical chemistry, however interesting the methods or physical/chemical concerns may be. Rather, we stress the development of new physical methods, the establishment of theoretical principles, the use of physical methods to obtain important biological results, and the use of physical hypotheses to explore and understand biological phenomena.

We are pleased to open this volume with Linus Pauling's recollections of his early days at the California Institute of Technology. In the remainder of the volume we present articles covering the four areas mentioned above.

Several articles discuss new methods that hold special promise for novel biological applications. The articles by Chiu and Wall & Hainfeld concern new developments in electron microscopy. The article by Avison, Hetherington & Shulman gives perspective on new applications of NMR in metabolic studies. Gray & Langlois explore the use of flow cytometry and sorting in chromosome studies.

Theory is represented in articles by Goad on computational analysis of gene sequences, by Honig, Hubbell & Flewelling on electrostatic interactions, by Levitt on ion channel analysis, and by Engelman, Steitz & Goldman on the prediction of some membrane protein structures from amino acid sequences. Each concerns an application of a theoretical approach to a specific, relevant issue in biology.

Applications of physical methods are reviewed in Wiegand & Remington's article on crystallographic studies of citrate synthase and in Coronado's discussion of reconstitution of ion channels.

The use of perspectives from biophysics and biophysical chemistry in understanding biological phenomena is a major component of the reviews of the nuclear matrix by Nelson, Pienta, Barrack & Coffey, of antifreeze glycoproteins by Feeney, Burcham & Yeh, of current views of the *lac* permease by Kaback, of halorhodopsin by Lanyi, of issues in muscular contraction by Hibberd & Trentham, and of recombinant lipoproteins by Atkinson & Small.

In future volumes we hope to continue the coverage of each of these areas that we conceive to be the province of modern biophysics and biophysical chemistry. We welcome our readers' suggestions for specific topics and authors.

THE EDITORS

Annual Review of Biophysics and Biophysical Chemistry
Volume 15, 1986

CONTENTS

SOME RELATED ARTICLES APPEARING IN OTHER *ANNUAL REVIEWS*

From the *Annual Review of Biochemistry*, Volume 55 (1986)

From the *Annual Review of Microbiology*, Volume 39 (1985)

From the *Annual Review of Physical Chemistry*, Volume 36 (1985)

Fluorescence Correlation Spectroscopy and Photobleaching Recovery, Elliot L. Elson

Theory of Hydrophobic Effects, Lawrence R. Pratt

From the *Annual Review of Physiology*, Volume 48 (1986)

Electrical Regulation of Sperm-Egg Fusion, Laurinda A. Jaffe and Nicholas L. Cross

Mimicry and Mechanism in Phospholipid Models of Membrane Fusion, R. P. Rand and V. A. Parsegian

Intracellular pH Regulation by Vertebrate Muscle, C. Claire Aickin

Intracellular pH Regulation in Epithelial Cells, Walter F. Boron

Mechanisms and Consequences of pH-Mediated Cell Regulation, William B. Busa

ATP-Driven H^+ Pumping into Intracellular Organelles, Gary Rudnick

From the *Annual Review of Plant Physiology*, Volume 37 (1986)

Biophysical Control of Plant Cell Growth, Daniel Cosgrove

Plant Chemiluminescence, Fred B. Abeles

From the *Annual Review of Cell Biology*, Volume 1 (1985)

Protein Localization and Membrane Traffic in Yeast, Randy Schekman

Acetylcholine Receptor Structure, Function, and Evolution, Robert M. Stroud and Janet Finer-Moore

Stabilizing Infrastructure of Cell Membranes, V. T. Marchesi

From the *Annual Review of Genetics*, Volume 19 (1985)

Selected Topics in Chromatin Structure, Joel C. Eissenberg, Iain L. Cartwright, Graham H. Thomas, and Sarah C. R. Elgin

Ann. Rev. Biophys. Biophys. Chem. 1986. 15 : 1–9

EARLY DAYS OF MOLECULAR BIOLOGY IN THE CALIFORNIA INSTITUTE OF TECHNOLOGY

Linus Pauling

Linus Pauling Institute of Science and Medicine, 440 Page Mill Road, Palo Alto, California 94306

I think that it might be said, and supported to some extent, that molecular biology originated in the California Institute of Technology.

I have been thinking about why chemistry and other sciences developed so rapidly in California from about 1920 on. I attribute much of the progress to the influence of Gilbert Newton Lewis, Dean of the College of Chemistry of the University of California in Berkeley, and Arthur Amos Noyes, Chairman of the Division of Chemistry and Chemical Engineering in the California Institute of Technology. These two men may have developed similar ideas because of their earlier association when Gilbert Newton Lewis was Vice-Director of the Research Laboratory for Physical Chemistry in the Massachusetts Institute of Technology, which Noyes had founded and directed until he left MIT to take a full-time job in the California Institute of Technology (then Throop College of Technology).

I had an opportunity in 1929 to observe the difference between life in Pasadena or Berkeley, for a chemist, and life at Harvard University, which may have been reasonably representative of other places. In Berkeley and in Pasadena a staff member in chemistry or a graduate student could obtain from the stockroom, without charge, whatever chemicals and apparatus he might want to work with. If he needed some item that was not too expensive, it would be ordered for him. I learned that at Harvard the situation was different. Every item obtained from the stockroom was charged to the professor or the student. In fact, the prices were high; the professors and students often went to the trouble of going to Boston to buy the items from a dealer.

1

I am sure that things have changed now, when so much research at the California Institute of Technology, the University of California, and Harvard is supported by grants of one sort or another. Probably in all institutions charges are made for items required in research. On thinking back I may say that I am sure, as I consider my impecunious situation from 1922 to 1925, that my research would have been hampered if I had had to think about the relative costs of alternative investigations that I might have wanted to undertake.

Another important point relates to freedom of inquiry. I remember some decades ago when I learned that in one university it was almost impossible for a student working toward a doctorate in chemistry to shift from one professor to another or from one subdepartment to another, such as from analytical chemistry to physical chemistry. I was in the Division of Chemistry and Chemical Engineering of the California Institute of Technology. There were no departments or subdepartments. Moreover, a student working for a PhD degree often carried on part of his research, perhaps for one year or two years, with one member of the Division and then shifted to another for another part of his training. As far as I was aware, there was no restriction on shifts of this sort.

The same spirit operated when, in 1929, the Division of Biology began to be active. Thomas Hunt Morgan and most of his group that was involved in the discovery of the gene had just come to Pasadena from Columbia University. From the beginning, probably under the influence of A. A. Noyes, there was close cooperation between the biologists and the chemists. Biochemistry was taught in the Division of Biology, and a good bit of biochemistry was in fact practiced in the Division of Chemistry and Chemical Engineering. Students in biochemistry were expected to take courses in chemical thermodynamics in the Division of Chemistry and Chemical Engineering. Students sometimes worked with a professor in the Division of Biology and obtained a PhD degree in chemistry. I myself had two students who carried out their doctoral research with me and obtained PhDs in biology, as well as two who obtained PhD degrees in physics and two who obtained PhD degrees in geology. The Chemistry seminars were attended by biologists and the Biology seminars were attended by chemists.

I had been receiving grants for several years from the Rockefeller Foundation, initially for support of my work on the crystal structure of the sulfide minerals, then for work on the magnetic properties of hemoglobin, and then for more general aspects of chemistry in relation to biology such as the determination by X-ray diffraction of the structures of amino acids and peptides to obtain information that we hoped would lead ultimately to the determination of the structure of proteins. In 1946 George Beadle, who was Chairman of the Division of Biology, and I submitted to the Rockefeller

Foundation a joint application for a grant for research in the borderline field between chemistry and biology. Earlier the Rockefeller Foundation had given a large sum of money to the California Institute of Technology to permit us to strengthen organic chemistry. We were successful in getting grants for chemistry and biology from the Rockefeller Foundation, and as far as I remember there was never any dispute between Beadle and me about how the available funds should be divided.

From 1929 on I became very interested in the work that the outstanding investigators in biology were conducting. Thomas Hunt Morgan was studying the phenomenon of self-sterility in the sea squirt. These animals are hermaphroditic. The sperm of one sea squirt will fertilize the eggs of almost any other sea squirt, perhaps 99.9% of them (I am not sure of this number, after 50 years), but the sperm will not fertilize the eggs of the same individual. I, of course, thought that this phenomenon must have a molecular basis, but I did not see how to go about studying it. There may be a good understanding of it now—I do not know.

Albert Tyler, who was a graduate student when he came to Pasadena with Morgan, was working on the entry of the sperm of the sea urchin into the ovum. In the course of his investigation he was able to show that the sea urchin produces two substances that combine with one another. I think that he may have mentioned to me in the early 1930s that they interacted similarly to an antigen and the homologous antibody.

I had been pleased to receive a grant from the Rockefeller Foundation to support my investigations of the crystal structure of the sulfide minerals (molybdenite, pyrite, chalcopyrite, sulvanite, enargite, binnite, and others). I still am interested in this problem, which in my opinion has not yet been solved in a thoroughly satisfactory manner (9). I did not object at all, however, when Warren Weaver, who had been an instructor in physics in the California Institute of Technology before I arrived in 1922 and was then head of one division of the Rockefeller Foundation, pointed out to me that the Rockefeller Foundation was not very interested in the sulfide minerals. He suggested that biochemistry, for example questions dealing with proteins, came closer to their interest. I was beginning to get interested in these biological questions and beginning to develop the feeling that a structural chemist might be able to contribute to their solution. Accordingly I applied for another grant, saying that I believed that I could answer the question of how the oxygen molecules are attached to the hemoglobin molecule in oxyhemoglobin by studying the magnetic properties of hemoglobin and its derivatives. I had had practically no experience in either the protein field or the field of magnetic properties, except that one of my students had measured the magnetic susceptibility of one substance at my suggestion, but the Rockefeller Foundation gave us the money. Charles Coryell then came to work with me, and he and I were able to interpret the results of the magnetic measurements of hemoglobin, carbon monoxyhemoglobin, and oxyhemoglobin to settle pretty well the question of how the oxygen molecules are attached. One of the oxygen atoms of O_2 is attached to the iron atom by a bond with some double-bond character, with predicted Fe-O-O angle of 136°, later verified by X-ray diffraction (8, 12).

It turned out that we had also made a significant contribution to experimental techniques for studying proteins, especially the heme proteins. By measuring the magnetic susceptibility we could determine equilibrium constants for many reactions involving addition of molecules or ions to ferrohemoglobin or ferrihemoglobin, and we could also measure rates of reactions. Theorell came from Sweden to work with us for one month, and then on his return he applied the magnetic technique to the study of the cytochromes.

We made a remarkable, completely unexpected discovery. I knew that there were two kinds of compounds of iron, either ferric or ferrous, those with high paramagnetic susceptibility and those with low paramagnetic susceptibility (zero in the case of the ferrous compounds, such as ferrocyanide). I had not anticipated that the iron atoms in hemoglobin would change from one type, with high spin and high magnetic moment, to the other type, with low spin or zero spin, when a ligand is added; but this is what we found (12).

By interpreting the magnetic measurements, Coryell and I were able to get a great deal of information about the two acidic groups associated with each iron atom in hemoglobin and about other aspects of the structure and properties of the substance (1). Also, I had a student named Robert C. C. St. George make magnetic measurements of various isocyanides combining with heme and also with hemoglobin. The equilibrium constants differed in a striking way, permitting us to conclude that the heme iron is buried within the globin molecule, fitting sufficiently tightly into a cavity that a larger group is partially inhibited from combining with the iron atoms (16). This result was substantiated, of course, when the structures of myoglobin and hemoglobin were determined later.

In addition to Coryell and Robert C. C. St. George, a number of other graduate students, postdoctoral fellows, and visitors to the California Institute of Technology worked on the magnetic properties of compounds of hemoglobin, including Fred Stitt, Richard W. Dobson, Charles D. Russell, and Allen Lein. Lein, who was at CIT for a year on sabbatical leave, studied the combining power of myoglobin for alkyl isocyanides and was able to show that, as with hemoglobin, the iron atom and its ligand are buried within the protein (3).

In 1936 I presented a seminar on the magnetic properties of hemoglobin at the Rockefeller Institute for Medical Research. There I met Alfred E. Mirsky, who had been studying the denaturation of proteins with M. L. Anson. In 1925 and 1931 they had reported that some denatured proteins could have the properties of the native protein restored, at least in part, by a careful treatment. Because of his interest in hemoglobin and his experience with it in the laboratory, I asked if he would be willing to come to Pasadena for a year or two. He agreed, and I asked the Director of the Rockefeller Institute for Medical Research to send him, which he agreed to do. In the same year Mirsky and I published a paper, "The nature of native, denatured, and coagulated proteins," in which the emphasis was laid on hydrogen bonds holding the molecule in its native configuration (5).

In the fall of 1937, while I was at Cornell University serving for one semester as George Fisher Baker Lecturer in Chemistry, Warren Weaver of the Rockefeller Foundation asked if I would talk with Dr. Dorothy Wrinch, who had come from England to the United States on a Rockefeller fellowship. She had developed a theory of the structure of proteins, the cyclol theory, which involved a complex of single-bonded rings. Her theory presented an alternative to the polypeptide chains that had been assigned to proteins by Emil Fischer, which Mirsky and I had accepted. My report to the Rockefeller Foundation, after Dorothy Wrinch had given her seminar at Cornell, was that the arguments and evidence seemed to me to have little weight. Then in 1938 Wrinch and Irving Langmuir published several papers in which they reported that an interpretation of the X-ray data of Dorothy Crowfoot (Hodgkin) for crystalline insulin provided proof that the insulin molecule actually had the structure of one of Wrinch's cyclols (2, 18). This stimulated Carl Niemann and me to marshall the evidence against the cyclol structures and for polypeptide chains in proteins, which we published in the *Journal of the American Chemical Society* in 1939 (14).

X-ray diffraction photographs of fibrous proteins had been prepared during the early 1930s by W. T. Astbury in England. Astbury's attempts to assign molecular structures to these proteins seemed unsatisfactory to me, in that he did not rely much on known values of bond lengths and bond angles, the planarity of the amide group, and the formation of the hydrogen bonds. During the summer of 1937 I spent a couple of months trying to find a structure that would satisfy the requirements of structural chemistry and that would account for the observed diffraction pattern of alpha-keratin (hair, horn, fingernail, muscle). I was unsuccessful in this effort, partially because I accepted a simple interpretation of the diffraction pattern and partially because I did not think hard enough—my structural assumptions were found later to be completely correct. At just this time Robert B. Corey came to Pasadena, sent for a year by the Rockefeller Institute for Medical

Research after the X-ray crystallographer with whom he had been working, R. W. G. Wyckoff, had left the Rockefeller Institute. Corey was also interested in proteins and had taken some X-ray diffraction photographs of proteins while he was with Wyckoff. I felt that there might be some new structural principle involved in peptides and proteins, and that, inasmuch as no correct structure determination had yet been made of any amino acid or peptide, it would be worth while to apply the X-ray diffraction method to some of these substances. Corey immediately began work in this field, and within a year he had succeeded in determining the structure of a cyclic dipeptide, diketopiperazine. During the following year he and Gustav Albrecht finished the determination of the first amino acid, glycine, and by 1948 the structures of a dozen amino acids and simple peptides had been determined at the California Institute of Technology. At that time no correct structure determination had been made at any other institution. Among Corey's co-workers, in addition to Albrecht, were Henri A. Levy, E. W. Hughes, Jerry Donohue, Verner Schomaker, David Shoemaker, Walter J. Moore, Kenneth N. Trueblood, Herman R. Branson, Sidney Weinbaum, R. A. Pasternak, Gene D. Carpenter, Harry L. Yakel Jr., and Richard E. Marsh.

By 1948 it was evident that no new structural principle had been discovered. The bond angles and bond lengths were essentially the same as those that I had assumed in 1937, the amide group was planar, and the hydrogen bond between nitrogen and oxygen had the length that I had accepted for it eleven years before. For some reason (perhaps the burden of work during the Second World War) I had not taken up again the problem of finding the structure of alpha-keratin until March 1948. On that day, in Oxford, England, where I was serving as George Eastman Professor in Oxford University, I decided to ignore the X-ray photograph and attack the problem as simply a problem in structural chemistry. With the assumption that all of the amino-acid residues are equivalent (ignoring the differences in side chains), I soon derived a structure, the alpha helix (10), that has turned out to be correct and to be a principal structural feature of many fibrous proteins and many globular proteins. Later Corey and I reported the discovery of the two pleated sheets, the parallel-chain pleated sheet and the antiparallel-chain pleated sheet, and we were able to show that silk fibroin has the antiparallel-chain pleated-sheet structure (4, 11).

Corey then made an effort to be the first person to determine the structure of a globular protein. He studied lysozyme by the heavy-atom isomorphous replacement method, but before he got very far Kendrew, in Cambridge, succeeded with myoglobin. Many other X-ray crystallographers began the attack on globular proteins, with such effectiveness that more than 200 globular protein structures have now been reported.

In the meantime my associates and I had been attacking the problem of the molecular basis of biological specificity. After my 1936 lecture at the Rockefeller Institute for Medical Research Dr. Karl Landsteiner invited me to come to his laboratory, where he asked how I might explain, on a molecular basis, the many observations that he had made in the field of immunochemistry. I was impressed by the specificity of the antigen-antibody reactions that he had been studying, and I decided that a molecular explanation of this specificity would probably apply to biological specificity in general. By 1940 I had concluded that biological specificity results from a detailed complementariness in structure of two molecules (6).

In 1940 Dan H. Campbell came as a Rockefeller fellow to the California Institute of Technology for one year; he returned to the University of Chicago, but then came back to CIT, where he was for many years professor of immunochemistry. He began studies similar to those carried out by Landsteiner, but more quantitative and with designs suggested by the complementariness theory, which I had published in 1940 in the *Journal of the American Chemical Society* (6). David Pressman carried out most of the quantitative studies with the assistance of graduate students and postdoctoral fellows, including L. A. R. Hall, Frank Lanni, Carol Ikeda, Miyoshi Ikawa, David H. Brown, John T. Maynard, Allan L. Grosberg, George G. Wright, Stanley M. Swingle, John H. Bryden, Arthur B. Pardee, and Leland H. Pence. These studies provided completely convincing evidence that the bond between an antibody molecule and the haptenic group of an antigen molecule involves a very close complementariness, to within a fraction of the diameter of an atom. We were able to show also that, in addition to the forces of van der Waals attraction and repulsion, the formation of hydrogen bonds and the interaction of positively charged and negatively charged groups are involved in the attractive forces (15). Precipitating antibodies were shown to have two combining groups, and blocking antibodies to have one combining group. By 1948 (7) I was able to write about the gene: "The detailed mechanism by means of which a gene or a virus molecule produces replicas of itself is not yet known. In general the use of a gene or virus as a template would lead to the formation of a molecule not with identical structure but with complementary structure. It might happen, of course, that a molecule could at the same time be identical with and complementary to the template on which it is molded. However, this case seems to me to be too unlikely to be valid in general, except in the following way. If the structure that serves as a template (the gene or virus molecule) consists of, say, two parts, which are themselves complementary in structure, then each of these parts can serve as the mold for the production of a replica of the other part, and the complex of two complementary parts thus can serve as the mold for the production of duplicates of itself." Eight years later the two mutually complementary strands were identified as polynucleotides, molecules of DNA—not in Pasadena, but in Cambridge, by Watson and Crick, with the help of Jerry Donohue. Donohue, the only person thanked by Watson and Crick in their first article (17), had spent several years in the chemistry department of the California Institute of Technology, and knew much more about hydrogen bonds and conjugated molecules, such as the purines and pyrimidines of DNA, than either Watson or Crick. This was the beginning of the DNA age in biology; biology had finally become a branch of chemistry.

This year, 1953, may also be taken as the time that marked the end of the early contributions of the Division of Chemistry and Chemical Engineering of the California Institute of Technology to the development of molecular biology. I shall mention only one more discovery. In 1945 I had the idea that there might be diseases of molecules, and that in particular sickle-cell anemia might be a disease of the hemoglobin molecule. By 1949 the idea had been verified by electrophoresis experiments carried out by Harvey Itano and Jon Singer, which resulted in the publication of our paper "Sickle-cell anemia: a molecular disease" (13). Harvey Itano and his collaborators then discovered several other abnormal human hemoglobins, some of them associated with diseases. Since then more than 300 human hemoglobins have been identified, and the branch of medicine called the hemoglobinopathies has developed.

I left the California Institute of Technology in 1963, and, although I know that much work in molecular biology is being carried out there by people in the Division of Biology as well as in the Division of Chemistry and Chemical Engineering, I cannot speak reliably about it. I consider molecular biology to be a part of structural chemistry, a field that was just beginning to be developed when I began working on the determination of the structure of crystals by X-ray diffraction at the California Institute of Technology in 1922. It did not enter my mind in 1922, or even in 1932 or 1942, that structural chemistry would develop in the marvelous way in which it has. I shall make no effort to predict what the future holds.

Literature Cited

1. Coryell, C. D., Pauling, L. 1940. *J. Biol. Chem.* 132:769–84
2. Langmuir, I., Wrinch, D. M. 1938. *Nature* 142:581–82.
3. Lein, A., Pauling, L. 1956. *Proc. Natl. Acad. Sci. USA* 42:51–54
4. Marsh, R. E., Corey, R. B., Pauling, L. 1955. *Biochim. Biophys. Acta* 16:1–16
5. Mirsky, A. E., Pauling, L. 1936. *Proc. Natl. Acad. Sci. USA* 22:439–44
6. Pauling, L. 1940. *J. Am. Chem. Soc.* 62:2643–53
7. Pauling, L. 1948. *Molecular Architecture and the Processes of Life; the 21st Sir Jesse Boot Lecture.* Nottingham, England: Sir Jesse Boot Foundation

8. Pauling, L. 1964. *Nature* 203:182
9. Pauling, L. 1978. *Can. Mineral.* 16:447–52
10. Pauling, L., Corey, R. B. 1950. *J. Am. Chem. Soc.* 72:5349
11. Pauling, L., Corey, R. B. 1951. *Proc. Natl. Acad. Sci. USA* 37:251–54
12. Pauling, L., Coryell, C. D. 1936. *Proc. Natl. Acad. Sci. USA* 22:210–14
13. Pauling, L., Itano, H. A., Singer, S. J., Wells, I. C. 1949. *Science* 110:545–48
14. Pauling, L., Niemann, C. 1939. *J. Am. Chem. Soc.* 61:1860–65
15. Pressman, D., Pardee, A. B., Pauling, L. 1945. *J. Am. Chem. Soc.* 67:1202–6
16. St. George, R. C. C., Pauling, L. 1951. *Science* 114:629–33
17. Watson, J. D., Crick, F. H. 1953. *Nature* 171:737
18. Wrinch, D. M., Langmuir, I. 1938. *J. Am. Chem. Soc.* 60:224–30

Ann. Rev. Biophys. Biophys. Chem. 1986. 15 : 11–28

HALORHODOPSIN: A LIGHT-DRIVEN CHLORIDE ION PUMP

Janos K. Lanyi

Department of Physiology and Biophysics, University of California, Irvine, California 92717

CONTENTS

PERSPECTIVES AND OVERVIEW

Studies of bacteriorhodopsin over the past fourteen years have contributed in a special special way to our present understanding of ionic pumps and membrane proteins in general. This is because bacteriorhodopsin, a light-driven proton pump found in the halobacteria, is a small membrane protein that contains the minimum features required for active proton transloca-tion; many other systems are larger proteins and show considerable structural complexity. Because bacteriorhodopsin is in a crystalline array, it was the first membrane protein for which a three-dimensional structure could be determined. The resulting characteristic bacteriorhodopsin struc-ture served as a test model for various strategies designed to predict and evaluate the structure of other membrane proteins. It has become evident in the last few years, however, that the halobacteria contain other retinal proteins that are potentially as interesting as bacteriorhodopsin. One of

11

0883–9182/86/0610–0011$02.00

these is halorhodopsin, which is also a small retinal protein, but which functions in the cytoplasmic membrane as an inward-directed light-driven chloride pump, rather than as an outward-directed proton pump. Recently, the halorhodopsin field has begun to expand rapidly, but without a much-needed concensus among the various laboratories on the interpretation of large amounts of fragmentary information. For that reason this review attempts to summarize the findings and concepts that have emerged so far in this system, particularly those that deal with the differences and similarities between halorhodopsin and bacteriorhodopsin.[1] Brief reviews of halorhodopsin and related findings are found in references 36, 37, 48, and 73.

TRANSPORT PROPERTIES OF HALORHODOPSIN

Halorhodopsin-containing cell envelope vesicles, prepared by sonication (42) of bacteriorhodopsin-depleted *Halobacterium halobium* cells of strains ET-15, R_1mR, or L-33, have been observed to take up protons upon illumination. Photoactivity in the membranes of these strains in the absence of bacteriorhodopsin was consistent with the observed photophosphory-lation in the intact cells (51). The proton influx resulted in a pH difference between the interior and the exterior of envelope vesicles, as demonstrated by uptake of radiolabeled N-methylmorpholine (49), the quenching of the fluorescence of 9-aminoacridine (18), and the partitioning of a spin-probe (31). Illumination also caused the generation of an electrical potential (negative inside) in this system, as demonstrated by the uptake of radiolabeled triphenylmethyl phosphonium (46, 49) and the quenching of the fluorescence of a cyanine dye (18). The rate of the light-driven proton inflow was increased in the presence of uncouplers; but agents that were observed to abolish the electrical potential, such as gramicidin and Na^+ or valinomycin and K^+, abolished the proton movement also (18, 46, 49). These observations can be explained only by passive proton transport into the vesicles driven by electrical potential, rather than by the active proton transport described earlier for bacteriorhodopsin-containing vesicles (61).

Quantitative estimations (18) showed that after a few minutes of illumination the steady-state pH difference and electrical potential in halorhodopsin-containing vesicles became about equal and of opposite sign. Thus, no steady-state protonmotive force was created by the illumination, in spite of the net transport of the protons. This result is consistent with the idea of passive proton flux.

---

[1] A preliminary version of this manuscript was circulated among workers in this field. The author is very grateful for the large number of comments received, and has attempted to incorporate the views expressed as far as possible.

The passive nature of the proton uptake is further illustrated by the consequences of dicyclohexylcarbodiimide treatment of the membranes. This agent is thought to block proton permeability through the F_o component of ATPase in many membranes, and inhibits ATP synthesis in intact *H. halobium* (6, 13, 20). In halorhodopsin-containing vesicles it inhibited the light-induced proton inflow but not the electrical potential. Adding uncoupler to the treated membranes restored light-dependent proton uptake (18) at rates higher than those in untreated control vesicles. The proton movement is thereby demonstrated to be a secondary effect of, rather than a necessary condition for, the membrane potential created by halorhodopsin.

The inside negative electrical potential generated by halorhodopsin was proposed (46, 49) to be due to active and outward-directed electrogenic transport of Na^+ for the following reasons: (*a*) Illumination of halorhodopsin-containing vesicles in NaCl caused uncoupler-insensitive export of sodium (46, 49). (*b*) The proton uptake and the membrane potential were not observed when the illumination was in ultrapure KCl solution, but were restored when NaCl was added, particularly after incubation in the dark to (presumably) allow the penetration of Na^+ into the vesicles (49). However, the ^{22}Na efflux data can be interpreted also as exchange rather than as net transport. When net loss of Na^+ was determined, either directly by measuring atomic absorption (18) or indirectly by following sodium gradient–dependent amino acid transport (47), either very low or no detectable sodium extrusion was seen. Likewise, the effects of added Na^+ on the pH and electrical potential can be interpreted otherwise, e.g. as having been caused by the activation of sodium/proton antiport, described earlier in this system (15, 41).

The arguments in favor of the only possible alternative model, active and inward-directed electrogenic chloride transport, are much more convincing. The first evidence presented below was obtained with halorhodopsin-containing envelope vesicles. All experiments were repeated with bacteriorhodopsin-containing vesicles as convenient controls, in which protonmotive force drives active outward sodium transport via the sodium/proton antiporter, accompanied by passive chloride efflux. The evidence with the vesicle system (62) is as follows.

1. Illumination caused net chloride uptake (not only exchange), as observed by determination of the total chloride content of dark-incubated and illuminated vesicles suspended in sodium sulfate or potassium phosphate containing 100 mM chloride. Over a period of 30–40 min the moles of chloride gained by the vesicles exceeded by over three orders of magnitude the moles of halorhodopsin present. The transport was against the electrical potential (inside negative) that existed during the transport, as

well as against the chloride concentration gradient (inside/outside = 5–10 after 10 min of illumination) that developed. The latter was determined from triphenyltin-dependent alkalinization 10 min after the terminating of illumination (62), since triphenyltin is an anion exchanger (64) and the newly created chloride gradients would drive the observed chloride/ hydroxyl exchange. A third kind of transport assay was based on volume increase due to chloride and sodium uptake during illumination, as measured either with light-scattering (62) or with a volume-sensitive ESR probe (53).

2. All secondary effects of the proposed chloride uptake, such as development of electrical potential, pH change, and light-scattering increase, were absent when chloride was not added. Chloride effectively restored these phenomena only when added to the vesicle exterior. Half-maximal effect was at about 40 mM chloride when the vesicles were in sodium sulfate or potassium phosphate. Preformed chloride gradients (inside/outside = 0.1–10) had little influence on the light-induced electrical potential. A role for Na^+ in the transport seems to be ruled out by the fact that in potassium phosphate the amount of chloride uptake was at least 20 times the amount of sodium adventitiously present inside the vesicles. Chloride and bromide were about equally effective (62), but iodide was much less so (23). This specificity agrees with the finding that only these anions were transported by halorhodopsin.

3. Agents that increase passive cation transport and consequently abolish the inside negative electrical potential, such as gramicidin and Na^+ or valinomycin and K^+, doubled or tripled the rates of chloride uptake and volume increase. The observed chloride uptake therefore cannot have been driven by the electrical potential as a secondary process; rather, it was the chloride transport that generated the electrical potential.

4. As discussed below, chloride and some other anions, but not cations, had specific effects on the halorhodopsin chromophore and its photoreactions. For the halides these effects were observed at concentrations consistent with the apparent affinity determined in the transport experiments. Thus, the halorhodopsin chromophore responds to anions as expected from its proposed role as a light-driven pump for chloride.

The above observations suggest very strongly that halorhodopsin transports chloride, and in an electrogenic fashion. Another line of evidence comes from experiments with partially and fully purified halorhodopsin. When membrane fragments containing halorhodopsin ("Tween-washed membranes") or proteoliposomes containing purified halorhodopsin were attached to planar lipid films separating two chambers (3, 4), a sandwich-like structure was formed that generated a current upon illumination. The appearance of these photocurrents depended on the presence of chloride, and was indifferent to the cation present. The following argument indicates that in these experiments the photocurrents were carried by chloride rather than by other ions: Without added ionophores the illumination generated only a transient capacitive current, which suggests that the ions were transported into the compartment formed between the Tween-washed membrane and the planar film. Only after addition of an uncoupler plus triphenyltin did the photocurrent become sustained (3), presumably because triphenyltin exchanged the chloride for hydroxyl ions and the uncoupler allowed the influx of protons, permitting a chloride current to flow between the two chambers. The fact that the photocurrent required a chloride-specific ionophore is an argument for a chloride current pathway. Thus, chloride must carry the current generated by halorhodopsin. Another report indicates that light-dependent passive proton transport by proteoliposomes containing purified halorhodopsin also requires chloride (10).

STRUCTURE OF HALORHODOPSIN

The apoprotein in halorhodopsin has been identified by bleaching with illumination in the presence of NH_2OH, reconstitution with radiolabeled retinal, reduction of the product with cyanoborohydride, and fluorography of SDS-acrylamide gels run with the membrane proteins (43, 66). One major labeled band has been observed, which appears as a double band in some reports. The labeled band runs somewhat ahead of bacterio-opsin. One (27, 66, 68) or both (D. Oesterhelt, personal communication) of these bands is halo-opsin. Several minor bands, labeled much less intensively, have been attributed to nonspecific binding of retinal to other proteins. Since bacteriorhodopsin runs anomalously on at least some SDS-acrylamide gels, a rigorous molecular weight cannot be obtained for halorhodopsin with the gel electrophoresis method. It is likely, however, that the molecular weight of halo-opsin, about 25 kd, is either similar to or about 1 kd lower than that of bacterio-opsin.

The halorhodopsin chromoprotein (retinal plus halo-opsin) can be purified by a number of methods (59, 71, 74, 75). All utilize a pretreatment step consisting of exposure to low salt and/or extraction of the membranes with Tween-20, detergent solubilization, and chromatography on hydro-xylapatite and phenyl- or octylsepharose columns. The main difference is the detergent used for solubilization: Lubrol PX (71), $C_{12}E_9$ (74), and Triton X-100 (59) are effective but difficult to remove; octylglucoside (75) is expensive but dialyzable in reconstitution experiments. Criteria for purity include the absence of extraneous bands on silver-stained SDS gels and a ratio of absorbance at 280 nm and 570 nm not higher than 1.6. The preparations are not contaminated (71) with sensory rhodopsin (8, 67), which is present in the strains used for the isolation. The purified halorhodopsin exhibits all of the chloride-dependent photochemical reactions of membrane-bound halorhodopsin, as described below (72), and it is functionally active in planar lipid film reconstitutions (4) and in proteoliposomes (10). Thus, we may conclude that the halorhodopsin transport activity resides in the 25-kd chromoprotein.

The amino acid composition of halo-opsin differs significantly from that of bacterio-opsin. The differences include the presence of a cysteine residue (2), the presence of two histidine residues, and the predominance of arginines over lysines in halo-opsin (3, 74). Bacteriorhodopsin, in contrast, contains no cysteine or histidine, and the same number of lysine as arginine residues (33). However, the two proteins are similar in containing a preponderance of hydrophobic residues. About 40% of the amino acid sequence is known from analysis of cyanogen bromide fragments (25; M. Ariki, J. K. Lanyi, unpublished results), and very little sequence overlap with bacterio-opsin is evident. Remarkably, in at least one of the fragments several of the arginine residues are clustered near each other.

Since halorhodopsin is an integral membrane protein a plausible secondary and tertiary structure can probably be constructed for it once its complete primary structure is available. The CD spectrum in the UV region suggests (70) that most of the protein, as with bacteriorhodopsin, is in a helical conformation. If this is so, this protein may contain up to seven helical segments looped across the membrane, as does bacteriorhodopsin (28). Most of the polypeptide chain is inaccessible to the aqueous environment; i.e. when in the membrane, the proteolytic digestion sites of halorhodopsin are limited to a short segment in either one or both of its terminal regions (68).

There is no evidence to suggest that halorhodopsin in the cytoplasmic membrane of the halobacteria exists in the extended two-dimensional arrays that are characteristic of bacteriorhodopsin in the purple membrane. However, CD spectra of some purified halorhodopsin preparations exhibit the same kind of band splitting in the visible region (70, 74) that is seen in the purple membrane but is not seen in monomeric bacteriorhodopsin. The band splitting is thought in the case of bacteriorhodopsin to reflect exciton interaction in the trimeric aggregates.

SPECTROSCOPIC PROPERTIES OF HALORHODOPSIN

Action spectra for the creation of electrical potential and pH difference in cell envelope vesicles indicate that the pigment responsible absorbs at 580–

585 nm (18, 19), somewhat redshifted from the absorption maximum of bacteriorhodopsin. A similar redshift has been observed for photophosphorylation in bacteriorhodopsin-depleted halobacterial strains (51, 78). The existence of a retinal pigment with such an absorption maximum can be demonstrated in envelope vesicles prepared from a bacteriorhodopsin-deficient strain that also lacks carotenoid pigments. The procedure is to bleach the membranes with lengthy illumination in the presence of NH_2OH and to follow the absorption changes upon reconstitution with added retinal. Difference spectra obtained in this way for two samples, with and without retinal, exhibited an absorption maximum near 588 nm (45), and the magnitude of this absorption band correlated well with reconstitution of light-driven passive proton uptake activity, as described in the previous section (45, 79). The extinction coefficient of the pigment at 580 nm was determined to be about 50,000 M^{-1} cm^{-1} (38, 45, 71), somewhat lower than that of bacteriorhodopsin. Subsequently it was found (8, 24) that the halobacterial membranes contain an additional retinal protein, "sensory rhodopsin" (8), that also bleaches and reconstitutes under these conditions. Thus, the pigment detected in these studies is a mixture of halorhodopsin and the third retinal protein (69). The use of mutants deficient in one or another rhodopsin (8, 67) has been helpful in sorting out absorption due to the two different pigments. Membranes containing exclusively halorhodopsin can also be produced by short periods of bleaching with light in the presence of NH_2OH, because halorhodopsin is more resistant to this treatment than the sensory rhodopsin (44).

The optical properties of purified preparations allow the easy determination of the absorption spectrum of halorhodopsin. However, in the presence of detergents (71) or lipids (10) the absorption maximum is blueshifted by 10–20 nm, although this effect is at least partially reversed upon removal of the detergent. This shift has led to divergent reports as to the absorption maximum of halorhodopsin. The most reliable estimate for the absorption maximum of the pigment in cell envelope vesicles (67) and in dialyzed purified halorhodopsin (71) is 578 nm. In estimations in which such ions as nitrate and thiocyanate were present but chloride was absent, the absorption maximum was blueshifted to about 565 nm (57–59, 72). The addition of chloride to these preparations restored the original absorption band. In 1 M nitrate, for example, the chloride-induced redshift gave a characteristic difference spectrum with a large maximum at 600 nm and a small minimum at 500 nm, arising from the different maxima and band shapes with and without chloride (59). The shift resulted in an approximately 14% increase in absorption at 580 nm upon addition of saturating concentrations of chloride (half-maximal effect at about 0.3 M chloride). Thus, halorhodopsin can bind chloride and other anions in the dark, and this binding results in relatively small but distinct changes in the chromophore.

Resonance Raman lines of halorhodopsin reveal major similarities and minor differences between the chromophore of this pigment and that of bacteriorhodopsin (1, 50, 65). Similarities of the retinal C–C and C=C stretch frequencies suggest, for example, that the electrostatic environments near the ionone ring are similar in the two systems. The differences include decreased coupling between the C=N stretch and N–H rock, which may originate from weaker hydrogen bonding of the Schiff base in halorhodopsin than in bacteriorhodopsin. Resonance Raman spectra of halorhodopsin reveal also that the retinal in this pigment is all-*trans*, as in bacteriorhodopsin (1, 65).

Upon flash illumination of halorhodopsin-containing membranes and purified halorhodopsin in molar NaCl solutions, the absorption of the halorhodopsin chromophore was depleted and absorption increased, owing to the most abundant photoproduct, near 500 nm (63, 76, 79). The absorption maximum of this photointermediate was at 520 nm (26, 63). The intermediate hR_{520} was formed in a few μsec (60), and was shown to decay with a 10–15 msec half-life (25, 63, 76, 79). In the absence of chloride, but at high ionic strengths provided by sulfate or other nonhalide anions, a different photoproduct was formed; it absorbed near 660 nm (63) or 640 nm (P. Hegemann, D. Oesterhelt, personal communication) and decayed on a 1–2 msec time scale. This redshifted species, hR_{640}, corresponds to the 630–640 nm intermediate obtained in the complete absence of added salt (11, 79). Conditions free of added salt are not recommended for use in general, however, as in many such cases purified halorhodopsin is rapidly denatured and its chromophore is lost. The apparent affinity for chloride to affect the photocycle is about 40 mM in vesicles (63), but about 10 mM in purified halorhodopsin (72). Only chloride, bromide, and iodide have been demonstrated to cause the appearance of the hR_{520} photoproduct (23, 72); other anions or cations have had no effect. A photoreaction scheme has been proposed that places hR_{640} after hR_{520} in the same photocycle (26), while other schemes (63, 73) put these intermediates in different photocyles.

The flash-induced difference spectrum, calculated from absorption changes within 1 msec after the flash in the presence of chloride, bears a remarkable resemblance to the difference spectrum obtained in the dark after adding chloride to halorhodopsin, as described above. It is an attractive idea, therefore, that light causes the reversible loss of chloride from halorhodopsin, and that it is this process that causes the observed changes in absorption (58, 59). However, this hypothesis predicts that the flash-induced depletion at 580 nm originates from relatively small changes in absorption (as mentioned above, a maximum of 14%); thus a given absorption change would correspond to a much greater degree of photo-conversion than if the absorption changes were to originate from depletion due to the production of hR_{520}, which absorbs much less at 580 nm than hR_{565}. Examination of the magnitude of depletion reveals that this prediction leads to unrealistically high quantum yields for the photoreaction of halorhodopsin as compared to that of bacteriorhodopsin (39). The hypothesis is therefore probably not tenable. However, a difference spectrum with a positive band near 590–600 nm and a smaller negative band at 500 nm is a recurring feature in bacterial rhodopsins. In bacteriorhodopsin this band is observed upon 13-*cis* to *trans* isomerization (7). The similarity among these difference spectra might not be pure coincidence.

The primary photoproduct, which absorbs at 600 nm, has a rise time of 5 psec, and its decay on a sub-μsec time scale may correspond to the rise time of hR_{520} (60). At $-196°$, illumination has been observed to cause the appearance of an absorption band at 620 nm (57) or 630 nm (73). Although this product could be driven back to hR_{578} with red illumination (57), it is doubtful that it is identical to the first intermediate of the photocycle, because it was stable for several hr upon warming to $-70°$, even though illumination at this temperature produced hR_{520} (J. K. Lanyi, Y. Mukohata, unpublished results). Photocycle schemes that include other intermediates of the photocycle have been published, e.g. species that absorb at 380 nm (73) and at 680 nm (65), but data supporting the existence of such photoproducts have not been presented. A later report (40) found no evidence for flash-induced absorption change centered at 380 nm.

Flash-induced absorption changes in bacteriorhodopsin in the near UV region have revealed two effects: *trans* to 13-*cis* isomerization of retinal (35) and deprotonation of a tyrosine (9, 29, 35). The former took place on a sub-μsec time scale, coincident with the appearance of the primary photoproduct K_{590}, and resulted in a difference spectrum in the near UV very similar to that produced during dark-adaptation of the pigment. The latter was seen on a slower time scale, similar to that of the rise and decay of the M_{412} photointermediate, and its difference spectra resembled those of tyrosine at neutral and alkaline pH. Flash-induced absorption changes for halorhodopsin in the UV have shown the kind of absorption changes associated with retinal isomerization, but not the kind that would originate from the deprotonation of tyrosine (40). Although their rise has not been resolved, the decay of these absorption changes coincides with the decay of hR_{520}. More recent results (J. K. Lanyi, unpublished observations) indicate that this kind of result can be obtained for hR_{640} also. It would appear, therefore, that in both photointermediates the retinal configuration is 13-*cis*. More direct demonstration would be obtained with resonance Raman spectroscopy of illuminated samples. When the conditions of the illumination were such that the samples produced a long-lived (deprotonated) photoproduct, hR_{410} (light) (see below), an increase in the 13-*cis* isomer content could be demonstrated by extraction and analysis of the retinal from the pigment (57). Illumination in the presence of NH_2OH produced oximes with increased 13-*cis* content (P. Hegemann, D. Oesterhelt, unpublished results).

It now appears that long incubations in the dark produce a dark-adapted form of halorhodopsin (22, 30) that was previously not seen (45). This form exhibits a blueshifted absorption band relative to light-adapted halorhodopsin (22, 30). The postulated dark-adapted form produces about half the hR_{520} and exhibits about half the transport activity of the light-adapted sample (31). These attributes are as expected if the dark-adapted form of the pigment contains 50% 13-*cis* retinal isomer, as is the case with bacteriorhodopsin. However, this has not been directly demonstrated as yet with extraction and analysis of the retinal. The suggested light/dark adaptation is similar, or perhaps identical, to spectral shifts obtained with red and blue illumination. Thus, red illumination is reported (65) to shift the halorhodopsin absorption maximum to 560–565 nm (red-adapted form), and blue illumination is reported to cause a shift back to 578 nm (blue-adapted form).

ANION EFFECTS ON HALORHODOPSIN CHROMOPHORE AND ITS DEPROTONATION

When cell envelope vesicles containing halorhodopsin were suspended in chloride-free salt, such as sulfate, at pH between 6 and 9, difference spectra with and without added chloride suggested that the absorption of the pigment at 580 nm decreased and its absorption at 410 nm increased as the pH was raised, and that this change was reversed by chloride (44). This phenomenon was seen more clearly with purified halorhodopsin; absolute spectra showed that at alkaline pH a new species, hR_{410} (dark), was produced at the expense of hR_{565}, and that the process was reversed by the addition of chloride (72). There can be little doubt that hR_{410}(dark) represents the deprotonated Schiff base, since the protonated Schiff base absorbs at wavelengths higher than 410 nm even without retinal-protein interaction (5). Titration curves with and without chloride in the dark revealed that chloride caused a pK_a shift for the deprotonation, from 7.4 to 8.9 (B. Schobert, J. K. Lanyi, unpublished results). Analysis of the chloride dependency of the apparent pK_a indicates that chloride does not dissociate from its binding site upon deprotonation, i.e. chloride must be bound to a permanently positively charged group, and thus not to the Schiff base itself (B. Schobert, J. K. Lanyi, unpublished results). For chloride the apparent affinity for this effect was about 40 mM in the vesicles (44), but about 10 mM in purified halorhodopsin (72). It appears that a large number of anions will raise the pK_a for the Schiff-base deprotonation in this way, with certain anions preferred in the order Cl, Br, I, SCN, N_3 > nitrate > others. This sequence of preference can be related to the hydrated radii of these ions, as described in the following section. The site of binding for these anions is referred to as Site I (B. Schobert, J. K. Lanyi, unpublished results). Occupancy of Site I causes a blueshift in the visible absorption maximum of halorhodopsin. In cell envelope vesicles Site I is accessible to the anions from the exterior side of the membrane (44).

Site I is thus distinct from a second kind of chloride binding site in halorhodopsin, in which occupancy causes the photoproduction of hR_{520} as described in the previous section. The latter is termed Site II, and its specificity is restricted to chloride, bromide, and, to a small extent, iodide (23, 58, 72). Binding to Site II produces a redshift of about the same magnitude as the blueshift for Site I. Diuretic drugs of the MK series have shown competitive inhibition with chloride for Site II as well as for transport (63). Occupancy of Site II does not change the pK_a of the Schiff base (B. Schobert, J. K. Lanyi, unpublished results). Access to Site II, as for Site I, is also from the exterior side of the membrane (10, 63).

Sustained illumination of halorhodopsin has been observed to cause the appearance of hR_{410}(light), at the expense of hR_{578} (57), particularly at alkaline pH (44). The light-dependent deprotonation proceeds with very low efficiency (27), however, and therefore flash illumination has not yielded appreciable absorption change at 410 nm. Once produced, however, the photoproduct hR_{410}(light) relaxed so slowly (over 1 hr) that it was observed to accumulate at pH above 7 even during 1–3 min of illumination with moderately intense yellow light. Thus, the hR_{410}(light) is not identical with hR_{410}(dark), which was rapidly reconverted to hR_{578} when the pH was lowered. The difference is probably that the dark product contains all-*trans* retinal, while in the illuminated product the retinal is in a 13-*cis* or another isomeric configuration. The rate of reprotonation after illumination was completely inhibited in the presence of low concentrations of Hg^{2+} (2), suggesting that the cysteine residue, with which it reacts, is near the Schiff base.

When azide was added, hR_{410}(light) was formed very rapidly during the illumination, and under these conditions its decay was about 200 msec (26; B. Schobert, J. K. Lanyi, unpublished results). Azide and a few other anions such as cyanate and sulfide (26) or even acetate at high concentration (J. K. Lanyi, unpublished results) are thought to act catalytically in accelerating these processes, as proton acceptor/donors that allow the Schiff-base proton to leave the protein and to return. Since the steady-state ratio of

hR_{410}/hR_{578} in the light is independent of azide concentration at any pH value, it seems clear that hR_{410}(light) is not produced by a branching reaction in the photocycle, but by a side reaction involving hR_{520} and/or hR_{640} (26).

The above results suggest very strongly that during the normal photocycle of halorhodopsin (i.e. without azide added) the Schiff base is not deprotonated. The reason is probably the presence of a kinetic barrier rather than a mismatch of the pK_a with the ambient pH, because the pK_a of deprotonation for the photoproduct(s) is about 4 pH units lower than for the parent species (26; B. Schobert, J. K. Lanyi, unpublished results).

When the binding of chloride to halorhodopsin was investigated by following NMR line broadening of ^{35}Cl (16), neither Site I nor Site II was detected. Rather, these measurements revealed a third binding site, with a binding constant for chloride of about 100 mM. From competition experiments it appears that the order of preference for anions by this site is reversed (compared to that of Site I): I^-, $SCN^- > Br^- > NO_3^- > Cl^-$. Site III has relatively small or no influence on the pK_a of deprotonation, and so far it has not been related to spectroscopic or transport properties.

MODELS FOR CHLORIDE TRANSLOCATION IN HALORHODOPSIN

From the foregoing it is evident that in some respects bacteriorhodopsin and halorhodopsin are similar retinal proteins. In both pigments the retinal-protein association appears to be via a Schiff base, the Schiff base is protonated, the charge distribution near the retinal gives large redshifts relative to the absorption band of a free protonated retinal Schiff base, the pigments contain all-*trans* retinal, and the main photoproducts contain 13-*cis* retinal. Furthermore, C–C stretching frequencies in the resonance Raman spectrum of halorhodopsin (1, 50, 65) are quite similar to those of bacteriorhodopsin. Since in halorhodopsin these frequencies are very sensitive to electrostatic perturbation of the ionone ring of retinal, close similarities between this pigment and bacteriorhodopsin are to be expected. Finally, the relaxation of the photochemical changes is on a similar time scale for the two pigments.

The fundamental difference between the two systems becomes evident, however, when the differences in amino acid composition and sequence, discussed above, are considered. These differences obviously are related to the fact that halorhodopsin is a chloride pump and bacteriorhodopsin is a proton pump, but at this time there are few clues as to how structure determines function in these systems. Another obvious difference is in the behavior of the pK_a of the Schiff base. In bacteriorhodopsin in the dark this

pK_a is above 12 (14), i.e. much above the value in free solution, presumably because the Schiff base in the protein is in an electronegative environment. The pK_a is lowered by 4–5 units upon illumination, most likely because the isomerization of the retinal around the $C_{13}=C_{14}$ double bond moves the Schiff base into a different environment. It is reasonable to expect that the lowering of the pK_a reflects the possibility of loss of the Schiff base proton to a strategically located acceptor group, and that this proton then migrates to the external membrane surface and is released. Replacement of the proton from the other side of the membrane would complete the proton translocation cycle in this pigment.

In contrast with bacteriorhodopsin, the Schiff-base pK_a of halorhodopsin is near neutrality, i.e. near what it would be without retinal-protein interaction. Various anions are able to raise the pK_a by as much as 2 units, presumably by binding in the immediate neighborhood of the Schiff base (Site I). It is possible to argue from the order of specificity of Site I for various anions, described in the previous section, that this site is at the aqueous phase: The relationship of hydrated size and observed binding constants for a large variety of anions suggests that the specificity of Site I derives entirely from the distance of closest approach for electrostatic attraction, which is equal to the Stokes radius (B. Schobert, J. K. Lanyi, unpublished observations). Upon illumination and the consequent isomerization of the retinal, the pK_a of the Schiff base was lowered by about 4 units (27; B. Schobert, J. K. Lanyi, unpublished observations), suggesting that this group is moved to another environment in the protein, as in bacteriorhodopsin. However, the deprotonation of the photointermediate(s) cannot compete with the recovery of the parent hR, because its rate is much lower.

Unlike Site I, which may be near the Schiff base, Site II is probably at a location removed from it. This is suggested by the fact that occupancy of Site II does not change the pK_a but causes a redshift in the absorption maximum. Furthermore, resonance Raman spectra of the pigment with and without chloride, but in the presence of large amounts of nitrate (a Site I anion), reveal that chloride has little effect on the N–H bending mode, but causes the expected shift of the C–C stretching frequency (50).

As in bacteriorhodopsin, the coupling between the light-induced absorption changes in the halorhodopsin chromophore and the transport is close. At saturating light intensities the turnover rate for chloride transport was near 20 sec^{-1} (53), only 2.5-fold below the approximate predicted maximal rate based on the spectroscopic turnover of the pigment. Since abolishing the membrane potential in the vesicles increases the transport 2–3-fold (62), the number of chloride ions transported per photocycle may be near 1.

Two kinds of models have been proposed for chloride translocation by halorhodopsin. Both include the isomerization of retinal as the principal driving force for the transport. In one of these models (56) the chloride is carried as the counter-ion to the positively charged Schiff base, which is displaced during the isomerization. Access to the Schiff base from the exterior is via a series of positively charged residues, which form a pathway for chloride movement in the protein. Release of the chloride in the isomerized state is to another series of positively charged residues, which leads to the other side of the membrane. This scheme, which is quite analogous to one proposed for bacteriorhodopsin, provides a mechanism for the chloride translocation that is compatible with bond rotational energies in the retinal. Thus, approach of a chloride to the positively charged Schiff base (in hR) is suggested to have the same consequences on the electron distribution on the retinal chain as leaving of the proton (in bR). Although not stated in these terms, the model states in effect that Site I is the chloride donor in the translocation, and another Site I–like group is the acceptor.

According to an alternative model (40), the chloride ion that is translocated is not the one bound to Site I near the Schiff base, but another, bound to Site II, remote from the Schiff base. Isomerization of the retinal and the accompanying movement of the protonated Schiff base alters the distribution of positive charges (perhaps via histidine tautomerization) in such a way that a new high-affinity binding site for chloride is created in the protein. The essential transport step in this model is the migration of the chloride from the initial site (Site II) to the newly created site. Release of the chloride on the cytoplasmic side occurs when the retinal re-isomerizes and the transient binding site loses affinity to the chloride. This model does not utilize Site I for the transport, although it allows Site I to be in the pathway of chloride migration to Site II. Thus, in this model the interaction of chloride and the Schiff base at Site I has no particular relevance for the central transport event.

PHYSIOLOGICAL ROLE OF HALORHODOPSIN IN THE HALOBACTERIA

The ionic circulation across the cytoplasmic membrane of halobacterial cells includes the following mechanisms: (a) proton extrusion via bacterio-rhodopsin or the respiratory chain; recirculation of the displaced protons to the cytoplasm through an ATPase (6, 13, 20), with the accompanying synthesis of ATP, and through a sodium/proton antiporter (15, 41); (b) massive net sodium efflux, via the electrogenic antiport, driven by protonmotive force; (c) net potassium uptake, probably driven (17, 77) by the electrical potential created, negative inside; (d) chloride uptake via both halorhodopsin and an undefined system driven by respiration (G. Wagner, unpublished observations).

Unlike the growth medium, which contains mostly NaCl (between 3 and 5 M), the cytoplasm of the halobacteria contains much more K^+ than Na^+ (12), although the internal anion is still largely chloride. Thus, water is retained in these cells mostly by KCl rather than by organic osmo-regulatory compounds, such as the sugars and amino acids found in many halotolerant organisms. Intuition suggests that during growth the net flux of ions should result in a K^+ uptake in excess of the Na^+ loss, and a Cl^- uptake equal to the difference, so as to provide a net gain of intracellular KCl commensurate with the gain in total intracellular volume (53). Furthermore, it is clear that the high internal chloride concentration is not in equilibrium with the large negative-inside electrical potential that accompanies the proton circulation and the sodium efflux. Thus, electrical potential–driven passive chloride movement can result only in chloride loss from the cells, rather than in the required uptake. For these reasons, it is necessary to postulate active inward chloride transport in these cells. It is not clear, as yet, whether or not the amount of halorhodopsin is sufficient for the chloride uptake required in all halobacterial strains. In any case, the halobacteria can grow aerobically in the dark, and therefore active chloride uptake mechanisms other than halorhodopsin must account for the ionic balance under dark conditions. The relationship of halorhodopsin to respiration-driven chloride uptake therefore appears to the be same as the relationship of bacteriorhodopsin to respiration-driven proton extrusion.

The electrogenic transport of chloride by halorhodopsin should contribute to the electrical potential across the cytoplasmic membrane that is established by proton transport. It has been suggested (18, 54) that the complex pH changes observed upon the illumination of whole cells of many halobacterial strains originate from the activities of both bacteriorhodopsin and halorhodopsin. According to this idea, the initial alkalinization observed would be passive proton uptake driven by the electrical potential from halorhodopsin, while the later acidification would be active proton extrusion by bacteriorhodopsin. At more acid external pH acidification predominates, while at more alkaline pH alkalinization is more apparent (54, 55); these observations agree with the pH optima of proton movements attributed to the two pumps in envelope vesicle suspensions (62). Furthermore, strains that lack bacteriorhodopsin, or strains in which bacteriorhodopsin is bleached, exhibit only the alkalinization (51, 52, 78), while heated cells in which halorhodopsin is inactivated exhibit only the acidification (51). It should be mentioned, however, that other explanations have been proposed for the alkalinization phenomenon in whole cells, including gating of the ATPase (6) and sodium/proton antiport (77).

The interaction of ionic fluxes that originate from the two light-driven pumps has been described in a cell-free system. A slowly sedimenting cell envelope fraction prepared from the R_1 strain (49) was used, which contained an appropriate ratio of halorhodopsin and bacteriorhodopsin. Illumination of this preparation caused proton uptake (external alkalinization) at pH higher than 5, but proton release at pH lower than 5 (18). A steady-state protonmotive force did develop, because even when protons were taken up the pH difference was smaller than the electrical potential. When an uncoupler was added, the net proton uptake increased and the electrical potential decreased, creating a balance that left no protonmotive force under these conditions. Such increased proton uptake upon addition of uncoupler to *Halobacterium halobium* envelope vesicles was observed long before the discovery of halorhodopsin (32). It appears, therefore, that in these experiments the pump for protons, i.e. bacteriorhodopsin, created an electrical potential by exporting protons, and chloride transport by halorhodopsin increased the electrical potential so that some of the protons were passively taken up. The data suggest that the two processes that generate the electrical potential are simply additive.

The light-dependent alkalinization of the medium by the cells has been associated with ATP synthesis, which implies that this proton influx occurs through the ATPase. Therefore the suggested role of halorhodopsin in this process predicts that the electrogenic chloride transport should be able to energize phosphorylation. Indeed, several bacteriorhodopsin-negative strains that contain halorhodopsin were found to photophosphorylate (51, 52, 78), although some others apprently did not (S. Helgerson, W. Stoeckenius, personal communication). The ATP was synthesized during the initial part of the illumination, i.e. during alkalinization. The action spectra for the ATP synthesis agreed with the absorption spectrum of halorhodopsin (51, 78). The electrical potential created by the chloride transport thus appears to be able to energize ATP synthesis. Under anaerobic conditions, however, the light-induced uptake of the protons will result in the continued acidification of the cytoplasm, unrelieved by a proton export process. During sustained illumination of the halorhodopsin-containing cells the ATP levels decline unless an electroneutral protonophore such as triphenyltin is added (54). These strains therefore cannot be grown anaerobically with illumination as the sole source of energy (78). In the strains that contain bacteriorhodopsin, however, the imported protons can recirculate and thus ATP synthesis can be sustained for at least five or six generations (21).

Literature Cited

1. Alshuth, T., Stockburger, M., Hegemann, P., Oesterhelt, D. 1985. *FEBS Lett.* 179:55–59
2. Ariki, M., Lanyi, J. K. 1984. *J. Biol. Chem.* 259:3504–10
3. Bamberg, E., Hegemann, P., Oesterhelt, D. 1984. *Biochim. Biophys. Acta* 773:53–60
4. Bamberg, E., Hegemann, P., Oesterhelt, D. 1984. *Biochemistry* 23:6216–21
5. Blatz, P., Mohler, J., Navangul, H. 1972. *Biochemistry* 11:848–55
6. Bogomolni, R. A., Baker, R. A., Lozier, R. H., Stoeckenius, W. 1976. *Biochim. Biophys. Acta* 440:68–88
7. Bogomolni, R. A., Baker, R. A., Lozier, R. H., Stoeckenius, W. 1980. *Biochemistry* 19:2152–59
8. Bogomolni, R. A., Spudich, J. L. 1982. *Proc. Natl. Acad. Sci. USA* 79:6250–54
9. Bogomolni, R. A., Stubbs, L., Lanyi, J. K. 1978. *Biochemistry* 17:1037–41
10. Bogomolni, R. A., Taylor, M. E., Stoeckenius, W. 1984. *Proc. Natl. Acad. Sci. USA* 81:5408–11
11. Bogomolni, R. A., Weber, H. J. 1982. *Methods Enzymol.* 88:434–39
12. Christian, J. H. B., Waltho, J. A. 1962. *Biochim. Biophys. Acta* 65:506–8
13. Danon, A., Stoeckenius, W. 1974. *Proc. Natl. Acad. Sci. USA* 71:1234–38
14. Druckmann, S., Ottolenghi, M., Pande, J., Callender, R. H. 1982. *Biochemistry* 21:4953–59
15. Eisenbach, M., Cooper, S., Garty, H., Johnstone, R. M., Rottenberg, H., Caplan, S. R. 1977. *Biochim. Biophys. Acta* 465:599–613
16. Falke, J. J., Chan, S. I., Steiner, M., Oesterhelt, D., Towner, P., Lanyi, J. K. 1984. *J. Biol. Chem.* 259:2185–89
17. Garty, H., Caplan, R. S. 1977. *Biochim. Biophys. Acta* 459:532–45
18. Greene, R. V., Lanyi, J. K. 1979. *J. Biol. Chem.* 254:10986–94
19. Greene, R. V., MacDonald, R. E., Perreault, G. J. 1980. *J. Biol. Chem.* 255:3245–47
20. Hartmann, R., Oesterhelt, D. 1977. *Eur. J. Biochem.* 77:325–35
21. Hartmann, R., Sickinger, H.-D., Oesterhelt, D. 1980. *Proc. Natl. Acad. Sci. USA* 77:3821–25
22. Hazemoto, N., Kamo, N., Kobatake, Y. 1984. *Biochem. Biophys. Res. Comm.* 118:502–7
23. Hazemoto, N., Kamo, N., Kobatake, Y., Tsuda, M., Terayama, Y. 1984. *Biophys. J.* 45:1073–77
24. Hazemoto, N., Kamo, N., Terayama, Y.,

25. Kobatake, Y., Tsuda, M. 1983. *Biophys. J.* 44:59–64
25. Hegemann, P. 1984. PhD Dissertation. Ludwig-Maximilian University, Munich
26. Hegemann, P., Oesterhelt, D., Steiner, M. 1985. *EMBO J.* In press
27. Hegemann, P., Steiner, M., Oesterhelt, D. 1982. *EMBO J.* 1:1177–83
28. Henderson, R., Unwin, P. N. T. 1975. *Nature* 257:28–32
29. Hess, B., Kuschmitz, D. 1979. *FEBS Lett.* 100:334–40
30. Kamo, N., Hazemoto, N., Kobatake, Y., Mukohata, Y. 1985. *Arch. Biochem. Biophys.* 237:In press
31. Kamo, N., Takeuchi, M., Hazemoto, N., Kobatake, Y. 1983. *Arch. Biochem. Biophys.* 221:514–25
32. Kanner, B. I., Racker, E. 1975. *Biochem. Biophys. Res. Comm.* 64:1054–61
33. Khorana, H. G., Gerber, G. E., Herlihy, W. C., Gray, C. P., Anderegg, R. J., Nihei, K., Biemann, K. 1979. *Proc. Natl. Acad. Sci. USA* 76:5046–50
34. Deleted in proof
35. Kuschmitz, D., Hess, B. 1982. *FEBS Lett.* 138:137–40
36. Lanyi, J. K. 1980. *J. Supramol. Struc.* 13:83–92
37. Lanyi, J. K. 1981. *Trends Biochem. Res.* 6:60–62
38. Lanyi, J. K. 1982. *Methods Enzymol.* 88:439–43
39. Lanyi, J. K. 1984. *Biochem. Biophys. Res. Comm.* 122:91–96
40. Lanyi, J. K. 1984. *FEBS Lett.* 175:337–42
41. Lanyi, J. K., MacDonald, R. E. 1976. *Biochemistry* 15:4608–14
42. Lanyi, J. K., MacDonald, R. E. 1979. *Methods Enzymol.* 56:398–407
43. Lanyi, J. K., Oesterhelt, D. 1982. *J. Biol. Chem.* 257:2674–77
44. Lanyi, J. K., Schobert, B. 1983. *Biochemistry* 22:2763–69
45. Lanyi, J. K., Weber, H. J. 1980. *J. Biol. Chem.* 255:243–50
46. Lindley, E. V., MacDonald, R. E. 1979. *Biochem. Biophys. Res. Comm.* 88:491–99
47. Luisi, B. F., Lanyi, J. K., Weber, H. J. 1980. *FEBS Lett.* 117:354–58
48. MacDonald, R. E. 1981. In *Chemiosmotic Proton Circuits in Biological Membranes*, ed. V. P. Skulachev, P. Hinkle, pp. 321–35. Reading, Mass: Addison-Wesley
49. MacDonald, R. E., Greene, R. V., Clark, R. D., Lindley, E. V. 1979. *J. Biol. Chem.* 254:11831–38
50. Maeda, A., Ogurusu, T., Yoshizawa, T.,

Kitagawa, T. 1985. *Biochemistry* 24: 2517–21

51. Matsuno-Yagi, A., Mukohata, Y. 1977. *Biochem. Biophys. Res. Comm.* 78:237–43

52. Matsuno-Yagi, A., Mukohata, Y. 1980. *Arch. Biochem. Biophys.* 199:297–303

53. Mehlhorn, R. J., Schobert, B., Packer, L., Lanyi, J. K. 1985. *Biochim. Biophys. Acta.* 809:66–73

54. Mukohata, Y., Kaji, Y. 1981. *Arch. Biochem. Biophys.* 206:72–76

55. Oesterhelt, D. 1975. In *Energy Transformation in Biological Systems. Ciba Found. Symp. 31*, pp. 147–67. Amsterdam: Excerpta Medica/North Holland

56. Oesterhelt, D., Schulten, K. 1985. *Eur. Biophys. J.* In press

57. Ogurusu, T., Maeda, A., Sasaki, N., Yoshizawa, T. 1981. *J. Biochem.* 90:1267–73

58. Ogurusu, T., Maeda, A., Sasaki, N., Yoshizawa, T. 1982. *Biochim. Biophys. Acta* 682:446–51

59. Ogurusu, T., Maeda, A., Yoshizawa, T. 1984. *J. Biochem.* 95:1073–82

60. Polland, H.-J., Frantz, M. A., Zinth, W., Kaiser, W., Hegemann, P., Oesterhelt, D. 1984. *Biophys. J.* 47:55–59

61. Renthal, R., Lanyi, J. K. 1976. *Biochemistry* 15:2136–43

62. Schobert, B., Lanyi, J. K. 1982. *J. Biol. Chem.* 257:10306–13

63. Schobert, B., Lanyi, J. K., Cragoe, E. J. Jr. 1983. *J. Biol. Chem.* 258:15158–64

64. Selwyn, M. J., Dawson, A. P., Stockdale, M., Gains, N. 1970. *Eur. J. Biochem.* 14:120–26

65. Smith, S. O., Marvin, M. J., Bogomolni, R. A., Mathies, R. A. 1984. *J. Biol. Chem.* 259:12326–29

66. Spudich, E. N., Bogomolni, R. A., Spudich, J. L. 1983. *Biochem. Biophys. Res. Comm.* 112:332–38

67. Spudich, E. N., Spudich, J. L. 1982. *Proc. Natl. Acad. Sci. USA* 79:4308–12

68. Spudich, E. N., Spudich, J. L. 1985. *J. Biol. Chem.* 260:1208–12

69. Spudich, J. L., Bogomolni, R. A. 1983. *Biophys. J.* 43:243–46

70. Steiner, M. 1984. PhD Dissertation. Ludwig-Maximilian University, Munich

71. Steiner, M., Oesterhelt, D. 1983. *EMBO J.* 2:1379–85

72. Steiner, M., Oesterhelt, D., Ariki, M., Lanyi, J. K. 1984. *J. Biol. Chem.* 259:2179–84

73. Stoeckenius, W., Bogomolni, R. A. 1982. *Ann. Rev. Biochem.* 52:587–616

74. Sugiyama, Y., Mukohata, Y. 1984. *J. Biochem.* 96:413–20

75. Taylor, M. E., Bogomolni, R. A., Weber, H. J. 1984. *Proc. Natl. Acad. Sci. USA* 80:6172–76

76. Tsuda, M., Hazemoto, N., Kondo, M., Kobatake, Y., Terayama, Y. 1982. *Biochem. Biophys. Res. Comm.* 108:970–76

77. Wagner, G., Hartmann, R., Oesterhelt, D. 1978. *Eur. J. Biochem.* 89:169–79

78. Wagner, G., Oesterhelt, D., Krippahl, G., Lanyi, J. K. 1981. *FEBS Lett.* 131:341–45

79. Weber, H. J., Bogomolni, R. A. 1981. *Photochem. Photobiol.* 33:601–8

Ann. Rev. Biophys. Biophys. Chem. 1986. 15:29–57
Copyright © 1986 by Annual Reviews Inc. All rights reserved

INTERPRETATION OF BIOLOGICAL ION CHANNEL FLUX DATA—
Reaction-Rate versus Continuum Theory

David G. Levitt

Department of Physiology, 6-255 Millard Hall, 435 Delaware Street SE, University of Minnesota, Minneapolis, Minnesota 55455

CONTENTS

PERSPECTIVE AND OVERVIEW

Ion channels (along with other membrane proteins) are the least understood class of biological proteins. The necessity of keeping ion channels in their membrane environment has severely limited physical structural

0883–9182/86/0610–0029$02.00

studies. Although X-ray and electron diffraction have provided low-resolution structures of the gap junction and acetylcholine receptor channels, these techniques are at present limited to membrane proteins that occur naturally in dense two-dimensional arrays; they cannot yet be applied to other biological ion channels (5, 13, 63, 80). The specter of molecular biology has recently entered this field in the form of cloning and sequencing of the acetylcholine receptor (74) and Na^+ channel (73). However, structure cannot be determined from the amino acid sequence alone. In fact, knowledge of the sequence increases the importance of functional studies because they are needed to guide molecular model-building.

The subject of this review is the permeation pathway of the open ion channel. Two approaches have been used to investigate this pathway. The first is performance of qualitative studies that can be interpreted directly (or by using very simple theory) in terms of structural features (e.g. maximum size of permeable ions, interaction between ions, voltage dependence of channel blockers, and number of ions in the channel). The second, which is the subject of this review, is the search for a detailed theoretical model that provides a quantitative description of the ion flux. The features and parameters of this model can then be interpreted in terms of the channel structure.

Current concepts about ion channels are heavily influenced by studies on gramicidin A, the only channel whose detailed structure is known. Knowledge of this structure has stimulated the recent development of a number of theoretical approaches to describe the flux in terms of the molecular interactions between the ion and the channel (76). Since these approaches require knowledge of the detailed molecular structure they cannot, at present, be applied to biological channels. What is needed is a simple theory with a minimum of adjustable parameters. The only theories that satisfy this condition are the reaction-rate and continuum theories. This review concentrates on the application of these theories to biological ion channels. Gramicidin is the subject of a number of recent reviews (33, 45, 76) and is discussed here only to illustrate the use or limits of these two elementary theories.

The first two sections review the two theories. They have been covered in a number of recent reviews and books (18, 44, 49, 77) and are well discussed by Cooper et al (18). The emphasis here is on the basic assumptions and limitations of the theories, particularly the weaknesses of the reaction-rate theory and the advantages of using the continuum approach where applicable. In addition a new result, the continuum theory for a one-ion channel, is briefly presented, with details in the Appendix.

The theories are first applied to the class of ion channels that can be occupied by at most one ion. This is an important class since it can be described by a theory that is both general and simple enough to allow inference about channel structure. The discussion is concerned primarily with the K^+-selective channel isolated from sarcoplasmic reticulum. The last section addresses the class of channels that contain more than one ion.

NERNST-PLANCK CONTINUUM THEORY

The Nernst-Planck continuum theory was the first quantitative approach to the modeling of ion transport (for review, see 44). The basic equation describing the flux (J) of an ion is:

$$J = -DA\left(\frac{dC}{dX} + \frac{zF}{RT}C\frac{dV}{dX}\right) = -DA\left(\frac{dC}{dX} + C\frac{dv}{dX}\right)$$

$$V = \Psi + \Phi, \qquad v = V/(RT/zF) = \psi + \phi.$$

1.

$D(X)$ is the ion diffusion coefficient, $A(X)$ is the area available for the ion, $C(X)$ is the ion concentration, and $V(X)$ is the total electrical potential, which is the sum of the applied external voltage (Ψ) and the intrinsic electrochemical potential (Φ) of the channel. In the second equality, the dimensionless potentials v, ψ, and ϕ are used. In Equation 1 the forces on the ion are divided into two major classes: the complicated short-range forces resulting from collisions with the channel water and wall, summarized by the mobility term (D); and the long-range forces that arise from the structural features of the channel and act through the intrinsic potential energy term (ϕ). The basic assumption of Equation 1 is that the ion remains in equilibrium with the short-range forces as it moves through the channel. This is equivalent to assuming that the noncollisional forces on the ion (ϕ) are of long range compared to its mean free path.

These assumptions should be satisfied in most ion channels because the forces arise primarily from long range electrostatic dipoles and monopoles acting in the high dielectric aqueous environment of the channel. This theory is especially well suited for those biological ion channels (see below) that have large-diameter mouths or vestibules in which the ion movement resembles bulk diffusion. A number of attempts have been made to theoretically calculate the size of the local energy barriers in the gramicidin channel (76) (Figure 1). Although accuracy is limited by uncertainty in the modeling of the intermolecular forces, particularly for the channel water, it is generally agreed that the barriers are small, of an order of several RT (56). These local energy barriers may be so small that the local ion mobility may be limited by the movement of the column of water that the ion must push and not by the local ion-wall interactions (24, 33, 45, 60).

In addition to the long range electrostatic forces that are probably dominant in ion channels, strong short-range forces must also be present. For example, the ability of a number of channels to discriminate between very similar ions (e.g. Rb^+ and K^+) requires the existence of short-range forces. Also, the forces involved in the dehydration of an ion when it enters a channel or passes through a narrow restriction should also be short-range. The movement of an ion in such short-range forces can still be described by Equation 1, but now the forces cannot be divided into long-range (ϕ) and short-range (D) components and D is reduced to an adjustable parameter with little physical significance.

The uses of the continuum and reaction-rate theory are illustrated by their application to the gramicidin A channel, the only channel for which the potential energy profile is known with some certainty (Figure 1). The small barriers in Figure 1 represent the local interactions between the ion and the carbonyl oxygens of the channel wall. (If the movement of the water is rate-limiting, then these barriers must be interpreted in terms of the forces on the channel water.) The long-range barrier in the center of the membrane is an electrostatic effect that results from the influence of the low dielectric lipid that surrounds the channel. The origin of the barriers at the mouth of the channel is uncertain, although it is probably related to dehydration of the ion. In applying the continuum theory to this channel, the smoothed potential profile (dashed line, Figure 1) would be used for the ϕ in Equation 1. The short-range local energy barriers influence the ion transport through

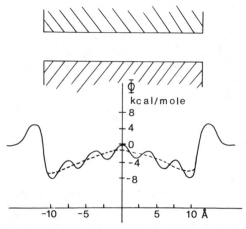

Figure 1 Potential energy profile in the gramicidin A channel estimated from a combination of experimental flux studies and theoretical molecular modeling of the known channel structure (*solid line*) (56). The *dashed line* is the long range component of the potential (ϕ) used in the continuum theory.

the local diffusion coefficient (D) in Equation 1. A major advantage of the continuum theory is that one can usually make a reasonable guess for the value of D. In large channels D should be close to the bulk solution value, and there is a continuum theory for estimating the decrease in D that occurs in narrow channels (61). Even in the very narrow gramicidin channel, D is only about $1/10$ the bulk value (24).

Assuming a steady state [which is valid for any biologically important time (17)], Equation 1 can be integrated exactly using an integrating factor:

$$Je^v = -DA \frac{d}{dX} (Ce^v); \qquad J \int_{-\infty}^{\infty} (e^v/DA) \, dX = C_1 e^{\psi_1} - C_2. \qquad\qquad 2.$$

The integration ranges from the bulk solution on side 1 (C_1, ψ_1) to the bulk solution on side 2 of the membrane ($C_2, \psi_2 = 0$). In the bulk solution the potential ϕ is zero. Equation 2 can be written in the form:

$$J = (C_1 e^{\psi_1} - C_2)/H, \qquad H = \int_{-\infty}^{\infty} (e^v/DA) \, dX. \qquad\qquad 3.$$

All the properties of the channel are summarized in the integral H. This equation provides a complete description of the channel transport.

The integral H in Equation 3 can be separated into the integrals over the bulk solutions (I_1 and I_2) and an integral for the channel proper (I_c): $H = I_1 + I_2 + I_c$. These three integrals correspond to the resistances of the three regions arranged in series. The integrals over the bulk solution (I_1 and I_2) correspond to the diffusion-limited access resistance of the channel (3, 4, 54, 61). If the voltage drop in the bulk solution is small (high concentration of inert electrolyte) and the value of ϕ in the bulk solution is assumed to be zero, then the integration over the bulk solution region can be performed using a hemispherical surface ($A = 2\pi X^2$) centered at the mouth of the pore and the bulk diffusion coefficient (D_0) for D:

$$I_1 = (e^{\psi_1}/D_0) \int_{-\infty}^{-a} dX/(2\pi X^2) = e^{\psi_1}/(2\pi a D_0), \qquad\qquad 4.$$

where a is the "capture radius" of the pore.

The only variables in Equation 3 that are under experimental control are the applied voltage (ψ_1) and the bulk concentrations. Since the flux simply scales with the concentration, no new information is obtained by varying the concentration and all the information about the channel is contained in the flux(or current)-versus-voltage relation. Attwell (6) has shown (assuming that the applied voltage varies linearly within the channel) that the function (exp ϕ)/DA in the channel can be uniquely determined given a perfect knowledge of the current-voltage (I-V) relation. Of course, the I-V curve has limited accuracy and this limits the reliability of the knowledge about (exp $\phi)/DA$. This is the only situation of which I am aware in which the structural features of the channel can be quantitatively and unambiguously related to the experimental flux measurements.

The major limitation of Equation 3 (and the reason it is rarely used to describe channel flux) is that it assumes that there is no interaction between ions. If, for example, there was already an ion in the channel, then the potential from this ion would have to be included in ϕ. This "independence principle" (46) is the basic condition for the validity of Equation 3. This is a severe limitation since in nearly every channel that has been studied, the independence principle has been shown to be invalid at some concentration. At low enough concentrations, all channels should be far from saturation and the independence principle (and Equation 3) will be valid.

The Nernst-Planck continuum equation can also be solved for the case of a channel that can be occupied by at most one ion. This is a more useful solution since a number of biological channels seem to be in this class (see below). The region of the channel that can hold at most one ion extends, by definition, from 0 to L. (The actual physical channel may extend out on both sides of this region.) The basic assumption of the derivation is given by the boundary conditions:

$$C(0) = C_1' P_0; \qquad C(L) = C_2' P_0, \qquad\qquad 5.$$

where P_0 is the probability that the channel is empty and C_1' and C_2' are the concentrations at the two ends of the unoccupied restricted region (which extends from 0 to L). Equation 5 is equivalent to the assumption that if the channel is not occupied the concentration at the entrance is equal to C_1', while if it is occupied the concentration is zero as a result of the ion-ion interaction. In addition, it is assumed that the rate-limiting process for ion transport is in the one-ion region; therefore the ions external to this region are in equilibrium with the bulk solution with which they are in contact:

$$C_1' = C_1 e^{-(v_1 - \psi_1)}; \qquad C_2' = C_2 e^{-(v_2)}, \qquad\qquad 6.$$

where v_1 and v_2 are the total potentials at the ends of the one-ion region. The details of the solution to Equation 1 using Equations 5 and 6 as boundary conditions are given in the Appendix. For the case where only one type of permeable ion is present at equal concentrations on both sides of the membrane ($C_1 = C_2 = C$),

$$J = C(e^{\psi_1} - 1)/[H(1)(1 + LCM)]$$

$$M(\psi_1) = \int_0^1 A e^{-v}[e^{\psi_1} - (e^{\psi_1} - 1)H(x)/H(1)]\, dx \qquad\qquad 7.$$

$$H(x) = L \int_0^x e^{v(x)}/(DA)\, dx,$$

where x ($= X/L$) is a dimensionless variable ranging from 0 to 1. This expression for the flux (J) has the simple Michaelis-Menten dependence on concentration with a dissociation constant $K = (LM)^{-1}$. The application of this solution to the experimental results on biological channels is described below under One-Ion Channels.

If a channel can be occupied by more than one ion then the continuum solution becomes very complicated. For example, for a channel that contains two ions and has a uniform area, one must use in place of Equation 1:

$$\frac{\partial P(x, y)}{\partial t} = 0 = -\frac{\partial u_x P}{\partial x} - \frac{\partial u_y P}{\partial y}$$

$$u_x P(x, y) = -(D/L) \left[\frac{\partial P(x, y)}{\partial x} + P(x, y) \frac{\partial v(x, y)}{\partial x} \right],$$

8.

where $P(x, y)$ is the joint probability that an ion is at x and y, $u_x(x, y)$ is the velocity of an ion at x when there is an ion at y, and $v(x, y)$ is the total potential energy of the channel (59). This equation has been solved for a channel that is similar to gramicidin for two special forms of the function $v(x, y)$ (59). In the first case it was assumed that the channel was so narrow that the ion and water could not pass each other anywhere in the channel. This means that the two ions must move as a coupled pair separated by a constant distance determined by the number of water molecules between them. In the second case it was assumed that there were large vestibules at each end of the channel where the ion and water could pass each other. These two special cases represent a hybrid of the reaction-rate and continuum approach with the rate of entry and exit from the channel described by reaction-rate theory. Even for these simplified cases, the solutions can only be obtained in numerical form. In the general case, Equation 8 would be solved by replacing it with a discrete finite difference equation. This is essentially the same procedure as used in the reaction-rate approach. However, Equation 8 provides the advantage that the transitions between the discrete states are defined not by the energy barriers of the reaction-rate model but in terms of the diffusion coefficient (D) and the potential energy $v(x, y)$.

Recently, another continuum solution has been presented in which Equation 1 is combined with strong electrolyte (e.g. Debye-Huckel) theory (61). If a channel contains fixed charges then, for an accurate solution, one must include in the potential ϕ (Equation 1) the influence of both the fixed charge and the counter ions that are attracted to it. The potential from the counter ions was determined by a self-consistent field approach similar to that used in the Debye-Huckel theory. This approach introduces a weak form of ion interaction because, as the bulk ion concentration increases and screens the field of the fixed charge, the flux tends to saturate. This is a very simple model. The only adjustable parameters are the geometric shape of the channel, the valence and locations of the fixed charges, and the valence and diameter of the ions. A solution has been presented for a channel that contains a single fixed charge and a shape thought to resemble that of the acetylcholine receptor channel, and there is general qualitative and some quantitative agreement between the theoretical predictions and the experimental measurements. This model has the major defect that it does not allow for the strong ion-ion interactions that, for example, are apparently present in one-ion channels.

The strong electrolyte continuum solution should be especially useful for modeling the bulk solution and large vestibule regions of the channel. For example, it can be used to provide a rigorous solution for the access limitation for the case where there are significant voltage gradients in the regions leading up to the pore mouth. Recently, this theory has been used to describe the equilibrium distribution of ions in the charged vestibule of a channel (J. A. Dani, manuscript in preparation).

REACTION-RATE THEORY

Although reaction-rate theory has a long history, its modern usage was popularized by Eyring (for review, see 44). The basic assumption of this theory is that there is a critical barrier that the reactants must overcome before the reaction can occur and that the system at this critical barrier is in equilibrium with a well defined initial state of the reactants (38). When this theory is applied to transport in ion channels, the initial state is represented by the ion in a local energy well, the critical state is when the ion is at the top of the local energy barrier, and the rate (k) of the ion crossing this barrier is described by the absolute reaction-rate theory:

$$k = v e^{-\Delta F/RT} = v e^{-\Delta \phi} e^{-\Delta \psi};$$

$$v = kT/h.$$

9.

ΔF, $\Delta \phi$, and $\Delta \psi$ are the difference between the peak and the well of the free energy, the intrinsic potential, and the applied voltage of the system, respectively. The preexponential (v) is a universal frequency factor. For example, for the gramicidin channel (Figure 1), $\Delta \phi$ is the height of each of the energy barriers and $\Delta \psi$ is the applied voltage difference between the peak of the barrier and the well.

The basic assumption of the theory requires that the ion jumps directly from the well to the peak of the barrier without any intermediate thermal collisions. This assumption is applicable, for example, to the breaking of a covalent bond where the reactants are vibrating in a potential well and the rate of the reaction depends on the probability that the reactants have enough energy to jump out of the well. However, the assumption is not generally valid for ion channel transport (18). Certainly, those regions of the pore that are wide enough to allow the ion to be hydrated will not have the sharp energy barriers required by this theory. Even for narrow channels, such as gramicidin, the theoretical calculations indicate that the barriers are only several RT high and are not steep enough to unequivocally define an initial state of the reactants. Although it is possible to develop more accurate theories for the preexponential factor, they cannot be used for interpreting studies on biological channels because they require detailed knowledge of the chemical structure of the channel wall (76). Thus, even though Equation 9 is commonly used to describe the rate of jumping over the barriers, the concept that "the preexponential factor v is entirely arbitrary" (21) is either stated or implied. This illustrates one of the serious weaknesses of this theory: The absolute values of the energy-barrier heights have no meaning. Only the product $ve^{-\Delta\phi}$ can be unambiguously determined from the theory. Furthermore, since one can always get a better fit by adding more barriers, the number of barriers is also arbitrary.

There are two special cases for which simple analytic solutions for the reaction-rate theory can be obtained for a channel with an arbitrary number, depth, and peak height of barriers. The first is the case where the channel either obeys the independence principle or can be occupied by at most one ion (54) (see One-Ion Channels). The second is the case where the channel is nearly saturated, containing at most one vacancy (48). This latter case is especially useful for interpreting flux ratio data. Although an analytical solution can be obtained for the general multi-ion channel, the resulting algebraic expressions are extremely complicated.

In order to simplify the solution, the generally accepted approach is to fit the experimental data using the smallest possible number of states. For example, for the gramicidin channel (Figure 1), which can be occupied by two ions, the simplest state diagram is:

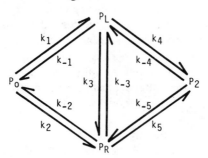

Here, P_0 is the probability the channel is unoccupied, P_L and P_R are the probabilities the channel is occupied by one ion in either the left or right well of Figure 1, and P_2 is the probability the channel is occupied by two ions. Each k in the diagram is the rate constant for a transition between states. The flux (J) through the channel is equal to $k_3 P_L - k_{-3} P_R$. Using this minimum number of energy barriers further decreases the validity of Equation 9. For example, the rate constant k_3 describes the movement of the ion from the left to the right well, a distance of at least 15 Å, during which the ion suffers hundreds of collisions with the wall and water. Clearly, the absolute reaction rate expression is not applicable to this case (45). The more rigorous way to justify and interpret the rate constants in this diagram is as an approximation to the continuum theory. For example, one can regard the above state diagram as a first-order approximation to the general continuum solution (Equation 8) with the joint probability $P(x, y)$ replaced by the four states and the differential Equation 8 replaced by the finite difference equation using the above rate constants. Using an approach due originally to Kramers (see 15), the rate constants can be related via continuum theory to the shape of the barriers (18, 55).

In order to simplify the theory and limit the number of adjustable parameters, it is almost always assumed that the rate constants have the form of Equation 9, i.e. an exponential dependence on the applied voltage difference. In fact, the voltage dependence is related to the shape of the barrier. The voltage dependence is not completely arbitrary since it must satisfy the microscopic reversibility constraint (81). [One advantage of the exponential dependence (Equation 9) is that it automatically satisfies this constraint.] The assumed voltage dependence can significantly affect the interpretation of the data. For example, Eisenman et al (29) assumed an exponential dependence for the rate constants and concluded that the above state diagram (referred to as three-barrier, two-site) could not fit the observed current-voltage relations for the gramicidin A channel. Therefore, they introduced a more complicated four-barrier, three-site model. However, Urban et al (81; reported in 45), treating the potential dependence of the rate constants as arbitrary (subject to the restrictions of microscopic reversibility), used a voltage dependence that was more linear than an exponential function and satisfactorily fit the data using the three-barrier, two-site model. Since there is a natural tendency to regard the barriers and sites as representing physical features of the channel, the assumed form of the voltage dependence of the rate constants significantly influences the physical model one derives from the experimental data.

Since there is no channel in which the potential profile is accurately known, one cannot determine the true voltage dependence of the rate constants. However, carrier-mediated ion transport has a similar state diagram and can be described either by the reaction-rate or continuum approach. Recently (27), the experimental voltage dependence of the nonactin-mediated K^+ transport was compared with the predictions of the two theories and it was shown that the results are quantitatively fit by the continuum theory and cannot, for any set of parameters, be fit by the reaction-rate model (using the exponential voltage dependence, Equation 9).

There are two basic approximations in the use of the reaction-rate theory as it is applied to channel flux. The first is the exponential voltage dependence of the rate constants (discussed above), and the second is the replacement of the general continuum solution by a difference equation with a minimum number of states. It has been shown that this second approximation can also lead to significant errors (59). The reaction-rate model (with the correct voltage dependence determined from continuum theory) was compared with a hybrid model in which reaction-rate constants were used to describe the entrance and exit steps and the continuum model was used for modeling the movement of ions within the channel. In the state diagram shown above for the gramicidin channel, an ion can cross the membrane (go from state P_L to P_R) only if the right end is unoccupied. As the concentration is increased and the channel becomes saturated with two ions, the conductance decreases to zero. In contrast, in the hybrid continuum model when the channel contains two ions, the ion at the left end, for example, can be moved deeper into the channel by an applied voltage, electrostatically displacing the ion at the right end. It has been shown that the reaction-rate approach is quite applicable for a channel that has a large central barrier and strong ion-ion interaction (e.g. gramicidin), while it can cause serious error for a channel with weaker ion-ion interaction, underestimating the conductance by a factor of 4 or more at high concentrations and applied voltages. Since this latter channel type may resemble biological multi-ion channels (see section on multi-ion channels, below), the reaction-rate theory may underestimate the flux in such biological channels at high (saturating) concentrations.

ONE-ION CHANNELS

This section refers to the class of channels that behave as if they can be occupied by at most one ion at any time. It is important to make this distinction, because channels in this class can be described by a relatively simple theory that allows some general, model-independent conclusions about channel structure. The criteria for distinguishing the different channel classes have been recently discussed by Hille (44) and Begenisich & Smith (10).

Theoretical and Experimental Criteria

The general operational definition of a one-ion channel is provided by the following features of the solution, which are valid for all one-ion channels (53). Although these results were obtained by Lauger (53) using the reaction-rate approach and expressions in the form of Equation 9 for the rate constants, the solution is completely general (arbitrary shape and number of barriers and arbitrary preexponential factors) and, therefore, is also valid in the continuum limit. These results could also be derived using the general continuum theory for a one-ion channel described in the Appendix.

For a one-ion channel, the flux can be written in the form:

$$J = A(C_1 e^{\psi_1} - C_2)/(B + DC_1 + EC_2), \qquad 10.$$

where A, B, D and E are functions of the applied voltage. If $C_1 = C_2 = C$, Equation 10 can be written in the Michaelis-Menten form with K and V_m functions of voltage:

$$J = V_m C/(K + C). \qquad 11.$$

In the limit of small applied voltage, the conductance (G_0) can also be written in the form:

$$G_0 = G_m C/(K + C). \qquad 12.$$

This prediction that the conductance reaches a maximum value (G_m) as the concentration of the permeable ion is raised provides the most direct test that the channel can be occupied by at most one of these ions. If the channel could be occupied by more than one ion, a second ion should enter the channel as the concentration is raised above K, increasing or decreasing the conductance. This test has its limitations. If the ionic strength is not held constant as the concentration is changed there may be accompanying changes in the surface potential, which could distort the conductance-versus-concentration curve. Also, if the channel has a very high affinity site that is saturated at concentrations so low that the conductance cannot be distinguished from zero, the binding of the second ion might mimic Equation 12.

The permeability ratio (P_b/P_a) is defined from the expression for the zero current membrane potential in asymmetric solutions:

$$\Psi_1 = -(RT/zF) \ln [C_a^1 + C_b^1 P_b/P_a)/(C_a^2 + C_b^2 P_b/P_a)]. \qquad 13.$$

It can be shown that if the channel obeys the independence principle, then the permeability ratio is a constant, independent of voltage and concentration, and is equal to the ratio of the ion conductances measured at the same applied voltage (43, 46). If the channel is of the one-ion class, the permeability ratio is a function of the voltage but not of the concentration (53). Since changing the concentration of one of the ions changes the potential (Equation 13), the permeability ratio may be an indirect function of the concentration. If the concentration is changed by multiplying all the concentrations by the same factor, the potential and therefore the permeability ratio will remain constant. For the special case where all the peak heights of one ion species, a, differ from those of the second ion species, b, by the same constant factor [peak energy offset condition (42, 43)] the one-ion channel permeability ratio becomes an absolute constant that satisfies the condition

$$P_a/P_b = G_m^a K^b/(G_m^b K^a).$$ 14.

If the channel is of the multi-ion type, the permeability ratio becomes a direct function of both the voltage and concentration.

If two permeable species of ions (a and b) are present with identical solutions on both sides of the membrane, the conductance in the low voltage limit for a one-ion channel can be written as:

$$G_0 = (K^b G_m^a C_a + K^b G_m^b C_b)/(K^a K^b + K^b C_a + K^a C_b),$$ 15.

where K and G are determined from Equation 12 when only one ion species (either a or b) is present. Neher (72) developed this relation to show that the gramicidin channel was not in the one-ion class: When either Tl^+ or Na^+ was present alone the conductance could be described by Equation 12, but when both ions were present the conductance could not be described by Equation 15. Subsequent investigations (30, 58) revealed that Tl^+ had a high-affinity binding component and that the conductance that was fitted by Equation 12 probably resulted from the binding of the second or third Tl^+ ion in the channel. This illustrates the general principle that the last two criteria for a one-ion channel (Equation 15 and the concentration independence of the permeability ratio) are more stringent than the requirement that the conductance-versus-concentration relation is in the form of Equation 12.

Another set of criteria comes from studies using ion-channel blockers. For a one-ion channel, the blocker should satisfy the classical kinetic equations for a competitive inhibitor. In addition to this quantitative kinetic prediction, there are a number of qualitative tests that are useful. For example, Hille (44) has demonstrated that the voltage dependence of the channel block depends on all the ions (both blocking and nonblocking) that must be transported in going from the open to the blocked state. Thus, if the voltage dependence is more than the maximum e fold change per 25 mv that is expected for a monovalent blocker, then the channel must be of the multi-ion type. For a one-ion channel the flux can be written in the form:

$$J = BP_0(C_1 e^{\psi} - C_2),$$ 16.

where P_0 is the probability that the channel is unoccupied and B is a function of the applied voltage. It can be seen from this equation that increasing the concentration of C_1 (which will decrease P_0) must decrease the flux in the direction from side 2 to side 1. However, in multi-ion channels just the opposite can be observed (10). Similarly, an increase in the concentration of the conducting ion in the presence of a blocking ion should increase the flux in a one-ion channel; while for a multi-ion channel, under some conditions, this increase in concentration can decrease the flux (26).

The first test used to demonstrate multiple-ion occupancy was the measurement of the flux-ratio exponent. It can be shown that under general conditions the minimum number of ions that can occupy the channel is equal to the flux-ratio exponent (48, 60). Thus, measurement of a flux-ratio exponent greater than 1 implies that the channel is of the multi-ion type.

Current-Voltage Relationship

The current-voltage (I-V) relation for a one-ion channel depends on the number and shape of the barriers. Since many biological ions channels have linear I-V curves in symmetrical solutions (over a range of ± 100 mv) it is important to consider the conditions under which the one-ion channel has a linear I-V curve. For the continuum model, a linear I-V curve is obtained if the energy profile (ϕ) is a constant for the entire channel. The closer the reaction-rate model comes to meeting this "constant field" condition, the more linear the I-V curve. Thus, in general, the more uniform the spacing and height and the greater the number of barriers, the more linear the I-V curve (i.e. the less the conductance depends on the voltage). For example, the conductance at 75 mv is 17%, 8%, 4%, and 1% greater than at 0 mv for 2, 3, 4, and 6 equally spaced uniform energy barriers, respectively (53). The I-V curve tends to be more linear at high, saturating ion concentrations where most of the biological measurements are made (53). The uniform barrier condition is not optimal since the I-V curve can be made more linear by small changes in the barrier spacing and height (20). The optimal conditions for linearizing the I-V curve have not been quantitatively described.

The one-ion version of the continuum model (see Appendix) also has a linear I-V curve for a constant field. In addition, the I-V curve remains linear if there is a narrow potential-energy well of arbitrary depth and position in the channel. This last condition will be used to model the biological channel data.

The constant field assumption was introduced by Goldman in 1943 (39) and became one of the basic tenets of electrophysiology. This assumption has fallen out of favor in recent years with the recognition that the dielectric forces and the structure of the channel wall must produce variations in the field within the channel. However, some biological channels (see Sarcoplasmic Reticulum K^+ Channel, below) seem to have large vestibules leading up to a relatively short restrictive region, a geometry for which the dielectric image forces become negligibly small (61). Also, for a one-ion channel a sharp energy well can be added to a constant field background (see Appendix) without affecting the I-V curve. These two extensions of the theory, coupled with the experimental observation that most biological channels have linear I-V curves (the sine qua non of the constant field) suggest that this assumption deserves to be revived.

Nearly all the experimental studies referred to in this review rely, in part, on the measurement of the change in ion conductance or permeability as the ion concentration is changed. Unfortunately, experiments of this type have an inherent drawback. In one approach the solutions contain only the ion(s) of interest so that the ionic strength varies as the concentration is changed. If the membrane is charged, changes in ionic strength will change the surface potential and the effective concentration at the mouth of the channel. Even if the membrane surface charge is known, it is difficult to correct for it because the potential at the mouth of the channel depends on the detailed channel and membrane structure (11, 22, 69). Thus, this approach can be used at low concentrations only if one is sure that the membrane is not charged (e.g. if the channel has been reconstituted in an uncharged lipid bilayer). An alternative approach is to keep the ionic strength constant by the use of an "inert" ion. This approach is not completely satisfactory because studies with most cation channels have shown that a truly "inert" ion may not exist (25, 75). Nevertheless, for a channel in a charged membrane or one of unknown charge, this is the only approach that can be used at low concentrations.

Sacroplasmic Reticulum K^+ Channel

The sarcoplasmic reticulum (SR) K^+ channel is probably the simplest and best characterized (in terms of its transport properties) biological ion channel. A large and varied set of studies has established the following features of this channel (68):

1. It meets all the above criteria of one-ion channels. The only exception is Tl^+, which may be associated with multiple occupancy (34, 35). K^+ has the highest maximum conductance (240 pS) and lowest affinity ($K_m = 54$ mM) and Li^+ has a low maximum conductance (7.7 pS) and high affinity (19 mM) (19). Cs^+ has such a high affinity (~ 4 mM)

and low conductance (15 pS) that under most conditions it can be treated as a nonconducting, blocking ion (21).

2. The conductance varies by less than 5% over a voltage range of ± 100 mV for saturating concentrations of K^+ (20). The I-V curve is also nearly linear for Na^+ and Li^+, but is highly nonlinear and asymmetrical for Cs^+ (21).

3. Studies with organic cations of different sizes indicate that the channel has a limiting restrictive region with a cross-sectional area of about 20 $Å^2$. Very large cations (e.g. glucosamine) added to the *trans* side (with respect to the side to which the vesicles that fuse with the bilayer are added) can reach a site at which the applied potential has fallen by 65% of the total transmembrane potential. This 65% of the voltage drop occurs over a distance of only 6–7 Å. Molecules that have a trimethylammonium group on each end separated by nine carbons can fold so that both charged groups reach this site (19, 66).

4. Streaming potential studies show that the restrictive region is short, at most three water molecules in length (67).

A schematic diagram summarizing these features is shown in Figure 2. The rest of this section addresses the question of what additional information can be added to this picture by the use of the quantitative kinetic theories.

Until now, the SR K^+ channel has been modeled exclusively by the reaction-rate theory; two sets of energy-barrier profiles that provide a quantitative fit to the experimental data are shown in Figure 2a (21). As emphasized above, these diagrams were constructed using Equation 9 with the realization that the absolute value of the preexponential factor, and therefore the barrier heights, are arbitrary. The barrier in the center was placed to roughly correspond to the site that the blocking ions reach. If only the single central barrier were used, the I-V curve would be highly nonlinear (regardless of the voltage dependence assumed for this barrier). The additional barriers in Figure 2a, which would have to be about the same size as the central barrier, were added to reproduce the observed linear I-V relation.

None of the qualitative studies on the SR K^+ channel provide any support for the existence of barriers near the ends of the channel. It has been suggested that the barriers may be related to dehydration of the ion, but this seems unlikely because ions as large as glucosamine (7 Å diameter) can traverse 65% of the voltage drop.

The above example illustrates a frequent result of the interpretation of ion channel data by the reaction-rate theory. A theoretical model is developed that can quantitatively fit most of the experimental data and can be used to summarize and parameterize the experimental results for the different ions. However, the linear I-V curve requirement leads to the prediction of large barriers in apparently unphysical locations. The overall effect of the use of the reaction-rate theory is often a muddying of the view of the channel.

The data for this channel can also be interpreted in terms of the one-ion modification of the continuum theory (see Appendix). As discussed above, a central narrow energy barrier cannot fit the observed linear I-V curve, even if the transport over the barrier is described by continuum theory. However, a continuum model that has no barriers but rather a sharp energy well (Figure 2b) can satisfactorily fit the experimental data and may provide a more physically realistic model of the channel. If it is assumed that the function $B(x) = e^{\phi}/A(x)$ (where ϕ is the potential energy in the channel and

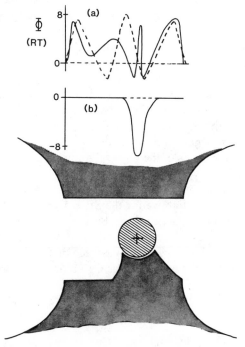

Figure 2   Potential energy profile for K^+ determined from fitting the reaction-rate theory (*a*) or the continuum theory (*b*) to the experimental flux data for the sarcoplasmic reticulum K^+ channel. Although the two curves in (*a*) provide equally good fits, the solid curve is believed to be more physically realistic (21). A schematic model of the rate-limiting region of the pore is shown at the bottom [modified slightly from (21)]. Its length is 10 Å and all other dimensions are to scale. A Cs^+ ion is shown at the binding site.

A is the effective cross-sectional area) is a constant ($= B_0$) except at some region (at an arbitrary position) where there is a sharp energy well where B is very small, then the I-V curve is exactly linear and the conductance (at any voltage) can be written in the form:

$$G = G_m C/(K + C)$$

$$G_m/K = (zF)^2 D/(RTN_{av}LB_0); \qquad K = B_w/L.$$

17.

(For derivation see Appendix, Equation 10A.) Applying this expression to the data for K^+ ($G_m = 240$ pS; $K = 54$ mM) and assuming an effective diffusion coefficient (D) in the channel of 10^{-5} cm^2/sec and a channel length (L) of 10 Å yields, from Equation 17, a B_0 of 8.7×10^{-13}. Since $B_0 = e^\phi/A$, one can then estimate the radius at, for example, the end of the one-ion channel region. Assuming that $\phi = 0$ at the channel end, this value of B_0 corresponds to a radius of about 6 Å. This is, of course, a very rough calculation but it at least shows that this feature of the model is physically realistic. One can also explain, semiquantitatively, the lower conductance for Li$^+$ just from the fact that Li$^+$ is more tightly hydrated and has a bulk diffusion coefficient about half that of K$^+$. If, for example, one assumes that the diffusion coefficient and effective area for Li$^+$ in the channel (61) is 1/3 that of K$^+$, then, from Equation 14 and using the experimental K for Li$^+$, the maximum Li$^+$ conductance will be 4% that of K$^+$, as is observed experimentally. Although these calculations are very rough, they demonstrate that one can begin to interpret the constants determined from the continuum theory in terms of the physical model, which is not possible for the reaction-rate theory.

The continuum theory seems to be consistent with all the other observations summarized above for the SR K$^+$ channel. The only aspect examined here is the apparent block by Cs$^+$ and the nonlinear asymmetrical I-V curve found for this ion (21). In order to explain the observations for Cs$^+$ it is necessary to assume that the channel has both a sharp potential well and a barrier. Since Cs$^+$ blocks primarily from the *cis* side, the barrier is placed on the *trans* side of the well. For the high (saturating) concentrations of pure Cs$^+$ where the flux was measured, the flux can be described by (Appendix, Equation 12A):

$$J = DB_w e^{-\alpha\psi_1}(e^{\psi_1} - 1)/(L^2 B_p),$$

18.

where α is equal to the fraction of the voltage drop that is in the region between the barrier and the well. A fraction $\alpha = 0.3$ fits the observed I-V conductance curve in symmetrical Cs$^+$ solutions. When both Cs$^+$ and another ion are present, only the Cs$^+$ on side 2 (*trans*) is effective in blocking the channel (see Appendix, Equation 13A).

As portrayed by the continuum theory, the rate-limiting region of this channel has a simple structure. It consists of a relatively featureless channel for which permeability is determined primarily by the same factors that influence the bulk diffusion coefficient, containing a localized binding site that has a high affinity for Cs^+ and a low affinity for K^+. The schematic in Figure 2 is drawn to be consistent with this picture. If the binding site is charged then the constant-field condition requires that the decrease in ϕ as the ion approaches the site is balanced by a decrease in A (cross-sectional area of a tapering channel) so that $B = (\exp \phi)/A$ remains constant.

Since as the number of barriers increases the reaction-rate and continuum models become identical, it is possible to simulate the behavior of this continuum model with a reaction-rate model. The model with the fewest barriers that summarizes the main features of this continuum model is a channel with a single binding site separated from each of the bulk solutions by a single barrier (a two-barrier, one-site model). Coronado et al (20), using an exponential voltage dependence (Equation 9), rejected this model because it could not fit the observed I-V curve. Even when an arbitrary voltage dependence is used, the I-V curve cannot be made linear because of the microscopic reversibility condition. If this model did match the I-V curve, it would still have the disadvantage that the barriers do not correlate with any known structural feature in the channel but are only a way of simulating ion diffusion through a wide-diameter channel.

The fact that a channel meets the one-ion criteria discussed above has important implications about its structure. Two conditions must be met. First, either there should be only one high-affinity site or there must be a very high interaction energy between the sites to prevent them both from being occupied. Second, since the maximum conductance is limited by the rate of ion movement through the entire one-ion region of the channel (not just the binding site region), the effective length of the one-ion region (L in Equation 17) must be independent of concentration. This is a very restrictive condition because as the ion concentration is raised one would expect that ions would penetrate further into the occupied channel, shortening the "effective" length of the one-ion region. The condition is satisfied only if the channel consists of two regions: a wide region at each end of a narrow, rate-limiting segment. As the concentration is raised, ions could enter further into the wide region of the occupied channel; this would not change the effective length of the channel, which is equal to the length of the narrow region. Electrostatic repulsion from the ion that occupied the binding site would have to be strong enough to prevent ions from entering the narrow region at the highest concentrations that are studied (about 1 M K^+ for this channel). The linear I-V curve found for this SR K^+ channel requires that the rate-limiting (narrow) region corresponds roughly to the region over which the potential drop occurs (about 10 Å). This is consistent with the picture of the channel shown schematically in Figure 2. Wide outer regions lead up to a more narrow region that contains the ion binding site. If the binding site were near the center of the narrow regions, then electrostatic repulsion would need only to keep the second cation from coming closer than about 5 Å from the ion in the binding site. Although this requirement seems physically reasonable, it needs to be quantitatively evaluated using rigorous electrostatic theory (61).

As this discussion illustrates, a channel cannot be expected to be strictly one-ion because as the concentration is raised to high enough levels (or another ion has a high enough affinity for the channel), a second ion should be able to enter an occupied channel. For example, in this SR channel K^+ above 1 M (20) and Tl^+ (34, 35) seem to be able to bind to the occupied channel. This creates a problem: We must decide whether deviations from strict one-ion behavior are relatively insignificant or are a sign of the basic multi-ion nature of the channel. In addition, the channel may be one-ion for some ions and not for others.

Other Possible One-Ion Channels

Most of the flux measurements for the acetylcholine receptor channel can be described by one-ion kinetics. The conductance-versus-concentration curve saturates about as one would predict from Equation 12, recognizing that the ionic strength varies and that the membrane or channel may be charged (23, 28, 47, 78, 79). In addition, in one study the permeability ratio seemed to be relatively independent of concentration (1). The channel is usually modeled by the two-barrier, one-site reaction-rate theory which, by definition, is a one-ion model. Because this simple model does not have the correct I-V dependence, more complicated models [e.g. fluctuating barrier models (28)] have been proposed. However, as the above discussion has emphasized, this two-barrier reaction-rate model is only a crude approximation and one would not expect it to have the observed I-V dependence. Several molecular models of the channel have been constructed using the known amino acid sequence (7, 32, 40, 85). Although these models have been constrained to be consistent with the known low-resolution structural data determined from electron microscopy and electron diffraction, they are not consistent with the structure predicted above for a one-ion channel. The proposed models have a long region of relatively uniform cross section with a surface lining consisting of a large number of fixed positive and negative charges; the models do not have the short restricted region that one would expect. In addition, the fixed charges should attract counter ions, which would reduce the electrostatic repulsion that is necessary to limit the occupancy to one ion. Resolution of this issue will have to wait for more definitive permeability and/or structural studies.

The Na^+ channel has some characteristics of a one-ion channel. The batrachotoxin-activated Na^+ channel reconstituted into neutral lipid bilayers has a Na^+ conductance–versus-concentration curve that is well described by Michaelis-Menten kinetics (Equation 12) for concentrations up to 1 M Na^+. Dissociation constants (K) of 8 mM (70) and 37 mM (36) have been reported. These are much smaller than the 368 mM reported for the intact frog node (42). Most of this difference can probably be accounted for by the presence of competing and blocking ions (e.g. K^+ and Ca^{2+}) in the node experiments (36). In addition, the interaction between Na^+ and the blocking ion Ca^{2+} is of the simple one-ion type (36). In contrast, two sets of measurements on the squid axon suggest that the channel is of the multi-ion type: (a) The Na^+/K^+ permeability ratio is a function of concentration (8, 14) and (b) the blocking by organic cations has the complicated kinetics of a multi-ion channel (26). The interpretation of these conflicting results should be clarified by the many studies on the reconstituted Na^+ channel that are now in progress.

MULTI-ION CHANNELS

The definition of the multi-ion channel is that it does not meet the criteria discussed above for a one-ion channel.

Inward Rectifier K^+ Channel

The inward rectifier K^+ channel is the first and still best studied multi-ion channel. Since this channel has been reviewed recently (10, 37, 50, 51), only one aspect is emphasized here: the relation of the reaction-rate model to the channel structure.

The observed flux ratio when the internal K^+ is 300 mM and the external K^+ is 30 mM is about 3.0 (9). This means that for these concentrations the channel is occupied by at least three and possibly four K^+ ions (see section on theoretical and experimental criteria under One-Ion Channels, above). If this represents the maximum (saturated) number the channel could hold, then as the concentration is increased the flux should stay constant or decrease (48). However, two studies have recently reported that the flux continues to increase as the K^+ concentration (equal on both sides) is raised from 300 to 500 mM (16) or 1000 mM (65). This means that the channel must be able to hold at least one more ion and thus can hold a minimum of four and probably five or six K^+ ions.

A reaction-rate description of this channel would require at least a five-barrier, four-site potential profile. The continuum description would require a solution of a partial differential equation similar to Equation 8 except that the probability would have to be a function of at least four variables. Although the reaction-rate theory has successfully provided a qualitative explanation of the large number of multi-ion effects that are seen with this channel, the kinetic description of the channel is so complicated that one cannot expect to obtain quantitative structural information by the use of either theory.

Some guesses about the channel structure can be made based on the observation that the channel can be occupied by four or more ions. Electrostatic calculations indicate that a narrow uncharged channel cannot be occupied by more than two ions (57, 71). In order to hold more than two ions, the channel probably needs to be charged and it is likely to have some regions with a large diameter so that the dielectric image charge effect (and resulting electrostatic repulsion between K^+ ions) is small. However, one cannot conclude that there are three or four localized binding sites in the channel. Because of the long range of the coulomb potential, a channel lined by negative charges will have a diffuse, nonlocalized potential and the K^+ ion may experience a relatively constant field profile in its passage through the channel.

Ca^{2+} Selective Channel

A longstanding question is whether it is possible for a channel to have a high ion selectivity. The problem is most clearly seen when the channel obeys the independence principle (i.e. at concentrations so low that the probability that the channel is occupied is small). In this case, it can be shown that the permeability is determined solely by the height of the energy barriers and is independent of the well depths (31, 43). Therefore, high selectivity requires a very discriminating energy barrier. However, there is no known physical or molecular property that can produce such a selective barrier. In contrast, an energy well or binding site can have a high specificity because it can structurally resemble the complexing sites in carriers or enzymes that are known to be highly selective. Thus, a selective channel would seem to require that a binding site be the locus of the rate-limiting process for the flux. However, a single binding site in the one-ion channel will not serve this function because, in order to be selective, it must have a high affinity (low K, Equation 12). Therefore it must also have a low conductance because the maximum conductance is proportional to K (53) (see Equation 14). In addition, since the permeability ratio for a one ion channel depends only on the energy barrier heights (53), a selective permeability would again require the existence of a selective energy barrier.

For these reasons, it has been argued that the highly selective Ca^{2+}

permeability of cell membranes involved a carrier rather than a channel mechanism. However, recent studies have clearly established that a Ca^{2+} channel does exist and there is good evidence that it achieves its high selectivity through a multi-ion mechanism (2, 41). The authors of these studies have proposed that there are two high-affinity interacting sites for Ca^{2+}. At micromolar concentrations, when only one Ca^{2+} site is occupied, the channel behaves as predicted for a one-ion channel, blocking the conductance of other ions but having a very low conductance because of the high affinity. As the concentration is raised into the millimolar range the second Ca^{2+} site becomes occupied, and thus the other ion is displaced and the conductance is increased. The experimental data for this channel can be qualitatively described by a two-site, three-barrier model. On the basis of the above theoretical discussion of the mechanism of channel selectivity and this one example, one is tempted to conclude that any highly selective channel must be multi-ion.

High-Conductance Ca^{2+}-Activated K^+ Channel

Although there are some variations in selectivity and gating properties among the channels in this class, the channels have enough features in common to support the idea that they all have the same basic structure (51). It has recently been shown that two of these channels are clearly of the multi-ion type; the conductance-versus-concentration relationship for a channel isolated from T-tubules (69) and reconstituted in uncharged lipid bilayers indicates that at least two K^+ ions can occupy the channel (with apparent dissociation constants 2.5 and 500 mM), and patch clamp studies on chromaffin cells (64, 82, 83) have revealed a channel of this class that has the complicated ion-ion interactions that are diagnostic of a multi-ion channel. In contrast, patch clamp studies on a Ca^{2+}-activated K^+ channel in muscle indicate that it is of the one-ion type (12). In particular, the permeability ratio of Rb^+ and NH_4^+ to K^+ is constant over a 10-fold range of concentration. This is presumably a good indicator that the channel is not multi-ion (see section on theoretical and experimental criteria under One-Ion Channels, above). This may imply that there are fundamental differences in the permeability pathways of different channels within this class.

The T-tubule channel demonstrates a selectivity for K^+ that is unique among the class of K^+-selective channels (50, 51). It has a measurable conductance only to K^+, Tl^+, and possibly NH_4^+ (52). Remarkably, it shows no detectable conductance to Rb^+, an ion that is very similar to K^+. The chromaffin cell channel also has a low Rb^+ conductance (84) but Rb^+ permeability (defined from Equation 13) similar to that of K^+ (12). This combination of a low conductance and a high permeability for Rb^+

is what would be expected for a channel with a high-affinity binding (blocking) site for Rb^+ (50).

Another interesting feature of Ca^{2+}-activated K^+ channels is that they have maximum conductances of about 600 pS, the highest recorded for any channels. This high conductance is somewhat misleading because it is consistent with a channel that has the same physical dimensions as the sarcoplasmic reticulum K^+ channel (Figure 2). Since the maximum conductance is proportional to the binding constant (K; see Equation 17), the threefold higher maximum conductance measured for the T-tubule channel is just what would be predicted for the approximately threefold larger K (threefold lower affinity) that is experimentally observed.

One result of this high conductance is that it could produce large voltage gradients in the regions leading up to the mouth of the channel. From an analysis of the "flickery" block of the channel produced by Na^+, Yellen (82) determined the voltage dependence of the rate constants for blocking and unblocking of the channel by Na^+ and showed that it could be explained by the voltage drop at the mouth of the channel. This is the most interesting illustration that the gradients of voltage and concentration in the region leading up to the channel mouth must be considered in any quantitative modeling of the channel flux. Although these effects have been examined in considerable detail for the gramicidin A channel (3, 4), the interpretation of the results is still controversial (45). One difficulty has been a lack of rigorous theoretical treatment for this problem. Yellen (82) used a theory developed by Lauger (54) that assumes electroneutrality in the bulk solution at the channel mouth. Although this electroneutrality cannot be rigorously correct, it may be a good approximation. Andersen (3) developed a different theory, modeling the mouth of the channel as an equivalent one-dimensional capacitor with an adjustable capacitance. Levitt's (61) recent approach avoids the electroneutrality assumption and should provide a more accurate solution.

SUMMARY

Although the reaction-rate theory may provide a useful mathematical description of the channel flux, it presents a misleading physical picture of the channel structure. There is a tendency to regard the barriers in the model as actual physical structures, whereas they are actually only mathematical artifacts that allow one to reduce a complicated differential equation with an infinite number of states to a finite difference equation with a minimum number of states. I argue that the energy profile in the permeation pathway of most biological channels should vary relatively smoothly with only a few localized energy barriers or wells. In these smoothly varying regions, the resistance to ion movement is similar to bulk diffusion and cannot be accurately modeled by one or two energy barriers. For the one-ion channel, the continuum approach is as general and at least as simple as the reaction-rate theory and may provide a more physical interpretation of the data. Thus for the SR K^+ channel, the structure suggested by the reaction-rate theory seems inconsistent with some experimental data, while the continuum-theory model is not only consistent with, but complements, the structure suggested by other data. Multi-ion channels have such complicated kinetics that one can only expect the theories to provide a qualitative description of the experimental data. They can be modeled by either the reaction-rate model or a finite difference approximation to the continuum model.

ACKNOWLEDGMENTS

I am grateful to Drs. Begenisich, Clay, Dani, French, Jakobsson, Miller, and Oxford for providing papers prior to publication.

APPENDIX: DERIVATION OF CONTINUUM SOLUTION FOR ONE-ION CHANNEL

A. General Solution

The solution is obtained for the differential Equation 1 subject to the boundary conditions of Equations 5 and 6. The equations are written in terms of the dimensionless variable $x = X/L$ where L is the length of the one-ion region. Equation 1 is integrated from 0 to x (see Equation 2):

$$J = [C(0)e^{v_1} - C(x)e^{v(x)}]/H(x).$$

$$1A.$$

$$H(x) = L \int_0^x D^{-1}B(x)e^{\psi(x)} \, dx, \qquad B = e^{\phi}/A$$

where C and J are in units of ions/unit volume and ions/sec.

At $x = 1$ ($X = L$), using Equations 5 and 6:

$$J = P_0[C_1 e^{\psi_1} - C_2]/H(1) \qquad \qquad 2A.$$

where P_0 is the probability the channel is not occupied.

Substituting Equation 2A for J into Equation 1A and solving for $C(x)$,

$$C(x) = e^{-v(x)}P_0[C_1 e^{\psi_1} - (C_1 e^{\psi_1} - C_2)H(x)/H(1)].$$

The probability the channel is occupied is equal to $1 - P_1$ where P_1 is the probability the channel contains an ion.

$$P_0 = 1 - P_1 = 1 - L \int_0^1 AC(x) \, dx = 1 - LP_0 C_1 M(\psi_1, C_2/C_1)$$

$$P_0 = [1 + LC_1 M]^{-1}. \qquad\qquad 3A.$$

$$M(\psi_1, C_2/C_1) = \int_0^1 B^{-1} e^{-\psi(x)} [e^{\psi_1} - (C_2/C_1)(e^{\psi_1} - 1) H(x)/H(1)] \, dx.$$

Substituting Equation 3A for P_0 into Equation 2A,

$$J = (C_1 e^{\psi_1} - C_2)/[H(1)(1 + LC_1 M)]. \qquad\qquad 4A.$$

In the limit as $\psi_1 \to 0$, the conductance (G_0) can be written in the form $(C_1 = C_2 = C)$:

$$G_0 = I/\Psi_1 = zeJ/(RT\psi_1/zF) = (G_{max}/K_m)C/(1 + C/K_m) \qquad 5A.$$

$$G_{max}/K_m = (zF)^2/[N_{av} RT H_0(1)]; \qquad K_m = (LM_0)^{-1}$$

where H_0 and M_0 are for $\psi = 0$ and N_{av} is Avogadro's number.

If two ions (a, b) are present, the flux can be written as:

$$J = J_a + J_b; \qquad J_x = P_0(C_x^1 e^{\psi_1} - C_x^2)/H_x(1)$$

$$P_0 = [1 + LC_a^1 M_a(\psi_1, C_a^2/C_a^1) + LC_b^1 M_b(\psi_1, C_b^2/C_b^1)]. \qquad 6A.$$

The bionic potential follows directly from the condition that $J = 0$:

$$\psi_1 = \ln [(C_a^2 + C_b^2 P_b/P_a)/(C_a^1 + C_b^1 P_b/P_a)] \qquad\qquad 7A.$$

$$P_b/P_a = H_a(1)/H_b(1).$$

Since H is a function of the applied voltage, so are the permeabilities (P) in Equation 7A. If ϕ satisfies the constant offset condition $\phi_a(x) = \phi_a^0 + \phi(x)$, and D and A are of the form $D_a(x) = D_a^0 D(x)$, $A_a(x) = A_a^0 A(x)$, then the voltage dependence cancels out of the permeability ratio.

B. Generalized Constant Field Assumption

This is the assumption that $B(x)$ is a constant (B_0) except for a very localized potential energy well where B is very small $(|\phi| \gg 1, \phi < 0)$. It is also assumed that in the rate-limiting region of the channel, the applied voltage varies linearly $(\psi(x) = \psi_1(1-x))$, D is a constant, and $C_1 = C_2 = C$. For these conditions, H and M can be explicitly evaluated:

$$H(x) = (L/D) \int_0^x e^\psi B \, dx \simeq (LB_0/D) \int_0^x e^\psi dx = LB_0(1 - e^{-\psi_1 x})e^{\psi_1}/(D\psi_1)$$

$$H(1) = LB_0(e^{\psi_1} - 1)/(D\psi_1) \qquad\qquad 8A.$$

$$M(\psi_1) = \int_0^1 B^{-1} \, dx \simeq \int_0^1 B_0^{-1} \, dx + \int_{\text{well}} B^{-1} \, dx = B_0^{-1} + B_w^{-1}$$

where B_w^{-1} is defined as equal to the integral of B^{-1} over the localized potential energy well. Substituting Equation 8A into Equation 4A, the flux is:

$$J = D\psi_1 C/[LB_0(1 + LC(B_0^{-1} + B_w^{-1}))]. \qquad\qquad 9A.$$

The I-V curve is linear and the conductance (at any voltage) can be written in the Michaelis-Menten form (Equation 12) with:

$$G_{\max}/K_m = (zF)^2 D/(RTN_{\text{av}}LB_0); \qquad K_m = [L(B_0^{-1} + B_w^{-1})]^{-1}. \qquad 10A.$$

The data for the SR K$^+$ channel indicate that $B_0^{-1} \ll B_w^{-1}$ and this is assumed throughout the text. The permeability ratio (Equation 7A) is independent of voltage:

$$P_a/P_b = D_a B_0^b/(D_b B_0^a). \qquad\qquad 11A.$$

In order to fit the results for Cs$^+$ (high affinity, low conductance, asymmetric I-V curve) it is necessary to assume that there is a localized energy well that dominates the integral for M, a localized potential barrier that dominates the integral for H, and a constant background $B(= B_0)$:

$$H(x) = \begin{cases} LB_0 e^{\psi_1}(1 - e^{-\psi_1 x})/(D\psi_1) & x < x_b \\ LB_b e^{\psi_b}/D & x > x_b \end{cases}$$

$$M = e^{-\psi_w}/B_w \qquad\qquad 12A.$$

$$J = DC(e^{\psi_1} - 1)/[LB_b e^{\psi_b}(1 + LCe^{-\psi_w}/B_w)]$$

$$J \to DB_w(e^{\psi_1} - 1)/(L^2 B_b e^{\alpha\psi_1}) \quad \text{as} \quad C \to \infty,$$

where x_b and x_w are the positions of the barrier and well, respectively; ψ_b and ψ_w are the values of the applied potential at these positions, B_b is the integral of B over the barrier region, and B_w^{-1} is the integral of B^{-1} over the well; and α is the distance between the well and the peak. It has been assumed in Equation 12A that $x_w > x_b$ and $C_1 = C_2 = C$. If Cs$^+$ and another ion are present, then the flux is described by Equation 6A and

$$C_{\text{Cs}}^1 M_{\text{Cs}}(\psi_1, C_{\text{Cs}}^2/C_{\text{Cs}}^1) = C_{\text{Cs}}^2 e^{-\psi_w}/B_w. \qquad\qquad 13A.$$

It can be seen that only the ions on side 2 (the side closest to the energy well) contribute to M and, by reducing P_0, block the channel.

Literature Cited

1. Adams, D. J., Dwyer, T. M., Hille, B. 1980. *J. Gen. Physiol.* 75:493–510
2. Almers, W., McCleskey, E. W. 1984. *J. Physiol.* 353:585–608
3. Andersen, O. S. 1983. *Biophys. J.* 41:135–46
4. Andersen, O. S. 1983. *Biophys. J.* 41:147–65
5. Anholt, R., Lindstrom, J., Montal, M. 1985. In *Enzymes of Biological Membranes*, Vol. 3, ed. A Martonosi, pp. 335–402. New York: Plenum
6. Attwell, D. 1979. In *Membrane Transport Processes*, Vol. 3, ed. C. F. Stevens, R. W. Tsien, pp. 29–41. New York: Raven
7. Bash, P. A., Langridge, R., Stroud, R. M. 1985. *Biophys. J.* 47:43a
8. Begenisich, T. B., Cahalan, M. D. 1980. *J. Physiol.* 307:217–42
9. Begenisich, T., DeWeer, P. 1980. *J. Gen. Physiol.* 76:83–98
10. Begenisich, T., Smith, C. 1984. *Curr. Top. Membr. Transp.* 22:353–69
11. Bell, J. E., Miller, C. 1984. *Biophys. J.* 45:279–87
12. Blatz, A. L., Magleby, K. L. 1984. *J. Gen. Physiol.* 84:1–23
13. Brisson, A., Unwin, P. N. T. 1985. *Nature* 315:474–77
14. Cahalan, M., Begenisich, T. 1976. *J. Gen. Physiol.* 68:111–25
15. Chandrasekhar, S. 1943. *Rev. Mod. Phys.* 15:1–89
16. Clay, J. R. 1985. *Biophys. J.* 47:221a
17. Cole, K. C. 1968. *Membranes, Ions and Impulses.* Berkeley/Los Angeles: Univ. Calif. Press
18. Cooper, K., Jakobsson, E., Wolynes, P. 1985. *Prog. Biophys. Mol. Biol.* 46:51–96
19. Coronado, R., Miller, C. 1982. *J. Gen. Physiol.* 79:529–47
20. Coronado, R., Rosenberg, R., Miller, C. 1980. *J. Gen. Physiol.* 76:425–46
21. Cukierman, S., Yellen, G., Miller, C. 1985. *Biophys. J.* In press
22. Deleted in proof
23. Dani, J. A., Eisenman, G. 1984. *Biophys. J.* 45:10–12
24. Dani, J. A., Levitt, D. G. 1981. *Biophys. J.* 35:501–8
25. Dani, J. A., Sanchez, J. A., Hille, B. 1983. *J. Gen. Physiol.* 81:255–81
26. Danko, M., Smith-Maxwell, C., McKinney, L., Begenisich, T. 1985. *Biophys. J.* In press
27. Dijk, C. V., de Levie, R. 1985. *Biophys. J.* 48:125–36
28. Eisenman, G., Dani, J. A. 1985. In *Proc. Int. Sch. Ionic Channels*, ed. R. Latorre. New York: Plenum. In press
29. Eisenman, G., Hagglund, J., Sandblom, J., Enos, B. 1980. *Upsala J. Med. Sci.* 85:247–57
30. Eisenman, G., Sandblom, J., Neher, E. 1977. In *Metal-Ligand Interaction in Organic Chemistry and Biochemistry*, ed. B. Pullman, N. Goldblumn, pp. 1–36. Dordrecht, Holland: Reidel
31. Eyring, H., Lumry, R., Woodbury, J. W. 1949. *Rec. Chem. Prog.* 10:100
32. Finer-Moore, J., Stroud, R. M. 1984. *Proc. Natl. Acad. Sci. USA* 81:155–59
33. Finkelstein, A., Andersen, O. S. 1981. *J. Memb. Biol.* 59:155–71
34. Fox, J. A. 1983. *Biochim. Biophys. Acta* 736:241–45
35. Fox, J., Ciani, S. 1985. *J. Membr. Biol.* 44:9–23
36. French, R. J., Krueser, B. K., Worley, J. F. 1985. See Ref. 28
37. French, R. J., Shouklima, J. J. 1985. *J. Gen. Physiol.* 85:669–98
38. Glasstone, S., Laidler, K. J., Eyring, H. 1941. *The Theory of Rate Processes.* New York: McGraw-Hill
39. Goldman, D. E. 1943. *J. Gen. Physiol.* 27:37–60
40. Guy, H. R. 1984. *Biophys. J.* 45:249–61
41. Hess, P., Tsien, R. W. 1984. *Nature* 309:453–56
42. Hille, B. 1975. *J. Gen. Physiol.* 66:535–60
43. Hille, B. 1975. In *Membranes*, Vol. 3, ed. G. Eisenman, pp. 255–323. New York: Marcel Dekker
44. Hille, B. 1984. *Ionic Channels of Excitable Membranes.* Sunderland, Mass: Sinauer
45. Hladky, S. B., Haydon, D. A. 1984. *Curr. Top. Membr. Transp.* 21:327–72
46. Hodgkin, A. L., Huxley, A. F. 1952. *J. Physiol.* 116:449–72
47. Horn, R., Patlak, J. 1980. *Proc. Natl. Acad. Sci. USA* 77:6930
48. Kohler, H., Heckmann, K. 1979. *J. Theor. Biol.* 79:381–401
49. Lakshminarayaniah, N. 1984. *Equations of Membrane Biophysics.* New York: Academic
50. Latorre, R., Alvarez, O., Cecchi, X., Vergara, C. 1985. *Ann. Rev. Biophys. Biophys. Chem.* 14:79–111

51. Latorre, R., Miller, C. 1983. *J. Memb. Biol.* 71:11–30
52. Latorre, R., Vergara, C., Moczydlowski, E. 1983. *Cell Calcium* 4:343–57
53. Lauger, P. 1973. *Biochim. Biophys. Acta* 311:423–41
54. Lauger, P. 1976. *Biochim. Biophys. Acta* 455:493–509
55. Lauger, P. 1982. *Biophys. Chem.* 15:89–100
56. Lee, W. K., Jordan, P. C. 1984. *Biophys. J.* 46:805–19
57. Levitt, D. G. 1978. *Biophys. J.* 22:209–19
58. Levitt, D. G. 1978. *Biophys. J.* 22:221–48
59. Levitt, D. G. 1982. *Biophys. J.* 37:575–87
60. Levitt, D. G. 1984. *Curr. Top. Membr. Transp.* 21:181–97
61. Levitt, D. G. 1985. *Biophys. J.* 48:19–32
62. Deleted in proof
63. Makowski, L., Caspar, D. L. D., Phillips, W. C., Baker, T. S., Goodenough, D. A. 1984. *Biophys. J.* 45:208–18
64. Marty, A. 1983. *Pflügers Arch. Gesomte Physiol. Menschen Tiere* 396:179–81
65. May, K. W., Oxford, G. S. 1985. *Biophys. J.* 47:221a
66. Miller, C. 1982. *J. Gen. Physiol.* 79:869–91
67. Miller, C. 1982. *Biophys. J.* 38:227–30
68. Miller, C., Bell, J. E., Garcia, A. M. 1984. *Curr. Top. Membr. Transp.* 21:99–132
69. Moczydlowski, E., Alvarez, O., Vergara, C., Latorre, R. 1985. *J. Membr. Biol.* 73:273–82
70. Moczydlowski, E., Garber, S. S., Miller, C. 1984. *J. Gen. Physiol.* 64:665–86
71. Monoi, H. 1983. *J. Theor. Biol.* 102:69–99
72. Neher, E. 1975. *Biochim. Biophys. Acta* 401:540–44
73. Noda, M., Shimizu, S., Tanabe, T., Takai, T., Kayano, T., et al. 1984. *Nature* 312:121–27
74. Noda, M., Takahashi, H., Tanabe, T., Toyosato, M., Kikyotani, S., et al. 1983. *Nature* 302:528–32
75. Oxford, G. S., Yeh, J. Z. 1985. *J. Gen. Physiol.* 85:583–602
76. Polymeropoulos, E. E., Brickmann, J. 1985. *Ann. Rev. Biophys. Biophys. Chem.* 14:315–30
77. Starzak, M. E. 1984. *The Physical Chemistry of Membranes.* New York: Academic
78. Suarez-Isla, B. A., Wan, K., Lindstrom, J., Montal, M. 1983. *Biochemistry* 22:2319
79. Tank, D., Huganir, R., Greengard, P., Webb, W. 1983. *Proc. Natl. Acad. Sci. USA* 80:5129
80. Unwin, P. N. T., Zampigi, G. 1980. *Nature* 283:545–49
81. Urban, B. W., Hladky, S. B., Haudon, D. A. 1980. *Biochim. Biophys. Acta* 602:331–54
82. Yellen, G. 1984. *J. Gen. Physiol.* 84:157–86
83. Yellen, G. 1984. *J. Gen. Physiol.* 84:187–99
84. Yellen, G. 1984. *Ionic permeation and blockade in calcium-activated potassium channel of chromaffin cells.* PhD Thesis, Yale Univ., New Haven, Conn.
85. Young, E. F., Ralston, E., Blake, J., Ramachandran, F., Hall, Z. W., Stroud, R. M. 1985. *Proc. Natl. Acad. Sci. USA* 82:626–30

Ann. Rev. Biophys. Biophys. Chem. 1986. 15:59–78

ANTIFREEZE GLYCOPROTEINS FROM POLAR FISH BLOOD

R. E. Feeney and T. S. Burcham

Department of Food Science and Technology, University of California, Davis, California 95616

Y. Yeh

Department of Applied Science, University of California, Davis, California 95616

CONTENTS

0883–9182/86/0610–0059$02.00

PERSPECTIVES AND OVERVIEW

In polar or subpolar areas of the ocean, fish may live in icy waters as cold as $-1.9°C$ (the freezing temperature of sea water). The blood of marine fish is hypoosmotic to ocean water and therefore fish would not be expected to survive near the freezing temperature of polar sea water. Early work by Gordon et al (34) and by Scholander and co-workers (57, 58) indicated that the blood serums of Arctic fish had lower freezing temperatures than did the serums of fish not adapted to the cold. They reported nondialyzable substances in the serum that helped to lower the freezing temperature. Sodium chloride, urea, and amino acids account for the observed $-1°C$ freezing point depression of serum from many marine temperate zone fish (34, 50). When scientists visited Antarctica (19, 22), they could see fish swimming among ice crystals and even resting upon anchor ice. The freezing temperature of the blood of these fish was determined and found slightly lower than the sea temperature of $-2°C$. Upon dialysis, the serum was found to have a significantly lower freezing temperature than could be accounted for by the sodium chloride and other dialyzable molecules in their serum (19, 23, 50). In contrast, temperate zone fish serum, upon dialysis, freezes at $0°C$. Therefore, polar fish synthesize a macromolecular "antifreeze" that lowers their serum freezing temperature below that of sea water; since the antifreeze is composed of macromolecules, it contributes very little to the osmotic pressure of their serum. In all organisms in which they have been found, these macromolecular antifreezes are glycoproteins or proteins (17, 20, 30, 60).

PROTEINS AS ANTIFREEZE AGENTS IN FISH BLOOD

Structures and General Properties of Antifreeze Glycoproteins

Because of the limited space for this review, initial studies on the chemistry of the antifreeze glycoproteins (17–19, 43) are not discussed. Only recent experiments on physical and chemical properties are discussed. Information available prior to 1977 has been previously summarized (27, 55, 64). For more recent physiological studies consult References 15, 27a, and 29. Some of the relevant properties of antifreeze glycoproteins are summarized in Table 1. Since current results support a mechanism of action at the ice-solution interface (7; T. S. Burcham, D. T. Osuga, Y. Yeh, R. E. Feeney, manuscript submitted for publication; W. L. Kerr, R. E. Feeney,

Table 1   Important observations on activities of antifreeze glycoproteins

Observation	Reference
1. Depresses the freezing temperature but not the melting point	Feeney & Hofmann (24)
2. Depresses the freezing temperature ~500 times as based on molecular weight	DeVries et al (17); Feeney & Yeh (27)
3. Depresses the freezing temperature additively with salt	DeVries et al (17)
4. There is no evidence for unusual interaction with liquid H_2O	Tomimatsu et al (61)
5. Functions in the presence of ice	Feeney & Hofmann (24); Burcham et al (9)
6. Is found to be trapped between crystal grains	Tomimatsu et al (61)
7. Ice is normal by X-ray and Raman	Tomimatsu et al (61)
8. Small AFGP "inactive" with "rough" ice made from deeply supercooled solutions; larger AFGP potentiates small AFGP with ice from supercooled solutions	Burcham et al (9)
9. Growing edge of ice crystals disrupted; growth axis of ice crystals change	Tomimatsu et al (61); Knight et al (42); W. L. Kerr, R. E. Feeney, D. T. Osuga, D. S. Reid, manuscript submitted
10. Surface secondary harmonic generation shows AFGP adsorbed to ice surface	Brown et al (7)
11. Surface energy of growing edge of ice crystal lowered	W. L. Kerr, R. E. Feeney, D. T. Osuga, D. S. Reid, manuscript submitted
12. Data for activities fit a kinetic model for adsorption	T. S. Burcham, D. T. Osuga, Y. Yeh, R. E. Feeney, manuscript submitted

D. T. Osuga, D. S. Reid, manuscript submitted for publication), particular attention is given to the following four indications of this mechanism: (a) The ice-crystal habit may affect activity, (b) the generation of surface second harmonic of light gives direct evidence for adsorption of the protein onto the ice surface, (c) interfacial studies show decreases in surface energies, and (d) kinetic modeling calculations fit a reversible adsorption mechanism.

STRUCTURE OF ANTIFREEZE GLYCOPROTEINS

Structure of the Glycotripeptide Unit and its Polymers

The most ubiquitous form of macromolecular antifreeze is the antifreeze glycoprotein (AFGP) of polar fish. AFGP is found in almost every fish living in the Antarctic; the Antarctic fish families of Nototheniidae, Channichthyidae, and Bathydraconidae all contain AFGP (1). Antifreeze glycoproteins have also been found in five Arctic fish, notably of the family Gadidae: *Boreogadus saida, Eleginus gracilis, Gadus morhua, Gadus ogac,* and *Microgadus tomcod* (28, 36, 50, 56, 63). All the above sources yield the same glycotripeptide monomer: The AFGP are polymers of H_2N[Ala-Ala(β-galactosyl(1 → 3)α-N-acetylgalactosamine)Thr]$_n$Ala-Ala-COOH and consist typically of eight distinct components [n = {50, 45, 35, 28, 17, 12, 6–4}] termed "AFGP 1" through "AFGP 8," respectively (17, 18, 43, 50, 62; T. S. Burcham, D. T. Osuga, B. N. N. Rao, C. A. Bush, R. E. Feeney, manuscript submitted for publication). The structure of an Ala-Ala-(disaccharide)Thr is shown in Figure 1.

STRUCTURES OF THE LOW MOLECULAR WEIGHT PEPTIDES   An exception to the general monomeric structure is found in the low molecular weight components, AFGP 6–8, which have proline replacing some alanine in the

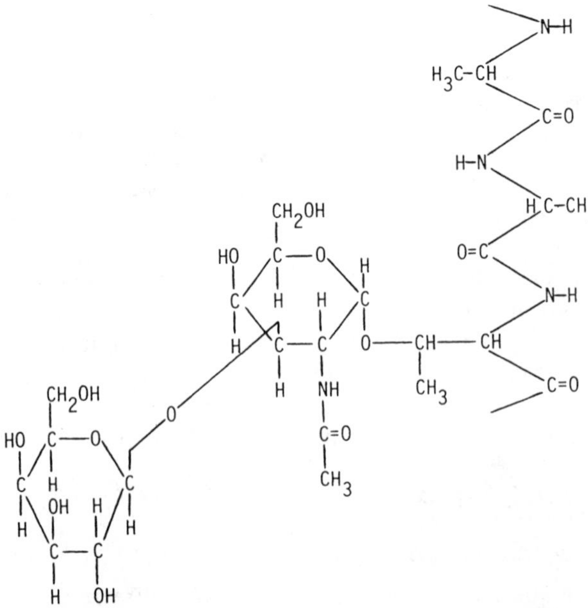

Figure 1   Polymer unit of AFGP 1–5. [Reproduced with permission from (25).]

first position (46). The low molecular weight components are heterogeneous with respect to the position of the proline(s), both within and among fish species. For example, AFGP 8 from the Antarctic fish *Pagothenia borchgrevinki* consists of three homologous glycopeptides that vary in the number of prolines and exist in the approximate proportion of 7:2:1 (46). In contrast, the AFGP 8 from the polar cod, *B. saida*, appears to be homogeneous with respect to sequence (50).

Conformation

The antifreeze glycoproteins appear to have a somewhat unusual conformation. Bush et al (12) and C. A. Bush & R. E. Feeney (manuscript submitted for publication) used high-field proton and ^{13}C NMR and found that AFGP cannot be a true random coil or an extended rigid rod. Instead, the results depict AFGP as a flexible-rod conformation having locally defined structure, particularly that of a three-fold left-handed helix (11, 32), but having enough segmental mobility to destroy any long-range order. Note that a three-fold left-handed helix would place all the carbohydrate residues on one side of the rod-shaped molecule; this is an important consideration for a proposed mechanism of AFGP activity. Theoretical conformational analysis of the AFGP has shown that a three-fold left-handed helical structure of AFGP is energetically stable (4).

STRUCTURE OF OTHER PROTEIN ANTIFREEZES

There are other antifreeze proteins that contain little or no carbohydrate. The most prevalent of these are very similar in structure to the polar glycoproteins in that they contain up to 60% alanine, but they also contain hydrophilic amino acids, such as Asp, that may have a role analogous to the carbohydrate of AFGP in substituting for the sugars. They are, however, less active on a molar basis than the glycoproteins (51; T. S. Burcham, D. T. Osuga, Y. Yeh, R. E. Feeney, manuscript submitted for publication). Nevertheless, it must be emphasized that there are some proteins that have been partially characterized, and are more like the structure of most proteins (37). Since much less information is available on the physical properties of these proteins than on the glycoproteins, the nonglycoproteins are not discussed in this review.[1]

---

[1] While this manuscript was in proof, data were presented on X-ray measurements for the high-alanine antifreeze protein containing no carbohydrate (Yang, D. S. C., Hew, C. L. 1985. *Abstr. Am. Crystallographic Assoc. Annual Meeting, Stanford, Calif.*).

GENERAL CHARACTERISTICS OF ANTIFREEZE GLYCOPROTEINS

Measurement of Antifreeze Activity

Activity can be measured with any apparatus that can determine the freezing and melting temperature of an AFGP solution with sufficient accuracy ($\pm 0.005°C$). One such apparatus is a sealed capillary that contains the solution whose freezing temperature is to be measured and some form of nucleus, typically an ice-crystal seed. Freezing in a bath is observed. A microscopic method that requires very small amounts of solution employs a microscope equipped with a cold stage, temperature sensor, and video screen monitor (24, 42).

An instrument commonly used to determine solution freezing temperature, a freezing point osmometer, measures the freezing temperature by first supercooling the solution and then nucleating it with a submerged rapidly vibrating wire. A thermistor immersed in the solution detects the rapid increase in temperature due to the release of the heat of fusion and measures the temperature of freezing by the temperature of the plateau during freezing (41).

Antifreeze Activity

On a molar basis, AFGP 1–5 (24) lowers the freezing temperature more than NaCl does (the lowering is ~ 500 times the lowering expected if its function were colligative). AFGP solutions melt at or near 0°C. This difference between the melting and freezing temperatures, or "thermal hysteresis," is the activity of AFGP (24; T. S. Burcham, D. T. Osuga, Y. Yeh, R. E. Feeney, manuscript submitted for publication). AFGP solutions exhibit hyperbolic functions of activity versus concentration (Figure 2) and strictly additive lowerings with most colligatively functioning substances, e.g. NaCl. The low molecular weight component, AFGP 8, has ~ 20% of the activity of the same weight concentration of the high molecular weight components, AFGP 1–5, and only ~ 5% of the activity based on molecular weight.

POSSIBLE MECHANISMS OF FUNCTION

Since the antifreeze glycoproteins do not function colligatively, they must act in some way to change the properties of the ice-water system. They might act on (a) the solution phase, (b) the ice solid phase, or (c) the ice-solution interface.

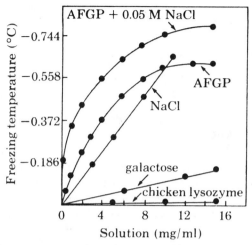

Figure 2   Freezing temperatures of solutions of sodium chloride, galactose, lysozyme, a mixture of AFGP 3–5 from *Pagothenia borchgrevinki*, and a mixture of glycoprotein and sodium chloride. [Reproduced with permission from (17).]

Solution Alterations

The binding of water by dissolved polyglycols has been extensively studied (31). In all instances the thermodynamic change leads to abnormal freezing and melting characteristics; usually substantially lowered freezing points are observed. Since the antifreeze glycoprotein contains such high amounts of carbohydrate distributed along the polymer, binding of large amounts of water to these groups might be expected.

Formation of Solid Solutions

The formation of solid solutions results in changed freezing and melting temperatures. With carbohydrate side chains along the length of the polymer the antifreeze protein might form a solid solution with ice, but these solid solutions might also be expected to affect the melting temperature.

Inhibition at the Ice-Solution Interface

Crystal growth may be inhibited by adsorption at a crystal surface. Inhibition of growth of calcium oxalate crystals by a protein has been reported (48, 49). The adsorption of AFGP on an ice surface could inhibit ice growth by producing a barrier between the ice surface and water molecules, thereby reducing the rate of crystal propagation (13). Crystal growth can occur only if the solution temperature is lower than some critical value at which the barrier can be overcome.

EXPERIMENTAL INVESTIGATIONS OF MECHANISMS OF FUNCTION

Solution Studies

CONFORMATION AND STRUCTURE   No direct evidence shows that solution conformation of AFGP is related to antifreeze activity. No changes in conformation of AFGP with temperature that might be related to function have been observed by ultracentrifugation, quasi-elastic light scattering, or CD measurements (2, 11). In order to determine whether conformational changes would occur under conditions where the protein would be functioning, measurements obviously needed to be done below 0°C and in the presence of ice crystals. Such measurements by quasi-elastic light scattering showed no unusual changes (2).

Burcham et al (9) attempted to demonstrate that molecular weight (or number of residues) of the polymer alone influences activity, by forming a covalent dimer of the weakly active AFGP 6 ($M_r \sim 4600$) to give a larger product (M_r 9200) similar in size to the highly active AFGP 5. This product was no more active than the original monomer. However, the dimer differed from AFGP 5 in containing prolines and a single run of four alanines at the connecting link.

With one exception, all covalent modifications of the carbohydrate residues of AFGP resulted in loss of activity (17, 33). Those causing losses of activity included O-acetylation of the hydroxyls, oxidative degradation by periodate, oxidation of the C-6 hydroxyls to the carboxyl group by bromine, and β-elimination of the disaccharides with alkali. However, removal of the acetyl group from the acetylated AFGP restored activity. The one modification that did not affect activity was the oxidation of the C-6 hydroxyls of the Gal and GalNAc to the aldehyde group using galactose oxidase. Formation of the bisulfite adduct to this AFGP polyaldehyde resulted in complete loss of activity that could be regained by removing the sulfite.

Other enzymatic changes involving the carbohydrate chains also have caused loss in activity. Addition of sialic acid to the galactose reduced activity to less than 10% (26); subsequent enzymatic removal of the sialic acid resulted in up to 75% recovery of activity.

AFGP activity was lost when 2 moles of borate were bound per mole of disaccharide side chain, and these losses and bindings were pH-dependent (3). The number of moles of borate bound per disaccharide unit was only about 0.5 at pH 8.0 but approached 2 at pH 9.0. When the effects of pH and borate on the conformation of the AFGP were studied, no significant changes were observed in the diffusion coefficients as determined by quasi-elastic light scattering or equilibrium centrifugation, or in the S_{20} values (3).

BINDING OF WATER TO THE POLYMER   Viscosity, translational diffusion, isopiestic determination, and NMR all show that the amount of water bound to AFGP is not significantly different than the amount of water bound to other glycoproteins (5, 12, 25, 27).

Studies of the Ice Phase

DISTRIBUTION OF AFGP BETWEEN ICE AND LIQUID   Duman & DeVries (21) found that the concentration of AFGP 1–5 in the liquid phase was unchanged upon partial freezing of the solution and concluded that AFGP 1–5 are incorporated into the ice phase. AFGP 8 was retained partially in the ice phase. While incorporation of AFGP 1–5 into the ice phase was found in Raman studies, AFGP 8 could be completely excluded from the ice phase when the surface of the growing ice was continually washed by circulation of AFGP 8 solutions (61). The crystal structure of the ice from AFGP solutions appeared to be identical with that of pure ice (53, 61).

Although the ice formed from AFGP 4 solutions appeared to be ordinary ice, disorientation of the crystal and poorly defined ice-solution interfaces have been observed (61). When a 1% solution of AFGP 4 was in contact with an oriented â-axis single crystal of ice, the crystal could not continue growing with further undercooling, even at $-4.5°C$ when growth would be very slow and even with continuous sweeping of the ice-solution interface with fresh glycoprotein solution. The ice that formed showed different optical polarization properties over the numerous regions of secondary crystallites (61). Comparative Raman spectroscopic studies of AFGP frozen into the ice phase and AFGP remaining in the liquid phase showed some structural differences and may be useful in a mechanistic interpretation (61). Difference spectra were most pronounced in the COH vibrational region.

Interfacial Studies

ICE CRYSTAL HABIT AFFECTS THE ACTIVITY OF AFGP 7–8   Initially, the low molecular weight AFGP 7–8 were considered to have very little or no activity. It was then shown (45, 54) that the apparent inactivity of these AFGP components was a result of the solution's nucleation temperature when the activity was measured. When the solution nucleation temperature was below $-4°C$, AFGP 7–8 had no activity, while at temperatures above $-3°C$, AFGP 7–8 had at most 5% of the molal activity of AFGP 1–5 (47). When the solution is nucleated at minimal supercooling, the activities of

AFGP 7–8 and 1–5 are purely additive, as long as the total activity is less than the maximum possible activity obtainable from such a mixture (9, 47). When the activity of the low molecular weight AFGP 7–8 is tested at temperatures less than $-4°C$ in the presence of very small amounts of the high molecular weight AFGP 1–5, the activity of the low molecular weight AFGP 7–8 is retained; this effect is termed "potentiation" (9, 47, 52). The effect is quite unlike the absence of activity of the low molecular weight AFGP 7–8 when examined alone at deep supercooling, as explained above. The specificity of the potentiation was shown by the inability of the antifreeze protein containing no carbohydrate (but containing $\sim 60\%$ alanine) to affect the activity of AFGP 8 at deep supercooling, although the nonglycoprotein functioned with AFGP 8 near $0°C$ or with the larger AFGP 4 at deep supercooling (51).

There is evidence that the loss of activity of the low molecular weight AFGP 7–8 may be due to the ice crystal habit resulting from the degree of supercooling and the attendant capacity of growth (9), and not due solely to the rate of ice crystal growth during the measurement of freezing temperature as was proposed earlier (54, 59). The ice crystal form was implicated as the cause of the loss of AFGP 7–8 activity when the activity of the AFGP 7–8 was examined with two different ice surfaces, one surface formed at $-1°C$ and the other at $-6°C$; but activity was measured in the 0 to $-1°C$ region. The results of this experiment are given in Table 2. Note that potentiation is also observed in this activity measurement. The loss of activity of the low molecular weight AFGP appears to be due to its inability to adequately cover ice growth sites. Potentiation could be due to a cooperativity between the high and low molecular weight AFGP at the ice surface.

The loss of AFGP 7–8 activity when the solution is nucleated at deep supercooling or in the presence of ice made at deep supercooling could be an effect of crystal surface quality on the inherent energy requirements for

Table 2   Results of visual determinations[a]

Sample	Freezing temperature (°C) in the presence of ice made at	
	$-1°C$	$-6°C$
AFGP 1–4 (4 mg ml^{-1})	-0.39	-0.36
AFGP 7–8 (10 mg ml^{-1})	-0.30	-0.05
AFGP 1–4 (4 mg ml^{-1}) + AFGP 7–8 (10 mg ml^{-1})	-0.51	-0.51

[a] Reproduced with permission from (9).

crystal growth. If the crystal surface is "rough," as in the case of ice made at −6°C, the energy barrier for more water molecules to be incorporated into the crystalline phase is lower than that of a "smooth" crystal, which instead depends on surface nucleation in order to form a new crystal layer (39). These effects, together with the probable difference in desorption coefficients for the high and low molecular weight AFGP, may indeed explain the loss of AFGP 7–8 activity at deep supercooling.

DIRECT EVIDENCE FOR AFGP ADSORPTION ONTO AN ICE SURFACE   At sufficiently high light intensities, a polarizable material begins to respond nonlinearly to incoming light, and the signal transmitted by such a system is a superposition of the fundamental and harmonic frequencies (6). If the polarizability of the material possesses inversion symmetry with respect to the light wave oscillations, then only the odd harmonics of the polarizability will be induced by intense incident light. Since a surface does not possess inversion symmetry with respect to its normal, the reflected light will also induce polarizability at even harmonics. The dominant component will be the second harmonic, an effect termed "surface second harmonic generation" (SSHG) (14). The preference at the ice-water interface for a tetrahedral structure of liquid water, even allowing for strained bonds, and the extremely small linear polarizability of either water or ice at the fundamental (1064 nm) and second harmonic (532 nm) lead to the observation that the pure ice-water interface exhibits no discernable SSHG (7). An adsorbed layer on an ice crystal should cause surface inversion asymmetry and thus a potentially higher SSHG signal. Figure 3 displays the normalized second harmonic intensity changes measured when pure water above an equilibrium ice surface is replaced by AFGP and control solutions. In every case, whether for a normal glycoprotein (ovomucoid) or for chemically inactivated AFGP, the control substances did not show an increase in SSHG (7). However, for every AFGP component and at all concentrations examined, active AFGP gave significant SSHG increases, and the amount of SSHG observed increased with bulk AFGP concentration. Thus, the AFGP appear to adsorb at an ice surface and the amount of AFGP at the ice surface is a function of solution AFGP concentration (7). The tendency of SSHG to increase with concentration parallels the increase of AFGP activity with concentration and is thought to be the result of a dynamic equilibrium between AFGP and the ice surface. The saturation in activity at high AFGP concentrations is probably the result of surface-site saturation.

EXAMINATION OF CHARACTERISTICS OF ICE CRYSTAL GROWTH   The growth patterns of ice crystals under different conditions can provide valuable information pertinent to understanding how AFGP functions (25). We have initiated two different studies under controlled temperature gradients: (a) direct macroscopic observations of columns of growing ice and (b) microscopic observations. Direct observations of growth in horizontal tubes showed differences in the growth patterns between solutions of AFGP 1–4 and water, as did observations of vertical columns. Both showed rougher and more opaque ice with AFGP. Microscopic observations of a growth front showed even greater effects (40; W. L. Kerr, R. E. Feeney, D. T. Osuga, D. S. Reid, manuscript submitted for publication). In the presence of AFGP 8, roughness and partial angularity of growth were seen, while with AFGP 1–4 the ice grew transversely to the temperature gradient. Furthermore, the front on the gradient direction developed angularly, i.e. crystals grew stepwise at right angles to the advancing front. On melting the reverse occurred. The different direction of growth with AFGP agrees with various observations that an axis approaching the c-axis is the preferred one of growth (42). However, the observed transverse growth did not appear to be a "10$\bar{1}$x" direction as described by Knight et al (42) for very low concentrations of AFGP.

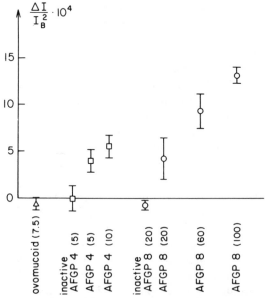

Figure 3 Surface second harmonic generation normalized intensity versus glycoprotein solution. Intensity increase is measured in photons detected per 2000 laser pulses and is normalized with respect to the square of the baseline intensity. Five sets of data are displayed: ovomucoid, AFGP 4, AFGP 8, inactive AFGP 4 (acetylated), and inactive AFGP 8 (periodate oxidized). Numbers in parentheses are concentrations in mg ml^{-1}. [Reproduced with permission from (7).]

We have since conducted grain boundary experiments where clearly discernable boundaries between the crystals and the solution can be seen (W. L. Kerr, R. E. Feeney, D. T. Osuga, D. S. Reid, manuscript submitted for publication). These experiments are conducted by first allowing for crystals of pure ice to grow into its own melt. After equilibration within the horizontal temperature gradient, the remaining water is then removed and AFGP solution is introduced. When equilibrium is reestablished, the new grain boundaries are photographed.

The basic objective of this series of microscopic studies is to measure directly the change in the curvature of the grain boundaries between the two cases, pure melt and AFGP solution, and to calculate surface energies therefrom. Preliminary calculations have shown a substantial decrease in surface energy in the presence of AFGP 1–5 and only a very small decrease in the presence of a control protein ovomucoid (W. L. Kerr, R. E. Feeney, D. T. Osuga, D. S. Reid, manuscript submitted for publication).

THEORY OF KINETIC ADSORPTION MECHANISM

In order to develop a theory to describe the mechanism by which AFGP lowers freezing temperature, we have to consider the preponderance of experimental data that favor a kinetic surface mechanism. The direct evidence that the AFGP is adsorbed onto the ice surface (7, 40), the existence of a freezing-melting temperature hysteresis (24), and the fact that the growth orientation of ice grown in the presence of AFGP differs from that of pure ice-crystal growth (40, 42, 53, 55), while the activity of the low–molecular weight AFGP depends on the growth habits of the prepared ice surface (9), all point toward a kinetic mechanism of AFGP activity where surface adsorption is implicated. Qualitatively, such a surface kinetic mechanism may be described in the following manner.

The Pure Ice-Water System

Ice and its liquid phase, water, can coexist in equilibrium at 0°C. If one plots the energy diagrams, the levels of free energy for the two coexisting phases must be the same. Equal exchange takes place from one phase into the other via an activation barrier, which, for the pure system, is diffusion controlled. The total height of this barrier depends not only on translational diffusion but on rotational diffusion such that a liquid molecule of water must be in the proper orientation to hydrogen bond onto the ice surface. When water is supercooled the effective activation barrier decreases in height. Both enthalpic and entropic factors are significant in reducing the effective activation barrier height. Essentially, when a pure system is supercooled the free energy level of the liquid rises compared to that of the liquid that was at equilibrium with the ice phase. This condition allows ice to form much more readily. At $-40°C$, the temperature corresponding to the homogeneous nucleation temperature of water, there is no activation barrier at all. When ice is already present, however, the kinetic barrier is much less than the barrier toward nucleation, and the height of this barrier becomes a function of crystal orientation. For example, normal ice growth favors the $\{10\bar{1}0\}$ â-axis direction while the ĉ-axis, $\{0001\}$, grows very slowly. Other orientations have not been observed to grow at all at atmospheric pressure.

In the Presence of AFGP

The presence of AFGP on the surface of ice alters both the thermodynamics of the surface and the activation barrier height. Thus, the total degree of freezing temperature depression is a result of thermodynamic and kinetic effects. These two effects can be distinguished by the exhibition of the freezing-melting temperature hysteresis. On a thermodynamic basis, the combined freezing point depression comes about from the sum of the colligative depression contribution, the solution nonideality contribution, and the multicomponent total surface free energy. Feeney & Yeh (27) discussed these in an earlier review. Incorporating AFGP molecules onto the surface increases the lattice-misfit factor between the pure and the impure surfaces (38). This increase is sufficient to cause a lowering of the freezing point of the heterogeneous system. On a kinetic basis, any adsorbed molecules on an ice surface will prevent a water molecule from occupying the same site. Site competition therefore provides a mechanism by which the freezing temperature is lowered, but the melting temperature is still the thermodynamic melting point.

Raymond & DeVries (55) attempted to fit experimental data on freezing temperature depression to AFGP concentration using surface thermodynamic considerations. In a recent review DeVries (16) further discussed a kinetic adsorption-inhibition idea. His hypothesis is based on the assumption that AFGP does not leave the surface of the ice, and consequently inhibition of certain growth direction can take place. This view seems to be too restricting on two bases (T. S. Burcham, D. T. Osuga, Y. Yeh, R. E. Feeney, manuscript submitted for publication): First, the suggestion that growth can only take place by two-dimensional nucleation on the inhibited surface is not true for nonfaceted surfaces such as the â-axis of an ice crystal. Secondly, there is now experimental evidence that the AFGP–ice surface interaction is dynamic (7). Burcham et al (9; T. S. Burcham, D. T. Osuga, Y. Yeh, R. E. Feeney, manuscript submitted for publication) postulated a reversible adsorption mechanism where only the adsorbed state can effectively prevent water molecules from bonding to the existing ice surface.

The amount of AFGP adsorbed onto the crystal surface is proportional to the antifreeze activity of these molecules. The activation energy barrier is then greater than that of the pure system. If the AFGP molecules are desorbed from that surface, then growth of the crystal can proceed in a normal way. Since the water molecules will choose a path of minimal resistance while AFGP is adsorbed, new growth axes may indeed develop. The ice may simply not grow at all along certain directions. Experimentally, high concentrations of AFGP forced a growth direction parallel to the ĉ-axis (40, 55), while at very low concentrations of AFGP a new direction of $\{10\bar{1}x\}$ was observed (42).

Reversible Adsorption of AFGP

We start by describing the interaction between AFGP and ice as a simple reversible adsorption onto an ice surface, which can be represented as:

$$AFGP + ICE \underset{k_d}{\overset{k_a}{\rightleftharpoons}} AFGP - ICE, \qquad\qquad 1.$$

where AFGP represents an antifreeze glycoprotein molecule, ICE represents an appropriate lattice site on the ice crystal, AFGP-ICE represents an antifreeze glycoprotein molecule adsorbed at the surface of an ice crystal, k_a is the rate of adsorption, and k_d is the rate of desorption. The AFGP-ICE component of the above scheme represents ice-crystal growth inhibition on the part of the antifreeze glycoprotein, although it is not specified whether the actual adsorption or desorption processes or the actual adsorbed state, as implied, causes the ice-crystal growth inhibition. (The latter has experimental evidence in its favor, i.e. the SSHG experiments described above.)

Since the above scheme represents a reversible process, one can draw an analogy between it and Langmuirian adsorption (44). The actual specific lattice site in the above scheme has not been specified, nor has the actual ligand unit of the antifreeze glycoprotein molecules, but since AFGP is a repeating polymer of identical units the actual lattice site on the ice crystal and the actual ligand on AFGP will more than likely be the entire monomer unit or some part of it. This model also assumes that the lattice sites on ice are identical (a reasonable assumption given the crystal structure of ice) and that there is no AFGP-AFGP interaction on the ice surface. The steady-state rate equation for Equation 1 is:

$$k_d\theta = k_a c(1 - \theta), \qquad\qquad 2.$$

where θ is the fraction of the total sites covered on the ice surface by an AFGP molecule. If $k_d/k_a = K$, the desorption coefficient, then Equation 2 rearranges to:

$$\theta = \frac{c}{K+c}.$$

3.

Note also that θ can be considered to be the probability that adsorption will occur, given a concentration of c. That is, if one takes a free site on the ice crystal to have a statistical weight of 1, the statistical weight for an adsorbed AFGP molecule is c/K.

At this time, a direct estimate of the fraction of an ice surface covered by AFGP, θ, has not been reported for an AFGP solution. Although the SSHG experiment gives strong evidence that surface AFGP concentration increases with the bulk concentration, the amount of AFGP adsorbed has yet to be quantified using this technique. If one assumes that θ is proportional to the observed antifreeze activity, then the observed plateau should correspond to the saturation of all the lattice sites on the ice crystal surface. If we assign the value of ΔT to be the difference between the activity of the melting and freezing of an AFGP solution, and ΔT_m to be the maximum observed activity for a specific component of AFGP, then θ can be defined as:

$$\theta = \frac{\Delta T}{\Delta T_m}.$$

4.

Inserting Equation 4 into Equation 3 and rearranging, we get:

$$\Delta T = \frac{\Delta T_m c}{K+c}.$$

5.

From this, we can see that ΔT_m represents the binding capacity of ice for a particular antifreeze component, and K represents the interaction between AFGP and ice.

The application of Equation 5 to AFGP 4 is given in Figure 4 (*top*). Figure 4 (*bottom*), however, shows that AFGP 8 can exhibit a sigmoidal activity-versus-concentration curve. This type of profile suggests cooperative interaction, and data accordingly must be analyzed using the Hill equation (Equation 6):

$$\Delta T = \frac{\Delta T_m c^n}{K'+c^n}.$$

6.

K' is a cooperative equilibrium constant, and n represents the number of cooperative units. We caution that the AFGP-ice-water system is more complex than a two-component enzyme-substrate system. However, it is possible to deduce a cooperativity number from data like that of Figure 4

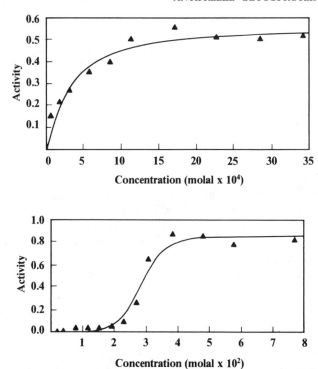

Figure 4   *Top*: Plots generated by the fitting routine for AFGP 4 at bath temperature of −2.0°C. *Bottom*: Plots generated by the fitting routine for AFGP 8 at bath temperature of −3.5°C. Points are actual data points, and curves drawn are the nonlinear least-squares best fit for the normal activity plot. [Reproduced from (8).]

(*bottom*) if one is careful not to suggest the concreteness of those numbers. Table 3 lists data from several species of AFGP. It is possible to conclude that the conditions of the ice surface at the given supercooling require the collective function of n small AFGP 8 molecules, while only $n = 1$, noncooperativity, is observed for AFGP 4.

SUMMARY AND CONCLUSIONS

Existing experimental evidence strongly suggests that the mechanism of activity of the Antarctic AFGP molecules is the inhibition of ice growth by competitive adsorption onto the growth sites of ice. The data further suggest the blocking of the formation of large critical nuclei for ice growth. Experiments showing that the longer polymers (AFGP 1–5) have different growth-prevention properties with different types of ice than the shorter polymers (AFGP 6–8) provide additional evidence that crystal size and habits are linked to function. Four main observations have been used in AFGP studies: (a) The ice crystal habit (size) affects the activity, (b) AFGP is on the surface of ice crystals, as shown by surface second harmonic generation (SSHG), (c) the presence of AFGP lowers the surface energy at the ice-solution interface, and (d) kinetic calculations of the inhibition of ice-crystal growth fit adsorption isotherms. In particular, AFGP 4 fits a Langmuirian adsorption curve.

On the detailed mechanistic side, there is a need to quantify the competitive rates of water-molecule attachment to ice as well as the rates of adsorption-desorption of AFGP on the ice surface. Toward this end, experiments are currently being conducted to examine the differential growth rates of ice crystals freely growing into a solution of AFGP. Direct observation of the adsorption and desorption of AFGP on the ice surface is possible in principle using the SSHG technique. A nonperturbative probe method that can differentiate adsorbed AFGP from the solution phase molecules is needed. Finally, a mechanistic description of molecular function is never complete unless the detailed molecular binding to the surface is elucidated. This task has not been accomplished yet, and thus there is still a question as to which molecular group is actually adsorbed onto the surface and for how long. Theoretically, as we have seen, the departure from Langmuirian adsorption is pronounced for small AFGP molecules under certain conditions. More refinements of the ideas of

Table 3 Summary of apparent desorption coefficient maximum activity, and goodness-of-fit parameters[a]

Sample[b]	Bath temperature (°C)	ΔT_m	K' ($\times 10^4$)	n	$K'^{1/n}$ ($\times 10^4$)	Ω^c	R^2	SE	Σ^2
AFGP 4	−2.0	0.58	2.9	1	2.9	1.0	0.99	0.035	0.011
	−3.0	0.69	3.2	1	3.2	1.0	1.00	0.029	0.007
	−4.5	0.65	1.9	1	1.9	1.0	0.99	0.062	0.035
AFGP 8	−2.5	0.97	68	1	68	1.0	0.99	0.067	0.054
	−3.5	0.85	4.1E-9	8	283	4.2	0.99	0.053	0.034
	−4.5	0.29	4.2E-16	12	523	7.7	0.97	0.018	0.004

[a] The abbreviations used for the goodness-of-fit parameters are: R^2, square of correlation coefficient; SE, standard error or standard deviation of regression; Σ^2, final sum of squares of the residuals. Data from Burcham (8).

[b] The molecular weights used in calculating the concentration are as follows: AFGP 4 = 17,500; AFGP 8 = 2600.

[c] The "interaction factor", Ω in this table, is only proportional to the true interaction factor. Since the nth root of K' is taken, $\Omega = (abc...)(a^{-1}b^{-2}c^{-3}...)$, where $a, b, c,...$ are defined as the microscopic interaction factors.

intermolecular or intramolecular cooperative interaction of AFGP on the ice surface are needed.

ACKNOWLEDGMENTS

The preparation of this review and much of the research described herein were supported in part by NIH Grant GM23817. The authors would like to thank Chris Howland for editorial assistance and Diana Melbourn for typing the manuscript.

Literature Cited

1. Ahlgren, J. A., DeVries, A. L. 1984. *Polar Biol.* 3:93
2. Ahmed, A. I., Feeney, R. E., Osuga, D. T., Yeh, Y. 1975. *J. Biol. Chem.* 250:3344
3. Ahmed, A. I., Yeh, Y., Osuga, D. T., Feeney, R. E. 1976. *J. Biol. Chem.* 251: 3033
4. Avanov, A. Ya., Lipkind, G. M., Kochetkov, N. K. 1982. *Bioorg. Khim.* 8:616
5. Berman, E., Allerhand, A., DeVries, A. L. 1980. *J. Biol. Chem.* 255:4407
6. Bloembergen, N., Pershan, P. S. 1962. *Phys. Rev.* 128:606
7. Brown, R. A., Yeh, Y., Burcham, T. S., Feeney, R. E. 1985. *Biopolymers* 24:1265
8. Burcham, T. S. 1985. *Chemical and physical properties of the antifreeze glycoproteins.* PhD thesis. Univ. Calif., Davis. 265 pp.
9. Burcham, T. S., Knauf, M. J., Osuga, D. T., Feeney, R. E., Yeh, Y. 1984. *Biopolymers* 23:1379
10. Deleted in proof
11. Bush, C. A., Feeney, R. E., Osuga, D. T., Ralapati, S., Yeh, Y. 1981. *Int. J. Pept. Protein Res.* 17:125
12. Bush, C. A., Ralapati, S., Matson, G. M., Yamasaki, R. B., Osuga, D. T., et al. 1984. *Arch. Biochem. Biophys.* 232:624
13. Cabrera, N., Vermilyea, D. A. 1958. In *Growth and Perfection of Crystals*, ed. R. H. Doremus, pp. 393. New York: Wiley. 609 pp.
14. Chen, C. K., Heinz, T. F., Ricard, D., Shen, Y. R. 1981. *Phys. Rev. Lett.* 46: 1010
15. DeVries, A. L. 1983. *Ann. Rev. Physiol.* 45:245
16. DeVries, A. L. 1984. *Philos. Trans. R. Soc. Lond. Ser. B* 304:575
17. DeVries, A. L., Komatsu, S. K., Feeney, R. E. 1970. *J. Biol. Chem.* 245:2901
18. DeVries, A. L., Vandenheede, J., Feeney, R. E. 1971. *J. Biol. Chem.* 246:305
19. DeVries, A. L., Wohlschlag, D. E. 1969.

Science 163:1073
20. Duman, J. G. 1982. *Cryobiology* 19:613
21. Duman, J. G., DeVries, A. L. 1972. *Cryobiology* 9:469
22. Feeney, R. E. 1974. *Professor on the Ice.* Davis, Calif.: Pacific Portals. 164 pp.
23. Feeney, R. E. 1974. *Am. Sci.* 62:712
24. Feeney, R. E., Hofmann, R. 1973. *Nature* 243:357
25. Feeney, R. E., Osuga, D. T., Reid, D. S., Yeh, Y. 1982. *Protein Nucleic Acid Enzyme* 27:1645
26. Feeney, R. E., Osuga, D. T., Ward, F. C., Rearick, J. I., Glasgow, L. R., et al. 1978. *Abstr. Div. Biol. Chem., ACS Meet., Miami Beach, Fla.*, Abstr. 13
27. Feeney, R. E., Yeh, Y. 1978. *Adv. Protein Chem.* 32:191
27a. Fletcher, G. L. 1981. *Can. J. Zool.* 59: 193
28. Fletcher, G. L., Hew, C. L., Joshi, S. B. 1982. *Can. J. Zool.* 60:348
29. Fletcher, G. L., King, M. J., Hew, C. L. 1984. *Can. J. Zool.* 62:839
30. Fourney, R. M. 1984. *Regulation of antifreeze polypeptide biosynthesis in the winter flounder* Pseudopleuronectes americanus. PhD thesis. Memorial Univ. Newfoundland, St. John's, Canada. 208 pp.
31. Franks, F., ed. 1972–1982. *Water. A Comprehensive Treatise*, Vols. 1–7. New York: Plenum
32. Franks, F., Morris, E. R. 1978. *Biochim. Biophys. Acta* 540:346
33. Geoghegan, K. F., Osuga, D. T., Ahmed, A. I., Yeh, Y., Feeney, R. E. 1980. *J. Biol. Chem.* 255:663
34. Gordon, M. S., Amdur, B. H., Scholander, P. F. 1962. *Biol. Bull.* 122:52
35. Deleted in proof
36. Hew, C. L., Slaughter, D., Fletcher, G. L., Joshi, S. B. 1981. *Can. J. Zool.* 59:2186
37. Hew, C. L., Slaughter, D., Joshi, S. B., Fletcher, G. L., Ananthanarayanan, V. S. 1984. *J. Comp. Physiol. B* 155:81

78   FEENEY, BURCHAM & YEH

38. Hobbs, P. V. 1974. *Ice Physics*. Oxford: Clarendon. 837 pp.
39. Jackson, K. A. 1958. In *Liquid Metals and Solidification*, pp. 174–86. Cleveland: Am. Soc. Metals. 348 pp.
40. Kerr, W. L. 1985. *Function of antifreeze glycoproteins at the ice-solution interface.* MS thesis. Univ. Calif., Davis. 185 pp.
41. Knauf, M. J., Burcham, T. S., Osuga, D. T., Ahmed, A. I., Feeney, R. E. 1982. *Cryo-Lett.* 3:221
42. Knight, C. A., DeVries, A. L., Oolman, L. D. 1984. *Nature* 308:295
43. Komatsu, S. K., DeVries, A. L., Feeney, R. E. 1970. *J. Biol. Chem.* 245:2909
44. Langmuir, I. 1918. *J. Am. Chem. Soc.* 40: 1361
45. Lin, Y., Duman, J. G., DeVries, A. L. 1972. *Biochem. Biophys. Res. Comm.* 46: 87
46. Morris, H. R., Thompson, M. R., Osuga, D. T., Ahmed, A. I., Chan, S. M., et al. 1978. *J. Biol. Chem.* 253:5155
47. Mulvihill, D. M., Geoghegan, K. F., Yeh, Y., DeRemer, K., Osuga, D. T., et al. 1980. *J. Biol. Chem.* 255:659
48. Nakagawa, Y., Abram, V., Kezdy, F. J., Kaiser, E. T., Coe, F. L. 1983. *J. Biol. Chem.* 258:12594
49. Nakagawa, Y., Margolis, H. C., Yoko-yama, S., Kezdy, F. J., Kaiser, E. T., Coe, F. L. 1981. *J. Biol. Chem.* 256:3936
50. Osuga, D. T., Feeney, R. E. 1978. *J. Biol. Chem.* 253:5338
51. Osuga, D. T., Feeney, R. E., Yeh, Y., Hew, C.-L. 1980. *Comp. Biochem.*

Physiol. 65B:403
52. Osuga, D. T., Ward, F. C., Yeh, Y., Feeney, R. E. 1978. *J. Biol. Chem.* 253: 6669
53. Raymond, J. A. 1976. *Adsorption inhibition as a mechanism of freezing resistance in polar fishes.* PhD thesis. Univ. Calif., San Diego. 181 pp.
54. Raymond, J. A., DeVries, A. L. 1972. *Cryobiology* 9:541
55. Raymond, J. A., DeVries, A. L. 1977. *Proc. Natl. Acad. Sci. USA* 74:2589
56. Raymond, J. A., Lin, Y., DeVries, A. L. 1975. *J. Exp. Zool.* 193:125
57. Scholander, P. F., Flagg, W., Walters, V., Irving, L. 1953. *Physiol. Zool.* 26:67
58. Scholander, P. F., VanDam, L., Kanwisher, J. W., Hammel, H. T., Gordon, M. S. 1957. *J. Cell. Comp. Physiol.* 49:5
59. Schrag, J. D., DeVries, A. L. 1983. *Comp. Biochem. Physiol.* 74A:381
60. Tomchaney, A. P., Morris, J. P., Kang, S. H., Duman, J. G. 1982. *Biochemistry* 21: 716
61. Tomimatsu, Y., Scherer, J. R., Yeh, Y., Feeney, R. E. 1976. *J. Biol. Chem.* 251: 2290
62. Vandenheede, J. R., Ahmed, A. I., Feeney, R. E. 1972. *J. Biol. Chem.* 247: 7885
63. Van Voorhies, W. V., Raymond, J. A., DeVries, A. L. 1978. *Physiol. Zool.* 51: 347
64. Yeh, Y., Feeney, R. E. 1978. *Acc. Chem. Res.* 11:129

Ann. Rev. Biophys. Biophys. Chem. 1986. 15 : 79–95

COMPUTATIONAL ANALYSIS OF GENETIC SEQUENCES[1]

Walter B. Goad

Theoretical Biology and Biophysics, Los Alamos National Laboratory, University of California, Los Alamos, New Mexico 87545

CONTENTS

PERSPECTIVES AND OVERVIEW

Since the late 1970s investigators in increasing numbers, representing a broad spectrum of current biological interests, have turned to nucleic acid sequencing for the analysis of key problems. The cumulative volume of sequence data now totals about six million nucleotide bases, and its exponential growth shows no sign of abating. Investigators working with sequence data, excepting, possibly, those who spend a great deal of time poring over a few particular sequences or portions of sequences, find the

[1] The submitted manuscript has been authored by an employee of the University of California, operator of the Los Alamos National Laboratory under Contract No. W-7405-ENG-36 with the US Department of Energy. Accordingly, the US Government retains an irrevocable, nonexclusive, royalty-free license to publish, translate, reproduce, use, or dispose of the published form of the work and to authorize others to do the same for US Government purposes.

help of a computer indispensable. It seems fair to consider publication of sequence data incomplete until it is available in computer-accessible form, especially since it is in this form that it becomes most amenable to the normal scientific processes of criticism, correction, and refinement. Two cooperating genetic sequence data banks, GenBank® in the United States and the European Molecular Biology Laboratory Nucleotide Sequence Library in Europe, compile all reported sequence data into computer databases that are available on magnetic media and, in the United States, through on-line access.

Much computational processing of sequence data can be accomplished by applying the kinds of computational tools and processes widely used in text processing, perhaps specialized to sequences, but not necessarily so. Thus many investigators use ordinary text editors or word-processing programs to assemble and manipulate sequence data, but specialized editors have also been developed. At GenBank, for example, the SEQEDIT program formats and manipulates sequence data in conformity with the structure of the GenBank database while communicating with the user through keyboard and cursor of a screen terminal in much the same way as any number of screen-oriented word-processing programs. A number of other programs have been developed specifically to assist in assembling sequences from individually sequenced restriction fragments. As in the case of restriction sites, many sequence patterns of biological interest are essentially lexically defined, and can be dealt with by text-processing methods. However, the specifically biological problems of identifying evolutionary and functional relatedness between sequences have led to development of computational tools that go beyond anything from the text-processing area. I first sketch the use of text-processing tools in sequence analysis and then go on to methods for identifying homology and other relationships within and between sequences. I have tried to include the most active areas of actual application in molecular genetics, although the important area of secondary structure prediction is omitted. A review of the mathematics of sequence comparison has been published by Waterman (25).

TEXT-PROCESSING METHODS IN SEQUENCE ANALYSIS

The key operation in text-processing is comparison of symbols represented in the computer by sets of binary digits; within strings of digitally encoded symbols computers can accurately and rapidly locate occurrences of a particular string. Thus restriction sites, open reading frames, or particular oligonucleotides are readily located with one of the many text-editing programs in common use. (However, the user must take care that a string being sought is not interrupted by an end-of-line character in the computer's memory.) Many other commonly needed operations (such as translation between nucleotide and amino acid sequences, tabulation of codon usage, and splicing of sequence fragments to assemble a representation of a sequence from separately sequenced fragments or from the exons of a gene interrupted by intervening sequences) represent straightforward application of match-and-manipulate operations on symbol strings that have been highly developed for general text and symbol-string processing, and adapted to sequence processing by a number of molecular biologists. The prototype of many subsequent program packages offering these services was developed by Korn et al (14); Korn & Queen (13) have cataloged a great many of its successors, and a number of others are found in reference 22.

Blattner et al (2) are exploring an alternative to writing a new program in one of the general computer languages for each newly conceived task. They have developed a high level command language, with elements that are operations on sequences, and computer programs that interpret expressions in the language representing arbitrarily complicated combinations of the permissible operations.

String-matching operations need not be restricted to finding exact matches of a given string. Strings can be defined by any rule that can be implemented with a finite series of operations to determine whether a particular string is or is not acceptable. The elementary operations needed are the counting and comparing of symbols and the entering of tables specifying symbol equivalence or translation. For example, many string-matching programs can locate occurrences of an A and T separated by two bases of any identity if the program is asked to search for the string "A??T." Quite general capabilities are included in programs that implement parsing of expressions given in a regular, completely defined syntax. GREP, found in most computer systems operated under the UNIX operating system, is such a program. One can, for example, readily find all occurrences of "TATA" followed, after between N and M intervening bases, by another short sequence, itself ambiguously specified by a variable number (with specified limits) of "wild cards" (any of a set of possible bases at particular positions). Naturally, the more complicated the rule, the more computation is required to carry it out; but certainly searches for short predefined patterns of a few tens of bases or less, in sequences of any length that one is prepared to process, can effectively be carried out by string-matching methods.

MEASURES OF PATTERN MATCH/MISMATCH

In a search for all occurrences of an ambiguously defined pattern it may well be that some occurrences will be considered better answers to the inquiry than others; if a variable number of "wild cards" is allowed, for example, pattern matches that require fewer of them may, for some purposes, be ranked above those with more. It is useful then to devise a biologically relevant measure to rank the possibilities and have the computer classify them, perhaps calling attention only to those ranking above some threshold of interest.

It is essential to adopt such a measure when there are many ways of generating some degree of pattern match, as is the case when the question is whether two sequences that are not identical nevertheless resemble each other according to some criterion. Among a number of criteria homology is perhaps the most fundamental, and here a cogent basis for constructing a measure of the distance between sequences lies in considering the number of mutational events by which one sequence may have been derived from another. A particular change can of course result from many different sequences of mutational events. Accordingly Ulam (24) explicitly, and Needleman & Wunsch (16) implicitly, suggested the following measure: First, each possible mutational event, formulated in terms of its consequences as base substitution, base insertion, or base deletion, is weighted according to some estimate of its relative frequency in the system at hand. Then, the combination of events sufficient to carry the one sequence into the other for which the sum of weights is smallest is determined; its value is used as a numerical measure of how far the two sequences have diverged from any possible common ancestor. Ulam conjectured, and Sellers (17) proved, that this measure satisfies minimal mathematical requirements for a distance as we intuitively use it when we describe what it means for objects to be close to or distant from one another relative to their distance from other objects: The distances among three sequences satisfy the triangle inequality. Thus we can be sure that if, by the measure, sequence A is close to B and A is distant from C, then B is certainly also distant from C, no matter how complicated the pattern of differences that compose the distances.

To implement such a measure we must have a way to determine for any pair of sequences what combination of base mismatches, insertions, and deletions sufficient to transform the one sequence into the other minimizes the total of weights assigned to each process. One could in principle devise a rule by which the computer could generate every possible combination, compute the measure for each, and select those for which the measure equals its smallest value. In most cases a number of combinations yield the same value for the measure. In short, one could in principle utilize procedures for matching ambiguous strings as outlined above. However, the number of possibilities is very large, growing factorially with length of the sequences. A procedure of this kind that is certain to include an optimum among the possibilities examined is beyond the capabilities of existing computers for most sequences of interest.

Practicable algorithms have been developed by proceeding step-by-step along the two sequences, and either severely restricting the possibilities for substitution, deletion, and insertion or, alternatively, examining at each step only the possible ways of extending optimal combinations already constructed at earlier steps. In their groundbreaking package of routines for sequence analysis, Korn et al (14) implemented a computationally economical algorithm exploiting the first alternative, although they did not explicitly include the computation of a measure. The idea and basic method of the second alternative, in which the number of operations needed to compare sequences of lengths N and M is proportional to $N \times M$, was discovered by Needleman & Wunsch (16), and proved by Sellers (17), to yield the overall optimum. Algorithms based on the latter method are sometimes called "rigorous" because the result satisfies with certainty a precisely defined criterion. The method itself is an example of a class of methods that has come to be called "dynamic programming." As described below, a number of algorithms seek computational economy by combining the two methods, striking various compromises between speed and rigor.

ALIGNMENT: THE NEEDLEMAN-WUNSCH-SELLERS ALGORITHM

Various notations can denote a particular set of replacements, insertions, and deletions by which two sequences differ; a particularly useful one is the alignment. In the alignment below,

tcaga tcagat

tc gacttca at,

two insertions and two deletions in the first sequence carry it into the second. The transformation can also be accomplished by six replacements as represented by the alignment

tcagatcagat

tcgacttcaat.

Which of these or of many other possible alignments is optimum depends on the measure of sequence difference, i.e. how insertion/deletion of four bases is weighted relative to six base replacements.

Implicit in any particular set of weights is a hypothesis of the relationship between the sequences being compared, a hypothesis that the comparison algorithm is used to test. From among the large number of possible ways two sequences can be aligned, the algorithm selects those that conform best to the hypothesis. Even for the two short sequences in the illustration above, a number of different alignments would be chosen as different hypotheses were tested. Suppose, for example, that the sequences come from longer sequences being compared for potentially encoding closely related proteins. In most contexts the replacement of a codon by another encoding the same amino acid is unimportant, so the replacement of tca by tcg, which both code for serine, would be given little weight in the second alignment. The replacement of gat, which codes for aspartic acid, by aat, which codes for asparagine, retains a hydrophilic residue at this point in the translated peptide, and so might not be given much weight in some structural contexts. By contrast, replacement of tca, which codes for serine, by ttc, which codes for leucine, replaces a polar residue by a hydrophobic one and, again depending on context, might be given so high a weight as to reject the alignment.

In models in which base insertion and deletion result from misalignment of some kind during DNA replication, it is plausible that the deletion or insertion of many bases will occur as frequently as the deletion or insertion of one. To reflect this, the distance between two sequences should give little more weight to the insertion or deletion of a number of contiguous bases than to the insertion or deletion of a single base. Fitch & Smith (8) have found this factor important in obtaining alignments among globin genes that make biological sense.

The Needleman-Wunsch-Sellers (NWS) algorithm economically yields the optimum alignment according to any measure that is additive, i.e. formed from the arithmetic sum of contributions from each of a set of elementary sequence differences that collectively carry one sequence into another. Thus the measure must be broken down into contributions from elementary differences that are simply added: If n base replacements are taken to add n times the distance contributed by one replacement, then replacement of each base is an elementary event; but if replacements are considered three at a time, or if deletion of n bases contributes less than n deletions of one base, then three-base replacements and n-base deletions are among the set of elementary differences.

Economy of computation is achieved by inductively constructing an optimal alignment step-by-step, starting at one end of the two sequences and proceeding to the other. The procedure can be defined by considering the step in which base N from one sequence and base M from the other are added to the alignment. We first find which optimal alignments of fewer than N with fewer than M bases can be extended to include bases N and M, by matching base N with M, if they are the same, or by introducing an elementary difference, if they are not. For this we need all of the optimal alignments obtained at earlier stages that can reach bases N and M when extended by either a match or an elementary difference. They must have been computed and remembered. Then each of the possible extensions is a candidate for the optimal alignment of the first N bases with the first M. The candidates include the optimal alignment of the first N-n bases with the first M extended by an n base deletion from the first sequence, and the optimal alignment of the first N with the first M-n extended by an n base deletion from the second sequence for each value of n that contributes less distance than any combination of shorter deletions. Also included as candidates are the shorter alignments that can be extended by base match or replacement. The distance measure for each alignment is computed. The best candidate is then the optimal alignment including bases N and M.

The NWS procedure can be visualized by means of a matrix in which the first row is filled with one sequence and the first column is filled with the other. Then in each element, under base N from the first row and to the right of base M from the first column, is placed the value of the measure for the optimal alignment up to and including N and M, and the value N' and M' from which a shorter optimal alignment was optimally extended to N, M. Values of the distance measure are used to elect the optimal extension, and N', M' simplify construction of the resulting alignment after computation. The computation must be carried out in such a way that each element (N, M) is dealt with after the values in the submatrix above and to the left of N, M have already been filled in. Visualization in terms of a "path matrix" introduced by Sellers (18) is discussed in detail by Goad & Kanehisa (10; Figure 1). Programming of the computer is usually simplest when each row (or column) is dealt with from beginning to end, then the next row (or column), and so on. However, to make use of the parallel processing capability of the CRAY supercomputer, it is necessary to proceed by antidiagonals (T. F. Smith, A. Warnock, unpublished).

The simplest set of elementary differences from which to build a distance measure is replacement, insertion, and deletion of one base; replacement then can be assigned a weight r, and insertion and deletion a weight g (for "gap"). Note that base match contributes zero. This set limits candidates for extension at each element (N, M) in the matrix in the NWS algorithm to extensions from $(N-1, M)$, $(N, M-1)$, and $(N-1, M-1)$, thus making computation more economical than with more complicated measures.

Applied to random sequences, replacements and gaps occur with relative frequencies approximately in inverse proportion to their weights. Except where one or the other of the sequences being compared is expected to predominate (as might, for example, be the case when an organism subjected to a frame-shift mutagen is the source of one of the sequences), it seems reasonable to approximately balance the two weights. At Los Alamos, values close to $r = 1$, $g = 2$ are routinely used when, for computational economy, this simple measure is employed. Except where deletions of more than a few bases are involved, the resulting alignments of homologous sequences seem biologically sensible.

As noted above there is good reason to weigh insertion/deletion of n bases at less than n times the weight for inserting one base. The next simplest model assigns a weight $g = a + bn$ for insertion or deletion of n bases. Fitch & Smith (8) found that a model of this kind is needed to reproduce alignments derived from extensive evidence in the globins, and that $a = 2$, $b = 0.5$ yield quite satisfactory results. To assign weights for

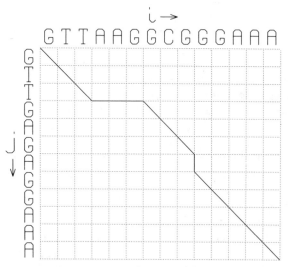

Figure 1   Matrix representation of sequence alignment. One sequence occupies the horizontal rows, the other the vertical columns. In each element, a diagonal line segment signifies that the base occupying that row is aligned opposite the base occupying the column; a horizontal line segment signifies deletion of the row base, a vertical segment deletion of the column base. Any contiguous set of segments proceeding from left to right and top to bottom forms a path representing a possible alignment. The path shown represents the alignment

GTTAAGGCG GGAAA

GTT   GAGAGGAAA.

replacement according to codon change, the set of elementary differences can be expanded to replacement of one, two, and three bases, typically with weights of r and $2r$ for the first two, and a table of weights for codon exchange for the third. The values for codon exchange can be zero for codons specifying identical amino acids; or values can be based on indices of chemical difference among the amino acids or on substitution frequencies (4). In any case, when codon identity is introduced at this level sequences similar in amino acid coding are found whether or not frame shifts have to be introduced to disclose the similarity; while the alternative of first translating a nucleotide sequence and then performing the NWS algorithm in terms of amino acid comparison uses reading frames predetermined at translation.

As discussed above, more complicated relationships can also be identified by adopting the appropriate measure: Given a rule for determining how frequently any particular series of codons corresponds to one or another of a set of structural motifs in the folded protein, correspondence or conflict between motifs in two sequences being compared can be made to override individual codon differences in the overall measure. These possibilities are added to the set of elementary differences for computing the measure and are weighted appropriately relative to individual codon differences. As another example, the distance computation might be based on the sequence-dependent parameters of DNA structure brought to light by Dickerson et al (5) instead of on base difference. For any computable function of sequence, as in the case of codon-dependent distance, such measures could be used to discover sequences that are close according to functional criteria, whether or not there are extensive base differences between them.

LOCAL SIMILARITY

As formulated above, the NWS algorithm gives the best overall alignment of two sequences; but it may fail to discover subregions of the two sequences that match strikingly better than average. For example, when a group of bases early in one sequence strongly matches a group much farther along in the other, the matching regions will be missed if the distance contributed by deleting the bases necessary to reach the match exceeds the total score for alignment without the long deletion.

Addressing this problem, Sellers (17) first introduced the important distinction between globally and locally optimal alignment. He offered an extension of the NWS procedure to yield all subalignments that are locally optimal, but optimal by a rather weak criterion. Goad (9), seeking a stronger and thus more selective way of defining and then computing

locally optimal alignments, took the following route: In global alignment
the measure of overall distance between sequences consists of a sum over
the changes required to carry one sequence into another, each with a weight
reflecting its biological significance in the context of the investigation at
hand. To reflect local relatedness at each point along an alignment, an
analogous quantity representing a weight per unit length along the
sequence is needed. This weight per unit length has to be computed from
changes along some finite segment of the alignment that includes the point
in question. If the speed and rigor of NWS-like algorithms are to be
retained, the measure must remain a simple sum; no quotients or products
can be included. In particular, we cannot use a sum of weights divided by
the number of bases to which the sum applies.

Within these limitations, a solution is to compute a second measure, a
weighted sum of base matches, as the NWS algorithm proceeds and, in
effect, to delimit locally optimal subalignments by the regions over which
the weighted sum of matches exceeds the weighted sum of differences. The
relative weights define a threshold for the number of matches per difference,
and the algorithm yields all subalignments for which the threshold is
exceeded.

Different ways of implementing these considerations are described by
Smith & Waterman (20) and by Goad & Kanehisa (10). Both carry out the
NWS algorithm with a measure combining weighted matches and weighted
differences with opposing signs. If matches are taken as positive and
differences negative, the measure is called a "similarity score" by Smith &
Waterman. They define a locally optimal alignment, or "local similarity" by
a region that, as the algorithm proceeds through the alignment matrix (see
Figure 1) from left to right, top to bottom, begins where the similarity score
rises above zero (the threshold) and ends at its maximum. Any continuation
beyond the maximum would make a net negative contribution by adding a
region for which the score would, overall, be below threshold. Goad &
Kanehisa carry out the algorithm twice, once in each direction along the
sequences, and retain as local similarities those regions for which the
measure is above zero for alignment in both directions. These methods
represent different ways of removing the artifact of having to carry out the
algorithm in one direction or another along the sequences by constructing
an alignment that is bounded in both directions by regions with differences
that exceed a threshold.

In addition to avoiding the possibility that interesting related local regions
may be missed, it is desirable that all subregions that are related be systematically isolated from those that have only a random relationship or that
have a degree of relationship below any other threshold that might be of
interest. Sellers (19) has shown that if, extending Goad & Kanehisa's pro-

cedure, alignment in alternating directions is repeated until the alignments for which the measure is above zero in both directions are unchanged by further sweeps through the matrix, local similarities are found that satisfy a strong and conceptually satisfactory condition: If any such locally optimal alignment is subdivided, each subdivision is itself locally optimal. Thus the algorithm systematically yields all subsequences for which the local degree of closeness exceeds a threshold that can be set to distinguish a significant relationship as defined by the investigator.

If two sequences are to be compared that are known to be related, or if for some other reason one is only interested in an alignment that starts with the pairing of bases at predetermined points in two sequences and continues to predetermined ends, the global NWS algorithm may be best suited to the task. In most cases, however, one form or another of the local similarity algorithm will be the method of choice. A long, close alignment will be discovered by either algorithm, but only the local algorithm guarantees that all similarities above some threshold of relationship will be located. The most computationally economical form is that of Smith & Waterman (20); its disadvantage is that a significant region of similarity downstream (with respect to the direction of alignment) of a region of greater similarity score and separated from it by a region of less than threshold similarity may be omitted. However, two sequences may be made up of closely related intervals separated by intervals that have either diverged (as in an intervening sequence) or been derived from another source altogether. In these circumstances the somewhat more laborious form of the algorithm given by Sellers (19) yields more definitive results: the set of all locally close subalignments, distinguished from unrelated portions of the sequence by a strong and logically compelling criterion.

It remains to put the set of optimal subalignments into an overall picture of sequence relationship. In a path matrix as in Figure 1, each local alignment would be represented by a path occupying, or ranging not far from, a segment of one of the matrix diagonals. If, starting from the upper left and proceeding toward the lower right, only one assembly of the local alignments into an overall alignment is possible, overall alignment is straightforward. In the general case an optimal overall alignment can be automatically generated by treating each subalignment as an element in a reduced matrix (analogous to a base match in the full matrix) and applying a modified NWS algorithm to the reduced problem. Where local alignments overlap, separate entries into the reduced matrix are made of their overlapping and non-overlapping parts. Since the overlaps represent repeats (not exact ones, in general), in the computer program being implemented at Los Alamos we include a reference to them in the output. An algorithm is needed for generating all arrangements of the locally optimal subalignments into overall alignments. For the similar but in many ways more difficult problem of assembling suboptimal RNA secondary structures, Williams & Tinoco (27) have now developed a practicable method.

DATABASE SEARCHES—THE QUEST FOR COMPUTATIONAL ECONOMY

Increasingly, it is of interest to ask if in some large body of sequence data, relatives of a given sequence are to be found. The computational investment for a "rigorous" search (i.e. a search that uses some form of the NWS algorithm) can be quite large. With the Smith-Waterman version of the NWS algorithm (8) run on the first of the commercially available supercomputers, the CRAY I, programmed to utilize the computer's parallel processing facility at close to optimum efficiency, about 2 hr are needed to compare a typical sequence of 2000 bases against the present GenBank database, 5 million bases, and its complement. It would take 2500 hours to compare a year's accumulation of data, in 1985 expected to be around 3 million bases, with the database. Given that many variations in measure and threshold are of interest, and that most investigators work on less productive computers, there is ample motivation for speeding up such searches.

Once a region to be aligned is identified, computation can be saved by limiting the region of the path matrix treated. Methods for this have been developed by Fickett (7) and Ukkonen (23). However, as the results of large-scale searches of the GenBank database show (21) the NWS algorithm for most sequence pairs yields nothing of interest. Large economies are achieved by devising a preliminary screening for sequences and sequence regions that are good candidates for more extended treatment.

There are rapid ways of pointing to short strings that match within two sequences. The most rapid procedure applies a device reinvented many times by designers of computer programs that involve searching for a particular piece of data, the "hash function" (12). For sequence searches a very simple hash function suffices: A table is prepared of the location in one sequence of each of a set of oligonucleotides (or amino acids). Typically, the table consists of a set of numbers representing the base numbers at which, for example, the tetranucleotide aaaa is to be found, followed by a set of base numbers of aaac, and so on. Another similar table is prepared giving the location in memory of the first of the aaaa base numbers, followed by the location in memory of the first of the aaac base numbers, and so on, ordered in a systematic way for each possible tetranucleotide. A few times N operations are required to prepare this table for a sequence of N bases. In a second sequence of M bases, for each of its M tetranucleotides a few address computations suffice to find the locations, if any, of matching tetranucleotides in the first sequence. Thus a table of matching tetranucleotides or, more generally, oligonucleotides can be prepared in a few times $(N + M)$ operations, orders of magnitude fewer than are required to execute the NWS algorithm, which requires a multiple of N times M operations.

The information about matching oligonucleotides can be used in various ways. Karlin et al (11) use the total number of matching oligonucleotides, perhaps for each of a number of oligonucleotide lengths, as a measure of relationship. Wilbur & Lipman (26) and Lipman & Pearson (15) use this information, generally starting with quite short oligonucleotides, to locate matching strings of all lengths in the two sequences, and proceed to focus on regions so identified. M. I. Kanehisa (unpublished) first locates matching oligonucleotides, then tests these for extendability (without, however, demanding exact match), and then subjects regions that meet a preliminary criterion of local closeness to the NWS procedure. W. B. Goad (unpublished) also employs a hash table to locate regions that are candidates for the NWS procedure, using the density of matching oligonucleotides along or near a diagonal of the path matrix as an indicator.

These strategies can achieve great increases in speed. Lipman & Pearson (15) have developed a program that provides effective protein searches with the IBM Personal Computer. Kanehisa's search algorithm, as routinely used at Los Alamos, reduces search times approximately 100-fold over full use of the NWS algorithm. It must be appreciated, however, that significant increases in speed entail a sacrifice of certainty that no close regions will be missed, and of precision in the measure of closeness. Quantitative characterization of the sacrifice will require more experience with database searches and further development of the statistics of sequence comparison.

STATISTICS OF SEQUENCE COMPARISON

As for any piece of circumstantial evidence, it is helpful to know how frequently a particular alignment would occur by chance alone. Such information can be generated empirically by tabulating the frequency with which various alignments occur in comparisons either within a collection of sequences sufficiently large that most can be assumed unrelated, or between a sequence of interest and randomly permuted versions of it. The two frequencies are informative in slightly different ways: The first includes whatever background correlations the sequences may have in common, stemming for example from codon usage in the case of protein-coding sequences of nucleotides, but at the upper end of similarity values it is not clear where relatedness begins to appear. In the second method, it is hard to insure that random permuting preserves background correlations. In principle, the needed information can also be obtained from probability theory; however, problems that involve the statistics of inexact string matchings are formidable and no comprehensive solution is available. Nevertheless, recent progress has provided a means of organizing the empirical results that allows economic construction of largely satisfactory results.

For the overall alignment of protein sequences there is a body of experience to draw on. Dayhoff et al (3), Doolittle (6), and Lipman & Pearson (15) have tabulated the frequencies of values of various similarity measures within large collections of data. They recommend somewhat different thresholds for significance, although they seem to agree on a threshold at about 50% similarity where there are very few insertions/deletions, with a similarity measure allowing for replacement of one amino acid by another of high chemical similarity or high replacement frequency in closely related proteins. Lipman & Pearson also offer an algorithm and a computer program for generating statistics with random permutations of the sequences of interest.

Researchers have less experience with local similarities, and more interest in relatively short fragments that may have been drawn from a common source or selected for common function. Smith et al (21) have tabulated the value of the similarity measure for the best local alignment in each of over 20,000 comparisons of pairs of eukaryotic sequences. They have organized the results to provide a very useful guide to statistical significance.

The key to Smith et al's organization of data is that the best local alignment in each comparison tends to correspond to a larger value of the similarity measure the longer the product of the lengths of the two compared sequences, approximately in proportion to the logarithm of the product. This would be expected of local alignments due to chance alone: If p is the probability that any two bases drawn from the two sequences match, the probability that any given n-base string from one sequence matches a particular n-base string from the other is p^n; for each n-base string in one sequence, there are N-n strings in the other (of length N) with which to compare it, and thus $(N-n) \times (M-n)$ pairs that have a possibility of matching between sequences of lengths N and M. Thus the frequency with which an n-base match would occur, at least at large enough values of n that no more than a few string matches would be likely, decreases exponentially with n but increases with $N \times M$. Then as $N \times M$ increases one expects to find string matches for larger values of n, roughly in proportion to log $(N \times M)$. Arratia et al (1), motivated by the empirical results, have confirmed this. To be precise, suppose sequences of lengths N and M are repeatedly drawn from urns containing the bases in appropriate proportions. Arratia et al have calculated (in a certain limit), the expected length of the longest matching subsequence between each pair of sequences drawn. They found that the expected average value of n grows approximately linearly with log $(N \times M)/$log $(1/p)$ or, equivalently, with LOG $(N \times M)$, where LOG means logarithm to the base $1/p$.

Locally optimal alignments as found by the NWS algorithm, allowing both mismatches and insertion/deletion, represent a more complicated relationship than matching strings, but for reasonably close alignments the number of matching bases is the leading term in the similarity measure. Therefore Smith et al (21) plotted the similarity value for the best local alignment between each pair of sequences versus the value of LOG $(N \times M)$ for the pair. The result, Figure 2, shows that for most sequences the best similarity score falls in a band rising linearly with LOG $(N \times M)$; strikingly enough, only about 5 in 1000 local alignments are scattered outside it. Thus there is a well-defined band containing more than 99% of

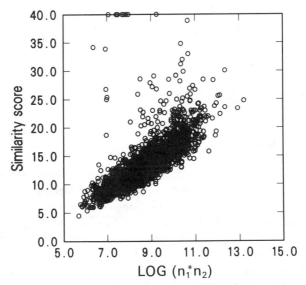

Figure 2 [Reproduced with permission from Smith et al (21).] Values of the similarity score (similarity measure) for a large set of sequences compared among themselves. The sequences represent an unselected sample from the vertebrate and eukaryotic virus sections of the GenBank database. Similarity score = (matches) -0.90 (mismatches) -2.0 (insertions/deletions). Each of the 20,706 data points shown results from comparing a pair of sequences; the score for the best locally optimal similarity found in the comparison is plotted against LOG $(n_1 \times n_2)$. The logarithm is computed to the base $1/p$, where p is the probability of base match if one base is drawn at random from each of the two sequences; n_1 and n_2 are the sequence lengths. All points falling above the upper horizontal axis are plotted on the axis.

the local alignments that can be assumed due to chance alone. For any local alignment found by applying the NWS algorithm to two sequences, the value of LOG $(N \times M)$ for the two sequences is computed. It can readily be seen whether the alignment's similarity score falls close to or within the band of statistically insignificant sequences.

An implicit assumption is that the frequency of chance alignments depends on similarity score alone. We do not know whether within the collection of random alignments of given similarity score some alignments, with various numbers of mismatches and insertions/deletions balanced by the appropriate number of base matches, occur much less frequently than others. But without this information, we simply neglect the possibility of discovering that some alignments indicated by only the distribution of similarity scores to be statistically insignificant are in fact significant. That is, the criteria of significance recommended by Smith et al (21) are conservative.

We can then formulate a rule of thumb, at least for results obtained with the similarity measure employed by Smith et al: weights of $+1$ per match, -1 per mismatch, and -2 per insertion/deletion. If a local alignment is found between two sequences of the order of 1000 bases, it is statistically significant if, with no mismatches or deletions, it contains more than about 25 matches; with mismatches and/or deletions, more matches are required at about 1 more per mismatch and 2 more per deletion. For longer or shorter sequences these numbers change only logarithmically, as detailed in reference 7.

CONCLUSION

The large and growing body of genetic sequence information forms an enormous scientific resource, describing an ever widening sample of the living world at the level at which heredity and development is controlled. As has happened so often in the past, new scientific data has brought into being new mathematical methods to exploit it, and I have tried to sketch what seems to me the center of the territory.

In the present era, mathematics in a scientific context usually includes computational science. For sequence data, the mode is symbol processing rather than numerical calculation. It seems clear that biology is doing for this area of mathematics and computational science what physics has done for numerical analysis, providing a substrate for steady advance that brings forth new ideas and new tools in both disciplines.

Literature Cited

1. Arratia, R. A., Waterman, M. S. 1984. *Adv. Math.* 55:13
2. Blattner, F. R., Schroeder, J. L. 1984. *Nucleic Acids Res.* 12:615
3. Dayhoff, M. O., Barker, W. C., Hunt, L. T., Schwartz, R. M. 1978. In *Atlas of Protein Sequence and Structure*, Vol. 5, Suppl. 3, p. 9. Washington, DC: National Biomedical Research Foundation
4. Dayhoff, M. O., Schwartz, R. M., Orcott, B. C. 1978. In *Atlas of Protein Sequence and Structure*, Vol. 5, Suppl. 3, p. 348. Washington, DC: National Biomedical Research Foundation
5. Dickerson, R. E. 1983. *J. Mol. Biol.* 166:441
6. Doolittle, R. 1981. *Science* 214:149
7. Fickett, J. W. 1984. *Nucleic Acids Res.* 12:175
8. Fitch, W. M., Smith, T. F. 1983. *Proc. Natl. Acad. Sci. USA* 80:1382
9. Goad, W. B. 1979. *Los Alamos Natl. Lab. Rep. LASL P-F80-5*
10. Goad, W. B., Kanehisa, M. I. 1982. *Nucleic Acids Res.* 10:247
11. Karlin, S., Ghandour, G., Ost, F., Tavare, S., Korn, L. J. 1983. *Proc. Natl. Acad. Sci. USA* 80:5660
12. Knuth, D. M. 1973. *The Art of Computer Programming*, Vol. 3, *Sorting and Searching*, p. 506. Reading, Mass: Addison-Wesley
13. Korn, L. J., Queen, C. L. 1984. *DNA* 3:421
14. Korn, L. J., Queen, C. L., Wegman, M. N. 1977. *Proc. Natl. Acad. Sci. USA* 74:4401
15. Lipman, D. J., Pearson, W. R. 1985. *Science* 227:1435
16. Needleman, S. B., Wunsch, C. D. 1970. *J. Mol. Biol.* 48:443
17. Sellers, P. H. 1974. *SIAM J. Appl. Math.* 26:787
18. Sellers, P. H. 1980. *J. Algorithms* 1:359
19. Sellers, P. H. 1984. *Bull. Math. Biol.* 46:501
20. Smith, T. F., Waterman, M. S. 1981. *J. Mol. Biol.* 147:195
21. Smith, T. F., Waterman, M. S., Burks, C. 1985. *Nucleic Acids Res.* 13:645
22. Soll, D., Roberts, R. J., eds. 1984. *The Applications of Computers to Research on Nucleic Acids II*. Oxford, Washington, DC: IRL Press (*Nucleic Acids Res.* 1984 12:1–855)
23. Ukkonen, E. 1984. *Inf. Control.* In press
24. Ulam, S. M. 1972. *Ann. Rev. Biophys. Bioeng.* 1:277
25. Waterman, M. S. 1984. *Bull. Math. Biol.* 46:473
26. Wilbur, W. J., Lipman, D. J. 1983. *Proc. Natl. Acad. Sci. USA* 80:726
27. Williams, A., Tinoco, I. 1985. *Nucleic Acids Res.* In press

Ann. Rev. Biophys. Biophys. Chem. 1986. 15 : 97–117
Copyright © 1986 by Annual Reviews Inc. All rights reserved

CITRATE SYNTHASE:
Structure, Control, and
Mechanism

Georg Wiegand and Stephen J. Remington[1]

Max-Planck Institut für Biochemie, 8033 Martinsried bei München,
Federal Republic of Germany

CONTENTS

PERSPECTIVES AND OVERVIEW

Citrate synthase is a component of nearly all living cells and plays a key role
in the central metabolic pathway of aerobic organisms, the citric acid cycle.
The citric acid cycle functions as a source of two reducing equivalents for
the electron transport chain and as a source of intermediates required for
the biosynthesis of amino acids for ketogenesis, lipogenesis, and gluco-
neogenesis (for a review see 36). The flux through the citric acid cycle is
maintained by activated acetic acid (49, 50) originating as acetyl-Coenzyme
A from the degradation of carbohydrate, fatty acids, and amino acids (47,
60, 86). Citrate synthase catalyzes an important step within the cycle, the

[1] Dr. Remington's present address: Institute of Molecular Biology, University of Oregon,
Eugene, Oregon 97403-1229.

0883–9182/86/0610–0097$02.00

Claisen condensation of acetyl-Coenzyme A with oxaloacetate to form citrate; and it is the only enzyme in the cycle that can catalyze the formation of a carbon-carbon bond.

Citrate acts not only as a substrate for the citric acid cycle within the mitochondria, but also as an acetyl donor for acetyl-Coenzyme A synthesis by ATP-citrate-lyase outside the mitochondria after transport through the mitochondrial membrane.

In eukaryotic cells citrate synthase occurs almost exclusively in mitochondria. An exception is in germinating plants where citrate synthase is found in glyoxysomes as well as in a component of the glyoxylate cycle (1, 6, 83, 84, 89, 90). The mitochondrial citrate synthase is coded by nuclear DNA, translated in the cytoplasm as a precursor, and transported into the mitochondria (2, 29, 59) where it is localized on the inner membrane (1, 84).

The enzyme found in animals, plants, fungi, archaebacteria (15), and gram-positive bacteria consists of two identical subunits with a combined molecular weight of about 100,000. Citrate synthase from algae and gram-negative bacteria consists of four to six identical subunits (also with a subunit molecular weight of about 50,000) and is NADH-dependent (15, 68, 91, 95).

Since the isolation of citrate synthase from pig heart mitochondria (61, 62), a number of excellent studies on the function and mechanism of the various enzyme species have been undertaken. These have been reviewed (5, 63, 71, 73, 77–80, 94–95).

It is clear, however, that knowledge of the spatial structure is required for a full understanding of the function and mechanism of this enzyme. As the three-dimensional structure of citrate synthase has recently been elucidated (66, 97–99), we are able to interpret the functional data on the mitochondrial enzyme in light of the molecular structure. The discussion in this review is primarily confined to the enzymes from pig and chicken heart.

INTRODUCTION

The stereochemistry of the enzymic condensation reaction has been elucidated by Eggerer et al (18). The acetyl group of acetyl-Coenzyme A adds to the si-face (28) of the keto moiety of oxaloacetate under inversion of the configuration of the methyl group (18, 37, 43, 67).

$$O=\underset{\underset{\displaystyle COOH}{|}}{\overset{\overset{\displaystyle CH_2COOH}{|}}{C}} + CH_3-\overset{\overset{\displaystyle O}{\|}}{C}-S-CoA + H_2O \rightleftharpoons HO-\underset{\underset{\displaystyle COOH}{|}}{\overset{\overset{\displaystyle CH_2COOH}{|}}{C}}-CH_2COOH + HS-CoA$$

The overall reaction may be conceptually divided into three chemical reactions: (a) enolization of the thioester group of the acetyl-Coenzyme A by proton abstraction from the acetyl moiety forming an enolate anion (17); (b) condensation of the enolate anion with the carboxyl group of oxaloacetate forming a citryl-thioester (19); and (c) hydrolysis of the citryl-thioester forming citrate and Coenzyme A (12, 20).

The overall reaction has a turnover number of about 6000 mol s^{-1} per active site (81) and an apparent equilibrium constant of 8380 liter mol^{-1} at pH 7.2 (87). The reaction has been shown to proceed by means of an ordered mechanism (33). Oxaloacetate binds first, increasing the binding constant for acetyl-Coenzyme A by a factor of at least 20. If acetyl-Coenzyme A is allowed to bind first, the on-rate for oxaloacetate decreases substantially (35).

Citrate synthase is highly specific towards its substrates. The condensation reaction occurs only with the physiological substrates oxaloacetate and acetyl-Coenzyme A. In the lyase reaction (the reverse of the condensation) only the citryl moiety of citryl-Coenzyme A is cleaved. Only citryl- or malyl-Coenzyme A can be hydrolyzed by the enzyme. With a much smaller V_{max} three other nonphysiological substrates are turned over (see Mechanism, below). The specificity for the second substrate, acetyl-Coenzyme A, is similarly high. Only the 3'-phosphate group seems unnecessary; when it is missing the K_m is decreased (85).

The sequences of the enzyme from pig heart and *Escherichia coli* have been determined (8, 9, 58). The pig heart enzyme consists of two identical subunits of 437 amino acids with molecular weight 48,969 (9). Each subunit has an active site with one binding site for oxaloacetate and acetyl-Coenzyme A (33, 66, 93). The active sites appear to be functionally independent from each other; cooperativity has not been observed (33, 93). However, the structural studies showed that each active site is formed from amino acid side chains from both subunits. Citrate synthase is protected against a variety of denaturing agents [for example urea (75, 100) or sodium dodecylsulfate (G. Wiegand, unpublished observations)] and proteolytic cleavage by oxaloacetate. Some polyvalent carboxylic acids seem to have control functions (see Control, below). The elucidation of the spatial structure forms the basis for explanations of these functional properties.

MOLECULAR STRUCTURE

Structures were determined for three crystal forms (space groups P $4_1 2_1 2$, C2, and P $4_3 2_1 2$) containing the following enzyme-ligand complexes

(66, 99):

1. citrate synthase from pig heart muscle with no ligands (P $4_1 2_1 2$),
2. pig heart citrate synthase with bound citrate (C2 and P $4_1 2_1 2$),
3. pig heart citrate synthase with bound citrate and Coenzyme A (C2),
4. pig heart citrate synthase with bound oxaloacetate and S-acetonyl Coenzyme A (P $4_3 2_1 2$),
5. chicken heart muscle citrate synthase with bound citrate and Coenzyme A (C2), and
6. chicken heart citrate synthase with bound (3R, S)-3, 4 dicarboxy-3-hydroxybutyl-Coenzyme A (a citryl-Coenzyme A analog) (C2).

The molecular structures were interpreted and crystallography was refined with the help of the amino sequence of the pig heart enzyme (9). In all crystal forms, except the crystal that contains oxaloacetate and S-acetonyl-Coenzyme A, one subunit is in the asymmetric unit and the dimeric molecule obeys an exact twofold symmetry relationship. In crystals that contain bound oxaloacetate the dimer is the asymmetric unit and the subunits are related by a local diad. The structure of citrate synthase is quite different in the three crystal forms. The molecule was shown to have one "open" and two "closed" conformations as shown in the space-filling diagrams of Figure 1. The two closed conformations superficially appear similar, but the packing of interior side chains is somewhat different for the two conformations.

Subunit Structure

Each subunit consists of 20 α-helical segments, which make up 75% of the 437 amino acid residues. CD measurements indicated 40% to 60% α-helical secondary structure (103). The residual segments are in extended irregular structure, except for a very small β-sheet of 13 residues, which connects α-helix C with an extended segment and is localized on the surface of the protein. Figure 2a shows the structure of the dimeric molecule of pig heart with bound citrate. Figures 2b and 2c show aspects of a single subunit of citrate synthase: the dimer interface as a ribbon trace and a schematic diagram of helix positions showing a view of the dimer interface. The subunit consists of two domains. The larger domain contains 15 helices (A to M and S to T). Four antiparallel pairs of helices, FF′, GG′, MM′, and LL′ (the prime indicates the twofold related helices), mediate the contact between the subunits, forming a twisted eight-helical sandwich reminiscent of the β-sheet.

Perhaps the most remarkable features of the secondary structure of the core are the bent antiparallel helices S and I, which are wrapped over either side of the interface sandwich (see Figures 2b and 2c). Helix I is kinked by a

(a)

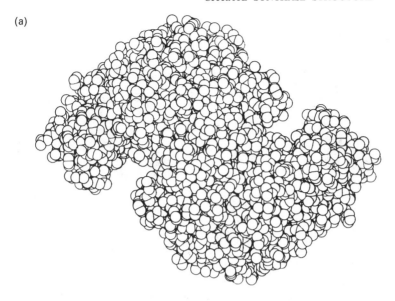

(b)

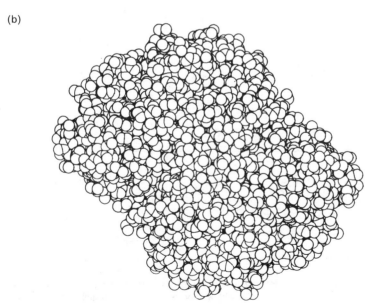

Figure 1   (*a*) Space-filling representation of the open form dimer of citrate synthase. (*b*) Space-filling representation of both closed forms of the dimer. The point of view is down the twofold axis.

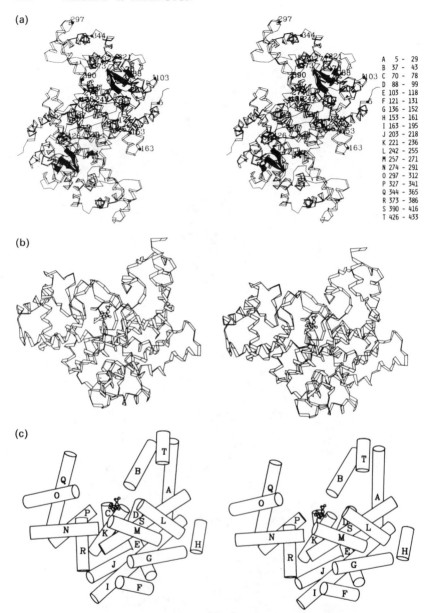

(a)

A	5 -	29
B	37 -	43
C	70 -	78
D	88 -	99
E	103 -	118
F	121 -	131
G	136 -	152
H	153 -	161
I	163 -	195
J	203 -	218
K	221 -	236
L	242 -	255
M	257 -	271
N	274 -	291
O	297 -	312
P	327 -	341
Q	344 -	365
R	373 -	386
S	390 -	416
T	426 -	433

(b)

(c)

Figure 2   (*a*) Stereo α-carbon drawing of the dimer in the open form. The point of view is down the twofold axis. Helices are figured as ribbon models. The starting residue of each helix is indicated by its number. The insert defines the starting and binding residues of the helical segments A–T. (*b*) Ribbon diagram of the dimer interface of one monomer. (*c*) Schematic diagram showing helix packing for the same view as in (*b*). Picture produced by ARPLOT (45).

proline at position 183, whereas helix S seems to be smoothly bent and has no corresponding proline residue. The axes through the ends of the bent helices S and I defined by the first six and last six α-carbons are inclined at an angle of about 40°. The interactions between the two helices appear to be determined predominantly by the packing of large hydrophobic side chains. Other kinked or bent helices are the A, P, and R helices. Both the A and P helices contain a proline residue. In the buried helix R two water molecules are incorporated into the hydrogen-bonding scheme on the helix axis. Stereo drawings of various structural details are shown in Remington et al (66).

The smaller domain consists of the five helices N, O, P, Q, and R. As in some other enzymes (7), the somewhat rigid globular domains move relative to one another, effecting large conformational change. The enzyme responds to the binding of substrates and Coenzyme A by domain movement and some internal rearrangement of amino acid side chains. Owing to the different conformations and consequently different inter-molecular contacts, the enzyme crystallizes in different crystal forms depending upon the particular combination of ligands. Figure 1a shows the enzyme in the open form, which is obtained in the absence of ligands or in the presence of citrate. In Figure 1b the effect of Coenzyme A or oxaloacetate is manifest; the molecule adopts a closed form in the presence of these ligands. The conformational difference is approximately described by an 18.5° rotation of the small domain relative to the large domain, so the small domain of one subunit makes contact with the large domain of the other subunit. In Figure 3 the result of this movement is shown in more detail. In Figure 3a the active site region is depicted for the open form. There is a large cleft between the small domain of one subunit and the large domain of the other subunit. After the conformational transition the small domain makes contact with the large domain of the other subunit near residue 164' (Figures 3b and 3c).

Active Sites

CITRATE/OXALOACETATE BINDING SITE   In the cleft between the two domains of one subunit lies the active site. In the closed forms of the enzyme, binding sites are observed both for citrate or oxaloacetate and for Coenzyme A. In the C2 crystal form, citrate (Figure 4) is bound deep in a cleft between the large and small domain and in the presence of Coenzyme A it is completely inaccessible to solvent. The citrate molecule is hydrogen-bonded or salt-bridged to three histidine residues (His 174, His 238, and His 320) and three arginine residues (Arg 329, Arg 401, and Arg 421 from the other subunit); every functional group of the citrate molecule is involved in a specific interaction with the enzyme. These are displayed as thin bonds in

(a)

(b)

(c)

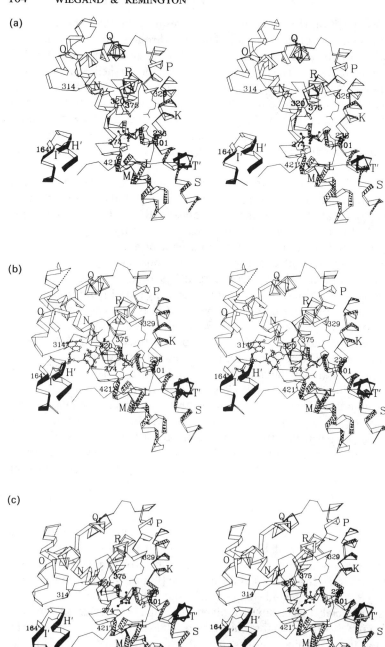

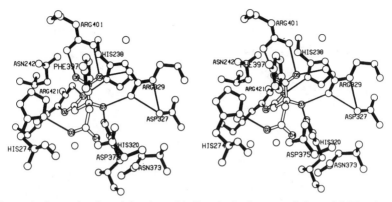

Figure 4   Stereo drawing of the citrate binding site in the monoclinic model. The citrate molecule is drawn with open bonds. Thin bonds represent the important hydrogen bonds or salt bridges. Arginine 421 is from the other subunit of the dimer.

Figure 4. Thus the active site is mainly polar, but nonpolar interactions probably help determine the extreme substrate specificity of citrate synthase and may serve as triggers for the conformational changes as well. (For example, Phe 397 forms an unusual edge-on interaction with the citrate molecule. See Figure 4.) The active site accommodates oxaloacetate much as it accommodates citrate (99). Helices K, L, and S of the large domain are involved in binding citrate (Figure 3b). The axes of helices K and L point at bound citrate with opposite polarity so that their electric fields may assist in citrate binding and polarization (30).

COENZYME A BINDING SITE   In the closed forms of the enzyme, Coenzyme A has a well defined binding site. Residues from the small and large domains of both subunits participate in binding (acetyl-)Coenzyme A. The most remarkable feature of the Coenzyme-A binding is the recognition of the adenine moiety by the enzyme. A section of the main chain (residues 314–320) between helices O and P wraps edgewise around the ring, making three

---

Figure 3   Parts of the enzyme that are involved in binding of substrates or products. Helices N–R form the flexible Coenzyme A binding domain. The dotted drawn helices belong to the rigid large domain of the related subunit. Residues that are involved in substrate binding are drawn and indicated by number: 314–320 ATP recognition loop, Asp 375, His 274, His 320, His 238, Arg 329, Arg 401, Arg 421′, and Arg 164′. Residues 291–294, indicated by the dotted line, are disordered and invisible in the electron density map of both closed forms. (*a*) Open enzyme form without substrates. (*b*) Closed enzyme form with bound products Coenzyme A and citrate. (*c*) Closed form with bound oxaloacetate. The point of view is the same as in Figure 1. The helices K, L, M, S, I′, H′, and T′ of the large domains have the same orientation in all pictures.

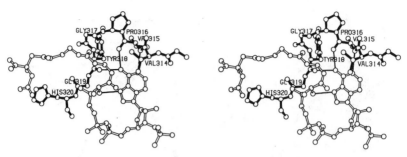

Figure 5   Stereo drawing of a model-built citryl-CoA based on the crystallographic analysis of bound citrate and Coenzyme A, and the adenine recognition loop 314–320. Citryl-CoA is indicated by open bonds while the thin bonds represent some of the important interactions. Note that the main-chain hydrogen bonds to the adenine ring.

specific main-chain hydrogen bonds with N-10 and N-1, shown in Figure 5 as the hypothetical intermediate citryl-Coenzyme A. An internal hydrogen bond between the N-7 of the adenine moiety and the hydroxyl O-52 of the pantothenyl moiety holds the Coenzyme A in a compact conformation. The ribose pucker is C-2′ endo, which allows the equatorial C-2′ hydroxyl to interact through a water molecule with the amide oxygen of the panto-thenyl amide. This further stabilizes the compact conformation. The glycosyl torsion angle is anti and the C-4′–C-5′ torsion angle is gauche-gauche (for nomenclature see 65). The three charged phosphate moieties each form a salt bridge to an arginine residue, the 5′-diphosphate interacts with Arg 324 and Arg 46, and the 3′-phosphate interacts with Arg 164 from the other subunit (Figure 3b). The sulfur of the pantothenic arm is not well ordered in the crystal form that contains citrate and Coenzyme A; there is density for at least two conformations, one of which places the sulfur atom near Asp 375, which is in van der Waals contact with C-5 and O-4 of the product citrate (Figures 4 and 5).

Conformational Change

As previously described, the conformational change can be considered a rigid body rotation of the small domain relative to the large domain of about 18°. This relative motion involves movements of up to 15 Å on the part of some atoms. The hinge points are well defined and are near the axis of the rotation that describes the conformational change. One of the hinge points is the main chain at His 274 and Gly 275 that contributes to the formation of the binding site of citrate or oxaloacetate; i.e. one of the two connecting strands between the two domains is directly involved in forming the active site. The observations that the addition of oxaloacetate cracks crystals of the open form of the enzyme and that the binary complex of enzyme and oxaloacetate crystallizes in a closed form (66, 99) suggest that oxaloacetate induces the conformational change. In addition, oxaloacetate prevents the enzyme from denaturing due to elevated temperature (100) or denaturants such as urea (75) or palmitoyl-CoA (100). An Arrhenius plot was shown to be nonlinear with a breakpoint at 22°C (40). Koshland (39) explains the nonlinearity of the Arrhenius plot by a conformational change during the enzymic catalysis.

The dramatic conformational change requires no changes of the helical secondary structure (66, 76). On the basis of the atomic coordinates of the refined crystal structures of the open and one of the closed enzyme forms (66), Lesk & Chothia (44) reported that the mechanism of domain closure involves small shifts and rotations of helices within the two domains and at their interface. Large motions of distant segments of the structure are the cumulative effect of the small relative shifts in intervening pairs of packed segments. These shifts are accommodated by small structural changes in side chains. Therefore it seems reasonable that small changes in side-chain orientation resulting from the binding of oxaloacetate induce the large conformational change. Citrate synthase crystallizes with citrate in the open form at pH 7.4. In this form the citrate binding site is quite different from that of the closed form (Figure 3a). The residues His 230 and Arg 329, which are in intimate contact with citrate in the closed form, are too distant to interact with citrate in the open enzyme form. In this enzyme conformation 28% of the surface of the citrate molecule is accessible to solvent. In the closed form with bound Coenzyme A, citrate is completely inaccessible to solvent. Oxaloacetate binds similarly to citrate in the closed form (Figure 3c) so that it must bind before acetyl-Coenzyme A binds. This is consistent with an ordered mechanism, as shown by Johansson & Pettersson (33).

This conformational change implies that the citrate/oxaloacetate binding site has several different configurations and that the (acetyl-)Coenzyme A binding site does not even exist until the enzyme adopts a closed conformation. The consequences are discussed under Mechanism, below.

CONTROL

A number of important intermediates or products of metabolic processes have structural features in common with the substrates or products of citrate synthase. Citrate synthase activity may therefore be affected by di- and tricarboxylic acids, ATP, or acyl-Coenzyme A. The effects have been studied with in vitro systems and purified enzymes, and they have been discussed in several reviews (5, 78–81, 94, 95). The relevance of these findings for the physiological situation is unclear owing to differences in concentration. The concentration of citrate synthase was about 10 μg ml^{-1} for the in vitro tests. In the mitochondrial matrix, however, a protein concentration of 270–560 mg ml^{-1} was reported. This corresponds to 27%–56% by weight (25, 31, 42, 64, 70). The protein concentration in crystalline protein samples was of similar magnitude (56, 97). We therefore expect extensive protein-protein contacts in mitochondria (25, 82). In addition, the structure and activity of water must be profoundly altered in mitochondria as it is in protein crystals (14, 83).

In such a dense packing of protein molecules the diffusion parameters of substrate molecules may be altered (69, 52), but the protein structure and flexibility may also be affected. In the crystalline state none of the crystal forms of citrate synthase show enzymatic activity (66) even though the enzyme from redissolved crystals is active. Domain motion is probably hindered by the crystal packing, and substrate turnover is all probably inhibited. Although crystalline order is not expected to occur in the mitochondrial matrix, the highly associated proteins probably influence the domain motion of citrate synthase. Such packing effects may alter in particular the action of di- and tricarboxylic acids on citrate synthase. Citrate bound to the open form of citrate synthase is less well ordered and makes fewer contacts with the enzyme than citrate bound to the closed C2 form, and may therefore be more weakly bound (66). The equilibrium between closed and open forms may be shifted by the protein matrix of which citrate synthase is a component. The affinity of possible effectors may be influenced in this way. Indeed, measurements on mitochondria made permeable after toluol treatment and stabilized by polyethylene glycol showed that the K_m for acetyl-Coenzyme A in the presence of oxaloacetate was 77 μM, nine times higher than in dilute solution. Oxaloacetate showed no change in V_{max} or K_m. The increase of the K_m of acetyl-Coenzyme A was not due to faster diffusion. The inhibition by tricarballylic acid was reduced in the permeabilized mitochondria as was the inhibition by ATP (54, 55, 101, 102). The formation of ATP-Mg^{2+} complexes may be responsible for the observed effects. Mg^{2+} suppresses the inhibition of the free enzyme by γ-phosphate groups of ATP (32). These groups are involved in the binding of acetyl-Coenzyme A to citrate synthase.

The effect of microenvironmental compartmentation on the steady-state rate of citrate synthase has also been investigated in a system where malate dehydrogenase and citrate synthase were immobilized by bonding to Sepharose (38, 57). The steady-state rate was increased by 10%–20% in this system, although the protein concentration was very low (1 mg protein per g Sepharose matrix) compared to that in mitochondria. Citrate synthase and malate dehydrogenase have shown a tendency to associate in the presence of polyethylene glycol (26). This interaction is probably specific, as it occurs only with mitochondrial malate dehydrogenase and not with S-MDH (26). Citrate synthase and mitochondrial malate dehydrogenase are bound to the inner mitochondrial membrane (16). Complex formation of citrate synthase and m-MDH is enhanced by palmitoyl-Coenzyme A. In contrast, oxaloacetate inhibits the citrate synthase m-MDH association (21). Palmitoyl-Coenzyme A interacts preferentially with the small domain of citrate synthase (9, 46, 98), while oxaloacetate induces domain closure.

These observations emphasize the specific nature of the interaction of these two enzymes. In general, the mitochondrial microenvironment is very likely to have a pronounced influence on the kinetic and regulatory properties of citrate synthase.

The regulatory role of palmitoyl-Coenzyme A has been a matter of intense interest. Palmitoyl- and stearoyl-Coenzyme A have been shown to inhibit citrate synthase with K_i of 4.2×10^{-6} M and 3.0×10^{-6} M, respectively. 3'-Dephospho-stearoyl-Coenzyme A was inactive even at 500 times higher concentration (100). This inhibition may be related to the fact that these detergents in stoichiometric amounts make citrate synthase susceptible to limited proteolysis. Palmitoyl-Coenzyme A and sodium dodecyl sulfate were observed to act similarly, but basic detergents were inactive. 0.1 M sulfate or phosphate inhibited the action of palmitoyl-Coenzyme A or sodium dodecyl sulfate (G. Wiegand, unpublished observations). These experiments indicate specific interactions as also shown by recent spin-label experiments (27): The binding of long-chain acyl-Coenzyme A analogs carrying a spin label was different from that of short-chain analogs. It was shown that 6-doxyloctanoyl-Coenzyme A binds stoichiometrically with two molecules per dimeric enzyme. It is not displaced by oxaloacetate. The long-chain analog 6-doxylstearoyl-Coenzyme A seems to bind in two different ways, as only 50% is displaced by oxaloacetate. The rest is displaced upon formation of the ternary complex with oxaloacetate and Coenzyme A (13, 27).

In summary, the regulation of citrate synthase at the molecular level is much less understood than the enzymatic mechanism. For the regulation in vivo the influence of the mitochondrial matrix on structural and functional properties of citrate synthase is probably of utmost importance.

MECHANISM

Irreversible modifications to various amino acid side chains formed the basis for early studies of the mechanism of enzymatic activity of citrate synthase. Acetic anhydride (74) and succinic anhydride (103) both inhibited the enzyme, whereby the latter caused the dissociation of the dimer into the two subunits. Photooxidation (96) and reactions with diethylpyrocarbonate or 2,3-butanedione (3, 51) indicated that arginine and histidine were important for the binding of oxaloacetate. Reagents that attack free amino groups or tryptophan had no effect on enzymatic activity (51). The crystallographic studies verified that arginine and histidine are indeed located in the active site and are involved in the binding of citrate and oxaloacetate (66, 99); in direct contact with either ligand are three arginine and three histidine side chains. However, due to the extreme substrate specificity of citrate synthase [only propionyl-CoA (92), fluoroacetyl-CoA, and fluorooxalooacetate can replace the natural substrates (11, 22, 23, 88)], little more could be determined by such modifications. For this reason studies of various inhibitors were undertaken. Since the overall reaction can theoretically be divided into three steps, inhibitors for the various partial reactions were sought. These are discussed where appropriate in light of the implications of the structural studies for the catalytic mechanism.

In the following we argue that the three structural states that have been observed for citrate synthase represent or are related to structures that catalyze distinct steps along the reaction pathway. Different functional groups on the enzyme direct the course of the reaction and prevent decoupling, for example to avoid the premature hydrolysis of acetyl-Coenzyme A before the condensation reaction has occurred.

Enolization and Proton Abstraction

The first step of the reaction is the enolization of acetyl-CoA and the abstraction of a hydrogen from the methyl group, which considerably favors the next step, the condensation reaction (17). Lipmann (47) classified the reactions of acetate into head, tail, and head-tail activated reactions; it is the latter class into which the mechanism of citrate synthase falls. The hydrogens of the methyl group of acetyl-CoA are nonexchangeable in aqueous solutions and a basic catalyst is required for abstraction. However, even in the presence of citrate synthase the protons from this methyl group are not exchanged with solvent protons (10, 53). Only in the presence of the enzyme and oxaloacetate or the analog L-malate can proton exchange be detected (17). The stereochemistry of this reaction has been studied and it has been determined that the hydrogens of the methyl group undergo inversion of configuration during the proton abstraction and condensation reactions. It was therefore postulated that a basic group on the enzyme accepts a proton from the methyl group of acetyl-CoA (18, 67), and that this group is activated upon binding of oxaloacetate by an induced fit process. The hypothesis was confirmed by the high resolution X-ray analysis of the three crystal forms. In the C2 crystal containing bound citrate and CoA, histidine 274 forms a hydrogen bond or salt bridge (see Figure 4) with the carboxylate of citrate (which would originate from acetyl-CoA in the full reaction); His 274 is undoubtedly the residue that accepts the proton from the incoming acetyl group. We note that if His 274 is protonated the positive charge will stabilize a negative charge at the enol oxygen.

$$B: \qquad \overset{O}{\underset{\|}{}} \qquad\qquad BH^+O^- $$

$$CH_3-\overset{\|}{C}-S-CoA \leftrightarrow CH_2=\overset{|}{C}-S-CoA$$

There is no other candidate for such a stabilizing positive charge.

This histidine residue is part of the hinge region of the enzyme and may well be activated by the open-closed conformational change (66). However, in all three crystal forms the side chains of residues His 274, Arg 401, and His 238 seem to be rigidly attached to the large domain and do not move during the conformational change. Using a least-squares procedure to superimpose these residues, it can be shown that within the limits of error of the structure determination the side-chain conformations of the three residues are identical for the three crystal forms (S. Remington, unpublished observations). Indeed, the side chain of His 274 is firmly bound to the N-terminal of helix L by hydrogen bonds to residues 242 and 244. The activation may therefore be related to the requirement that oxaloacetate be bound, forming a tight pocket for the acetyl moiety of acetyl-Coenzyme A.

Johansson & Pettersson (33) have shown that in the binary complex of citrate synthase and acetyl-Coenzyme A the acetyl group is highly mobile, but it is immobilized in the ternary complex. Also, oxaloacetate increases the binding constant for acetyl-CoA by a factor of at least 20 (33, 34, 93), presumably by inducing a number of additional contacts. In the crystal structure containing oxaloacetate and the inhibitor S-acetonyl-Coenzyme A, the electron density for the inhibitor is weak and uninterpretable in spite of near stoichiometric occupancy (99). This inhibitor differs from acetyl-Coenzyme A by a single methylene group, but is unable to bind in a well defined conformation. Studies have been done with spin-labelled analogs of acetyl-Coenzyme A (93). When the methyl group of acetyl-CoA is replaced with the large spin label N-oxy-pyrrolidine, oxaloacetate has no effect on the binding constant nor does it immobilize the label. ATP behaves similarly; it is a competitive inhibitor of acetyl-Coenzyme A with a binding constant of 2.0 mM^{-1} regardless of the presence or absence of oxaloacetate (34). Succinyl- and acetoacyl-CoA bind more weakly to the enzyme than either acetyl-CoA or CoA (72, 85). On the other hand, oxaloacetate does increase the binding constant for propionyl-Coenzyme A (150 mM^{-1}) by the same factor (about 20) as for acetyl-Coenzyme A (170 mM^{-1}) (33, 34, 92). Most remarkably, an analog of the transition state, carboxymethyl-Coenzyme A (which resembles enoyl-acetyl-Coenzyme A), shows an increase in affinity by a factor of 3000 in the presence of oxaloacetate (4).

In summary, the enolization activity of citrate synthase arises through an induced fit process requiring specific residues on the enzyme, the presence of oxaloacetate (or malate), and the specific configuration of acetyl-Coenzyme A. The thioester is important for the formation of the ternary complex and there is very little additional room in the binding pocket. The crystallographic studies verify this conclusion; in the C2 crystal form of the enzyme both citrate and the pantothenic arm of CoA are in an extremely tight binding pocket and are inaccessible to the solvent continuum. There are, however, three water molecules in the vicinity of these groups so that some rearrangement to accommodate slightly larger molecules such as propionyl-Coenzyme A seems possible.

Condensation Reaction

The observation by Eggerer et al (18) that the methyl group of acetyl-CoA undergoes inversion of configuration upon reaction with oxaloacetate indicates that proton abstraction and condensation are concerted reactions. The group most likely to be the proton acceptor, His 274, opposes the si-face of bound oxaloacetate in the $P4_32_12$ crystal form, with a narrow channel centered on the carbonyl carbon of oxaloacetate (the site of nucleophilic attack). It therefore seems likely that proton abstraction and condensation proceed via a pentacoordinate carbon intermediate. One problem with this interpretation is that L-malate is also capable of stimulating proton abstraction (17) from the methyl group of acetyl-CoA, but the corresponding nucleophilic attack does not occur. In the absence of structural data on the mode of binding of malate we can only speculate that enolization follows a somewhat different pathway in this instance.

Structural studies (66, 99) showed that citrate and oxaloacetate are in a predominantly polar environment. Six electron-withdrawing groups (three arginines and three histidines) are arranged in three dimensions about the anion. Kurz et al (41) have studied the polarization of the carbonyl bond of oxaloacetate by C-13 NMR and have concluded that the bond order is substantially reduced when this compound is liganded to citrate synthase. Thus the carbonyl carbon would carry a partial positive charge and would be highly susceptible to nucleophilic attack. We have already indicated that this carbon is centered at the end of a narrow tunnel and that the acetyl group of AcCoA must approach through this tunnel. Histidine 320, which is hydrogen-bonded to the carbonyl oxygen in the crystal that contains oxaloacetate, is in an ideal position to protonate the oxygen as the condensation reaction occurs. In summary, the condensation reaction can be understood in terms of a simple polarization (activation) of the two substrates and their coming together in the proper orientation. The product of the reaction, citryl-Coenzyme A, is probably formed in one concerted step.

Hydrolysis Reaction

Eggerer & Remberger (19) have shown that citryl-Coenzyme A is cleaved by the enzyme into citrate and Coenzyme A as well as oxaloacetate and acetyl-Coenzyme A. This observation led to the original proposal that citryl-Coenzyme A was an intermediate in the reaction. Buckel & Eggerer (12) proposed that the hydrolysis step occurred via an intramolecular attack resulting in citric anhydride and Coenzyme A. The crystal form containing both citrate and Coenzyme A has been studied at 1.7 Å resolution. This crystal form represents the conformation of citrate synthase with bound products, which would be the step immediately following the reaction. In this closed form of the enzyme the citrate molecule is very tightly bound to the enzyme with salt bridges to each of the carboxylates (66). As ocaloacetate is bound in the same manner as citrate (99) and the binding pocket is quite narrow, such an intramolecular attack seems unlikely. However, an acid group from the enzyme (Asp 375) would be in van der Waals contact with the carboxylate of citrate that would originate from acetyl-Coenzyme A. An attack by this aspartate to form a mixed anhydride would be analogous to the intramolecular reaction proposed by Eggerer and associates. This would release Coenzyme A. On the other hand, Remington et al (66) provided evidence from the electron density map that aspartyl(375)-Coenzyme A was present to some extent in this crystal form; they suggested that transesterification was a step in the hydrolysis reaction, releasing citrate. Regardless of the function of this aspartate residue, the presence of several bound water molecules in the immediate vicinity of the proposed thioester bond indicates that the final step of the hydrolysis could occur without the necessity of opening the active site to the solvent continuum. Indirect lines of evidence that imply the importance of this residue include the sequence of *E. coli* citrate synthase (58) where this aspartate is conserved. In fact, every residue that directly interacts with citrate or oxaloacetate is conserved between pig heart and *E. coli* citrate synthases even though the overall homology is quite low.

Crystallographic analyses of the mode of binding of (R,S)-3,4-dicarboxy-3-hydroxybutyl-Coenzyme A (66) have provided solid support for the hypothesis that the closed form of the enzyme with bound citrate and CoA is the hydrolysis form. This compound is a close analog of citryl-Coenzyme A; the only difference is that the thioester oxygen of citryl-CoA is replaced by two hydrogen atoms. This compound inhibits both the hydrolysis and the lyase reactions and binds extremely tightly to the enzyme (4). The crystallographic analysis was unusually easy to interpret: The citryl-CoA analog binds in exactly the manner of citrate and CoA; the only changes observed on the part of the enzyme were small positional shifts of Asp 375 and His 274.

Conformational Change and the Reaction Pathway

New kinetic evidence has recently become available that supports the idea that various conformational states of citrate synthase have different enzymatic activities. The enzyme has been shown to have a biphasic rate dependence on enzymatic concentration and to exhibit hysteretic behavior (46, 48). For the hydrolysis of citryl-Coenzyme A an initial fast phase is followed by a slow steady-state rate for certain enzyme concentrations. This was interpreted in terms of the structural information available previously as an equilibrium between the closed form containing citrate and Coenzyme A, which represents the ligase activity, and the open form, which represents the hydrolase activity (48). The slow phase was thought to result from the gradual appearance of oxaloacetate, which would shift the equilibrium from the (predominant) hydrolytic form to the ligase form. Limited proteolysis of citrate synthase (46) was observed to lower the rate of the overall reaction by a factor of 20 but actually stimulated hydrolysis or C—C bond cleavage of citryl-Coenzyme A by a factor of 10 (presumably by interfering with this conformational change).

The demonstration that a third structural form exists in the presence of oxaloacetate (99) complicates the issue slightly but provides evidence for a satisfying model that is not different in substance from that just discussed. The model for catalytic activity proposed is that the three conformational states observed crystallographically represent the ligase activity, the hydrolase activity, and the product-release form. The closed forms are the catalytically competent forms, as the reaction requires a highly specific three-dimensional environment isolated from the solvent. However, to allow substrate/product entry/release, the enzyme must open up. The structure analyzed in the presence of oxaloacetate must be (or be very closely related to) the condensation (ligase) form of the enzyme. In this form the active site is not completely closed : Asp 375, His 320 and Arg 329 shift 0.8–1.3 Å away from their positions in the closed form containing citrate. We have already argued that the closed form containing citrate and Coenzyme A is the hydrolysis form, and is induced by the appearance of citryl-Coenzyme A. In this form Asp 375 moves about 1.3 Å (S. Remington, unpublished observations) and presumably is in position to catalyze the hydrolysis reaction. It is reasonable to assume that this step is necessary in order to prevent the enzyme from hydrolyzing acetyl-Coenzyme A before the condensation reaction is complete. The kinetic model of Löhlein-Werhahn et al (48) is applicable to this problem if we assume that the presence of either substrates, products, or intermediates strongly favors the closed forms of the enzyme over the open (product release) form of the enzyme, thus allowing concentration to be neglected. It should be noted,

CITRATE SYNTHASE STRUCTURE   115

however, that citrate alone does not appear to induce a conformational change. Crystals of both the open and closed forms of the enzyme can be grown from citrate buffer, with the deciding factor apparently being the pH of the solution (66).

In summary, the following order of events is proposed. Oxaloacetate binds to the enzyme, inducing the "nearly closed" conformation observed in the $P4_32_12$ crystal form. This form strongly polarizes and orients oxaloacetate and provides a binding site for acetyl-Coenzyme A with a narrow channel leading to oxaloacetate. As the pantothenic arm of AcCoA enters, a proton is abstracted from the methyl group and condensation occurs in a concerted reaction that results in the inversion of the configuration of the hydrogens of the methyl group. The appearance of citryl-Coenzyme A on the enzyme induces the hydrolytic "closed" form of the enzyme observed in the crystals containing citrate and Coenzyme A (the C2 form). The hydrolysis reaction occurs via Asp 375 and bound water. The enzyme is now free to open, releasing the products citrate and Coenzyme A.

Citrate synthase can be considered a classic enzyme in terms of the induced fit model. The enzyme possesses a domain structure with the active site situated between the domains. The large domain appears to be relatively rigid (99) while the small domain undergoes several conformational changes in response to the stereochemistry of bound substrates and/or intermediates. These conformational changes influence the various functional groups to order the reaction as it proceeds.

Literature Cited

1. Addink, A. D. F., Boer, P., Wakabayashi, T., Green, D. E. 1972. *Eur. J. Biochem.* 29:47
2. Alam, T., Finkelstein, D., Srere, P. A. 1982. *J. Biol. Chem.* 257:11181
3. Bayer, E., Bauer, B., Eggerer, H. 1981. *FEBS Lett.* 127:101
4. Bayer, E., Bauer, B., Eggerer, H. 1981. *Eur. J. Biochem.* 120:155
5. Beeckmans, S. 1984. *Int. J. Biochem.* 16:341
6. Beevers, H. 1979. *Ann. Rev. Plant Physiol.* 30:159
7. Bennett, W. S., Huber, R. 1984. *CRC Crit. Rev. Biochem.* 15:291
8. Bloxham, D. P., Ericsson, L. H., Titani, K., Walsh, K. A., Neurath, H. 1980. *Biochemistry* 19:3979
9. Bloxham, D. P., Parmelee, D. C., Kumar, S., Wade, R. D., Ericsson, L. H., Neurath, H., Walsh, K. A., Titani, K. 1981. *Proc. Natl. Acad. Sci. USA* 78:5381
10. Bove, J., Martin, R. O., Ingraham, L. L., Stumpf, P. K. 1959. *J. Biol. Chem.* 234:999
11. Brady, R. O. 1955. *J. Biol. Chem.* 217:213
12. Buckel, W., Eggerer, H. 1969. *Hoppe-Seyler's Z. Physiol. Chem.* 350:1367
13. Caggiano, A. V., Powell, G. L. 1979. *J. Biol. Chem.* 254:2800
14. Cooke, R., Kuntz, I. D. 1974. *Ann. Rev. Biophys. Bioeng.* 3:95
15. Danson, M. J., Black, S. C., Woodland, D. L., Wood, P. A. 1985. *FEBS Lett.* 179:120
16. D'Souza, S. F., Srere, P. A. 1983. *J. Biol. Chem.* 258:4706
17. Eggerer, H. 1965. *Biochem. Z.* 343:111
18. Eggerer, H., Buckel, W., Lenz, H., Wunderwald, P., Gottschalk, G., Cornforth, J. W., Donninger, C., Mallaby, R., Redmond, J. W. 1970. *Nature* 226:517
19. Eggerer, H., Remberger, U. 1963. *Biochem. Z.* 337:202

20. Eggerer, H., Remberger, U., Grünewälder, C. 1964. *Biochem. Z.* 339:436
21. Fahien, L. A., Kmiotek, E. 1983. *Arch. Biochem. Biophys.* 220:386
22. Fanshier, D. W., Gottwald, L. K., Kun, E. 1962. *J. Biol. Chem.* 237:3588
23. Fanshier, D. W., Gottwald, L. K., Kun, E. 1963. *J. Biol. Chem.* 239:425
24. Deleted in proof
25. Hackenbrock, C. R. 1968. *Proc. Natl. Acad. Sci. USA* 61:598
26. Halper, L. A., Srere, P. A. 1977. *Arch. Biochem. Biophys.* 184:529
27. Hansel, B. C., Powell, G. L. 1984. *J. Biol. Chem.* 259:1423
28. Hanson, K. R., Rose, I. A. 1963. *Proc. Natl. Acad. Sci. USA* 50:981
29. Harmey, M. A., Neupert, W. 1979. *FEBS Lett.* 108:385
30. Hol, W. G. J., van Duijnen, P. T., Berendsen, H. J. C. 1978. *Nature* 273:443
31. Hoppel, C. L., Tomec, R. J. 1972. *J. Biol. Chem.* 247:832
32. Jaffe, E. K., Cohn, M. 1978. *Biochemistry* 17:652
33. Johansson, C.-J., Pettersson, G. 1974. *Eur. J. Biochem.* 42:383
34. Johansson, C.-J., Pettersson, G. 1974. *Eur. J. Biochem.* 46:5
35. Johansson, C.-J., Pettersson, G. 1977. *Biochim. Biophys. Acta* 484:208
36. Klingenberg, M. 1979. *Meth. Enzymol.* 56:245
37. Klinman, J. P., Rose, I. A. 1971. *Biochemistry* 10:2267
38. Koch-Schmidt, A.-C., Mattiasson, B., Mosbach, K. 1977. *Eur. J. Biochem.* 81:71
39. Koshland, D. E. 1959. *J. Cell. Comp. Physiol.* (Suppl. 1) 54:245
40. Kosicki, G. W., Srere, P. A. 1961. *J. Biol. Chem.* 236:2560
41. Kurz, L. C., Ackerman, J. J. H., Drysdale, G. R. 1985. *Biochemistry* 24:452
42. Landriscina, C., Papa, S., Coratelli, P., Mazzarella, L., Quagliariello, E. 1970. *Biochim. Biophys. Acta* 205:136
43. Lenz, H., Buckel, W., Wunderwald, P., Biedermann, G., Buschmeier, V., Eggerer, H., Cornforth, J. W., Redmond, J. W., Mallaby, R. 1971. *Eur. J. Biochem.* 24:207
44. Lesk, A. M., Chothia, C. 1984. *J. Mol. Biol.* 174:175
45. Lesk, A. M., Hardman, K. D. 1982. *Science* 216:539
46. Lill, U., Schreil, A., Henschen, A., Eggerer, H. 1984. *Eur. J. Biochem.* 143:205
47. Lipmann, F. 1948, 1949. *Harvey Lect.* 44:99
48. Löhlein-Werhahn, G., Bayer, E., Bauer, B., Eggerer, H. 1983. *Eur. J. Biochem.* 133:665
49. Lynen, F. 1942. *Justus Liebigs Ann. Chem.* 552:270
50. Lynen, F., Reichert, E., Rueff, L. 1951. *Justus Liebigs Ann. Chem.* 574:1
51. Mahlen, A. 1975. *FEBS Lett.* 51:294
52. Makinen, M. W., Fink, A. L. 1977. *Ann. Rev. Biophys. Bioeng.* 6:301
53. Marcus, A., Vennesland, B. 1958. *J. Biol. Chem.* 233:727
54. Matlib, M. A., Boesman-Finkelstein, M., Srere, P. A. 1978. *Arch. Biochem. Biophys.* 191:426
55. Matlib, M. A., Shannon, W. A. Jr., Srere, P. A. 1977. *Arch. Biochem. Biophys.* 178:396
56. Matthews, B. W. 1968. *J. Mol. Biol.* 33:491
57. Mosbach, K. 1978. In *Microenvironments and Metabolic Compartmentation*, ed. P. A. Srere, R. W. Estabrook, pp. 381–98. New York: Academic
58. Ner, S. S., Bhayana, B., Bell, W. A., Giles, I. G., Duckworth, H. W., Bloxham, D. P. 1983. *Biochemistry* 22:5243
59. Neupert, W., Schatz, G. 1981. *Trends Biochem. Sci.* 6:1
60. Novelli, G. D., Lipmann, F. 1950. *J. Biol. Chem.* 182:213
61. Ochoa, S. 1955. *Meth. Enzymol.* 1:685
62. Ochoa, S., Stern, J. R., Schneider, M. C. 1951. *J. Biol. Chem.* 193:691
63. Parvin, R. 1969. *Meth. Enzymol.* 13:16
64. Pihl, E., Bahr, G. F. 1970. *Exp. Cell Res.* 63:391
65. Pullman, B., Saran, A. 1976. *Prog. Nucleic Acid Res. Mol. Biol.* 18:225
66. Remington, S., Wiegand, G., Huber, R. 1982. *J. Mol. Biol.* 158:111
67. Retey, J., Lüthy, J., Arigoni, D. 1970. *Nature* 226:519
68. Rubin, B. H., Stallings, W. C., Glusker, J. P., Bayer, M. E., Janin, J., Srere, P. A. 1983. *J. Biol. Chem.* 258:1297
69. Rupley, J. A. 1969. In *Structure and Stability of Biological Macromolecules*, ed. S. N. Timasheff, G. D. Fasman, pp. 291–352. New York: Dekker
70. Schnaitman, C., Greenawalt, J. W. 1968. *J. Cell Biol.* 38:158
71. Sheperd, D., Garland, P. B. 1969. *Meth. Enzymol.* 13:11
72. Smith, C. M., Williamson, J. R. 1971. *FEBS Lett.* 18:35
73. Spector, B. L. 1972. *Enzymes* 7:357
74. Srere, P. A. 1963. *Biochim. Biophys. Acta* 77:639
75. Srere, P. A. 1965. *Biochim. Biophys. Acta* 99:197
76. Srere, P. A. 1966. *J. Biol. Chem.* 241:2157
77. Srere, P. A. 1969. *Meth. Enzymol.* 13:3

78. Srere, P. A. 1971. *Adv. Enzyme Regul.* 9:221
79. Srere, P. A. 1972. *Curr. Top. Cell. Regul.* 5:229
80. Srere, P. A. 1974. *Life Sci.* 15:1695
81. Srere, P. A. 1975. *Adv. Enzymol.* 43:57
82. Srere, P. A. 1980. *Trends Biochem. Sci.* 5:120
83. Srere, P. A. 1981. *Trends Biochem. Sci.* 6:4
84. Srere, P. A. 1982. *Trends Biochem. Sci.* 7:375
85. Srere, P. A., Matsouka, Y., Mukherjee, A. 1973. *J. Biol. Chem.* 248:8031
86. Stern, J. R., Ochoa, S. 1949. *J. Biol. Chem.* 179:491
87. Stern, J. R., Ochoa, S., Lynen, F. 1952. *J. Biol. Chem.* 198:313
88. Stern, J. R., Shapiro, B., Stadtmane, R., Ochoa, S. 1951. *J. Biol. Chem.* 193:703
89. Tolbert, N. E. 1971. *Ann. Rev. Plant Physiol.* 22:45
90. Tolbert, N. E. 1981. *Ann. Rev. Biochem.* 50:133
91. Tong, E. K., Duckworth, H. W. 1975. *Biochemistry* 14:235
92. Weidman, S. W., Drysdale, G. R. 1973.

93. Weidman, S. W., Drysdale, G. R., Mildvan, A. S. 1973. *Biochemistry* 12:1874
94. Weitzman, P. D. J. 1981. *Adv. Microb. Physiol.* 22:185
95. Weitzman, P. D. J., Danson, M. J. 1976. *Curr. Top. Cell. Regul.* 10:161
96. Weitzmann, P. D. J., Ward, B. A., Rand, D. L. 1974. *FEBS Lett.* 43:97
97. Wiegand, G. 1976. *FEBS Lett.* 62:281
98. Wiegand, G., Kukla, D., Scholze, H., Jones, T. A., Huber, R. 1979. *Eur. J. Biochem.* 93:41
99. Wiegand, G., Remington, S., Deisenhofer, J., Huber, R. 1984. *J. Mol. Biol.* 174:205
100. Wieland, O., Weiss, L., Eger-Neufeldt, I. 1964. *Biochem. Z.* 339:501
101. Williamson, J. R., Olson, M. S. 1968. *Biochem. Biophys. Res. Comm.* 32:794
102. Williamson, J. R., Olson, M. S., Herczeg, B. E., Coles, H. S. 1967. *Biochem. Biophys. Res. Comm.* 27:595
103. Wu, J.-Y., Yang, J. T. 1970. *J. Biol. Chem.* 245:212

Fed. Proc. 32(549):1863 (Abstr.)

Ann. Rev. Biophys. Biophys. Chem. 1986. 15:119–61

RELATIONSHIPS BETWEEN CHEMICAL AND MECHANICAL EVENTS DURING MUSCULAR CONTRACTION

Mark G. Hibberd

Harvard Medical School, Boston, Massachusetts 02115

David R. Trentham

Physical Biochemistry Division, National Institute for Medical Research, Mill Hill, London, NW7 1AA, United Kingdom

CONTENTS

PERSPECTIVES AND OVERVIEW

The discovery of the sliding filament mechanism of striated muscle contraction (75, 84) set the stage for a new era of muscle research in which investigators sought to elucidate the molecular events that set the filaments

0883–9182/86/0610–0119$02.00

in motion. Extensive improvements in the spatial resolution of structural probes and in the time resolution of mechanical and biochemical measurements have not, as yet, yielded a satisfying or unifying theory by which specific structural, mechanical, and biochemical assignments can be made for each of the steps in the cross-bridge cycle. These uncertainties are particularly prominent in regard to the key reactions that produce force or shortening.

In this review, we focus mainly on the advances made possible by new experimental techniques that have allowed increasingly specific links to be made between biochemical and mechanical theories of contraction. We consider experimental evidence within the general framework of theories in which cross-bridges between the filaments act as independent force

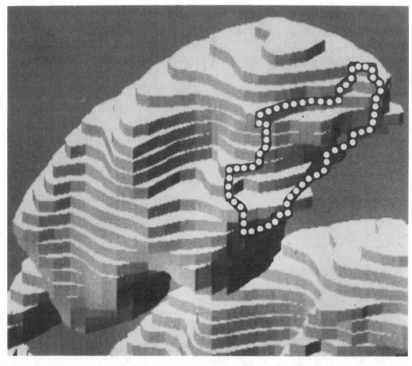

Figure 1   Three-dimensional reconstruction of myosin subfragment 1 obtained by image processing of electron micrographs. The view is from the axis of helical symmetry of actin-tropomyosin-subfragment 1. The part assigned as actin-tropomyosin was removed. The actin binding site of subfragment 1 is enclosed with a dotted line. (The length of the long axis of the enclosure is 4 nm.) A small portion of the image distal to the actin binding site lies beyond the picture to the lower left. Part of a second subfragment 1 is shown in the lower right-hand corner. (T. Wakabayashi, C. Toyoshima, A. Tomioka, M. Tokunanga, unpublished data.)

generators; the merits and shortcomings of many other theories have been discussed in detail by A. F. Huxley (68, 70, 71) and by Woledge et al (161). We are concerned primarily with the characterization and properties of intermediates within the cross-bridge cycle of activated skeletal muscle. Finally, we incorporate the mechanical and biochemical results into a scheme of the cross-bridge cycle, paying particular attention to the influence of force on the elementary biochemical steps.

There are detailed reviews of mechanics (70, 71), energetics (95, 161), and solution biochemistry (3, 86, 139, 148, 151, 157). Advances in space- and time-resolved structural studies, made possible in part by the use of synchrotron radiation, spin and magnetic resonance, and optical probes of the cross-bridges, have also been reviewed (4, 54, 57, 82, 85, 112, 140, 149). While structural aspects of the cross-bridge cycle are outside the scope of this review, a recent reconstruction of the myosin head is presented to help nonspecialists in muscle research visualize its shape and the likely extent of its spacial interaction with actin (Figure 1). As yet there is little definitive information on what structural component of the cross-bridge gives rise to its force-generating and elastic properties or what structural changes are associated with the transfer of stored chemical energy into external mechanical work.

At present there are very few points of unambiguous reference between the biochemical and mechanical data. Measurements of the mechanical properties of fibers do not contain information about the biochemical states of the cross-bridges except by inference. Similarly, the biochemical experiments with solubilized proteins are conducted without the mechanical constraints that are so crucial to the macroscopic properties of muscle. Neither is it valid to assume that the biochemical experiments give precise data applicable to the mechanical situation during shortening at maximum velocity (i.e. at zero load). The assumption is invalid because theoretical and experimental considerations indicate that maximum shortening velocity is determined by a balance between cross-bridges that drive the filaments in a shortening direction and those that, by virtue of their limited detachment rate, push the filaments in opposition to shortening (68, 71). Thus, while recent theories of the cross-bridge cycle all presume that the organization of proteins within the myofilament and the performance of work affect the rates of the elementary reactions, the results of experiments on solubilized systems cannot easily be transposed to those that occur in vivo.

Advances in understanding the thermodynamic basis for energy transduction in biological systems (64–67) have been extended to include specific issues related to muscle contraction (31–33, 67, 133, 161). Several key points have emerged from these contributions, two of which influence much current experimental work in this area. First, it is to be expected that the coupling of mechanical work or force production into a molecular transition will reduce the equilibrium constant of that step. The free energy change associated with the unrestrained reaction can be considered as transduced into a mechanical form, or alternatively the free energy associated with the mechanical work must be considered as part of the elementary reaction and therefore as a contributor to the overall equilibrium. Second, gross and basic free energy changes (133) depend on the actual concentrations of the chemical species involved in addition to the energy changes associated with the standard free energy changes of the molecular transitions. It is likely that the steps in the overall cycle at which appreciable actual free-energy changes occur are, or are closely coupled to, processes in which ligands bind or dissociate from the ATPase active site (160). Thus it is essential to know the rates of the elementary reactions and the concentrations of the intermediates as they exist in vivo in order to describe properly the thermodynamics of the cycle.

The general reader may not be familiar with the usage by physiologists of mechanical terms applied to muscle. An excellent summary is given in the Appendix of Reference 134 (pp. 144–45). In this review we frequently use the term "tension" for tension per cross-sectional area where the sense is otherwise clear (as in figure legends).

Mechanical Events of the Cross-Bridge Cycle

Hanson & H. E. Huxley (165) and A. F. Huxley (68) proposed theories incorporating cycling properties of cross-bridges into the sliding filament hypothesis. These theories still describe best the large body of mechanical and thermal data that was known at that time. We briefly detail the main properties of A. F. Huxley's theory (68) and outline how the essential ideas have been developed in the light of more recent data, so that we may relate them to the biochemical schemes devised during the 1970s and 1980s. We also refer the reader to much more detailed comparisons of the theory's predictions with experimental determinations of the mechanical and thermal properties of muscle (70–72, 120, 134, 161).

The main mechanical features of Huxley's model were as follows. A cycle of interactions was believed to occur between the myosin and actin filaments. In this cycle links from myosin would attach to actin, undergo movement through an elastic element, and then detach. Huxley accommodated the vectorial nature of force production in the theory by making the rates of attachment and detachment of force generating links (now called cross-bridges) sensitive to position. The rate constants for these two processes were believed dependent on the relative axial positions of the myosin-based projections and the actin-based attachment sites. Attachment would be faster than detachment in regions where cross-bridges generated positive tension, but detachment would be much faster from positions in which the cross-bridges exerted a force opposed to the normal shortening motion. Cross-bridges could not attach directly in positions that exerted negative tension, and could reach such positions only through the sliding of the filaments relative to each other during shortening. In the isometric state, cross-bridges would be attached in a variety of positive force–producing states, but during shortening tension would be inversely proportional to the velocity of shortening for three reasons. First, the finite time required for attachment would lead to a lower average force per attachment, particularly because the number of cross-bridges generating the most force would be lowest at high velocity. Second, fewer cross-bridges would be attached because of the net increase in the rate of detachment and also because at high velocities some myosin links would fail to attach to actin sites that passed by relatively rapidly. Third, tension would drop to zero at the maximum velocity of shortening when the sum of the negative forces rose to match the sum of the remaining positive forces in the cross-bridges. Some cross-bridges would be carried into a region where they would develop negative force owing to drag.

When rapid changes in load (21, 116) or length (5, 42, 76) are applied to the muscle, the velocity or tension responses occur in up to four phases. The Huxley theory does not make specific predictions about such transients. However, A. F. Huxley & Simmons (78) and Ford et al (43) have concluded that all phases of the transients result almost exclusively from the properties conferred on the muscle by the formation of cross-bridges. Podolsky et al (117) suggested modifications to the position sensitivity and the rates of the processes in Huxley's original model. Their model fit the transients well, but the modifications made the model less energetically efficient. Moreover, the adapted model predicted effects on stiffness (related to the number of attached cross-bridges) that do not seem to be supported by recent investigations. Specifically, the modifications predicted a rise in stiffness immediately following an abrupt release and during shortening. In the former instance, very little change in stiffness is apparent (41, 78) and in the latter case several investigators have shown that stiffness falls to between 20% and 40% of the isometric level during rapid shortening (10, 44, 88, 89).

A. F. Huxley & Simmons (76) suggested a different explanation of the two fastest phases of the tension transients. They proposed that phase one results from an almost simultaneous change in length of a linear elastic element and that the rapid recovery of tension, phase two, results from the rapid stepwise transitions between two or three attached cross-bridge states. These transitions would tend to reestablish the original length of the elastic element and therefore the original tension also. Activation energy barriers between the various states would depend in part on the stress supported by the cross-bridge. Therefore transitions would be faster when the force was low, such as after a release, and slower after a stretch. Huxley & Simmons ascribed the two slower transient phases (phases three and four) to attachment/detachment reactions similar to those described earlier by Huxley (68). The Huxley-Simmons theory predicts that stiffness will be essentially unchanged immediately following a step length change; this is in fact observed (41). When combined with the postulates of Huxley's original formulation, this theory would almost certainly account as well as the earlier theory for the steady-state mechanical and thermal data, and would also account for the fall in stiffness during shortening.

Other models of the transients (1, 2, 91) do not fit as well to the most precise data (42, 77). They are also inferior because they require stiffness changes after a quick length change unless net cross-bridge attachments and detachments are coincidentally balanced during the large force changes that occur after a step.

Together, the Huxley (68) and Huxley-Simmons (76) theories predict a number of testable features of shortening muscle. The link between the stress in a cross-bridge and the rate of its transitions from one position to another predicts that tension recovery following a release should be more rapid when the sudden length step is imposed during steady overall shortening, since the average tension in the cross-bridge is then presumably lower. Huxley's original proposal also predicted a reduced number of cross-bridges, a reduced average extension of each cross-bridge, and a wider range of cross-bridge extensions during shortening. All four features are substantially supported by recent measurements (44). Ford et al (44) also suggested that part of the rapid tension recovery during shortening may result from the very rapid detachment of negatively strained cross-bridges. However, they were unable to make measurements on a sufficiently rapid time scale to estimate and compare the proportion of this contribution with the Huxley-Simmons type of tension recovery process. It is likely that newly developed equipment with faster response time will be of value for making such measurements (16, 73, 74, 154).

Biochemistry of Actomyosin ATPase

In this section we give some of the background that provides the basis for the actomyosin ATPase mechanism with an appropriate degree of detail for our later correlations of the biochemistry and mechanics of fibers. It is also necessary to know the source and nature of solubilized proteins used in biochemical studies for these correlations. Protein preparations for study of the actomyosin ATPase have, for the most part, been isolated from rabbit back and leg fast-twitch skeletal muscles. Actin is prepared in filamentous form, frequently with the regulatory proteins and tropomyosin removed.

Myosin subfragment 1, the monomeric head of the myosin molecule, which contains both an actin binding site and an ATP hydrolytic site, is frequently used in place of myosin (100). However, other soluble forms of rabbit skeletal muscle myosin have been used. For the purposes of this review the various preparations are not distinguished but are referred to as isolated actomyosin (in contrast to actomyosin in fibers). The reader must refer to original texts to establish the protein preparations used.

Unless myosin is cross-linked chemically to actin (113), ATPase activities at physiological ionic strength are much less than those measured at low (i.e. at or below 20 mM) ionic strength. This is because the K_m of actin with respect to the ATPase increases markedly with ionic strength. Biochemical experiments are therefore frequently conducted at nonphysiological ionic strength; this must be accounted for in making mechanochemical correlations.

The framework for the biochemical mechanism of the actomyosin ATPase used here stems from the scheme of Lymn & Taylor (102). This scheme contains a conceptually simple view of how the chemistry of ATP hydrolysis might be related to cycling cross-bridges, as illustrated in Figure 2. Lymn & Taylor's proposal, somewhat updated by more recent findings, is as follows. The ATP, which binds tightly to myosin but only weakly to actomyosin, induces dissociation of the cross-bridge. Hydrolysis of ATP

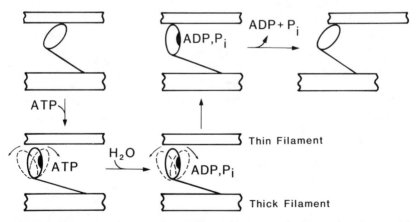

Figure 2   Cross-bridge cycle and its relation to the actomyosin ATPase based on model proposed by Lymn & Taylor (102). The attached states are characterized by two preferred orientations depending on whether or not nucleotide is bound to the myosin head. The dissociated heads are mobile (14, 24) with freedom of rotational motion, having relaxation times (0.1–1 μs) comparable to those of free heads (110, 131). This mobility is indicated by multiple orientations of dissociated heads. The heavy marking in the heads represents nucleotide bound to a site distinct from the actin binding site (112, 145).

occurs, followed by reattachment of the cross-bridge. At this point an actomyosin-products complex is formed in which the affinity of actin for myosin is relatively weak [$K_{diss} \approx 100$ μM (19)]. However, release of products results in the high-affinity actomyosin complex [$K_{diss} \approx 0.1$ μM (22)] and the cycle is complete. Figure 2 shows structural correlations that were proposed to accompany the biochemical events. It was suggested that the preferred orientation of the actomyosin cross-bridge to which ATP binds differs from that of the actomyosin-products complex. The rearrangement from one orientation to another during product release generates force and the consequent movement of the filaments.

During the past decade there have been several important new findings that we have incorporated into a more complex scheme of the actomyosin ATPase (Figure 3). An attached-state hydrolysis (step 3a) had been postulated for more than a decade by Inoue et al (86) and by Tonomura (150), but the constraints on the mechanism imposed by the available experimental data were insufficient to make the argument strong. Sleep & Hutton (137) provided the first convincing demonstration of step 3a. They showed that actin was able to bind almost as readily to M.ATP as to M.ADP.P$_i$ (steps 2 and 4). This meant that the researchers could add near-saturating amounts of actin to M.ATP and M.ADP.P$_i$ and monitor the reaction between AM.ATP and AM.ADP.P$_i$. Eisenberg and coworkers (3, 19, 31, 144) made several important contributions that confirm both the existence of attached state hydrolysis in solution (step 3a) and the concept that actin binds reversibly, rapidly, and relatively weakly to M.ATP and M.ADP.P$_i$. Whether the major flux during actomyosin ATPase activity is via step 3 or 3a depends on actin concentration and ionic strength. Flux through step 3a is favored at low ionic strength. In any event, flux through step 3a in fibers does not preclude relative filament sliding during shortening, because the high values of rate constants associated with steps

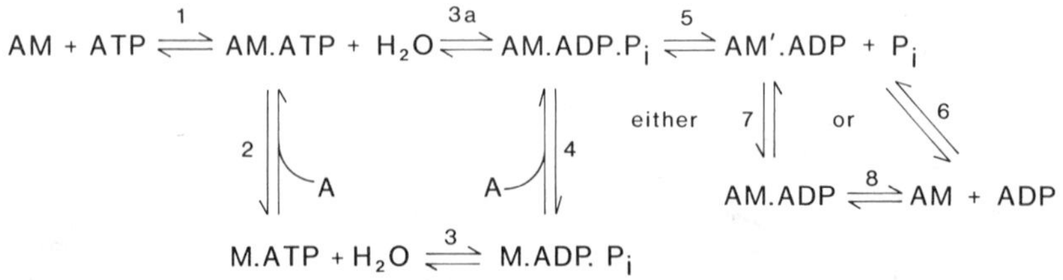

Figure 3 Actomyosin ATPase mechanism based primarily on data derived by Eisenberg, Sleep, and Taylor and their colleagues (3, 19, 31, 87, 102, 124, 137–139, 144, 148). AM denotes actomyosin. In the text, $K_i(= k_{+i}/k_{-i})$ is the equilibrium constant of the ith step, and k_{+i} and k_{-i} are the rate constants.

2 and 4 would permit rapid attachment and detachment of cross-bridges (31, 32).

The other major modification depicted in Figure 3 is the inclusion of an AM'.ADP state. This state was proposed to exist during actomyosin ATPase activity and to be distinct from AM.ADP, the state formed by adding ADP to actomyosin (138). As discussed later, whether AM.ADP is a state on the ATPase pathway is open to question. This uncertainty is reflected in Figure 3 by alternate pathways via either step 6 or steps 7 and 8.

Data that might allow one to list equilibrium and rate constants of each of the steps in Figure 3 are not available for the actomyosin ATPase in solution under physiological or even low ionic strength conditions. The principal reasons for this lack of data are that interpretations of equilibrium and rate measurements are model dependent and that the steps under investigation may be influenced by additional processes, so researchers are appropriately cautious. For example, the step controlling actomyosin dissociation on addition of ATP has a rate constant of $5000–18,000$ s^{-1} (111). However, this process may be an isomerization between AM.ATP states rather than actin dissociation per se, so that k_{+2} may be greater than $18,000$ s^{-1}. With this reservation one can make estimates of equilibrium and rate constants of the mechanism in Figure 3, which provide a useful base from which to correlate mechanical or biochemical data from fibers. Most numbers relate to rabbit skeletal muscle actin and subfragment 1 at pH 7, low ionic strength (≈ 20 mM), and 15–20°C when $k_{cat} \approx 10$ s^{-1} (155). $K_1 = 100–500$ M^{-1} and $k_{+2} = 5000–18,000$ s^{-1} with the product $K_1 k_{+2} = 1.5 \times 10^6$ M^{-1} s^{-1} (111, 160) and $k_{-1} > 6$ s^{-1} (137). The affinities of actin for M.ATP and M.ADP.P$_i$ are similar, with $K_2 = 30–100$ μM and $K_4 = 1–3 \times 10^4$ M^{-1} (19, 144).

For the hydrolysis steps, Rosenfeld & Taylor (124) found $k_{+3a} = 12$ s^{-1}, $k_{-3a} = 28$ s^{-1}, $k_{+3} = 30$ s^{-1}, and $k_{-3} = 20$ s^{-1}. Their data further suggest that ATP cleavage is the rate-limiting step of the actomyosin ATPase so that a mixture of AM.ATP and M.ATP comprises the major component of the steady-state complex during ATPase activity. This view was supported by the data of Biosca et al (7), but differs from that of Stein et al (141–143), who claim that the steady-state complex at full actin activation is an actomyosin-products complex. If Stein et al's claim is true, there is a problem of compatibility with the mechanism in Figure 3, since it has been found that the K_m for actin with respect to the ATPase is sixfold less than K_4^{-1} and the mechanism in Figure 3 predicts $K_m \approx K_4^{-1}$ (142; but see also 124). This paradox has yet to be resolved and is of particular relevance when the biochemical ATPase mechanism of shortening muscle is considered. Eisenberg & Greene (31) and Stein et al (142) postulated a six-state model of muscle contraction to reconcile their data. The additional AM.ADP.P$_i$ and

M.ADP.P$_i$ states that they propose are not included in the biochemical schemes of this review. Evidence from oxygen isotope exchange studies favors the Rosenfeld & Taylor interpretation. At high actin concentration the exchange between water and P$_i$ formed during ATP hydrolysis occurs by reversal of step 3a. The amount of exchange depends on k_{-3a}/k_{+5}; the greater this ratio, the more the exchange. According to Rosenfeld & Taylor's scheme, exchange should be minimal because $k_{-3a} < k_{+5}$, while according to the six-state model extensive exchange would occur (135). In fact, the amount of exchange that occurs is small (136). However, as pointed out by Stein et al (143), it is possible that exchange would also be suppressed if actin inhibited phosphate rotation in the active site (process X' in Figure 10, below). We regard this explanation as unlikely, in part because, as we show below, the rate constants of P$_i$ dissociation from AM.ADP.P$_i$ measured by oxygen isotope exchange and mechanical experiments are approximately the same.

Neither the rate constant controlling P$_i$ release, k_{+5}, nor that of ADP release from AM'.ADP have been measured directly. Sleep & Hutton (138) have discussed the kinetic and equilibrium relationships between AM'.ADP and AM.ADP and concluded that one, though not the exclusive, interpretation of their data gives $K_7 = 50$, $K_8 = 160$ μM, and $k_{+8} > 500$ s^{-1} ($= 400$ s^{-1} at 4°C) (55, 132, 159). Finally, it is important that the dissociation constant of actin from AM.ADP equals 1 μM, which is 30- to 100-fold smaller than K_2 and K_4^{-1} (46). This means that dissociation of M.ADP from AM.ADP is unlikely to occur.

The correlate of the myosin ATPase in solution occurs in fiber studies when actin and myosin filaments are stretched beyond overlap (e.g. 39, 62). The solution correlate of unstretched relaxed muscle fibers is a preparation of myosin and actin with its complement of regulatory proteins in a solvent containing EGTA (to lower Ca^{2+} concentration to $< 10^{-8}$ M). The basic myosin ATPase mechanism described in reference 151 still appears to be valid with the proviso that, as with the actomyosin ATPase, additional myosin and myosin-nucleotide states have been characterized under certain conditions. Several studies (see 30 and references therein) give indirect evidence for a second, perhaps regulatory, ATP binding site.

We have deliberately limited the scope of our discussion on the actomyosin ATPase mechanism. Thus this review does not consider the implications of myosin being a two-headed molecule. Several biochemical studies of the ATPase mechanism have addressed this point, and the potential role in contraction of interactions between the heads has been carefully considered (8, 86, 115). However, none of these studies has led to a generally accepted explanation for the existence of two heads. Many observations from biochemical studies also suggest that several additional intermediates exist during actomyosin-catalyzed hydrolysis of ATP (6, 22, 87, 111, 125, 130, 131, 152). Some of these observations have led to interesting theoretical descriptions of the mechanism of muscle contraction (e.g. 45, 129). However, our ability to make meaningful correlations between biochemical states of the actomyosin ATPase and mechanical properties of muscle is still severely limited by lack of direct experimental data, so we limit our discussion of the biochemistry to the more rudimentary mechanism of Figure 3.

PHYSIOLOGICAL AND BIOCHEMICAL ANALYSIS OF SKINNED FIBERS

Fortunately there is a good model experimental system that allows mechanical measurements to be made while the biochemical milieu of the myofilaments is altered. "Skinned fiber" preparations are chemically (146) or mechanically (59, 114) demembranated muscle cells with largely intact myofilaments. When placed in solutions that mimic the intracellular environment, the cells contract or relax accordingly. Thus chemical perturbations can be imposed on the fibers while force is monitored. However, caution is required in extrapolating from skinned fibers to intact muscle. Skinned fibers are swollen compared with intact cells, and they do not obey the constant lattice volume relation observed in shortening of the intact fibers (104, 105). The cross-bridges in skinned fibers do not exhibit the almost linear force-extension curve of cross-bridges in electrically stimulated muscle (51). However, the linearity can be improved by shrinking the skinned fiber lattice toward the value observed in vivo [by imposing osmotic constraints on the fiber value with high molecular weight polymers that are largely excluded from the filament lattice (52, 105)]. Nevertheless, demembranated muscle cells appear to have general contractile properties very similar to those of intact cells. Differences between skinned and intact fibers do not severely limit the usefulness of the skinned fiber preparation as long as appropriate qualifications are made.

A more difficult problem has been the diffusion limitation on the rate of addition of nucleotides and their analogues to the myofilaments. This limitation must be overcome if high time-resolution mechanical measurements are to be made during experiments analogous to the rapid mixing and quenching experiments that are performed with purified muscle proteins, which have a time resolution of about 1 ms. The synthesis of the inert photolabile precursor of ATP, P^3-1-(2-nitro)phenylethyladenosine 5'-triphosphate (Compound I) or "caged ATP" (90) has provided an effective solution. Caged ATP (and the chemically related compounds caged ADP and caged P_i) can be allowed to equilibrate within skinned fibers and then can be cleaved photochemically with near-UV light from a laser or xenon arc pulse. This photolysis yields ATP at a rate limited by a dark reaction, the decay of an *aci*-nitro intermediate, at 118 s^{-1} in a solution at physiological ionic strength, pH 7.1, and 20°C (49, 109):

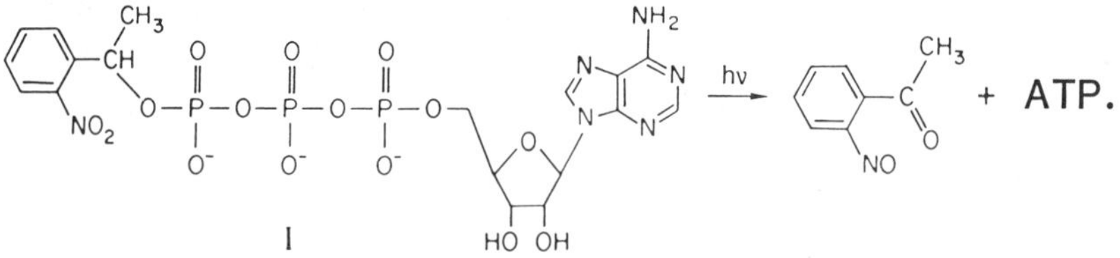

We now consider states and processes around the cross-bridge cycle. In presenting experimental evidence we draw primarily on our work and that of our colleagues using skinned fiber preparations from rabbit psoas muscle. Use of this muscle is appropriate for correlating mechanics with the biochemistry of actomyosin isolated from rabbit fast-twitch skeletal muscle.

Cross-Bridge Detachment

The kinetics of cross-bridge detachment have been studied by the liberation of up to 1 mM ATP from caged ATP within skinned muscle fibers in the rigor state (48–50). In the presence of a constant and saturating concentration of Ca^{2+}, tension falls rapidly in response to ATP liberation and then rises to a steady active level (Figure 4). With the exception of the first 3–4 ms, during which the reaction is limited by the rate of caged ATP photolysis, the tension records are well fitted by a simple scheme in which two rapid and sequential steps cause the cross-bridges first to detach and then to reattach rapidly and produce full active isometric tension. The first of these steps is strongly dependent on the ATP concentration, while the second process is also rapid and either independent of or only very weakly dependent on the ATP concentration released by photolysis (50). A more detailed discussion of this second step is presented below (see Activation of Muscle and Cross-Bridge Attachment). When the data are analyzed by fitting them to the two-step scheme, a second-order rate constant of 0.5–1×10^6 M^{-1} s^{-1} is found for ATP-induced detachment. This constant is of the same order as the corresponding rate constant for isolated actomyosin.

Analyses of the effects of ATP on maximum steady shortening velocity (37) and on tension responses to sinusoidal length oscillations (91) suggest less directly a value for the second-order rate constant between 5.7×10^5 and 3.6×10^6 M^{-1} s^{-1}.

The decrease in the tension or stiffness observed in this experiment only demonstrates cross-bridge detachment indirectly. At least two alternative explanations for the observed tension transients should be considered. First, cross-bridges might not detach at all, but directly enter a low-tension attached AM.ATP state. To be compatible with the stiffness records, such a hypothetical state would have to show a low in-phase stiffness and a very low quadrature stiffness (defined in the legend of Figure 4) that corresponds to the properties of relaxed muscle fibers (92) and not to the characteristics of cross-bridges identified in intact activated muscle fibers during rapid mechanical transients (42, 76).

A second possibility might be that cross-bridges bind ATP and then enter a "weakly bound" state (AM.ATP) in rapid equilibrium with M.ATP (12, 13, 142, 144). In this state the cross-bridges would behave much as relaxed cross-bridges before proceeding to form active force-generating complexes. This possibility is a direct extension of the first hypothesis. Representation of a weakly bound state such as AM.ATP may be shown schematically in an approach described by Hill (65). Figure 5 and its accompanying legend show the relationships of cross-bridge stiffness, chemical potential, and the probability distribution of cross-bridge strains with longitudinal displacement within the filament. Thus in the absence of geometrical constraints,

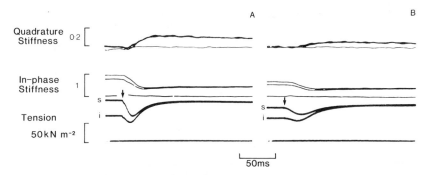

Figure 4   Stiffness and tension records following release of ATP (520 μM in *A*, 170 μM in *B*) by laser pulse photolysis of caged ATP in the presence of 30 μM free Ca^{2+}. Fibers were either held isometric (i) or prestretched by 0.74% and then held isometric (s). The in-phase sinusoidal component of tension, labelled "in-phase stiffness," is approximately proportional to the high-frequency stiffness of the fiber and is indicative of the number of attached cross-bridges. The (90°) out-of-phase component, or "quadrature stiffness," is related to viscosity or viscoelasticity. Quadrature stiffness is related to the number of active cross-bridges (49) that undergo force-producing transitions similar to those that account for the rapid tension recovery after a small length change (76, 92). The arrow marks the time of the 30 ns laser pulse. Baselines were recorded from the fiber bathed in a relaxing solution containing 5 mM ATP. Records of the isometric and stretched fibers were overlaid and superimposed by alignment of the relaxation records and the times of the laser pulse. For further details see Figure 4 of Reference 50; published with permission.

cross-bridges exert both positive and negative forces with a distribution that depends on their attachment and detachment rate constants about a point at which zero force is exerted.

Thus AM.ATP formed either directly or subsequent to detachment may produce a small but significant force. Simulations of tension transients such as those depicted in Figure 4A were made in which the best fits to experimental records were made without any constraint on the tension of

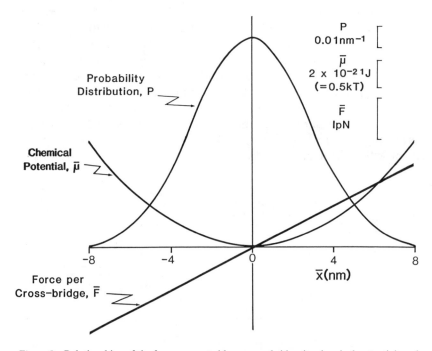

Figure 5 Relationships of the force generated by a cross-bridge, its chemical potential, and probability distribution of strains with its longitudinal displacement. The graph shows a linear dependence of force, $\bar{F}$, on longitudinal displacement within the myofilament, $\bar{x}$, of a cross-bridge from the point of its attachment with minimum chemical potential, $\bar{\mu}$. For a linear spring-like element associated with each cross-bridge, $\bar{F} = \kappa\bar{x}$, where κ is the stiffness defined as the instantaneous dependence of force on length (161, p. 21; see also 134, pp. 144–45). κ is taken to be 2.5×10^{-4} Nm^{-1} (76). $\bar{\mu}$ is given by $\bar{F}(\bar{x})\,d\bar{x} = \kappa\bar{x}^2/2 + \text{constant}$. The constant is taken here arbitrarily as zero. The probability, P, that the cross-bridge at equilibrium has a particular strain prior to attachment at a given value of $\bar{x}$ is $P = (\kappa/2\pi kT)^{1/2}[\exp(-\kappa\bar{x}^2 2kT)]$ where k is the Boltzmann constant. Note that the strain (i.e. relative elongation of the elastic component) prior to attachment arises through thermal motion of cross-bridges and that the actual distribution of attached states is affected by the association and dissociation rate constants of the cross-bridges.

the intermediate formed immediately after ATP generation by photolysis. The value for the best fit to the tension of this intermediate (presumably AM.ATP in equilibrium with M.ATP) was not significantly different from zero. However, this was not the case when ADP first had to be displaced from the fiber in rigor (as in Figure 11B, below). We have no convincing explanation for this, though it suggests that in appropriate circumstances an ATP-bound state may exert tension (Y. E. Goldman, J. A. Dantzig, M. G. Hibberd, D. R. Trentham, unpublished observations). The suggestion of a force-generating ATP-bound state has been inferred from mechanical studies with skinned frog fibers (40).

In spite of these uncertainties, changes in the equatorial and meridional X-ray reflections of insect flight muscle are consistent with the notion that rapid detachment of the cross-bridges followed by transient reattachment occurs when ATP is released from caged ATP into skinned fibers in rigor in the absence of Ca^{2+} (53).

STRAIN DEPENDENCE OF DETACHMENT KINETICS   Most currently favored theories of contraction explicitly include stress- and/or strain-dependent rate constants for certain key transitions (e.g. 33, 68, 76, 117, 118). Although estimates of the rate of cross-bridge detachment indicate that AM.ATP is a short-lived intermediate (48), it is possible that the detachment rate measured following caged ATP photolysis in rigor fibers was influenced by the low tension per cross-bridge (one-third of active tension) that existed prior to ATP liberation (48, 49). Figure 6 shows data with which this hypothesis was tested; the rate of relaxation was determined from tensions spanning the full isometric range. In the absence of Ca^{2+}, records beginning at different tension levels all converged rapidly after ATP liberation was initiated. The simplest explanation of such transients is that cross-bridge detachment is fast. After detachment the behavior of cross-bridges is independent of their initial rigor tension. Detachment is followed by a transient reattachment and then a slower detachment so that overall relaxation occurs in several phases (49). The cross-bridges reattach because of cooperative protein interactions, probably based on properties of the thin filament (9).

The initial rate of detachment of the cross-bridges in rigor can be estimated from the decay in the difference in tension records. This estimate depends upon the assumption that tension drops rapidly to zero once a cross-bridge detaches (or forms a weakly bound AM.ATP state with properties similar to those described in Figure 5). Tension differences for each adjacent pair of records in Figure 6 are illustrated. It is evident that the rate of convergence from the higher tension levels is about twice that for the lower tensions. The actual range of rates is probably greater than indicated because ATP release on photolysis of caged ATP may be partially rate limiting. The second-order rate constant for ATP-induced detachment in the absence of Ca^{2+} obtained with this approach is $\approx 10^6 \ M^{-1} \ s^{-1}$.

To provide an explanation of the Fenn effect (35, 121) A. F. Huxley postulated (68, 72) that the reaction that controls the energy output and work production of muscle is accelerated by a release (i.e. allowing the filaments to slide in the shortening direction). Such a reaction is inconsistent with the faster ATP-induced cross-bridge dissociation at high strain (Figure 6). The step that controls the shortening must be an attached transition that precedes ATP binding to actomyosin (48, 49).

It is interesting that the K_m (0.47 mM) with respect to the ATP dependence on maximum shortening velocity is high (37). Although several kinetic schemes might account for this high K_m (40), the only one consistent with the very rapid detachment observed in the photolysis experiments also places the rate limitation earlier in the cycle. The most recent biochemical evidence also favors such an interpretation (124, 143).

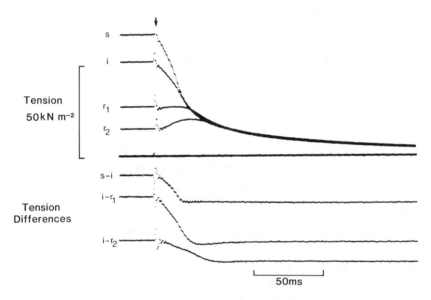

Figure 6 Tension records following release of 560 μM ATP by laser pulse photolysis of caged ATP in the absence of free Ca^{2+}. Fibers were either held isometric (i), prestretched by 0.38%, (s), or released by 0.76% (r_1) or 1.52% (r_2). In each case tension base lines were recorded one minute after relaxation. Records were overlaid by superimposing the full relaxation records and the time of the laser pulse. Algebraic differences between adjacent pairs of tension records are shown in the lower part of the figure. For further details see Figure 9 of Reference 49; published with permission.

Detached States and Their Relationship to Relaxed Muscle

As we saw in discussion of the biochemistry of the actomyosin ATPase, it is likely that detached states correspond to M.ATP and M.ADP.P$_i$. These two intermediates comprise the steady-state complex of the Mg-dependent myosin ATPase (151). Thus it is natural to suppose that M.ATP and M.ADP.P$_i$ predominate in relaxed muscle. This hypothesis is consistent both with the absence of structurally identifiable cross-links between the filaments (123) and with the low muscle stiffness. Radionucleotide binding measurements (103) have demonstrated that most myosin heads in relaxed skinned fibers from rabbit contain ADP.

At low ionic strength (< 100 mM) several aspects of cross-bridge attachment appear to be different. Both stiffness and X-ray diffraction measurements indicate that an increasing proportion of cross-bridges becomes attached as the ionic strength is lowered from 100 mM to 20 mM (12, 13). These cross-bridges are considered to be in rapid equilibrium with detached cross-bridges, since the stiffness is dependent on the rate of the length change used to estimate stiffness (13). At physiological ionic strength cross-bridge attachment is not evident; but stiffness measurements indicate that detachment could still occur because time resolution of the stiffness measurements is such that an attachment-detachment process at $> 10^4 \, \mathrm{s}^{-1}$ would not be detected. Nevertheless, if myosin heads are weakly associated with actin in relaxed muscle at physiological ionic strength for any time, the association does not appear to influence the myosin ATPase activity to an appreciable extent (38, 39) or to result in force generation.

The development of the caged ATP photolysis technique permits characterization of the chemical state of bound nucleotide in both active and relaxed fibers and correlation of kinetic parameters associated with ATP cleavage in fibers and isolated proteins. The data in Figure 7 suggest that the steady-state intermediate in relaxed fibers at full overlap and in fibers stretched beyond overlap is predominantly a myosin-products complex, probably M.ADP.P$_i$ (39). ATP cleavage also occurs rapidly after ATP liberation in activated fibers held isometrically but, as is discussed below, the steady-state intermediate may well be AM'.ADP. The available data indicate that the time courses of ATP cleavage in fibers and in isolated actomyosin are similar (36).

Activation of Muscle and Cross-Bridge Attachment

Activation of contraction in intact muscle fibers is regulated by the binding of Ca^{2+} to troponin on the thin filament (28, 29). Until recently, the steric blocking model (98 and references therein) was widely considered to account for the regulation of contraction and relaxation mediated by Ca^{2+}.

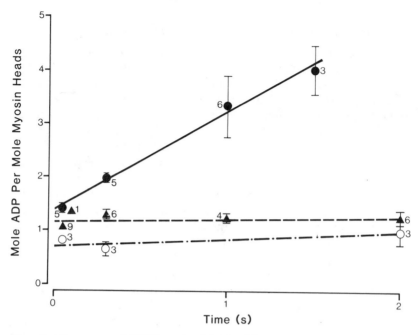

Figure 7 Time courses of ADP formation following ATP release from caged ATP in single fibers or bundles of fibers held isometrically. ● = activated fibers, 20 μM free Ca²⁺; ▲ = relaxed fibers, absence of free Ca²⁺; ○ = fibers stretched beyond filament overlap, 20 μM free Ca²⁺. Error bars show ±1 S.E. of the mean; for symbols without bars the S.E. falls within the symbols. The data points are the average of the number of experiments recorded beside the symbols. In the experiments [2-³H]ADP and [2-³H]ATP were isolated from fibers freeze-clamped at the indicated times after photolysis of tritiated caged ATP. For further details see Figure 6 of Reference 39; published with permission.

According to this hypothesis tropomyosin physically obstructs the myosin binding site on actin when Ca²⁺ is low, but moves away to allow access for myosin when Ca²⁺ binds to troponin (28). Although it is not explicit in the steric blocking hypothesis, it is implied from the model, in combination with biochemical data both from fibers and from solution studies, that the predominant intermediate in relaxed muscle is M.ADP.Pᵢ. Even with the time resolution improved by use of synchrotron radiation sources (82, 83, 98), the time resolution of the X-ray measurements cannot rule out the possibility of a rapid equilibrium between M.ADP.Pᵢ and A.M.ADP.Pᵢ. However, recent X-ray results strongly support the following temporal sequence when muscles are activated by electrical stimulation at 6°C: Tropomyosin moves with a half time of 17 ms after stimulation followed by a change in the equatorial reflection intensities due to movement of mass from the thick filaments to the thin filaments with a delay of about 10 ms, followed some 18 ms later by a rise in isometric tension (98).

An alternative mechanism of activation has been proposed (19, 20) in which it is suggested that tropomyosin's role is not to physically block, but rather to accelerate the transition from an attached but non–force producing state to one that is able to generate tension. The data are difficult to interpret because they are derived from equilibrium distributions of states with isolated proteins and we must therefore rely on inference for describing the temporal sequence of events; if it were not for the clear temporal separation between the tropomyosin and myosin movements during the onset of contraction (98) it would always be possible to suggest that tropomyosin motion succeeded rather than preceded the attachment of myosin to actin. However, it seems probable from experiments with regulated isolated actomyosin that once an activated isometric steady state is established in a muscle, Ca^{2+} influences the kinetics of nucleotide and P_i binding to attached states (E. W. Taylor, personal communication).

We now consider the kinetics of cross-bridge attachment. The experiment illustrated in Figure 4 shows that in the presence of Ca^{2+} full active isometric tension is achieved less than 100 ms after the liberation of ATP by photolysis within a rigor muscle fiber. This indicates that cross-bridges rapidly pass through the steps between the rigor state and the state that sustains tension with the usual mechanical properties of activated fibers (such as quadrature stiffness). Ferenczi et al (39) have shown that the steps include the hydrolysis reaction, and the analysis presented by Goldman et al (50) indicates that these intermediate states are not able to sustain the rigor tension that they supported prior to the initiation of the transient (see Cross-Bridge Detachment, above). The processes from AM to the first force-producing transition occur relatively rapidly and do not limit the overall rate of cycling in activated isometric muscle $(1–2 \text{ s}^{-1})$ (15, 39, 147). The apparent rate constant for the regeneration of force from the detached state intermediate, measured from records such as Figure 4, is $80–100 \text{ s}^{-1}$ (50) and is compatible with the known rate of activation of intact mammalian fibers (34, 122).

The inference from the photolysis experiments is that the predominant intermediates present in an active muscle are attached states, since detachment hydrolysis and reattachment with force generation are all much more rapid than the overall cycling rate. This general idea is not novel and has been deduced from several studies. For example, when muscle fibers are producing active tension they have 50–75% of the stiffness observed in rigor (51, 162), and the low angle equatorial X-ray diffraction intensities (considered to reflect the distribution of filament mass between the thick and thin filaments, and by inference the relative position of the myosin heads with respect to the filaments) suggest that 50–90% of the cross-bridges are at least close to, if not attached to, the thin filament during a full isometric contraction (58, 107, 108, 163).

It is important to discuss how electrical stimulation of intact muscle as a way of initiating contractile events is related to the principal approach described here involving rapid release of ATP from caged ATP. In the latter case, the kinetics of tension generation following ATP hydrolysis are associated with formation of AM'.ADP from AM.ADP.P_i. [This presumes that AM'.ADP (Figure 3) is a state that generates positive force; this point is discussed in Attached Cross-Bridge States]. The rate of tension generation is determined in part by K_4, since the rate of AM'.ADP formation depends on how much the rapid preequilibration step, step 4, lies in favor of AM.ADP.P_i. In the case of electrical stimulation, we infer from the steric blocking model that cross-bridges exist predominantly as M.ADP.P_i prior to activation, and that movement of tropomyosin permits AM.ADP.P_i and hence AM'.ADP formation. Insofar as the alternative view of regulation is correct (19, 20), activation would permit the transition from AM.ADP.P_i to AM'.ADP to occur. Either way the transition from AM.ADP.P_i to AM'.ADP probably occurs in muscle fibers only in the presence of Ca^{2+} [unless the usual regulatory inhibition is released by cross-bridges in rigor (9)], even though the transition may not be the initial process in muscle activation.

Attached Cross-Bridge States

We now focus on the mechanical and biochemical properties of the four attached states on the right-hand side of Figure 3, AM.ADP.P_i, AM'.ADP, AM.ADP, and AM, and the transitions between them.

AM.ADP.P_i The following lines of experimental evidence suggest that an attached state producing little or no net positive force is formed before the first tension-bearing state is reached (in terms of Figure 3 this state is AM.ADP.P_i): (a) During activation of intact muscle, stiffness increases before tension is produced (17, 18, 128). (b) The equatorial X-ray intensities change upon activation in a way that suggests a substantial transfer of mass from the region of the thick to that of the thin filament before tension development (80, 98). (c) Rapid equilibrium between M.ADP.P_i and AM.ADP.P_i is established during ATPase activity of isolated actomyosin (12, 19, 144). The mechanical properties of AM.ADP.P_i might therefore be expected to be similar to those of AM.ATP in its unstrained state (see Figure 5). (d) The attachment of cross-bridges initially exerting high force is unlikely (see Figure 5). A. F. Huxley & Simmons (76) therefore proposed that cross-bridges might attach with low force and then undergo transitions to states exerting high force. (*e*) The formation of such a reversible attached state that is uncommitted to the power stroke of the cycle helps explain the drop in the rate of energy liberation that is observed in muscle at high shortening velocities (63, 69). These data and inferences suggest that a discrete low-force AM.ADP.P$_i$ state is formed after hydrolysis and before the principal force-producing reaction in the cycle.

AM.ADP.P$_i$ TO AM'.ADP TRANSITION AND THE AM'.ADP STATE   Studies on isolated actomyosin (159, 160) suggest that product (i.e. ADP and/or P$_i$) release is probably the transition linked to force production in muscle, since it is associated with a large drop in free energy. If product release were to occur in muscle fibers without concurrent mechanical coupling, a substantial proportion of the free energy available from the hydrolysis of ATP would be lost, severely reducing the maximum possible efficiency of contraction. Stretching of the elastic component within the cross-bridge and relaxing of the elastic component with relative movement of the filaments may well be linked to more than one biochemical transition. The data below imply that stretching of the elastic component is associated with the transition from AM.ADP.P$_i$ to AM'.ADP with accompanying P$_i$ dissociation.

Since P$_i$ has a marked influence on isometric tension and mechanical transients elicited after a quick stretch (60, 93, 94, 126, 158, 164), it may be that P$_i$ can bind to a specific site on cross-bridges during contraction. P$_i$ in the range 1–20 mM also affects the form of the tension and stiffness transients observed after the release of ATP by caged ATP photolysis in skinned fibers in rigor (61) (Figure 8). The final steady-state levels of tension, in-phase stiffness, and quadrature stiffness were all lower than

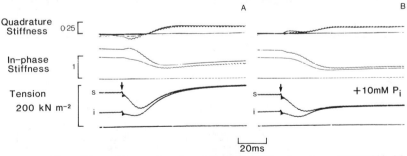

Figure 8   Effect of P$_i$ on stiffness and tension of single fibers following release of ATP (700 μM in *A*, 870 μM in *B*) on laser pulse photolysis of caged ATP in the presence of 100 μM free Ca^{2+}. Fibers were either held isometric (i) or prestretched (s). For further details of the protocol see the legend to Figure 4, and Figure 3 of Reference 61; published with permission (copyright 1985 by the AAAS).

those obtained in the absence of P_i, although stiffness is reduced proportionately less than tension. End-compliance effects would be likely to make the true value of stiffness in the presence of P_i greater than the measured value, since the final tension is lower. The relative reductions in tension and stiffness differ from those in glycerinated cardiac muscle preparations, which are proportionately equal (60). Nonlinearities in the stiffness records due to the more complex internal structure of the cardiac cells and the presence of substantial non–cross-bridge compliances may account for this difference.

The rate of convergence of tension records with prestretched and isometrically held fibers is not appreciably altered if P_i is present. Thus, if P_i is able to bind to cross-bridges in rigor, it presumably only does so in a way that does not interfere with ATP binding. The tension transient in the presence of P_i can be fitted with the same two-step reaction sequence that fits the data in the absence of P_i, although the apparent rate of the cross-bridge reattachment/force-producing step increases with P_i concentration (61). In Figure 8 the rate constant of the second step increases from $76 \, s^{-1}$ in the absence of P_i to $140 \, s^{-1}$ in the presence of 10 mM P_i. The scheme in Figure 3 can qualitatively accommodate these results if it is supposed that AM'.ADP is a force-generating state to which P_i binds relatively easily to form AM.ADP.P_i and then M.ADP.P_i with resulting lower tension.

The idea that P_i binds to AM'.ADP in activated fibers is supported by evidence that P_i accelerates relaxation following ATP release into rigor muscle in the absence of Ca^{2+} (61). In Figure 9A the record i shows a transient increase of tension that has been assigned to reattachment of cross-bridges during relaxation, because residual rigor cross-bridges are thought to relieve the inhibitory effect of the regulatory system (9, 49). Reattached cross-bridges are presumed to exist predominantly as

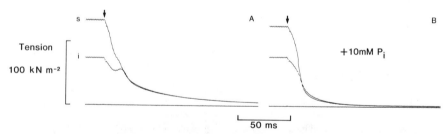

Figure 9   Effect of P_i on the tension of single fibers following release of approximately $600 \, \mu M$ ATP on laser pulse photolysis of caged ATP in the absence of free Ca^{2+}. Fibers were either held isometric (i) or prestretched (s). Baseline records are of fully relaxed fibers. For further details of the protocol see the legend to Figure 4 and Reference 61.

AM'.ADP. The absence of the transient rise and the more rapid approach to full relaxation in the presence of 10 mM P_i (Figure 9B) can be explained by less AM'.ADP formation (61).

Kawai et al (93) have shown that P_i in the millimolar range increases the characteristic frequency at which the oscillatory power output from rabbit skinned fibers is maximal; this confirms earlier observations made on insect fibrillar flight muscles (158). Kawai et al (93) suggest that this result is most easily incorporated into cross-bridge cycles in which P_i increases the rate of a work-producing reaction. Their hypothesis implies an expected increased rate of the overall ATPase in the presence of P_i, but this increase does not occur (M. R. Webb, M. G. Hibberd, Y. E. Goldman, D. R. Trentham, unpublished; J. A. Sleep, personal communication). Nevertheless, the increased magnitude and frequency for work production is unexpected and is as yet unexplained.

Oxygen isotope exchange   The inclusion of AM'.ADP was based in the first instance on experiments of Sleep & Hutton (138) that showed that ^{32}P-exchange could be measured between P_i and ATP during ATPase activity of isolated actomyosin. This implied that all the steps between and including ATP binding and P_i release were reversible. The exchange was not seen if P_i and ADP were added to actomyosin in the absence of ATP; this absence necessitated the proposal of an AM'.ADP state in addition to AM.ADP, the state formed on addition of ADP to actomyosin. P_i-ATP exchange of ^{32}P has also been observed in muscle fibers, but at the time of Sleep & Hutton's research too little was known about the ATPase mechanism to incorporate the data into detailed reaction schemes (47, 153).

The chemistry of P_i release in fibers has also been investigated by oxygen exchange techniques (62, 156) following extensive studies with isolated actomyosin (for a review see Reference 157). Figure 10 illustrates the rationale for two experimental approaches. For simplicity the two schemes only show detached state hydrolysis (c.f. Figure 3). Figure 10A shows what may happen if a fiber is incubated in ^{18}O-water during ATPase activity. In steps 3 and 3a (Figures 3 and 10) the oxygen atom of a water molecule is incorporated into P_i during ATP hydrolysis. Steps 4 and 5 enable P_i release to occur. The upper row of reactions in Figure 10A shows $(^{18}O_1)P_i$ formation. This sequence does not involve exchange. However, if P_i rotates in the active site (processes X and X'), the new intermediate in the second row can either dissociate as $(^{18}O_1)P_i$ or reform ATP with elimination of ^{16}O-water. In the latter case, "intermediate" oxygen exchange will occur and the P_i product will contain two, three, or four ^{18}O-atoms depending on how many times the exchange cycle is repeated. Thus, provided X or X' is rapid, the extent of exchange is related to the rate of ATP reformation

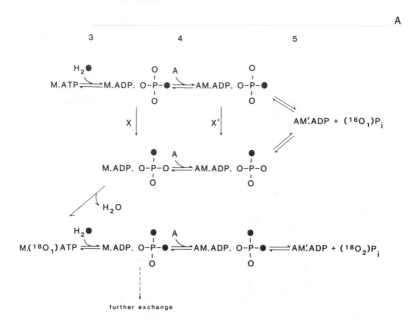

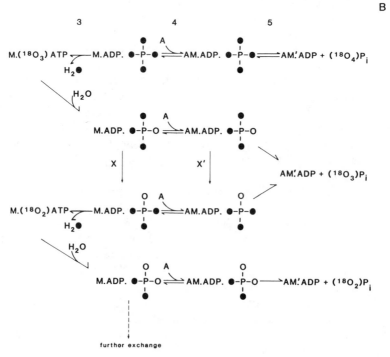

controlled by k_{-3} (and k_{-3a}; see Figure 3) compared to the rate of P_i release controlled by k_{+5}. Such experiments show that P_i release occurs 2.3-fold faster than ATP reformation at the active site in activated fibers held isometrically. The factor of 2.3 was derived from the distribution of ^{18}O in the major population of P_i formed during this intermediate oxygen exchange (62) and during the medium oxygen exchange (156) discussed in the next paragraph.

The kinetics of P_i association controlled by k_{-5} can be investigated in a different type of oxygen exchange experiment termed "medium" oxygen exchange (Figure 10B). $(^{18}O_4)P_i$, added to an activated fiber held isometrically, is able to bind to intermediates formed during ATPase activity and, if the rate constants are appropriate, ATP can be reformed at the active site with elimination of ^{18}O. This is shown in the top row of reactions in Figure 10B. Equilibration of the liberated ^{18}O-water with medium water is presumed to be rapid, so that hydrolysis of $M.(^{18}O_3)ATP$ results in $(^{18}O_3)P_i$ as shown in the second row of equations. As in the case of intermediate oxygen exchange, rotation of P_i in the active site through X and/or X' permits further exchange. The reaction that liberates $(^{18}O_3)P_i$ into the medium is shown as irreversible because in the experiments the P_i in the medium prior to exchange was much more abundant than that formed as a result of exchange. In such experiments the amount of oxygen exchange gives information about k_{-5}. The distribution of oxygen isotopes in the exchange P_i is related, as in the case of intermediate oxygen exchange, to the rate of reformation of ATP compared to the rate of P_i release.

In our analysis we presume that X and X' occur rapidly. In the case of the ATP synthase associated with submitochondrial particles, restricted rotation is likely to occur (56). As discussed in the section on the biochemistry of the actomyosin ATPase, it has been suggested that low values of X and X' could be responsible for the incompatibility of oxygen exchange data with the ATPase mechanism of Eisenberg and coworkers (143). However, there is no evidence that P_i rotation is restricted during myosin and actomyosin ATPase activity.

Figure 10   Oxygen exchange schemes to relate results of experiments described in the text to the ATPase mechanism of Figure 3. Numbers relate to steps of that mechanism. Filled circles and open circles represent ^{18}O and ^{16}O respectively. In A, ^{18}O-water is shown entering at step 3 during ATPase activity. $(^{18}O_1)P_i$ is formed without oxygen exchange. X and X' are processes describing rotation of P_i in M.ADP.P_i and AM.ADP.P_i, respectively. Formation of $(^{18}O_2)P_i$, $(^{18}O_3)P_i$, and $(^{18}O_4)P_i$ occurs as a result of "intermediate" oxygen exchange. In B, $(^{18}O_4)P_i$ in the medium binds to AM'.ADP, a state that occurs during ATPase activity and in which ADP is derived from unlabelled ATP. A sequence of reactions is shown in which $(^{18}O_3)P_i$ is formed by "medium" exchange. Further medium exchange requires process X or X' to occur.

Estimation of the rate of P_i release thus rests on the distribution of isotopes in intermediate and medium exchange experiments, on the value of $(k_{-3a}K'_4 + k_{-3})/(K'_4 + 1)$ (a function related to the rate of reformation of M.ATP and AM.ATP in which $K'_4 = K_4[A]$, where [A] is the effective actin concentration in fibers), and on the equilibrium constant between M.ADP.P_i and AM.ADP.P_i. Ferenczi (36) is measuring the kinetics of ATP hydrolysis at 12°C in fibers; a preliminary estimate at 20°C of $(k_{-3a}K'_4 + k_{-3})/(K'_4 + 1)$ is 20 s^{-1}. From this it follows that $k_{+5}K'_4/(K'_4 + 1)$, the rate constant of P_i dissociation, equals about 50 s^{-1} (from 2.3 × 20 s^{-1}). Estimation of k_{-5}, the rate constant of P_i association, depends on the fraction of cross-bridges that exists as AM'.ADP during steady-state ATPase activity. Analysis of medium oxygen data leads to $k_{-5} = 500[AM_0]/[AM'.ADP]$ M^{-1} s^{-1} (156; M. R., Webb, M. G. Hibberd, Y. E. Goldman, D. R. Trentham, unpublished), where $[AM_0]$ is the total cross-bridge concentration.

It is interesting to compare the values of these biochemically determined rate constants with those from the photolysis-induced tension transients. In the latter case (Figure 8), the overall rate of force generation equals 76 s^{-1}. In terms of Figure 3 this probably corresponds to the rate constant controlling P_i release $[= k_{+5}K'_4/(K'_4 + 1)]$ rather than that controlling ATP cleavage, which appears to be more rapid (though the rates are possibly comparable) (36; M. A. Ferenczi, personal communication). The 64 s^{-1} increase in the rate of approach to steady tension (from 76 s^{-1} to 140 s^{-1}) in the presence of 10 mM P_i, with the accompanying fall in steady tension to 50% of that in the absence of P_i, suggests that the second-order rate constant of P_i association, k_{-5}, equals 6400 M^{-1} s^{-1}. However, if 10 mM P_i is added to an activated isometric fiber, the drop in tension is only about 50% of the drop seen in Figure 8 (26; M. R. Webb, M. G. Hibberd, Y. E. Goldman, D. R. Trentham, unpublished work). These apparent discrepancies have been reconciled by J. A. Dantzig & Y. E. Goldman (personal communication), who have shown that the apparent steady tension in the presence of P_i (as in Figure 8B) continues to rise, with a rate constant of 1–3 s^{-1}, to about 75% of the steady isometric tension in the absence of P_i (as in Figure 8A). Nevertheless, there is uncertainty in estimating K_5. K_5 may be up to threefold greater and hence k_{-5} may be as much as threefold smaller than suggested by the data in Figure 8.

However, the above approach for calculating equilibrium and rate constants for a biochemical mechanism from mechanical data may be somewhat superficial. This raises a key problem in understanding muscle contraction: how to establish a valid approach for correlating mechanical and biochemically determined rate constants. We return to this issue in Further Analysis of Force Generation.

AM.ADP AND AM   Skinned muscle fibers in rigor with or without ADP added are convenient to study. Unfortunately, it is not known whether these two readily available preparations represent states that occur during the cross-bridge cycle of actively contracting muscle.

It has been suggested that ADP release from AM.ADP, the state formed by adding ADP to fibers in rigor, may correspond to the rate-limiting reaction during shortening, owing to the close correspondence of the observed or inferred ADP release rate from isolated actomyosin with the expected lifetime of cross-bridges at maximum shortening velocity (132). However, this step need not correspond to the rate-limiting step for ATPase activity nor to the overall turnover rate of ATP hydrolysis (37, 132). Photolysis experiments in which an attempt is made to determine the rate constant of ADP release in rabbit skinned muscle fibers are illustrated in Figure 11. In the presence of Ca^{2+}, a considerably slower tension transient on release of ATP is observed in the presence of ADP than in its absence. At first sight it appears that ADP release from AM.ADP limits the net rate of ATP-induced cross-bridge detachment. Based on the subtraction procedure (lower record of Figure 11B), an apparent rate constant for the initial detachment of rigor cross-bridges is about 30 s^{-1}. The effect is almost saturated at 100 μM ADP, and the apparent dissociation constant for ADP may be close to the value of 40 μM measured by binding radiolabeled ADP to skinned rabbit fibers (103). However, this rate constant may underestimate the rate of ADP release, especially in shortening muscle, for the following reasons. First, the resulting tension transient cannot be simulated with the simple two-state scheme that accounts for the transient in the

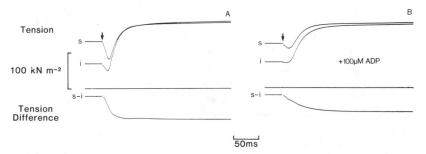

Figure 11   Effect of ADP on the tension of single fibers following release of 500 μM ATP in laser pulse photolysis of caged ATP in the presence of 30 μM Ca^{2+}. In *A*, the solution "rigor with Ca^{2+}" of Table 1 of Reference 50 was used but without creatine phosphate and creatine kinase. In *B*, the solution was as in *A* but with 100 μM ADP added. Fibers were either held isometric (i) or prestretched (s). Baseline records are of fully relaxed fibers. Tension difference records were obtained as described in Figure 6. (Y. E. Goldman, J. A. Dantzig, M. G. Hibberd, D. R. Trentham, unpublished.)

absence of ADP by inclusion of a simple ADP-limited initial detachment rate. This problem is still under investigation. Secondly, the overall shortening velocity would be seriously limited by a rate constant as slow as 30 s^{-1}. Thirdly, the apparent dissociation constant for ADP is at least an order of magnitude below that described for the effects of ADP on the steady shortening velocity in skinned fibers (26).

However, it has already been pointed out that AM.ADP may only form as a side-branch to the main pathway. J. A. Sleep (personal communication) favors this alternative since he finds that the rate of vanadate inhibition of the isolated actomyosin ATPase (presumably by binding to AM'.ADP) is tenfold faster than the rate of vanadate-induced actin dissociation from the AM.ADP complex, yet the inhibition and dissociation constants with respect to vanadate are about the same at 1 mM. In fact these data require distinctly different properties of actomyosin and of the species formed on ADP dissociation from AM'.ADP during ATPase activity. A similar need to propose an additional actomyosin state has been suggested by Konrad & Goody (96), and recent transient kinetic data provide evidence for an additional state in isolated actomyosin (22).

In general, this part of the ATPase cycle is not well characterized. Understanding of the biochemical nature and structure of the cross-bridge as it moves from strained AM'.ADP at the start of the working stroke to the end of the working stroke when ATP binds is one of the major goals in present studies of the mechanochemistry of muscle contraction.

Further Analysis of Force Generation

At least within the framework of existing kinetic data AM'.ADP can reasonably be presumed as an important state, which both exerts positive force and transduces that force into mechanical work. One can further presume that in an activated fiber held isometrically, addition of P_i reduces the population of cross-bridges in the AM'.ADP state with consequent reduction of force. In addition, from the oxygen isotope exchange studies we have been able to make inferences about the kinetics of P_i release from AM.ADP.P_i and about P_i binding to AM'.ADP. A next step is to see what can be derived from a formulation that aims to relate the biochemical kinetics of P_i binding to the mechanical behavior. In regions where the cross-bridge is strained on P_i dissociation we can expect the equilibrium constant to favor AM.ADP.P_i compared to AM'.ADP, but where it is unstrained the converse is true. It is likely to be an oversimplification to confine force generation to a single process (76). Nevertheless, it is useful to constrain our analysis in this way, if only to see if the data we now have linking biochemical and mechanical processes are reasonably compatible. In the following we consider the influence of P_i on force production (Figure

8), on the kinetics of reversible P_i binding as measured by oxygen exchange, and on steady-state ATPase activity of an activated fiber held isometrically.

Suppose cross-bridges in such a fiber exist principally in two states related kinetically as follows:

$$AM.ADP.P_i \underset{k_r'}{\overset{k_f}{\rightleftharpoons}} AM'.ADP + P_i \overset{k_a}{\rightarrow},$$

where $k_a \ll k_f$, so that k_a equals the catalytic center activity of the ATPase at low P_i concentration. We assume that $AM'.ADP$ is a force-generating state while $AM.ADP.P_i$ is not. (The effect of removing this assumption is discussed later.) The chemical potentials, μ, of $AM.ADP.P_i$ and $AM'.ADP$ at two different P_i concentrations are drawn in Figure 12. The dependence of force on displacement is presumed linear for $AM'.ADP$ as seems likely from mechanical studies (42), so that μ as a function of x is a parabola (see Figure 5). If we define $k_r = k_r'[P_i]$, thermodynamic considerations at any x give

$$k_r = k_f \exp[-(h-\varepsilon-hx^2)/RT], = k_f Q \exp h(x^2-1)/RT$$

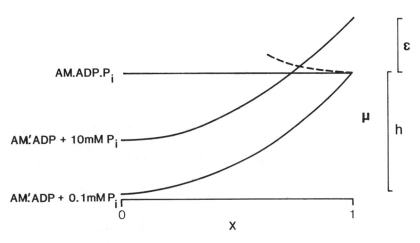

Figure 12 Hypothetical dependence of chemical potential, μ, of cross-bridge states $AM.ADP.P_i$ and $AM'.ADP + P_i$ on longitudinal displacement within the myofilament. The normalized distance, x, is drawn from the minimum of μ for the $AM'.ADP + 0.1$ mM P_i state to the point at which μ for the $AM.ADP.P_i$ and $AM'.ADP + 0.1$ mM P_i states are equal. As explained in the text, μ for $AM.ADP.P_i$ is arbitrarily made independent of x. The dashed line is the chemical potential dependence of $AM.ADP.P_i$ on x if the instantaneous stiffness, κ in Figure 5, was the same for $AM'.ADP$ and $AM.ADP.P_i$. 0.1 mM was chosen as an approximate mean concentration of P_i present in the medium during the first turnover following ATP release in a rigor muscle, as might be appropriate for Figure 9A.

since AM.ADP.P_i and AM'.ADP are in steady-state equilibrium, where for AM'.ADP $\mu = hx^2 + \varepsilon$ and $\varepsilon = RT \ln Q$. Q is the ratio of the concentration of P_i to that of a standard state, chosen here arbitrarily as 0.1 mM (see legend in Figure 12), and h is the chemical potential difference between AM.ADP.P_i and AM'.ADP + 0.1 mM P_i at $x = 0$.

Now let $F(x)$ be the force generated per cross-bridge at x and $[AM'.ADP]_x$ be the concentration of AM'.ADP at x. Then the total force, F, between points x_1 and x_2 is given by

$$F = \int_{x_1}^{x_2} \alpha[AM'.ADP]_x F(x)\, dx = \int_{x_1}^{x_2} \alpha[AM'.ADP]_x \kappa x\, dx,$$

where κ is stiffness, since $F(x) = \kappa x$ (Figure 5). The constant α relates the number of cross-bridges to their concentration. If we assume that the cross-bridges are uniformly distributed along x and this uniformity is represented by a concentration $[AM_0]$ that is independent of x, then

$$[AM'.ADP]_x = k_f[AM_0]/(k_f + k_r) = [AM_0]/[1 + Q \exp h(x^2 - 1)/RT].$$

Therefore

$$F = \int_{x_1}^{x_2} \alpha\kappa[AM_0]x\, dx/[1 + Q \exp h(x^2 - 1)/RT],$$

which on integration gives

$$F = \tfrac{1}{2}\alpha\kappa[AM_0]\{x_2^2 - x_1^2 - (RT/h)\ln[1 + Qe^{h(x_2^2 - 1)/RT}]/$$
$$[1 + Qe^{h(x_1^2 - 1)/RT}]\}.$$

This general expression for the force exerted by activated isometric cross-bridges can be simplified if we confine the range over which cross-bridges are located to that drawn in Figure 12. This is comparable to the simplification in A. F. Huxley's 1957 model for isometric muscle (68). Therefore $x_2 = 1$ and $x_1 = 0$. Then

$$F = \tfrac{1}{2}\alpha\kappa[AM_0]\{1 - (RT/h)\ln[1 + Q]/[1 + Q \exp -(h/RT)]\}.$$

For the situation in Figure 12, $Q = 1$ at 0.1 mM P_i and, if we take $h = 25$ kJ mol^{-1} so that $h/RT = 10.3$,

$$F = \tfrac{1}{2}\alpha\kappa[AM_0](1 - 0.097 \ln 2) = 0.47\alpha\kappa[AM_0].$$

F is less than $\tfrac{1}{2}\alpha\kappa[AM_0]$ because a significant fraction of cross-bridges in the region of $x = 1$ exist as AM.ADP.P_i and so exert no force.

At 10 mM P_i, $Q = 100$, so $F = 0.28\alpha\kappa[AM_0]$. Thus introduction of 10 mM P_i causes the steady tension to drop to 60% of the level at 0.1 mM P_i. Thus the number chosen for h/RT is compatible with the data (see Figure

9) and the ATPase thermodynamics (160). However, it is fivefold greater than the number derived from $\bar{\mu} \times$ Avogadro's Constant/RT when calculated from $\kappa = 2.5 \times 10^{-4}$ Nm^{-1} with $\bar{x}$ from 8 to 0 nm (Figure 5) (76).

It is apparent from the above analysis that the cross-bridges represented on the right-hand side of Figure 12 have the major role in force production. It is important to know whether a similar nonhomogeneity among the cross-bridges influences the biochemical data. Figure 13 shows how two functions of $[AM'.ADP]_x$ vary with x. We have seen in discussion of the transition from AM.ADP.P$_i$ to AM'.ADP that the rate of force production (76 s^{-1}) approximately equals the rate of P$_i$ release (50 s^{-1}). If we now assume that k_f is independent of x, k_f can be equated with $k_{+5}K'_4/(K'_4+1)$ and so $k_f \approx 60$ s^{-1}. Figure 13 also shows that when averaged over all x, the mean value for $k_r = 0.26\, k_f\, [AM_0]/[AM'.ADP]$. Recalling that $k_r = k'_r[P_i]$, at 10 mM P$_i$ we have $k'_r \approx 1560\, [AM_0]/[AM'.ADP]$ M^{-1} s^{-1}. Thus k'_r is 3.1-fold greater than k_{-5} calculated from medium oxygen exchange data. Better agreement is obtained by increasing h or decreasing ε.

Another way of viewing the problem is to note that $k_r = 100\, k_f$ at $x = 1$, $k_r = k_f$ at $x = 0.74$, and $k_r = 0.0045\, k_f$ at $x = 0$. Thus across the range from $x = 0$ to $x = 1$ all the cross-bridges respond very differently to P$_i$ binding, and only those cross-bridges to the right of the diagram are effective in promoting medium oxygen exchange. On the other hand, the dashed line in Figure 13 shows that only a small fraction of cross-bridges exist as AM'.ADP at $x = 1$. Overall the net flux giving rise to medium oxygen exchange occurs in the range $0.6 < x < 1$. Cross-bridge strain can give rise to a wide spectrum of rate constants, as is well illustrated by the 1000-fold range of rate constants when cross-bridges in rigor are dissociated by (β,γ-imido)ATP or PP$_i$ (127).

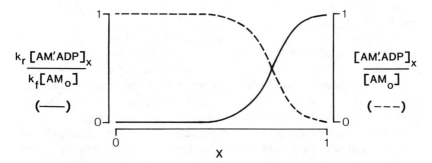

Figure 13   Dependence of $k_r[AM'.ADP]_x/k_f[AM_0]$ and of $[AM'.ADP]_x/[AM_0]$ on x in the presence of 10 mM P$_i$. Terms are defined in the text. $[AM'.ADP]_x/[AM_0] = (1+k_r/k_f)^{-1} = [1+Q \exp h(x^2-1)/RT]^{-1}$.   Note   that $k_r[AM'.ADP]_x/k_f[AM_0] = [AM.ADP.P_i]_x/[AM_0] = 1 - [AM'.ADP]_x/[AM_0]$. The lines are drawn for $h/RT = 10.3$ and $Q = 100$.

From the dashed line in Figure 13 we calculate that only 74% of the cross-bridges exist as AM'.ADP in the presence of 10 mM P_i. Thus we can expect a reduction in the steady-state ATPase activity by a fraction that depends on how k_a varies with x.

If the chemical potential of AM.ADP.P_i is allowed to change with x (dashed line in Figure 12), there is relatively little change in the dependence of the functions plotted against x in Figure 13. The two curves in Figure 13 will intersect at a slightly higher value of x, indicated by the shift in the point of intersection of μ for AM.ADP.P_i and AM'.ADP + 10 mM P_i in Figure 12. The result is that the mean values of $k_r (= 0.26\ k_f[AM_0]/[AM'.ADP]$ in Figure 13) and k_r' are little affected.

Another general aspect of force generation is that within the overall thermodynamic constraints there is room for either the forward or reverse rate constant of a process to vary when the equilibrium constant is changed, for example by stretching the elastic element within the cross-bridge. In the above discussion we have assumed that the forward rate constant associated with AM'.ADP formation, k_{+5}, in contrast to k_{-5} is independent of x. There may be important consequences if this is true. Thus, if k_{+5} does decrease with increasing x, it may lead to a less efficient muscle, since the cross-bridges will be kinetically restricted from reaching those AM'.ADP states that exert most force. If the muscle is shortening, the reduction of k_{-5} as x decreases may allow useful work, since the flux of the ATPase will continue more readily in a forward direction.

Shortening Muscle

It is important to extend biochemical studies in fibers to shortening muscle, but transient kinetics and isotope exchange studies are only just beginning. In several interesting experiments steady-state tensions and shortening velocities of skinned fibers have been measured as functions of ATP, ADP, and P_i concentrations (23, 26, 37).

The rate limitation imposed by an external load on the transformation of attached force-generating states in isometric muscle is absent in muscle at maximum velocity of shortening, so we may expect different cross-bridge states to be predominant. Several studies have suggested that these are detached states. In the context of our previous discussion, such states also include those such as AM.ATP and AM.ADP.P_i that dissociate rapidly and reversibly. Most evidence for the distribution of cross-bridges between attached and detached states comes from stiffness and X-ray diffraction measurements.

Stiffness is only 10–35% of the isometric level (10, 44, 88, 89) and by X-ray diffraction the intensity ratio of the 11 and 10 reflections is lower than during isometric contraction (80, 119). However, the correspondence between stiffness and X-ray data is not as close as it is for isometric muscle and at any given velocity the X-ray results seem to indicate up to twice as many attached cross-bridges as the stiffness data suggest. This difference probably arises because X-ray reflections may reveal more than just the tension-producing attached cross-bridges. Cross-bridges in the vicinity of the thin filament and weakly bound cross-bridges may contribute to the X-ray reflections but not to either tension or stiffness. In addition, there is no direct evidence that the 11 : 10 ratio is directly related in shortening muscle to the fraction of cross-bridges that are attached [as it is in isometric contractions (163)], and there are theoretical reasons for supposing that the assumption of linearity could be incorrect (101). The stiffness measurements seem most likely to underrepresent the number of attached bridges during shortening (see Reference 44 for a detailed discussion), owing to: (a) truncation of the instantaneous stiffness by rapid phase-two recovery or by rapid detachment during the mechanical step used to estimate stiffness; (b) nonlinearities in the force-extension curve of the cross-bridges and/or (c) a velocity-dependent increase in the number of cross-bridges that have a lower stiffness; and (d) our uncertainty whether a myosin molecule with one head attached and one detached contributes as much or only half as much to the stiffness as a myosin molecule with both heads attached. Overall, it is probable that a substantial reduction in the proportion of attached cross-bridges occurs as the shortening velocity increases.

We can take advantage of the different proportion of attached and detached cross-bridges in shortening and isometric fibers to probe the kinetics of shortening muscle. An experiment of Brenner (Figure 14) shows how a transient kinetic approach may be used to probe the kinetics of the rate-determining step. An activated skinned rabbit fiber was allowed to shorten at a low load to reduce the number of attached cross-bridges that characterize the isometric steady state and allow them to accumulate in detached (or weakly bound, i.e. AM.ATP or AM.ADP.P_i) states. The fiber was then rapidly restretched to its original length and held isometric, allowing force to redevelop. Brenner suggests that the single exponential rise in force results from the transition between low force– and the normal high force–generating states of isometric contractions. The rate constant of this transition and the catalytic center activity of the isolated actomyosin ATPase when measured in the same solvent over a 5–30°C temperature range are equal, and thus it is inferred that they are related to the same process (11).

Unfortunately the nature of the process that produces the tension rise in Figure 14 is ambiguous, even if it is the same process in isolated actomyosin ATPase, since the rate determining step of the latter is not known (see Biochemistry of the Actomyosin ATPase). The process could be ATP hydrolysis or, as favored by Eisenberg & Hill (32), an AM.ADP.P$_i$ isomerization. A third possibility raised by the oxygen isotope exchange experiments is that the tension rise could be equated with P$_i$ release from AM.ADP.P$_i$, in view of the rate of approximately 50 s^{-1} for the process as determined by oxygen exchange experiments with isometric fibers. However, in that case the rapid transients (>300 s^{-1}) measured in quick release experiments (2) would have to be related to interconversion of force-generating intermediates beyond the first formed AM'.ADP state in the cycle [i.e. successive force-generating states would be needed, as indeed Huxley & Simmons suggested (76)].

Direct measurements of the rates of P$_i$ and ADP release during shortening have not been made. A recent study of the effect of these hydrolysis products on the steady shortening velocity has demonstrated that P$_i$ has no significant effect on the shortening velocity, while ADP reduces the rate of shortening with a K_m of about 200 μM (26). Cooke & Pate (26) interpreted this effect of ADP as follows: ADP effectively competes with ATP for cross-bridges in the AM state, and therefore reduces the net rate of ADP release. The slower release rate reduces the shortening velocity because more AM.ADP cross-bridges are translated into positions where they exert negative force as the filaments slide, since it is assumed that detachment may only occur after ADP release and subsequent binding of ATP to AM.

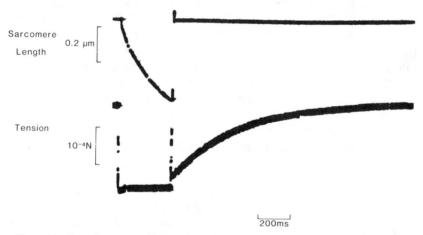

Figure 14 Rate of recovery of isometric tension after a lightly loaded isotonic contraction followed by a rapid stretch of the fiber back to its isometric length (11). The record, kindly provided by B. Brenner, is of an experiment with an activated fiber at 5°C in the presence of 1 mM MgATP at physiological ionic strength. Isometric tension prior to the isotonic contraction is shown at the left of the record. (Published with permission, © 1985 by Chapman & Hall.)

An alternative explanation might be that cross-bridges in the AM.ADP state could detach and reattach to another actin site, exerting a lower force at their new location; the cross-bridge would in effect "slip" to a new location on the thin filament. Cross-bridge slipping of this kind has been proposed as an explanation for several other aspects of the effects of nucleotides on cross-bridge mechanics (25, 37, 40, 99).

However, much still needs to be learned about the nature of biochemical intermediates in shortening muscle. One useful approach may be through the photolysis technique, which releases pulses of ATP, ADP, or P_i into shortening muscle fibers. In Figure 15 we show the feasibility of this method. A skinned fiber was allowed to shorten at a relatively high velocity but in the presence of a low initial concentration of ATP. At the arrow, the ATP concentration was raised to approximately 800 μM while the fiber continued to shorten at a constant rate. The high final tension achieved was as expected from the known steady-state ATP dependence of the force-velocity relations for skinned fibers (23, 37); also the rapid rise of tension without a substantial lag or initial drop is consistent with the suggestion that the tension transient arises as a consequence of detachment of cross-bridges bearing negative tension, because of the greater availability of ATP

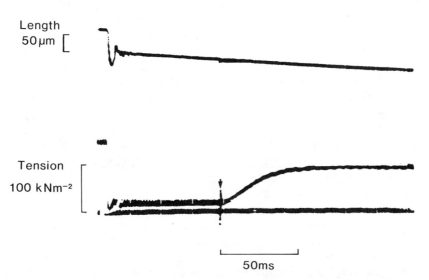

Figure 15 Dependence of tension on rapid changes of ATP concentration in a shortening fiber. An isotonic contraction was initiated in an activated fiber held isometric in the presence of 100 μM ATP with a creatine phosphate/creatine kinase ATP-regenerating system. At the time marked by the arrow a laser pulse caused the release of 700 μM ATP from caged ATP. The baseline of the tension record is of the fiber when relaxed. (M. G. Hibberd, Y. E. Goldman, D. R. Trentham, unpublished.)

to react with AM. The idea of cross-bridges bearing negative tension is important in A. F. Huxley's 1957 theory (68). However, we emphasize that Figure 15 illustrates an approach only and we wish to be cautious in its interpretation.

Cross-Bridge Cycle

The major features of Huxley's 1957 theory of the cross-bridge cycle (68, 78) and the proposed mechanism of force generation (76) have been largely supported by recent mechanical investigations in shortening muscle. These theories were formulated without knowledge of the biochemical transitions that have since been discovered and that presumably underlie the mechanical and energetic processes of contraction.

Introduction of biochemical transient kinetic data into cross-bridge theories has been strongly influenced by Eisenberg & Hill (32). Eisenberg et al (33) showed explicitly how force generation and the chemical kinetics of the actomyosin ATPase could be linked. They were successful in reproducing the main features of the dependence of isometric transients on the size of the imposed step-length change, and they also accounted for a large body of biochemical data. Eisenberg & Greene (31) revised this version of the model to account for more recent biochemical data but, as they pointed out, the newer formulation predicts little dependence of the number of attached cross-bridges on shortening velocity, and therefore fails to match the most recent mechanical measurements. Furthermore, the Hill-Eisenberg class of theory is generally less successful in reproducing the force-velocity relation and energetic properties of muscle than is the Huxley-Simmons formulation. Additionally, there are several aspects of the energetic properties of muscle that are not inherent in either set of theories (see Reference 161, pp. 277–308 for detailed comments). Most current theories are essentially two-dimensional in nature. Three-dimensional properties of the filament lattice have not generally been introduced into models, although it is clear that quite substantial radial forces are generated by the cross-bridges as they attach in rigor or during contraction (97, 105, 106).

The theme of this review has been to work within the framework of several well established theories and these include the tilting cross-bridge model of H. E. Huxley (79). Figure 16 provides a description of the main features that relate the mechanical properties of cross-bridges to their biochemistry. We emphasize the schematic nature of Figure 16 and in particular that the location of the elastic element of the cross-bridge is unknown. Mechanical work is produced through the cyclic interaction of myosin with actin coupled to the hydrolysis of ATP, with the sequential release of P_i and ADP. The overall vectorial nature of the work production results from the stress and strain dependence of key transitions in the cycle.

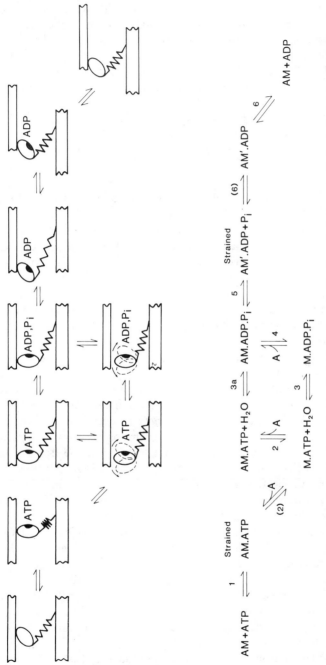

Figure 16 Correlation of mechanical and biochemical states in shortening muscle. The figure includes a negatively strained state, AM.ATP, and a positively strained state, AM′.ADP. All other states are drawn unstrained.

A large proportion of the free energy of ATP hydrolysis (which would otherwise be lost if the proteins were free in solution) may be converted in two or more rapid transitions into the extension of an elastic element in the cross-bridge, with the consequent development of force or relative sliding motion of the filaments. Biochemical evidence suggests that for isolated proteins the largest drop of free energy during product release is associated with release of ADP from AM'.ADP during the formation of the AM state (138). Strong evidence for the AM'.ADP state has been obtained in studies that involve vanadate trapping of intermediates (27).

Figure 16 does not show the rapid force-generating transitions of the Huxley-Simmons model (76). In the figure the strained and unstrained states relate only to the possible average strain of the whole cross-bridge ensemble. In each biochemical state, individual cross-bridges are likely to be distributed about a mean strain, because the periodicities of the actin sites and myosin projections that form the cross-bridges differ slightly (81). The mean strain in any state will also be governed by the rate of filament sliding, such that the strain is reduced at higher velocities. The cross-bridges must be considered to be in a constant dynamic flux rather than statically fixed in one position (as might be implied by Figures 2 and 16). With these factors in mind, we now consider the complete cross-bridge cycle in terms of Figure 16 as a summary of the detailed discussions presented earlier.

Once the muscle is activated, $M.ADP.P_i$ attaches to actin and enters the force-producing state(s) after P_i release. Prior to such product release, the cross-bridge may still detach without having performed work. If the external load applied to the muscle is smaller than that generated by the cross-bridges, the filaments slide in the shortening direction. ADP release from AM'.ADP is sufficiently slow to prevent detachment of the cross-bridges before they have passed through their normal working distance as the filaments slide. During filament sliding the force in the cross-bridges is reduced as mechanical work is performed. As the shortening velocity rises, an increasing proportion of cross-bridges are brought into positions where they exert a force opposed to the sliding motion, thus limiting the maximum shortening velocity. At any given velocity, ADP release may occur either before or after any individual cross-bridge begins to develop negative force. Few AM or AM.ATP states should exist under physiological conditions, due to the rapid detachment of M.ATP induced by the binding of ATP. This rapid detachment also relieves the negative strain in the AM.ATP state formed initially upon ATP binding. M.ATP is then hydrolyzed to reform $M.ADP.P_i$, which may attach to a new actin site. M.ATP can also rebind to actin, but AM.ATP will not, on average, exert significant net force or perform net work.

If the load on the muscle is large a similar sequence of events results, with the following differences. Since filament sliding no longer shifts the equilibrium of attached states in favor of more rapidly detaching positions, ADP dissociation is presumed to be slower and the predominant intermediates during isometric contraction exert positive force.

As mentioned earlier, the greatest uncertainty in the cycle relates to transition between AM'.ADP and AM. We do know whether AM is the same species as the rigor complex (i.e. the state when ATP is removed from a muscle fiber) or whether the ADP-bound rigor complex (AM.ADP in Figure 3) is on the reaction pathway. If the physiological AM state is not similar to the rigor complex, it is of great importance to know the rate at which it is dissociated by ATP. Nevertheless there are some grounds for suggesting that the rigor state and AM.ADP lie in the pathway. Evidence that the AM state behaves kinetically like the rigor complex is provided by the internal consistency of the rate constant of the ATP-induced dissociation of the rigor complex, the K_m of ATP with respect to the maximum velocity of shortening, and the cross-bridge detachment rate (37). The strong correlation between shortening velocity and the dissociation rate constant at ADP from an ADP-rigor complex suggests that AM.ADP is also in the pathway (132). A final difficulty with the mechanism of Figure 16 is that, on purely thermodynamic grounds, significant phosphate exchange would not be predicted between P_i and ATP in isolated actomyosin if the rate of association of P_i to AM'.ADP, k_{-5}, were strongly force-dependent. However, the data of Sleep & Hutton (138) show that k_{-5} is at least $0.5[AM_0]/[AM'.ADP]$ $M^{-1} s^{-1}$ and therefore is within a factor of 1000 of k_{-5} measured in fibers (see Attached Cross-Bridge States). This ratio is less than might be expected from the discussion in Further Analysis of Force Generation. Thus we can see from these examples how much still must be understood.

SUMMARY

In this review we have attempted a synthesis of ideas from cross-bridge theories of muscle contraction with biochemical mechanisms of the actomyosin ATPase. This synthesis of ideas has been based on experimental approaches that permit mechanical and biochemical investigations on the same system. We have formulated an example of how biochemical processes may be influenced by strain in the cross-bridge and have highlighted how much has yet to be learned about the biochemistry (and protein structure) of the working stroke of the cross-bridge. Processes that do not appear to be related to the working stroke such as ATP-induced dissociation of actomyosin or protein-bound ATP hydrolysis appear to be similar kinetically in fibers and isolated actomyosin. But, as might be expected, this is not the case in those processes that involve force production and the performance of mechanical work. There appears to be a sound base from which the mechanochemistry of individual processes within the cross-bridge cycle can be analyzed in detail. There is a need for the development of spectroscopic techniques, particularly those that might detect the rate of P_i and ADP dissociation from cross-bridges into the medium. The combination of pulse photolysis of caged ATP and time-resolved structure analysis by use of synchrotron radiation (53) should lead to better understanding of the structure of cross-bridge states in relation to the chemistry and mechanics of transient intermediates.

ACKNOWLEDGMENTS

We would like to thank our colleagues Drs. M. A. Ferenczi, Y. E. Goldman, and M. R. Webb for helpful discussions during the preparation of this review; many colleagues for permission to report unpublished data; and Mrs. A. Humphrey-Gaskin for help in preparing the manuscript. This work is supported by the Medical Research Council, UK and the National Institutes of Health through grant number HL 15835 to the Pennsylvania Muscle Institute and the Muscular Dystrophy Association of America.

Literature Cited

1. Abbott, R. H. 1972. *Nature New Biol.* 239:183–86
2. Abbott, R. H., Steiger, G. J. 1977. *J. Physiol.* 266:13–42
3. Adelstein, R. S., Eisenberg, E. 1980. *Ann. Rev. Biochem.* 49:921–56
4. Amos, L. A. 1985. *Ann. Rev. Biophys. Biophys. Chem.* 14:291–313
5. Armstrong, C. F., Huxley, A. F., Julian, F. J. 1966. *J. Physiol.* 186:26P–27P
6. Biosca, J. A., Barman, T. E., Travers, F. 1984. *Biochemistry* 23:2428–36
7. Biosca, J. A., Travers, F., Barman, T. E., Bertrand, R., Audemard, E., Kassab, R. 1985. *Biochemistry* 24:3814–20
8. Botts, J., Cooke, R., Dos Remedios, C., Duke, J., Mendelson, R., Morales, M. F., et al. 1973. *Cold Spring Harbor Symp. Quant. Biol.* 27:195–200
9. Bremel, R. D., Weber, A. 1972. *Nature New Biol.* 238:97–101
10. Brenner, B. 1983. *Biophys. J.* 41:33a
11. Brenner, B. 1985. *J. Muscle Res. Cell Motil.* 6:659–64
12. Brenner, B., Schoenberg, M., Chalovich, J. M., Greene, L. E., Eisenberg, E. 1982. *Proc. Natl. Acad. Sci. USA* 79:7288–91
13. Brenner, B., Yu, L.-C., Podolsky, R. J.
1984. *Biophys. J.* 46:299–306
14. Bunghardt, T. P., Thompson, N. L. 1985. *Biochemistry* 24:3731–35
15. Cain, D. F., Infante, A. A., Davies, R. E. 1962. *Nature* 196:214–17
16. Cecchi, G., Colomo, F., Lombardi, V. 1976. *Bull. Soc. Ital. Biol. Sper.* 52:733–36
17. Cecchi, G., Colomo, F., Lombardi, V., Piazzesi, G. 1985. *J. Muscle Res. Cell Motil.* 6:103
18. Cecchi, G., Griffiths, P. J., Taylor, S. 1982. *Science* 217:70–72
19. Chalovich, J. M., Chock, P. B., Eisenberg, E. 1981. *J. Biol. Chem.* 256:575–78
20. Chalovich, J. M., Eisenberg, E. 1982. *J. Biol. Chem.* 257:2432–37
21. Civan, M. M., Podolsky, R. J. 1966. *J. Physiol.* 184:511–34
22. Coates, J. H., Criddle, A., Geeves, M. A. 1985. *Biochem. J.* 232:351–56
23. Cooke, R., Bialek, W. 1979. *Biophys. J.* 28:241–58
24. Cooke, R., Crowder, M. S., Thomas, D. D. 1982. *Nature* 300:776–78
25. Cooke, R., Pate, E. 1985. *Biophys. J.* 47:773–80
26. Cooke, R., Pate, E. 1985. *Biophys. J.* 48:789–98

27. Dantzig, J. A., Goldman, Y. E. 1985. *J. Gen. Physiol.* 86:305–27
28. Ebashi, S., Endo, M. 1968. *Prog. Biophys. Mol. Biol.* 18:125–83
29. Ebashi, S., Endo, M., Ohtsuki, J. 1969. *Q. Rev. Biophys.* 2:351–84
30. Eccleston, J. F. 1980. *FEBS Lett.* 113: 55–57
31. Eisenberg, E., Greene, L. E. 1980. *Ann. Rev. Physiol.* 42:293–309
32. Eisenberg, E., Hill, T. L. 1985. *Science* 227:999–1006
33. Eisenberg, E., Hill, T. L., Chen, Y. 1980. *Biophys. J.* 29:195–227
34. Elmubarak, M. H., Ranatunga, K. W. 1984. *Muscle Nerve* 7:298–303
35. Fenn, W. O. 1923. *J. Physiol.* 58:175–203
36. Ferenczi, M. A. 1985. *Biophys. J.* 47: 60a
37. Ferenczi, M. A., Goldman, Y. E., Simmons, R. M. 1984. *J. Physiol.* 350: 519–43
38. Ferenczi, M. A., Homsher, E., Simmons, R. M., Trentham, D. R. 1978. *Biochem. J.* 171:165–75
39. Ferenczi, M. A., Homsher, E., Trentham, D. R. 1984. *J. Physiol.* 352: 575–99
40. Ferenczi, M. A., Simmons, R. M., Sleep, J. A. 1982. In *Basic Biology of Muscles: A Comparative Approach*, Vol. 37, *Society of General Physiologists Series*, ed. B. M. Twarog, R. J. C. Levine, M. M. Dewey, pp. 91–107. New York: Raven. 406 pp.
41. Ford, L. E., Huxley, A. F., Simmons, R. M. 1974. *J. Physiol.* 240:42P–43P
42. Ford, L. E., Huxley, A. F., Simmons, R. M. 1977. *J. Physiol.* 269:441–515
43. Ford, L. E., Huxley, A. F., Simmons, R. M. 1981. *J. Physiol.* 311:219–49
44. Ford, L. E., Huxley, A. F., Simmons, R. M. 1985. *J. Physiol.* 361:131–50
45. Geeves, M. A., Goody, R. S., Gutfreund, H. 1984. *J. Muscle Res. Cell Motil.* 5:351–61
46. Geeves, M. A., Gutfreund, H. 1982. *FEBS Lett.* 140:11–15
47. Gillis, J. M., Marechal, G. 1974. *J. Mechanochem. Cell Motil.* 3:55–68
48. Goldman, Y. E., Hibberd, M. G., McCray, J. A., Trentham, D. R. 1982. *Nature* 300:701–5
49. Goldman, Y. E., Hibberd, M. G., Trentham, D. R. 1984. *J. Physiol.* 354: 577–604
50. Goldman, Y. E., Hibberd, M. G., Trentham, D. R. 1984. *J. Physiol.* 354: 605–24
51. Goldman, Y. E., Simmons, R. M. 1984. *J. Physiol.* 350:497–518
52. Goldman, Y. E., Simmons, R. M. 1978. *Biophys. J.* 21:86a
53. Goody, R. S., Guth, K., Maeda, Y., Poole, K. J. V., Rapp, G. 1985. *J. Physiol.* 364:75P
54. Goody, R. S., Holmes, K. C. 1983. *Biochim. Biophys. Acta* 726:13–39
55. Greene, L. E., Eisenberg, E. 1980. *J. Biol. Chem.* 255:543–48
56. Hackney, D. D. 1983. *Fed. Proc.* 42: 1923
57. Harrington, W. F., Rogers, M. E. 1984. *Ann. Rev. Biochem.* 53:35–73
58. Haselgrove, J. C., Huxley, H. E. 1973. *J. Mol. Biol.* 77:549–68
59. Hellam, D. C., Podolsky, R. J. 1969. *J. Physiol.* 200:807–19
60. Herzig, J. W., Peterson, J. W., Ruegg, J. C., Solaro, R. J. 1981. *Biochim. Biophys. Acta* 672:191–96
61. Hibberd, M. G., Dantzig, J. A., Trentham, D. R., Goldman, Y. E. 1985. *Science* 228:1317–19
62. Hibberd, M. G., Webb, M. R., Goldman, Y. E., Trentham, D. R. 1985. *J. Biol. Chem.* 260:3496–500
63. Hill, A. V. 1964. *Proc. R. Soc. London Ser. B* 159:319–24
64. Hill, T. L. 1968. *Thermodynamics for Chemists and Biologists*, Chapter 7. Reading, Mass: Addison-Wesley. 181 pp.
65. Hill, T. L. 1974. *Prog. Biophys. Mol. Biol.* 28:267–340
66. Hill, T. L. 1977. *Free Energy Transduction in Biology.* New York: Academic. 229 pp.
67. Hill, T. L., Simmons, R. M. 1976. *Proc. Natl. Acad. Sci. USA* 73:95–99
68. Huxley, A. F. 1957. *Prog. Biophys. Biophys. Chem.* 7:255–318
69. Huxley, A. F. 1973. *Proc. R. Soc. London Ser. B* 183:83–86
70. Huxley, A. F. 1974. *J. Physiol.* 243:1–43
71. Huxley, A. F. 1980. *Reflections on Muscle. The Sherrington Lecture XIV.* Liverpool: Liverpool Univ. Press. 111 pp.
72. Huxley, A. F. 1981. In *Advances in Physiological Sciences*, Vol. 5, *Molecular and Cellular Aspects of Muscle Function*, ed. E. Varga, A. Kovér, T. Kovécs, L. Kovécs, pp. 1–12. Elmsford, NY: Pergamon
73. Huxley, A. F., Lombardi, V. 1980. *J. Physiol.* 305:15P–16P
74. Huxley, A. F., Lombardi, V., Peachey, L. D. 1981. *J. Physiol.* 317:12P–13P
75. Huxley, A. F., Niedergerke, R. 1954. *Nature* 173:971–73
76. Huxley, A. F., Simmons, R. M. 1971. *Nature* 233:533–38
77. Huxley, A. F., Simmons, R. M. 1972. *Nature New Biol.* 239:186–87

78. Huxley, A. F., Simmons, R. M. 1973. *Cold Spring Harbor Symp. Quant. Biol.* 37:669–80
79. Huxley, H. E. 1969. *Science* 164:1356–66
80. Huxley, H. E. 1979. In *Cross-Bridge Mechanism in Muscle Contraction*, ed. H. Sugi, G. H. Pollack, pp. 391–405. Tokyo: Univ. Tokyo Press. 665 pp.
81. Huxley, H. E., Brown, W. 1967. *J. Mol. Biol.* 30:383–434
82. Huxley, H. E., Faruqi, A. R. 1983. *Ann. Rev. Biophys. Bioeng.* 12:381–417
83. Huxley, H. E., Faruqi, A. R., Kress, M., Bordas, J., Koch, M. H. J. 1982. *J. Mol. Biol.* 158:637–84
84. Huxley, H. E., Hanson, J. 1954. *Nature* 173:973–76
85. Huxley, H. E., Kress, M. 1985. *J. Muscle Res. Cell Motil.* 6:153–61
86. Inoue, A., Takenaka, H., Arata, T., Tonomura, Y. 1979. *Adv. Biophys.* 13:1–194
87. Johnson, K. A., Taylor, E. W. 1978. *Biochemistry* 17:3432–42
88. Julian, F. J., Morgan, D. L. 1981. *J. Physiol.* 319:193–203
89. Julian, F. J., Sollins, M. R. 1975. *J. Gen. Physiol.* 66:287–302
90. Kaplan, J. A., Forbush, B. III, Hoffman, J. F. 1978. *Biochemistry* 17:1929–35
91. Kawai, M. 1982. See Ref. 40, pp. 109–30
92. Kawai, M., Brandt, P. W. 1980. *J. Muscle Res. Cell Motil.* 1:279–303
93. Kawai, M., Wolen, C., Cornacchia, T. 1985. *Biophys. J.* 47:24a
94. Kentish, J. C. 1985. *J. Physiol.* 358:75P
95. Kodama, T. 1985. *Physiol. Rev.* 65:467–551
96. Konrad, M., Goody, R. S. 1982. *Eur. J. Biochem.* 128:547–55
97. Krasner, B., Maughan, D. W. 1984. *Pflügers Arch.* 400:160–65
98. Kress, M., Huxley, H. E., Faruqi, A. R., Hendrix, J. 1985. *J. Mol. Biol.* In press
99. Kuhn, H. J. 1978. *Biophys. Struct. Mech.* 43:169–78
100. Lowey, S., Slayter, H. S., Weeds, A. G., Baker, H. 1969. *J. Mol. Biol.* 42:1–29
101. Lymn, R. W. 1978. *Biophys. J.* 21:93–98
102. Lymn, R. W., Taylor, E. W. 1971. *Biochemistry* 10:4617–24
103. Marston, S. B. 1973. *Biochim. Biophys. Acta* 305:397–412
104. Matsubara, I., Elliott, G. F. 1972. *J. Mol. Biol.* 72:657–69
105. Matsubara, I., Goldman, Y. E., Simmons, R. M. 1984. *J. Mol. Biol.* 173:15–33
106. Matsubara, I., Umazume, Y., Yagi, N. 1985. *J. Physiol.* 360:135–48
107. Matsubara, I., Yagi, N., Hashizume, H. 1975. *Nature* 255:728–29
108. Matsubara, I., Yagi, N., Miura, H., Ozeki, M., Izumi, T. 1984. *Nature* 312:471–73
109. McCray, J. A., Herbette, L., Kihara, T., Trentham, D. R. 1980. *Proc. Natl. Acad. Sci. USA* 77:7237–41
110. Mendelson, R. A., Morales, M. F., Botts, J. 1973. *Biochemistry* 12:2250–55
111. Millar, N., Geeves, M. A. 1983. *FEBS Lett.* 160:141–48
112. Morales, M. F., Borejdo, J., Botts, J., Cooke, R., Mendelson, R. A., Takashi, R. 1982. *Ann. Rev. Phys. Chem.* 33:319–51
113. Mornet, D., Bertrand, R., Pantel, P., Audemard, E., Kassab, R. 1981. *Nature* 292:301–6
114. Natori, R. 1954. *Jikeikai Med. J.* 1:119–26
115. Offer, G., Elliott, A. 1978. *Nature* 271:325–29
116. Podolsky, R. J. 1960. *Nature* 188:666–68
117. Podolsky, R. J., Nolan, A. C. 1973. *Cold Spring Harbor Symp. Quant. Biol.* 37:661–68
118. Podolsky, R. J., Nolan, A. C., Zaveler, S. A. 1969. *Proc. Natl. Acad. Sci. USA* 64:504–11
119. Podolsky, R. J., St. Onge, R., Yu, L., Lymn, R. W. 1976. *Proc. Natl. Acad. Sci. USA* 73:813–17
120. Podolsky, R. J., Tawada, K. 1980. In *Muscle Contraction: Its Regulatory Mechanisms*, ed. S. Ebashi, K. Maruyama, M. Endo, pp. 65–78. Tokyo: Japan Sci. Soc. Press. 549 pp.
121. Rall, J. A. 1982. *Am. J. Physiol.* 242:H1–H6
122. Ranatunga, K. W., Wylie, S. R. 1983. *J. Physiol.* 339:87–95
123. Reedy, M. K., Holmes, K. C., Tregear, R. T. 1965. *Nature* 207:1276–80
124. Rosenfeld, S. S., Taylor, E. W. 1984. *J. Biol. Chem.* 259:11908–19
125. Rosenfeld, S. S., Taylor, E. W. 1984. *J. Biol. Chem.* 259:11920–29
126. Ruegg, J. C., Schadler, M., Steiger, G. J., Muller, C. 1971. *Pflügers Arch.* 325:359–64
127. Schoenberg, M., Eisenberg, E. 1985. *Biophys. J.* 48:863–71
128. Schoenberg, M., Wells, J. B. 1984. *Biophys. J.* 45:389–97
129. Shriver, J. W. 1984. *Trends Biochem. Sci.* 9:322–28
130. Shriver, J. W., Sykes, B. D. 1981. *Biochemistry* 20:6357–62
131. Shriver, J. W., Sykes, B. D. 1982. *Biochemistry* 21:3022–28
132. Siemankowski, R. F., Wiseman, M. O., White, H. D. 1985. *Proc. Natl. Acad. Sci. USA* 82:658–62

133. Simmons, R. M., Hill, T. L. 1976. *Nature* 263:615–18
134. Simmons, R. M., Jewell, B. R. 1974. In *Recent Advances in Physiology*, No. 9, ed. R. J. Linden, pp. 87–114. London: Churchill-Livingstone. 467 pp.
135. Sleep, J. A. 1983. *Biochem. Soc. Trans.* 11:152
136. Sleep, J. A., Boyer, P. D. 1978. *Biochemistry* 17:5417–22
137. Sleep, J. A., Hutton, R. L. 1978. *Biochemistry* 17:5423–30
138. Sleep, J. A., Hutton, R. L. 1980. *Biochemistry* 19:1276–83
139. Sleep, J. A., Smith, S. J. 1981. *Curr. Top. Bioenerg.* 11:239–86
140. Squire, J. 1981. *The Structural Basis of Muscular Contraction*. New York: Plenum. 698 pp.
141. Stein, L. A., Chock, P. B., Eisenberg, E. 1981. *Proc. Natl. Acad. Sci. USA* 78:1346–50
142. Stein, L. A., Chock, P. B., Eisenberg, E. 1984. *Biochemistry* 23:1555–63
143. Stein, L. A., Greene, L. E., Chock, P. B., Eisenberg, E. 1985. *Biochemistry* 24:1357–63
144. Stein, L. A., Schwarz, R. P. Jr., Chock, P. B., Eisenberg, E. 1979. *Biochemistry* 18:3895–909
145. Sutoh, K., Yamamoto, K., Wakabayashi, T. 1984. *J. Mol. Biol.* 178:323–29
146. Szent-Gyorgyi, A. 1949. *Biol. Bull.* 96:140–61
147. Takashi, R., Putnam, S. 1979. *Anal. Biochem.* 92:375–82
148. Taylor, E. W. 1979. *CRC Crit. Rev. Biochem.* 6:103–64
149. Thomas, D. D. 1978. *Biophys. J.* 24:439–62
150. Tonomura, Y. 1972. *Muscle Proteins,*

Muscle Contraction and Cation Transport, pp. 191–245. Tokyo: Univ. Tokyo Press. 433 pp.
151. Trentham, D. R., Eccleston, J. F., Bagshaw, C. R. 1976. *Q. Rev. Biophys.* 9:217–81
152. Trybus, K. M., Taylor, E. W. 1982. *Biochemistry* 21:1284–94
153. Ulbrich, M., Ruegg, J. C. 1977. In *Insect Flight Muscle*, ed. R. T. Tregear, pp. 317–33. Amsterdam: North-Holland. 367 pp.
154. Van den Hooff, H., Blange, T. 1984. *Pflügers Arch.* 400:280–85
155. Wagner, P. D., Weeds, A. G. 1979. *Biochemistry* 18:2260–66
156. Webb, M. R., Hibberd, M. G., Goldman, Y. E., Trentham, D. R. 1985. *Biophys. J.* 47:60a
157. Webb, M. R., Trentham, D. R. 1983. In *The Handbook of Physiology: Skeletal Muscle*, ed. L. D. Peachey, R. H. Adrian, S. R. Geiger, pp. 237–55. Bethesda, Md: Am. Physiol. Soc. 688 pp.
158. White, D. C. S., Thorson, J. T. 1972. *J. Gen. Physiol.* 60:307–36
159. White, H. D. 1977. *Biophys. J.* 17:40a
160. White, H. D., Taylor, E. W. 1976. *Biochemistry* 15:5818–26
161. Woledge, R. C., Curtin, N. A., Homsher, E. 1985. In *Energetic Aspects of Muscle Contraction. Monogr. Physiol. Soc.*, No. 41. London: Academic. 357 pp.
162. Yamamoto, T., Herzig, J. W. 1978. *Pflügers Arch.* 373:21–24
163. Yu, L. C., Hartt, J. E., Podolsky, R. J. 1979. *J. Mol. Biol.* 132:53–67
164. Altingham, J. D., Johnston, I. A. 1985. *J. Physiol.* 368:491–500
165. Hanson, J., Huxley, H. E. 1955. *Symp. Soc. Exp. Biol.* 9:228–64

Ann. Rev. Biophys. Biophys. Chem. 1986. 15 : 163–93
Copyright © 1986 by Annual Reviews Inc. All rights reserved

ELECTROSTATIC INTERACTIONS IN MEMBRANES AND PROTEINS

Barry H. Honig

Department of Biochemistry and Molecular Biophysics, Columbia University, New York, NY 10032

Wayne L. Hubbell and Ross F. Flewelling

Jules Stein Eye Institute and the Department of Chemistry and Biochemistry, University of California, Los Angeles, California 90024

CONTENTS

PERSPECTIVES AND OVERVIEW

The generation and utilization of electrical potentials is a central function of biological membranes. The electrical properties of pure lipid bilayers and biological membranes are well understood owing in large part to the ease of measuring changes in membrane potentials and currents. In contrast,

0883–9182/86/0610–0163$02.00

despite the fact that a great deal is known about the three-dimensional structure of globular proteins, their electrical properties are rather poorly understood. A goal of this article is to review the relationship between structure and electrostatics both in membranes and in globular proteins. In this way we hope to provide researchers familiar with one area with insights available from the other.

The article is organized into three main sections dealing with electrical aspects of lipid bilayers, globular proteins, and membrane proteins. There are fundamental principles common to all systems and these serve as unifying themes. Specific issues that are addressed include: (a) Structural origins of electric fields. Examples include fields due to surface charges and dipoles of oriented phospholipids in bilayers, and charged amino acids and helices in proteins. (b) The energetic principles associated with transferring ionic species from solvent into proteins and lipid bilayers. These issues arise in problems such as the transport of ions across pure lipid bilayers and through channels, the stability of charged amino acids on the surface and in the interior of proteins and the possible role of charged groups as voltage sensors. The relation of these issues to chelation is emphasized. (c) The molecular nature of electrically active conformational changes in membrane proteins. The possible chemical nature of the voltage sensor is discussed in light of evidence available from well characterized model systems and from the energetic considerations outlined in the previous paragraph.

The topics discussed in this article have been reviewed separately in numerous articles but have not previously been combined into a single review. For separate discussions of the individual topics the reader should consult articles by McLaughlin (123) and Andersen (5) on lipid bilayers, by Matthew (118) and Warshel & Russell (187) on proteins, and by Armstrong (10), French & Horn (64), and Bezanilla (22) on gating phenomena in excitable membranes.

LIPID BILAYERS

Total Membrane Potential Profile

The total membrane electric potential profile of a lipid bilayer, even in the absence of proteins, is a complicated sum of multiple components (Figure 1): the transmembrane potential ($\Delta\Psi$), surface potentials (ψ_{so}, ψ_{si}), and various internal potentials (e.g., ψ_o and ψ_i). Internal membrane potentials can be quite varied, but include most prominently the adsorption potentials and the membrane dipole potential.

TRANSMEMBRANE POTENTIAL   The transmembrane potential is the electric potential difference between the bulk aqueous phases. It arises from net charge separations across the bilayer, and in biological systems is typically on the order of 10–100 mV. Invariably the inside (cytoplasmic side) is negative relative to the outside. For a system at equilibrium the potential is directly related to the ion concentration gradient of membrane-permeable ions, as given by the Nernst equation. Under steady-state conditions, the transmembrane potential is determined both by the concentrations ($[X]$) and relative permeabilities (P_X) of the various membrane-permeable ions, and is often well described by the Goldman-Hodgkin-Katz equation (81). If Na^+, K^+, and Cl^- are the dominant permeable species, as in the squid axon, the transmembrane potential is given as follows:

$$\Delta\Psi = -\frac{RT}{F}\ln\left(\frac{P_K[K]_i+P_{Na}[Na]_i+P_{Cl}[Cl]_o}{P_K[K]_o+P_{Na}[Na]_o+P_{Cl}[Cl]_i}\right).$$

When the permeability of one ion dominates, this expression reduces to the simple Nernst equation for that ion.

SURFACE POTENTIAL   The membrane surface potential is the electrostatic potential at the membrane-aqueous interface relative to that in the corresponding bulk solution, and arises from a net charge on the membrane surface. In virtually all biological systems studied this potential is negative owing to the ubiquity of negatively charged lipids (primarily phos-

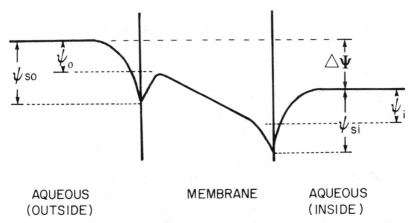

AQUEOUS   MEMBRANE   AQUEOUS
(OUTSIDE)   (INSIDE)

Figure 1 Total membrane electric potential profile as the sum of the transmembrane potential ($\Delta\Psi$), surface potentials (ψ_{so}, ψ_{si}), and internal (e.g. dipole or adsorption) potentials such as ψ_o and ψ_i. An intramembrane potential ($\Delta\psi_{ip}$) can also be defined depending on the situation. In the absence of surface and internal potentials, the intramembrane potential is simply the transmembrane potential; but in the general case it is often necessary to define it in terms of the surface potentials ($\Delta\psi_{ip} = \psi_{so}-\psi_{si}$) or adsorption potentials ($\Delta\psi_{ip} = \psi_s-\psi_i$). In many membrane electrical phenomena this intramembrane potential is the important consideration.

phatidylserine in mammalian membranes), although adsorbed molecules and proteins also contribute to the net surface charge density. Biological membranes typically have 10–20% of their effective lipid area composed of negatively charged lipids (about 1 electronic charge per 1000 Å^2), which together with protein contributions give surface potentials of -8 to -30 mV as measured by electrophoretic mobility (46).

The Gouy-Chapman model (40, 70), with modifications by Stern (168) sometimes used to account for ion adsorption, combines the Poisson and Boltzmann equations to treat surface potentials near charged membranes. By this model the surface potential is related to the surface charge density, σ_s, as follows (symmetric salt at 20°C, the ionic strength C in mol/liter, and σ_s in electronic charges per Å^2): $\psi_s = 51$ mV $\sinh^{-1} 136\sigma_s/C^{1/2}$). Several excellent reviews summarize the theoretical and practical aspects of surface potentials (e.g. 14, 74, 123). Surface potentials result in alterations of bulk ion surface concentrations, as given by the Boltzmann equation: $C_{surf} = C_{bulk} \exp(-ZF\psi_s/RT)$. Thus for a surface potential of -60 mV, the surface concentration of monovalent anions is depleted tenfold while that for monovalent cations is enhanced tenfold.

Recent extensions of the Gouy-Chapman-Stern model include more detailed consideration of ion size (4, 35), nonplanar geometries (127, 137), discrete charge effects (e.g., 30, 42, 133), and the distribution of surface charges out from the physical interface (e.g. 31, 36, 121, 122, 138). For early reviews of these points see Haydon (75) and McLaughlin (123, 124). Monte Carlo and molecular dynamics simulations are in good agreement with the Gouy-Chapman-Stern theory (24, 174–176, 183). Most small inorganic ions adsorb to lipid bilayers with at least some affinity and corrections for this must be made, especially at high salt concentrations (e.g. 54, 102, 125).

INTERNAL POTENTIALS   Internal potentials arise from either dipoles or charges in the low dielectric interior of the membrane. Two specific types of such potentials are the membrane dipole potential and adsorption potentials. As reviewed below, lipid bilayers possess a substantial dipole potential arising from the structural organization of polar molecules at the membrane-water interface. Its magnitude is typically several hundred millivolts, inside positive, and thus the dipole potential contributes significantly to the total membrane potential profile. Specific boundary potentials are also possible. These arise from the transfer of net charge from the water to the inner side of the membrane interface, and are distinguished from surface potentials in that their effects within the bilayer are not screened by changes in the ionic strength of the bulk medium. They can thus be generated by hydrophobic ion binding within the membrane (e.g. 6, 7), charge movement within the bilayer (e.g. 32, 33, 38, 50, 52, 54), or channel gating charges (see below).

It is important to keep in mind that as far as the electrostatic properties of molecules in membranes are concerned, it is actually what we call here the "intramembrane potential" (Figure 1) that is important, and not simply the transmembrane potential. Indeed these two potentials may be quite different. Surface charge asymmetry and internal dipole layers as well as adsorbed amphiphiles, hydrophobic ions, or proteins can modify the intramembrane potential substantially, with important consequences for membrane electrical phenomena (e.g. 68, 101).

Charged Molecule Interactions with Bilayers

TOTAL POTENTIAL ENERGY PROFILE   The free energy as a function of position z (from the membrane center) for small charged molecule interactions with lipid bilayers can be written as the sum of electrical and nonelectrical terms:

$$W_{TOT}(z) = W_{Born}(z) + W_{Image}(z) + W_{Dipole}(z) + W_{Neutral}(z). \qquad 1.$$

The Born, image, and dipole energy contributions contain all of the dominant electrical interactions of a charged molecule with the membrane. They represent the ways in which a charged molecule interacts with bulk dielectrics, dielectric interfaces, and any intrinsic dipole potential, respectively. The neutral energy term includes all of the other contributions to the free energy: hydrophobic, van der Waals, steric, and specific chemical interactions. This formulation, minus the dipole potential, was first proposed by Ketterer et al (96) to explain the properties of hydrophobic ion interactions with bilayers; it has been reviewed on several occasions (5, 76, 80, 104). The membrane dipole potential contribution has since been explicitly included in the total energy, and this more complete treatment has been successful in accounting for the detailed binding and translocation properties of hydrophobic ions (60, 61).

The dielectric constant enters into several of these terms and therefore deserves special attention. Pure water has a dielectric constant of about 80 at 20°C, and changes due to temperature or ionic composition are generally negligible. Most oils and aliphatic hydrocarbons have dielectric constants in the range 2–5, with a value of about 2 generally accepted for the hydrocarbon portion of lipid bilayers (49). Dielectric constants in the range 10–30 are reasonable for the region between bulk water and lipid hydrocarbon as inferred from low-frequency impedance measurements (12, 13) and spectroscopic probes (37, 56, 109, 156). Thus it is reasonable that the dielectric constant should vary with perpendicular distance from the interface; a bulk water value of 80 is attained at least a few water-molecule layers out from the interface, a bulk lipid dielectric value of 2 is attained below about the ester groups of the lipid, and in between, in the transition region of not more than 10 Å, a sigmoidal functional dependence is reasonable. Such a model has been introduced and found useful for estimating the effects of the dipole potential in the interfacial region (61).

Born energy   The free energy of transfer in moving a charged body from a region of dielectric strength ε_2 to a region of dielectric strength ε_1 is given by the Born (28) expression,

$$W_B = \frac{q^2}{2_r} \left(\frac{1}{\varepsilon_1} - \frac{1}{\varepsilon_2} \right). \qquad\qquad 2.$$

The Born model is relatively successful in accounting for hydration energies if appropriate radii are used (148; for reviews see 26, 155). With a bulk aqueous dielectric constant $\varepsilon_2 = 80$ and a bulk lipid membrane dielectric constant $\varepsilon_1 = 2$, the Born energy for transfer of a charge $q = Ze$ becomes (r in Å): $W_{Born} \approx 81\ Z^2/r$ kcal/mol. This poses the most significant barrier to ion transport across bilayers and, as described below, nature solves this fundamental electrostatic problem in several unique ways.

Image energy   In addition to the bulk electrostatic Born energy, a charged body in a lipid bilayer also feels the polarization ("image") forces arising from the dielectric interfaces. Image energy calculations applicable to lipid bilayers have been developed in the literature on several occasions (76, 135, 140, 141). It is not generally recognized, however, that none of these developments is in agreement with the others (59). The treatment of Haydon & Hladky (76) differs from that of Neumcke & Läuger (135) by a factor of two; the difference appears to arise from an error on the part of Haydon & Hladky to distinguish between the electrostatic potential they derive and the work done against such a potential in moving an ion from far away to a position near the interface. The Neumcke & Läuger (135) and Parsegian (140, 141) results are also not in agreement; the difference arises from their implicit reference-state assumption (59). However, the numerical differences between the two calculations are normally small, except in the case of large ion radii or high membrane dielectric. Therefore use of either expression gives an entirely adequate representation of the image energy contribution, especially given the various other assumptions in the model.

For a charge q in a dielectric slab of thickness d and uniform dielectric constant ε_1, surrounded on both sides by bulk dielectrics ε_2, and at distance x ($x = d/2 - z$) from one interface, the complete Born-plus-image energy solution has been given by Neumcke & Läuger (135). Under normal conditions ($\varepsilon_2 \gg \varepsilon_1$) a convenient Born-image energy expression for $x > 2r$ is given approximately by (61):

$$W_{B-I} = \frac{q^2}{2\varepsilon_{1r}} \left[1 - \frac{r}{2x} - 1.2 \left(\frac{r}{d} \right) \left(\frac{x}{d} \right)^2 \right] \qquad\qquad 3.$$

At the membrane center this becomes $W_{B-I}(0) \approx 83Z^2(1/r - 0.033)$. The effect of the image contribution is thus twofold: It reduces the Born energy by 10–15% in the middle of the membrane, or about 4 RT (2.5 kcal/mol) in activation energy, and it gives the energy function a more realistic profile near the interfaces.

Dipole energy   There are three possible sources for the membrane dipole potential: lipid headgroups, surface water molecules, and lipid carbonyls (e.g. 123, 139). The principal source is probably the lipid carbonyls that link the fatty acid chains to the glycerol backbone. This idea is supported by the fact that a wide range of lipid types with different head groups give remarkably similar dipole potentials, while ether-linked rather than ester-linked lipids have substantially reduced dipole potentials. Water and lipid headgroup dipoles may also contribute, although in an unknown way. Previously the dipole potential has been treated by a simple capacitor model. However, with the molecular sources noted above the dipole potential can be treated quasi-microscopically, for example by a two-dimensional lattice array of point dipole sources located at each membrane surface. Such a model has been successful in accounting for the detailed properties of hydrophobic ion interactions with bilayers; the best value for the dipole potential in phosphatidylcholine vesicles was about 240 mV and in the center of the bilayer $W_{Dip}(0) \approx 5.4Z$ kcal/mol (60, 61). The model is useful not only for quantitatively accounting for the intrinsic dipole potential, but also for investigating alterations in the dipole potential, for example by membrane adsorption of dipolar molecules or by changes in lipid type.

An impotant general theme to keep in mind is that as a consequence of the membrane dipole potential, lipid soluble anions invariably partition much more readily into and translocate much more readily across lipid bilayers than similarly structured cations. This is certainly true of hydrophobic ions and ion carriers, and may have broader relevance as well.

Neutral energy   The neutral energy contribution is the free energy of transfer for moving a small molecule from the aqueous phase to the membrane, apart from electrical contributions. It can best be accounted for by considering the free energy of transfer for a corresponding uncharged analog (for tabulations of molecular partition coefficients see 48, 110, 193). As a rough estimate, when the neutral analog of interest is nonpolar, the nonelectrical free energy of transfer for simple molecules is largely dominated by its unfavorable entropy in water. This contribution amounts to about 22 cal/mol per square angstrom of accessible surface area (150). For small molecules, free energies of transfer between octanol and water can be fit empirically to $W_{Neut} \approx 0.6 - 0.45r^2$ kcal/mol, where r in angstroms is an effective spherical radius (111). This amounts to about 0.5–1 kcal/mol for small ions, while larger hydrophobic ions fall in the range 5–10 kcal/mol.

APPLICATIONS   Equation 1 describes quite well the total potential-energy profile for charged molecule interactions with lipid bilayers. It was originally developed and is best suited for describing the properties of hydrophobic ions. However, it may also be applied to understanding how small ions, ion carriers, and perhaps polypeptides and larger macromolecules interact with membranes.

Small ions   Most small inorganic ions (e.g. Na^+, K^+, Ca^{2+}, SO_4^{2-}) are virtually lipid-impermeable. Shown in Figure 2 is the total potential profile for small ions based on Equation 1. By far the most important factor is the Born electrostatic energy barrier that the ions face when moving from the favorable high dielectric environment of water to the unfavorable low dielectric of the membrane. As given by Equation 2, the Born energy for a small ion gives rise to an energy barrier in the lipid bilayer of at least 40 kcal/mol. When this is compared to the average thermal energy (0.6 kcal/mol), it is apparent that small ions are virtually excluded from the lipid phase, with intrinsic membrane translocation rates of less than $10^{-10} s^{-1}$.

Hydrophobic ions   In marked contrast to small ions, hydrophobic ions readily permeate lipid bilayers. (Included in this class of molecules are many uncouplers of oxidative phosphorylation such as dinitrophenol and $FCCP^-$, metal ion complexes of ionophores such as valinomycin and nonactin, some spin label probes, and potential-sensitive dyes.) The primary barrier in moving from the aqueous to the lipid phase is still the unfavorable Born electrostatic energy. However, the salient feature of hydrophobic ions is that they have hydrophobic surfaces and possess large effective ionic radii, typically 4 Å or greater. Consequently not only is the Born energy considerably reduced, but the ions have a significantly more favorable neutral (hydrophobic) free energy of partitioning into the hydrocarbon. For example, the two structural analogs tetraphenylphosphonium (TPP^+) and tetraphenylboron (TPB^-) are representative of hydrophobic ions. Detailed analysis of their comparative binding and translocation properties shows that the cation TPP^+ binds only moderately strongly ($\Delta G_b^0 \approx -2.5$ kcal/mol) and translocates across the bilayer relatively slowly ($k \approx 10^{-2}-10^{-3} s^{-1}$ with $\Delta G^{0\ddagger} \approx 20$ kcal/mol), while the anion TPB^- binds very strongly ($\Delta G_b^0 \approx -7.5$ kcal/mol) and translocates very rapidly ($k \approx 10-100 s^{-1}$ with $\Delta G^{0\ddagger} \approx 14$ kcal/mol) (see 60 and references therein). These differences are due primarily to the dipole potential contribution. Overall the properties of hydrophobic ions are well accounted for by Equation 1.

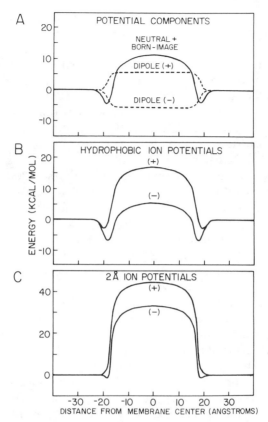

Figure 2   Theoretical membrane potential profiles combining Born, image, dipole, and neutral energy profiles as in Equation 1 of the text. (*A*) Combined Born-image and neutral energy profiles for hydrophobic ions are identical for ions that differ only by the sign of the charge, while the dipole potential profiles differ. (*B*) Total potential profiles for hydrophobic ions with effective 4 Å radius. Note the adsorption sites just within the bilayer, a moderately high barrier for cation translocation, and a relatively low barrier for anion translocation. (*C*) Total potential profiles for 2 Å ions, showing significant electrostatic barriers for translocation, although the barrier for anions is significantly less than that for cations.

Macromolecules   Interactions with membranes of macromolecules such as ion carriers, peptides, and proteins may also be analyzed in terms of Equation 1 to a first approximation. Additional important considerations are discussed below.

ELECTROSTATIC INTERACTIONS IN PROTEINS

Solvent Effects

The classic problem in electrostatics of proteins has been to account for the shape of titration curves. Tanford & Kirkwood (171) treated a protein as a spherical region of low dielectric constant surrounded by a solvent of high dielectric constant. They calculated the interaction of charged groups, represented by point charges, within the protein. The solvent was treated as a simple electrolyte behaving according to Debye-Hückel theory. The Tanford-Kirkwood theory and modified versions of it (see e.g. 89, 119, 120, and Matthew's review, 118) have successfully accounted for pH- and ionic strength–dependent phenomena in a number of proteins. In this light, and given the success of Poisson-Boltzmann type treatments of lipid bilayers (see above for a discussion of Gouy-Chapman theory) and nucleic acids (see e.g. 9), it is clear that these treatments constitute a useful basis for obtaining at the very least a qualitative understanding of the effects of solvent polarization on electrical phenomena in globular proteins.

The extent to which continuum electrostatic models are the method of choice when quantitative predictions are required is less certain. The range of applicability of macroscopic dielectric models can be extended by using numerical methods to treat complex shapes. However, the numerical algorithms can become quite time-consuming, particularly if an accurate representation of the protein-water boundary is required. Warwicker & Watson (189) used a finite difference method to study the electrical potential outside phosphoglycerate mutase due to dipoles within the protein. Their algorithm has since been used in a number of other applications (152, 153). Rogers et al (152) found that a spherical representation of a protein (based on the use of image charges) gave a significantly different extent of solvent screening in cytochrome c_{551} than did the Warwicker-Watson algorithm. The errors involved in mapping a protein into a sphere may be reduced if information as to the depth of each charge relative to the surface is retained (see e.g. 89, 129).

A second problem of macroscopic models is that because charges are represented as points, self energies, which are closely related to solvation free energies (28), are infinite. However, it has recently been shown that the problem of infinite self energies can be avoided by treating charged atoms as charged spheres rather than points. A formalism has been developed in which self energies are incorporated into a continuum electrostatic model; thus macroscopic concepts can be applied to problems where ionic solvation is important (69).

A further complication is the necessity of choosing appropriate values for the dielectric constant to represent both the protein and the solvent. The major uncertainty with respect to the solvent concerns the dielectric constant of the first water layer around the protein, which is likely to be different from the bulk value of 80. However, the success of Poisson-Boltzmann treatments of pure lipid bilayers suggests that the problem may not be too severe in many applications. On the other hand, for groups close to one another and in contact with the solvent the microscopic properties of the solvent clearly have to be accounted for.

Unfortunately, microscopic description of the solvent has not yet progressed to the point where the dielectric properties of water can be accurately reproduced. For example, despite recent progress (39, 126) molecular dynamics simulations cannot yet accurately account for vapor-water transfer free energies of individual ions. In contrast, Born-type models can reproduce solvation free energies although the reasons for this capacity are not clear (148). Microscopic models may encounter problems associated with the use of periodic boundary conditions, inadequacies in the potentials, or perhaps convergence difficulties. It is clear that the reliability of detailed simulations of ions in water will continue to improve and that much insight can be gained from current studies on model systems (see e.g. 23, 105). On the other hand, current simulations of water around proteins (72, 134, 184) are designed to study the structure of boundary waters and would not be expected to account for, say, solvent screening of electrostatic interactions.

An intermediate approach between continuum models and detailed atomic descriptions of solvent molecules involves the use of Langevin dipoles to simulate the dielectric properties of the solvent. Warshel and co-workers have applied this method to a variety of problems associated with ionic species both in proteins and in polar fluids (extensively reviewed in 187). While some successes have been reported, the range of applicability must still be determined for the method's adjustable parameters.

Dielectric Properties of Proteins

The dielectric constant of a protein is difficult to define, yet the concept has provoked widespread interest. Alternate definitions of the dielectric constant have been discussed in a number of recent articles (69, 118, 143, 187, 188). It is frequently assumed that proteins have a dielectric constant of about 4, based on the observed dielectric constant of crystalline polyamides (16). An effective dielectric constant of 5.5 is obtained for the interaction of two sodium ions in the gramicidin channel (182). Since the ions are separated by a single file of water molecules and may be partially screened by surface water, the dielectric constant of the channel-forming protein may be somewhat lower. In keeping with these results a recent theoretical analysis based on the Kirkwood-Fröhlich theory of polar fluids (67, 98) obtained a dielectric constant ranging between 3 and 5, with the precise value depending on the degree of flexibility of the polypeptide backbone (M. K. Gilson, B. Honig, unpublished results).

A number in the range 3–5 appears appropriate for the response of a protein to a weak field in which the protein deforms slightly but undergoes no significant conformational change. Thus, under the assumption that the protonation and deprotonation of ionizable groups near the surface has only small effects on protein structure, the standard use of a dielectric constant of 4 (118) in the calculation of titration curves appears reasonable. In contrast, for any calculation in which the protein's permanent dipoles are accounted for explicitly, the contribution of these dipoles should not be incorporated into the dielectric constant. In this case the high-frequency dielectric constant (due only to electron polarizability) of about 2–3 should be used. As an example, only the high frequency dielectric constant should enter in the calculation of electrostatic interactions between helix dipoles. Of course, all electrostatic interactions will be screened by the solvent as well as by the protein so that the effective dielectric constant for most interacting groups will be much higher than 3. The actual value depends on the separation between the groups, on their distance from the protein surface, and on the shape of the protein-water boundary. For example, two charged groups that are both 2 Å below the surface of a spherical protein but are 7 Å apart experience an effective dielectric constant of about 20 even if the optical dielectric constant is only 2 (see Figure 3; 69).

A number of attempts have been made to extract effective dielectric constants from experimental data. Rees (149) obtained a set of effective dielectric constants from the change in redox potential of the heme iron in ferricytochrome c that results from the chemical modification of surface lysines. Values ranging from 44 to 128 were obtained. Rogers et al (152) determined an effective dielectric of 19.5 for the propionate-iron interaction

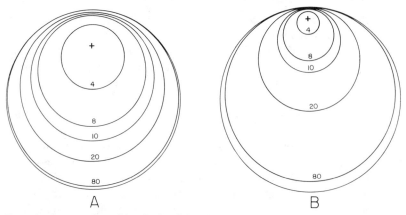

Figure 3   Contour plots of the effective dielectric constant for the interaction of a fixed charge, denoted by +, and a test charge in a spherical protein of radius 20 Å and dielectric constants embedded in water. (*A*) Fixed charge 12 Å from sphere's center. (*B*) Fixed charge 18 Å from sphere's center. (See reference 69 for details of the calculations.)

in cytochrome c_{551}, a value that was fairly well reproduced in calculations based on the Warwicker-Watson algorithm (189). Expressions in which the dielectric constant depends on the distance between interacting charges have been reported (79, 117, 187). The functions were derived to fit interactions that take place on the protein surface; thus the effective dielectric constant tends to be quite high, which reflects, primarily, screening by the solvent rather than the protein. This does not imply that folded proteins themselves have high dielectric constants. In fact the high effective dielectric constant is consistent with the model of a protein as a low dielectric cavity embedded in a high dielectric medium (69, 79, 118, 152; see also Figure 3).

Warshel and co-workers (187, 188) have argued against the description of a protein as a low dielectric cavity. Warshel has correctly emphasized that the self energy of charged groups is frequently ignored in treatments of protein electrostatics. Using a value of about 75 kcal/mol for the transfer energy of an ionized carboxylate or ammonium ion from the gas phase to water (86, 188), the transfer energy of these groups from water to a medium of dielectric constant 4 is calculated from the Born expression to be about 19 kcal/mol, leading to a pK change of about 14 units. Since the pKs of protein groups are generally quite close to those of the isolated amino acids in water, it was argued (188) that the protein cannot be viewed as having a low dielectric constant. It is possible, of course, that ion-solvent interactions are stronger than is indicated by accessibility calculations on crystal structures. In any case, for a completely buried group near the surface the increase in self energy is only half that of a deeply buried group [i.e. about 9 kcal/mol (6 pK units) for a dielectric constant of 4] while a group that is half buried has its self energy increased by only about 4 kcal/mol (3 pK units) (69). From these numbers it is clear that the formation of one or two hydrogen bonds with groups on the protein can compensate for the loss of aqueous solvation. The effect is identical in principle to chelation (see below) and may be viewed as equivalent to the generation of a high local dielectric constant. However this dielectric constant, which is defined in terms of the Born energy, is different from the one that screens charge-charge interactions (69, 118, 188). Thus, the unmodified pKs of ionizable groups are not inconsistent with a low dielectric constant for proteins.

One effect that can, in principle, be used to determine the dielectric constant of proteins is the spectral shift observed when certain chromophores are transferred from various solvents to their binding sites in the interior of proteins. Perhaps the most dramatic shifts are observed for the retinal chromophores of visual pigments and bacteriorhodopsin. These have been attributed to the effects of charged amino acids that are adjacent to the chromophore and that act on it with an extremely low effective dielectric constant (94, 131). The conformation of bacteriorhodopsin is not yet known to sufficiently high resolution to reveal the structural basis of the generation of such large electric fields.

Energetics

INTERACTIONS BETWEEN SURFACE CHARGES   Friend & Gurd (65) and Matthew (118) used a modified Tanford-Kirkwood theory to calculate the net electrostatic free energy resulting from the interaction of charged groups on the surface of a number of proteins. Net stabilization energies on the order of 10 kcal/mol were obtained (see also 89). These results have led to a modification of the traditional picture that assumes that the protein's net charge is uniformly distributed on the protein surface and thus must destabilize the native structure (170). Apparently, charges are situated on the surface in a nonrandom fashion that produces a net stabilization; this conclusion is supported by surveys of known structures (20, 186). While it is not possible to test whether the calculated energies are accurate, they agree reasonably well with experimental evidence for a number of pH-dependent phenomena (89, 120).

We emphasize that negative interaction energies alone do not imply that surface groups enhance protein stability. Some of the groups are only partially accessible to solvent in the native conformation and, as a result, their self energy should be higher (owing to a loss of aqueous solvation) than when in the unfolded state (unless, as discussed above, they form hydrogen bonds with other groups on the protein). One indication that surface charges do in fact contribute to protein stability is the observation that much of the calculated stabilization energy in myoglobin is due to specific and highly conserved ion pairs (66). This conclusion is further supported by the study of Barlow & Thornton (20), who found that ion pairs are conserved in the trypsin and lysozyme families as well. The first direct evidence that a salt-bridge can enhance protein stability was obtained by Fersht (57), who determined a value of 2.9 kcal/mol for the main-chain Ile-16 . . . Asp-194 salt-bridge in chymotrypsin. On the other hand, Hollecker & Creighton (85) found that the stability of a number of proteins was essentially unaffected by large variations in their net charge resulting from succinylation of lysines. However, stability was substantially affected for a number of groups, suggesting that they were involved in specific stabilizing interactions such as salt-bridges. To our knowledge, no attempt to account for these results theoretically has been reported.

Direct evidence on the contribution of surface charges to protein stability is becoming available from the studies of Bierzynski et al (25), Shoemaker et al (165), and Rico et al (151) on ribonuclease fragments. Both groups demonstrated that Glu-2 in the sequence contributes to the stability of the isolated N-terminal helix. The effect was attributed to either an interaction of the carboxylate ion with the helix dipole (165) or to a Glu-2 . . . Arg-10 salt-bridge. It should be emphasized here that salt-bridge formation is affected not only by purely electrostatic factors, but also by a decrease in conformational entropy of the side chains in question, at least a partial loss of aqueous solvation, and a possible gain in hydrophobic energy resulting from the interactions between the hydrocarbon portions of the Glu and Arg residues and hydrophobic segments of the helix. Thus, even if they are stabilizing in some cases, salt-bridges do not always provide a stabilizing interaction; this is indicated by the evidence (165) that a possible Glu-9 . . . His-12 salt-bridge does not contribute to helix stability.

THE HELIX DIPOLE   Theoretical and experimental evidence suggests that α-helices have significant dipole moments (185), which can be represented approximately by the field generated by two opposite charges of magnitude $e/2$ placed at either end of the helix (84, 164). Recent calculations on crystals of an uncharged helical undecapeptide suggest that dipole-dipole interactions provide significant electrostatic stabilization to the crystal energy (82). However, the proposal that the alignment of helix dipoles is a significant factor in determining protein structure (83, 164) is less likely to be correct. Calculations by Rogers & Sternberg (153) have shown that solvent screening diminishes helix-helix interaction energies significantly unless the helix termini are well removed from the solvent. This may occur in special cases, but most helix termini are fairly exposed.

Other proposals for the role of the helix dipole include the attraction and binding of charged substrates, stabilizing interactions with charged residues located near the helix termini, and the enhancement of reaction rates (83). Recently, the helix dipole has been implicated as a stabilizing factor that facilitates the binding of sulfate ions to the sulfate-binding protein of *Salmonella typhimurium* (144). The general applicability of these proposals is difficult to assess. Many of the calculations upon which they are based do not account for the effects of solvent screening and thus lead to a significant overestimate of interaction energies. Moreover, the observations that charged groups tend to appear at the termini of helices could also be due to specific interactions with hydrogen-bonding groups that are available from the peptide backbone at either terminus. Indeed it is sometimes difficult to distinguish the role of the helix dipole from that of the first or last turns of a particular helix. Nevertheless, the helix dipole does appear to be an important structural feature; it is likely to have a mechanistic role in at least some cases and has been clearly implicated as a voltage sensor (see below).

BURIED CHARGES AND CHELATION   The ability of proteins to remove charged groups from water is critical to a variety of biological functions such as catalysis, voltage gating, and ion transport. It is thus of considerable importance to understand the energetics of "burying" ions and ion-pairs. The basic problem, as discussed above, is that proteins must be able to compensate for the loss of the solvation energy of the ion in water. If proteins are simply considered to constitute a low dielectric medium, say of dielectric constant 4, then the Born expression predicts a transfer free energy from water to the protein interior of about 19 kcal/mol for a charged lysine or carboxylate. As is evident from crystal structures, the mechanism used by proteins to reduce the self energy of buried charges is to surround them with polar groups and, in some cases, with charges of the opposite sign. A novel example of the sequestering of an ion is provided by the sulfate-binding protein discussed above (144). The sulfate forms seven hydrogen bonds and may be further stabilized by the dipoles of three helices. The structures of two particularly interesting calcium-binding proteins, troponin-C (77) and calmodulin (15), have also been reported in the past year.

A simple calculation based on the Born equation was recently used to estimate the free energy of ion pairs in membrane proteins (86). Assuming that the protein can be treated as a medium of dielectric constant 4, the cost of completely burying an ion pair in the interior of a protein was predicted to be about 7 kcal/mol. However, using a bulk dielectric constant assumes nothing about the microenvironment of the ion pair. If it is further assumed that the ion pair makes a number of additional hydrogen bonds, as is observed in globular proteins (147), the contribution of the ion pair to the free energy of the protein can be significantly reduced and can even become negative. Of course, the protein must still invest the energy to bury the groups that "chelate" the ion pair (see below), but the transfer energies of polar but uncharged groups from water to organic solvents are quite small (192). A similar argument may be made for isolated ions. As pointed out above, burying a charged lysine in a medium of dielectric constant 4 is estimated to cost about 19 kcal/mol. A protein can clearly reduce this value significantly depending on the number and strength of the hydrogen-bonding interactions it provides. In this regard we note the recent study of Scheiner & Hillenbrand (157) that demonstrated the great sensitivity of hydrogen-bonding interactions to conformational factors.

The removal of ionic species from water by proteins may be viewed as an example of chelation. The underlying principles of this process have been understood for some time (see e.g. 90). Open-chain chelating agents extract ions from water by providing a high local concentration of liganding groups. Jencks suggested that some proteins may act in this way. He also suggested a mechanism in which a charged site was fixed in a protein in a poor solvating environment thus providing a strong driving force for ion binding and charge neutralization. Additional mechanisms have become evident from synthetic molecular receptors that bind a variety of ionic substrates (see e.g. 108). Specific examples include crown ethers (142), cryptands (107), and spherands (45). Crowns, cryptands, and natural ionophores undergo conformational reorganization during complex formation while liganding groups of spherands are organized prior to complex formation. Crowns bind ions more tightly than open-chain analogs as a result of the macrocyclic effect of reduced reorganization entropy during complex formation. The effect may have an enthalpic contribution as well (191). Cryptates have higher binding affinities than crowns owing primarily to the enthalpic effect discussed above (90) involving a strong interaction of cations with weakly solvated polydentate ligands (107). This enthalpic effect apparently compensates for the fact that the bound ions are more accessible to solvent in crowns than in cryptates. In spherands the ligands are essentially fully organized for cation complexation and consequently have much higher binding affinities than cryptands (45). Their large binding affinities are due both to preorganization, which is an entropic effect, and to an enthalpic compensation for electron-electron repulsion in the uncomplexed molecule. A detailed theoretical study of the inclusion of alkali cations into spherands has recently been reported (99).

The ion-binding properties of these chelating molecules have led to the generalization that "the larger the number of host ligating sites organized for binding during synthesis rather than during complexation, the greater the standard free energy change that accompanies complex formation" (45). Whereas the negative solvation free energy of the ion in water contains positive contributions (both entropic and enthalpic) due to organization of the water molecules, these contributions are reduced in chelates where the energy price is "paid" during synthesis. Similarly, ionic species can be buried in the interior of proteins in at least a partially preorganized site, but the driving force for the binding must ultimately derive either from the folding energy of the protein or, in analogy to simple chelation, from the placement of liganding groups in nearby positions along the primary sequence as is typically observed for calcium-binding sites. In such cases an entropic effect due to a high local concentration of ligands accounts for ion binding, and the investment of folding energy is not necessary.

PROTEINS IN MEMBRANES

Ion Carriers and Channels

It is well known that the barriers to ion transport through protein channels can be no more than a few kilocalories per mole (e.g. 78). The problem of accounting for the reduction of the approximately 30 kcal/mol electrostatic barrier posed by the pure lipid bilayer has been of considerable interest.

Qualitatively, the mechanism may be related to the ability of carriers (e.g. $FCCP^-$, DNP^-, valinomycin, and nigericin) to transport small ions across lipid bilayers (for reviews see 80, 103). Carriers basically function as hydrophobic ions and consequently their energetics can be understood as the sum of Born, image, and dipole energies together with an effective neutral energy, as given by Equation 1 above. In the charged state, carriers have effective ionic radii for transport of at least 5 Å, and from Equation 1 the free energy of activation is therefore not more than 12 kcal/mol (see Figure 2). Translocation rates would then be at least 10^4 s^{-1}, and this is observed experimentally (80, 103).

Related estimates have been made for channels under a variety of assumptions. Parsegian (140, 141) first approached the problem by treating the channel as an infinitely long narrow cylinder with an interior dielectric constant comparable to that of water and surrounded by a medium with dielectric constant 2. For a narrow channel (such as gramicidin) of radius ~ 2 Å, a barrier height of about 14 kcal/mol is estimated. This value can be further reduced by considering the finite thickness of membranes (91, 113) although the experimental value for gramicidin of about 3–4 kcal/mol (17) is difficult to reproduce with this simple model. It has been suggested that the problem can be resolved by treating the channel walls as a medium of higher dielectric constant than the surrounding bilayer (91, 113).

The difficulty with this approach is that one knows very little about the dielectric constants of water in the channel or of the surrounding walls. An approach based on a detailed molecular description of the channel would be preferred and, indeed, a number of recent calculations of potential energy profiles in the gramicidin channel have been reported (55, 97, 106, 116, 146). These studies have provided a great deal of useful information on the energetics of ion-peptide interactions, the state of water in the channel, and the flexibility of the polypeptide. They do not, however, directly address the question of the lowering of the activation energy for transport. This problem would require that the energy of the ion in the channel be compared to that of the free ion solvated in water, which, as noted above, has not yet been done in a satisfactory fashion. The problem is rather complex since, as discussed above, it is even difficult to obtain quantitative agreement for the solvation energies of individual ions with detailed simulations.

Despite the complexities associated with obtaining a quantitative description of channel energetics, the problem does not appear to pose any conceptual difficulties. Indeed it is generally agreed that narrow channels such as gramicidin compensate for the loss of aqueous solvation through interactions of the bound ion with the permanent dipoles of the polypeptide. Thus, the energetics of transport are closely related to the energetics of chelation, for which the principles were discussed above.

Voltage-Dependent Conformational States

In the native state, it is not unusual for a biological membrane to have a transmembrane potential of the order of 100 mV, leading to internal electric fields of up to 10^5 V/cm. Membrane proteins are thus exposed to intense electric fields, and many systems are now known in which the conformations are functionally coupled to the electric field for the purpose of switching (100), transport (93), energy transduction (167), or polypeptide insertion into the membrane (47). Voltage-dependent conformational changes involve the motion of a charge along a direction normal to the membrane surface or a change in the component of the dipole moment along this direction, and give rise to displacement currents in the membrane.

Displacement currents associated with the opening of voltage-dependent ionic channels in excitable membranes have been detected and referred to as "gating" currents (3, 10, 11, 64). Although they have so far only been observed in excitable membranes, displacement currents accompanying conformational changes are expected to be common and may be referred to in general as "conformational displacement currents."

In the following paragraphs, information on the molecular nature of electrically active conformational changes is reviewed, with particular emphasis on the chemical nature of the voltage sensor. The scope is necessarily narrow since virtually all of the data at this level have come from "simple" model systems and all of these channels are voltage-dependent (switches). Unfortunately, nothing is known regarding either the three-dimensional structure or the identity of the voltage-sensitive elements in any native, electrically active membrane protein. Development of working hypotheses may be guided by information from the model systems.

Figure 4 shows some general models of electrically active conformational changes that might be expected to occur in membrane proteins. A proper thermodynamic analysis would treat them in a unified way in terms of their dipole moment of transition (112). However, mechanistic details are of interest here and each model is discussed separately since they are very different in terms of protein conformational changes.

NET CHARGE DISPLACEMENT   Figure 4A illustrates net displacement of a protein fixed charge along a direction perpendicular to the bilayer concomitant with a conformational change. Consider the change to represent a switch that triggers a biological process of some sort and let the conformation a be the "off" state and b be the "on" state. For illustration, the "off" state is taken as the most stable state in the absence of intramembrane electric fields, although other situations could equally well obtain. For this simple two-state model, the fraction of the molecules in the "on" state, ξ, as a function of membrane potential is given by (54)

$$\xi = 1/[1 + e^{(\Delta W_{ab} + ZF\Psi_{ab})/RT}],$$

where ΔW_{ab} is the conformational free energy difference between the two protein states in the absence of electric fields, $\Delta \Psi_{ab}$ is the difference in electric potential between the locations of the charge in the two states, and Z is the equivalent valence of the displacement charge. For low noise, ξ should be small in the resting state (i.e. at $\Delta\Psi = 0$). As an example, the acetylcholine receptor has 10^{-7} or less of the channels open at any instant in the absence of acetylcholine (Jackson, personal communication). Even for a modest requirement of $\xi = 10^{-4}$ at $\Delta\Psi_{ab} = 0$, $\Delta W_{ab} \approx 6$ kcal/mol. This system will switch ($\xi = 0.5$) when the electrical energy is equal in magnitude and opposite in sign to the conformational energy of 6 kcal/mol. If the system is to switch at the reasonable value of 50 mV and the displacement charges are allowed to move the full thickness of the membrane, $Z \simeq 5$, i.e.

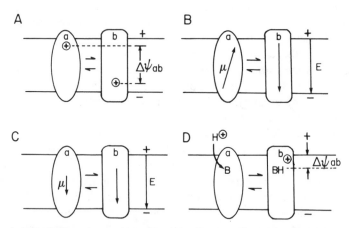

Figure 4   Electrically active conformational changes. In each case, a and b represent two conformations of a membrane protein. The conformations have an intrinsic energy difference, schematically represented by the different shapes. In the presence of an intramembrane electric field, they have additional energy differences due to interactions with the field. In (A), the electrical energy difference is due to the net displacement of bound charge; in (B), to dipole reorientation; in (C) to an induced structural dipole moment, and in (D) to an interfacial charge migration. In this latter case a proton is shown moving to an acceptor protein, but the charge could be any chemical species with a binding site in the protein, or a portion of the protein itself that migrates from the aqueous phase to the membrane or protein interior.

five charges must move. Even more must move if the displacement does not cover the full thickness of the membrane. Involvement of rather large electrical and conformational energies (compared to thermal energies) also insures a conductance that is a steep function of the potential, giving a sharp rather than graded change. The conformational energies involved are of the order of a few hydrogen bonds and are roughly comparable to the folding energies of globular proteins, which are typically in the range of 10–20 kcal/mol (145).

An example of a system that appears to employ fixed charges to operate a voltage-dependent switch is the polyene, antibiotic monazomycin. Although this is not a protein, it provides a relatively well-established example of this mechanism. The small molecule (MW $\approx$ 1400) is elongated in shape ($\approx$ 40 Å long); one surface bears hydrophilic hydroxyl groups and the opposing surface is entirely hydrophobic. A single positive charge contributed by a protonated amino group occurs at one end of the molecule (132). A *trans* negative membrane potential leads to the formation of oligomeric transmembrane channels composed of 5–6 monazomycin monomers arranged like the staves of a barrel with the hydrophilic surface facing the barrel interior and the hydrophobic surface facing the lipid bilayer. The mechanism of channel formation appears to proceed by a voltage-dependent insertion of small oligomers into the bilayer followed by aggregation into channels (8, 130). The channels thus formed are in equilibrium with a nonconducting oligomer pool.

It seems likely that monazomycin employs the ammonium group as a gating charge. This is rather curious, considering the enormous energy expenditure required to bury an ammonium ion in the low dielectric environment of the membrane interior. As a result, the kinetics of gating would be expected to be extremely slow. Muller & Andersen (130) have provided an explanation for this situation. It appears that monazomycin enters the membrane in a voltage-dependent fashion only after some fraction of the monazomycin has inserted in the bilayer. That is, the insertion step is autocatalytic. To "seed" the bilayer with an initial dose, very high potentials must be momentarily applied. Further monazomycin addition then proceeds under more moderate voltage. The displacement charges of the entering molecules are facilitated in their entry by the resident monazomycins, perhaps by a mechanism in which the ammonium ions move through a partially formed channel. Thus, monazomycin gating may involve the movement of the gating charge through its own nascent channel.

We emphasize that channels in biological membranes are formed from single polypeptides or very stable subunits and do not operate on the principal of monomer (or oligomer) aggregation (78). However, one could conceive of channel formation based on the movement of mobile subunits linked together at one surface by a flexible hinge (179).

There is little available evidence other than monazomycin to suggest that simple charged groups are employed in nature as voltage sensors. One further possibility is found in the F_0 segment of the ATPase complexes of energy-transducing systems. The very hydrophobic "c" subunit of the F_0 proton-conducting complex contains a single essential aspartic acid residue in the center of a very hydrophobic stretch of polypeptide that traverses the membrane (88). It is generally agreed that this carboxyl group faces the hydrophobic domain of the phospholipid bilayer, but it is not known to what extent it is unprotonated. This group is selectively blocked by hydrophobic carbodiimides (87, 88). Modification of the carboxyl group with a spin-labelled carbodiimide analog gives rise to a completely immobilized spectrum, indicating that the carboxyl group is in a very restricted environment (possibly chelated by protein groups) and is not free to move within the boundary layer of the protein (166). Since the membrane bound ATPase complex employs the transmembrane electrochemical potential to produce ATP, it is possible that these carboxyl groups are potential sensors (29).

If native membrane proteins employ the displacement of net charge to sense changes in potential, the charges are likely to be carboxylate and/or ammonium groups, since recent data give no indication of unusual charged amino acid side chains in electrically active proteins (136).

Whatever the chemical identity of the charge, it must be located within the region of electric potential drop across the membrane so that it can sense the potential change. A simple charged ammonium (or carboxylate) group has too high an energy relative to the deprotonated (or protonated) form to exist at equilibrium within the low dielectric environment of the bilayer. This problem could be overcome by chelation of the charge by protein groups as discussed above, perhaps by groups on the walls of the channel itself (58).

In considering relative merits of the ammonium versus carboxyl groups as displacement charges, it should be noted that the carboxyl group, being anionic, has an advantage over the ammonium since in crossing the bilayer it experiences a lower activation energy (by about 10 kcal/mol) due to the membrane dipole potential (61). This would only be a factor if the displacement charge crossed the dipolar region (Figure 1).

As pointed out above, reliable gating systems that are to switch from "off" to "on" with a steep function of potential must operate by moving the equivalent of several charges the full thickness of the membrane. Gross movement of the protein to accomplish the gating is generally considered an unattractive idea, since conformational changes known so far tend to be rather small in terms of atomic displacement. There is little direct experimental evidence that bears directly on this matter. The apparent activation volume of the gating process in the sodium channel of the squid giant axon is quite small (44). However, by itself this provides little information on the extent of conformational changes, since in multicomponent systems opposing volume changes in different components can cancel.

Several models have been proposed to provide for minimal conformational movements of the channel protein during gating. The most extreme of these proposals suggests that the gating charges do not move at all (43). However, this could not give rise to potential-dependent conformational changes since no electrical work could be done in the conformational change.

Finkelstein & Peskin (58) have proposed a model in which the entire membrane potential drops across a thin physical gate in the channel. Since the gate is only a fraction of the membrane thickness, the gating charges only have to move a small distance to gain the full electrical energy.

Armstrong (10) has proposed a novel model in which two helical segments bearing opposite charges on their surfaces are in contact and form interhelical salt-bridges. If the helices are arranged with their long axes parallel to the membrane perpendicular, an applied potential may slide one with respect to the other. A movement of only a few angstroms then leads to a net displacement of an entire charge through the full thickness of the bilayer.

DIPOLE REORIENTATION   Figure 4B shows a simple two-state conformational manifold with states that differ in free energy due to the rotation of a dipole moment in the presence of an intramembrane electric field. If, as above, we consider the b state to be "on" and the a state to be "off," the fraction of the molecules in the "on" state, ξ, is given by

$$\xi = 1/[1 + e^{(\Delta W_{ab} + \Delta \mu E)/RT}],$$

where $\Delta \mu$ is the change in the component of the dipole moment along the direction of the electric field due to the reorientation, ΔW_{ab} is the difference in the free energy between the a and b states in the absence of the electric field, and E is the (constant) intramembrane field. The most likely candidate for a dipole element is the α-helix with a dipole moment of about 3.5 Debye (D) per residue.

The same arguments regarding channel noise as made above in the discussion of monopole displacement charge can be applied here. Thus, in order to have $\approx 10^{-4}$ of the channels open in the resting state ($E = 0$ in this example), $\Delta W_{ab} \approx 6$ kcal/mol. For the switching potential to be 50 mV (across 40 Å of membrane interior), $\Delta\mu \approx 100$ D. This is indeed a very large change, corresponding to a 90° rotation of no less than 10 transmembrane α-helices. Such large movements would be very sensitive to bilayer fluidity and should be relatively easy to detect in proteins made of helical bundles.

A well-characterized model system gated by the dipole reorientation mechanism is provided by the antibiotic alamethicin, which forms voltage-dependent ion channels in membranes (27, 73). Alamethicin is a 20-peptide, linear protein for which the crystal (63) and certain solution structures are known (18, 27). The secondary structure in organic solvents is α-helical between proline-2 and proline-14. This segment is also completely hydrophobic and alone should give rise to a dipole moment of about 40 D. The C-terminal end of the molecule is considerably more polar and adopts a nonhelical configuration.

The generally accepted mechanism for the voltage-dependent channel formation is the reorientation of alamethicin dipoles in the membrane field followed by aggregation into oligomeric channels. However, there is not unanimous agreement on the details of the voltage-dependent process. Boheim et al (27) propose that alamethicin forms aggregates in the membrane at zero potential due to the dipole-dipole attraction of antiparallel helices in the low dielectric interior of the membrane. This requires that populations of alamethicin interact from opposite sides of the membrane. Application of a potential across the membrane induces a flip-flop of helices (i.e. entire alamethicin molecules) to a parallel arrangement, which is proposed to lead to the opening of a water-filled pore.

In the detailed model offered by Hall and co-workers (73), the 3-turn helical segment at the amino-terminal end of the molecule undergoes an approximately 180° rotation about proline-14 under an applied electric field. This produces a $\Delta\mu = 80$ D. Since approximately 10 monomers form the conductive channel, the total change in dipole moment is ≈ 800 D, in the appropriate range to yield reliable switching characteristics with a steep voltage dependence.

Jung et al (92) have synthesized a number of peptides of the series Boc-(L-Ala-Aib-Ala-Aib-Ala)$_n$-OMe. The structures with $n = 2$–4 showed voltage-dependent single channel conductance fluctuations. All of these structures are hydrophobic and form α-helices in organic solvents. Thus the conclusion was reached that only pure hydrophobic α-helices are required for voltage-dependent channel formation. The mechanism was suggested to be similar to that of alamethicin, i.e. a dipole flip-flop with the final channel formed by a bundle of α-helices.

A curious feature about the models for alamethicin and the synthetic peptides is that the channel itself is lined by the side chains of helices that are completely hydrophobic. It is difficult to imagine a hydrophilic pore lined with hydrophobic residues. These molecules are all α-helical in low dielectric medium, but their structure has not been confirmed in the bilayer under the influence of strong electric fields. It is conceivable that the structure of the pore formed under such conditions may be other than an α-helical bundle (181).

Melittin is another small peptide that apparently exhibits voltage-dependent channel formation in lipid bilayers. The peptide is a naturally occurring component of bee venom with 26 amino acids. A recent report of the crystal structure of the melittin tetramer (172) reveals that the molecule has two helical domains separated by a hinge region at proline-14. The C-terminal segment is highly polar with four positive charges, while the N-terminal domain is relatively nonpolar except for the presence of lysine-7.

Boheim et al (27) have emphasized the structural and functional similarity of melittin to alamethicin, i.e. melittin has a relatively hydrophobic N-terminal domain, a helix-breaking proline at position 14, a polar C-terminal domain, and formation of similar voltage-dependent channels in membranes. On the basis of these similarities, Boheim et al suggest that the mechanism of voltage dependence is also similar, i.e. helix reorientation in an electric field. Even though melittin and alamethicin have a number of structural features in common, there is one outstanding difference. The melittin N-terminal helical domain is clearly amphiphilic, as opposed to those in alamethicin and the hydrophobic peptides mentioned above. Indeed it is much easier to understand pore or channel formation by melittin because of this property.

Kempf et al (95) and Tosteson & Tosteson (177) suggest that the lysine-7 in the hydrophobic N-terminal helix acts as a positive gating charge, but this idea is untenable considering that acylation of all but two C-terminal arginines leaves the channel-forming properties generally intact (27). Under any condition, the Born energy of removing simple ammonium groups from water in the gating process would be too high to give reasonable kinetics for channel formation. Nevertheless lysine-7 must enter the membrane interior at some point, either in the neutral form or chelated by other peptide groups, to lower the electrostatic energy. It is interesting to note that in the melittin sequence two threonine residues are sufficiently close to form good hydrogen bonds with the lysine ammonium group. This would stabilize the charge to a considerable degree in the low dielectric interior of the membrane.

The extent to which native systems use the fixed-helix dipole reorientation mechanism to couple to transmembrane fields is completely unknown. If the mechanism is used, the reorienting helical segments are probably part of one molecule or a very stable subunit collection rather than independent units aggregating as in alamethicin. If some voltage-gated channels are indeed single molecules, as seems likely for the sodium channel from the electroplax (2), a structurally acceptable means of returning the polypeptide across the membrane without an opposing dipole moment must be found. This is necessary because all of the helices must be oriented in the same direction. Urry has suggested the "helical rack" model (181) in which the returning polypeptide chain is in the β configuration and actually forms one section of a β-barrel channel lining.

INDUCED STRUCTURAL DIPOLE MECHANISMS   Certain conformational changes in polypeptides occur with a substantial change in dipole moment, as illustrated in Figure 4C. A well-known example is the helix-coil transition studied by Schwarz (158). The formal energy analysis of this class of conformational change in the presence of an electric field is identical to that given above for the dipole reorientation. They are distinguished here under different sections since they involve quite different structural changes at the level of the polypeptide chain.

Urry (181) has proposed an interesting conformational transition that involves a large change in structural dipole moment and that leads to possible channel modulation. This is the so-called "β-spiral to β-helix" transition. The β-spiral configuration has a dipole moment of $\mu = -1.8$ D/Å and an occluded channel while the β-helix has a dipole moment of $\mu = 1.4$ and a channel sufficiently large to accommodate small ions. Thus a change in conformation sufficient to open a channel with a change in dipole moment of 3.2 D/Å is possible. For a channel length equal to the thickness of the membrane (about 30 Å) the change in dipole moment is nearly 100 D, sufficient to give an energy difference of ≈ 1 kcal/mol for transmembrane potentials of around 50 mV. One difficulty with this model for channels like the sodium channel of neuronal membranes is that 1 kcal/mol is not large enough to give a low channel conductance in the resting state with a steep voltage dependence in the physiological range of potentials. For sufficient energy, the gating event of several channels would have to be coupled.

INTERFACIAL CHARGE TRANSPORT   Figure 4D shows a conformational change with obligatory interfacial charge transport. The energy analysis of this case is similar to that for the movement of net charge within the membrane (see above). This model is considered specifically since at least two possible examples have been described.

The membrane protein rhodopsin undergoes a well-documented series of conformational changes initiated by the absorption of a photon of visible light by its covalently attached chromophore, 11-cis retinal. The conformational transition between the intermediates Meta I and Meta II takes place with the uptake of a proton from the aqueous solution. This event leads to a measurable change in the electrical properties of the membrane-solution interface (21, 32, 33, 114, 178). Cafiso & Hubbell (32, 33) demonstrated that the potential in the boundary region of the membrane was increased coincident with the proton uptake, giving rise to an apparent conformational displacement current. The results are consistent with a simple interfacial proton transport although the displacement current could arise as a result of a coupled conformational change. This implies that the two conformations should be in a potential-dependent equilibrium, although this has not yet been studied directly.

Bacteriorhodopsin shows apparently similar effects (34, 51, 173). In this system, the rate of decay of the M-intermediate in the photocycle is potential-dependent (71), although the origin of the voltage dependence is unknown. Tsuji & Neumann (180) have observed voltage-dependent changes in bacteriorhodopsin at moderate fields, which they attribute to the interaction of induced dipoles with the field.

FUTURE DIRECTIONS   The theoretical predictions of the models presented above or elaborations thereof are quite compatible with experimental data for a number of systems (53, 128, 159, 160). The task is now to determine which of the models is a realistic description of the actual gating process.

Several very interesting and important systems have now been isolated and/or their biochemistry is sufficiently understood to make them outstanding candidates for physical and chemical analyses directed at elucidating the voltage-dependent mechanisms. These include (a) the voltage-dependent sodium channel, which has now been isolated in reasonable quantities (1, 19, 74), been reconstituted in membranes (154, 169, 190), and had its complete sequence determined (136); (b) the voltage-dependent K^+ channel from sarcoplasmic reticulum, which has been extensively characterized (128) and appears to obey the simple two-state model discussed above (see, however, 62); (c) colicins of the type E_1, which display voltage-dependent insertion and conductance increases in membranes (41, 162); and (d) the procoat protein from the filamentous phage M13, which shows voltage-dependent insertion in bacteria (47).

In the near future many or all of these systems will be reconstituted into lipid vesicles in sufficiently high concentration for structural investigations by spectroscopic approaches. If procedures are established for maintaining stable potentials across such vesicles, it will be possible to directly test some of the mechanisms discussed above, particularly those invoking dipole reorientation or induction.

ACKNOWLEDGMENTS

The authors would like to thank Drs. B. Wallace, P. Boyer, M. Gilson, R. Fine, and A. Rashin for informative discussions, and Drs. A. Finkelstein, J. Fox, M. Gilson, J. Trudell, and M. Jackson for critical readings of the manuscript. Work reported by the authors was supported by NIH grants GM30518 to BHH and EY5216 to WLH, NSF grant PCM82-07145 to BHH, and the Jules Stein Professor Endowment to WLH.

Literature Cited

1. Agnew, W. S., Levinson, S. R., Brabson, J. S., Raftery, M. A. 1978. *Proc. Natl. Acad. Sci. USA* 75:2606
2. Agnew, W. S. 1984. *Ann. Rev. Physiol.* 46:517
3. Almers, W. 1978. *Rev. Physiol. Biochem. Pharmacol.* 82:96
4. Alvarez, O., Brodwick, M., Latorre, R., McLaughlin, A., McLaughlin, S., Szabo, G. 1983. *Biophys. J.* 44:333
5. Andersen, O. S. 1978. *Membrane Transport in Biology*, Vol. 1, ed. D. C. Tosteson, p. 369. New York: Springer-Verlag
6. Andersen, O. S., Feldberg, F., Nakadomari, H., Levy, S., McLaughlin, S. 1978. *Membrane Transport Processes*, Vol. 2, eds. D. C. Tosteson, Yu. A. Ovchinnikov, R. Latorre, p. 325. New York: Raven
7. Andersen, O. S., Feldberg, F., Nakadomari, H., Levy, S., McLaughlin, S. 1978. *Biophys. J.* 21:35
8. Andersen, O. S., Muller, R. U. 1982. *J. Gen. Physiol.* 80:403
9. Anderson, C. F., Record, M. T. Jr. 1982. *Ann. Rev. Biophys. Bioeng.* 33:191
10. Armstrong, C. M. 1981. *Physiol. Rev.* 61:644
11. Armstrong, C. M., Bezanilla, F. 1973. *Nature* 242:459
12. Ashcroft, R. G., Coster, H. G. L., Laver, D. R., Smith, J. R. 1983. *Biochim. Biophys. Acta* 730:231
13. Ashcroft, R. G., Coster, H. G. L., Smith, J. R. 1981. *Biochim. Biophys. Acta* 643:191
14. Aveyard, R., Haydon, D. A. 1973. *An Introduction to the Principles of Surface Chemistry.* Cambridge: Cambridge Univ. Press
15. Babu, Y. S., Sack, J. S., Greenhough, T. J., Bugg, C. E., Means, A. R., Cook, W. J. 1985. *Nature* 315:37
16. Baker, W. O., Yager, W. A. 1942. *J. Am. Chem. Soc.* 64:2171
17. Bamberg, E., Läuger, P. 1974. *Biochim. Biophys. Acta* 367:127
18. Banerjee, U., Tsui, F., Balasubramanian, T., Marshall, G., Chan, S. 1983. *J. Mol. Biol.* 165:757
19. Barchi, R. L. 1983. *J. Neurochem.* 40:1377
20. Barlow, D. J., Thornton, J. M. 1983. *J. Mol. Biol.* 168:867
21. Bennett, N., Michel-Villaz, M., Dupont, Y. 1980. *Eur. J. Biochem.* 111:105
22. Bezanilla, F. 1985. *J. Membr. Biol.* In press
23. Berkowitz, M., Karim, O. A., McCammon, J. A., Rossky, P. J. 1984. *Chem. Phys. Lett.* 105:577
24. Bhuiyan, L. B., Blum, L. 1983. *J. Chem. Phys.* 78:442
25. Bierzynski, A., Kim, P. S., Baldwin, R. L. 1982. *Proc. Natl. Acad. Sci. USA* 79:2470
26. Bockris, J. O'M., Reddy, A. K. N. 1970. *Modern Electrochemistry*, Vol. 1. New York: Plenum
27. Boheim, G., Hanke, W., Jung, G. 1983. *Biophys. Struct. Mech.* 9:181
28. Born, M. 1920. *Z. Phys.* 1:45
29. Boyer, P. D. 1984. In H^+-ATPase (ATP Synthase): Structure, Function, Biogenesis, The $F_0 F_1$ Complex of Coupling Membranes, ed. S. Papa, K. Altendorf, L. Ernster, L. Packer, pp. 329–38. Bari, Italy: Adriatica
30. Brown, R. H. Jr. 1974. *Prog. Biophys. Mol. Biol.* 26:343
31. Buff, F. P., Goel, N. S. 1969. *J. Chem. Phys.* 51:4983
32. Cafiso, D. S., Hubbell, W. L. 1980. *Biophys. J.* 30:243
33. Cafiso, D. S., Hubbell, W. L. 1980. *Photochem. Photobiol.* 32:461
34. Carmeli, C., Quintanilha, A. T., Packer, L. 1980. *Proc. Natl. Acad. Sci. USA* 77:4707
35. Carnie, S., McLaughlin, S. 1983. *Eur. J. Biochem.* 117:483
36. Cevc, G., Svetina, S., Zeks, B. 1981. *J. Phys. Chem.* 85:1762
37. Cevc, G., Watts, A., Marsh, D. 1981. *Biochemistry* 20:4955
38. Chance, B., Crofts, A. R., Nishimura, M., Price, B. 1970. *Eur. J. Biochem.* 13:364
39. Chandrasekhar, J., Spellmeyer, D. C., Jorgensen, W. L. 1984. *J. Am. Chem. Soc.* 106:903
40. Chapman, D. L. 1913. *Phil. Mag.* 25:475
41. Cleveland, M. B., Slatin, S., Finkelstein, A., Levinthal, C. 1983. *Proc. Natl. Acad. Sci. USA* 80:3706

42. Cole, K. S. 1969. *Biophys. J.* 9:465
43. Colombini, M. 1984. *J. Theor. Biol.* 110:559
44. Conti, F., Inoue, I., Kukita, F., Stuhmer, W. 1984. *Eur. Biophys. J.* 11:137
45. Cram, D. J., Lein, G. M., Kaneda, T., Helgeson, R. C., Knobler, C. B., et al. 1981. *J. Am. Chem. Soc.* 103:6228
46. Curtis, A. S. G. 1967. *The Cell Surface: Its Molecular Role in Morphogenesis.* New York: Academic
47. Date, T., Goodman, J. M., Wickner, W. T. 1980. *Proc. Natl. Acad. Sci. USA* 77:4669
48. Diamond, J. M., Wright, E. M. 1969. *Ann. Rev. Physiol.* 31:581
49. Dilger, J. P., Fisher, L. R., Haydon, D. A. 1982. *Chem. Phys. Lipids* 30:159
50. Drachev, L. A., Kalamkarov, G. R., Kaulen, A. D., Ostrovsky, M. A., Skulachev, V. P. 1981. *Eur. J. Biochem.* 117:471
51. Drachev, L. A., Kaulen, A. D., Khitrina, L. V., Skulachev, V. P. 1981. *Eur. J. Biochem.* 117:461
52. Drachev, L. A., Semenov, A. Yu., Skulachev, V. P., Smirnova, I. A., Chamorovsky, S. K., et al. 1981. *Eur. J. Biochem.* 117:483
53. Ehrenstein, G., Lecar, H. 1977. *Q. Rev. Biophys.* 10:1
54. Eisenberg, M., Gresalfi, T., Riccio, T., McLaughlin, S. 1979. *Biochemistry* 18:5213
55. Etchebest, C., Ranganathan, S., Pullman, A. 1984. *FEBS Lett.* 173:301
56. Fernandez, M. S., Fromherz, P. 1977. *J. Phys. Chem.* 81:1755
57. Fersht, A. R. 1972. *J. Mol. Biol.* 64:497
58. Finkelstein, A., Peskin, C. S. 1984. *Biophys. J.* 46:549
59. Flewelling, R. F. 1984. *Hydrophobic ion interactions with membranes: thermodynamic analysis and applications to the study of membrane electrical phenomena.* PhD thesis. Univ. Calif., Berkeley
60. Flewelling, R. F., Hubbell, W. L. 1986. *Biophys. J.* 49: In press
61. Flewelling, R. F., Hubbell, W. L. 1986. *Biophys. J.* 49: In press
62. Fox, J. 1985. *Biophys. J.* 47:573
63. Fox, R. O., Richards, F. M. 1982. *Nature* 300:325
64. French, R. J., Horn, R. 1983. *Ann. Rev. Biophys. Bioeng.* 12:319
65. Friend, S. H., Gurd, F. R. N. 1979. *Biochemistry* 18:4614
66. Friend, S. H., Gurd, F. R. N. 1979. *Biochemistry* 18:4620
67. Fröhlich, H. 1958. *Theory of Dielectrics.* London: Oxford Univ. Press
68. Gilbert, D. L., Ehrenstein, G. 1984.

Curr. Top. Membr. Transp. 22:407
69. Gilson, M. K., Rashin, A., Fine, R., Honig, B. 1985. *J. Mol. Biol.* 183:503
70. Gouy, M. 1910. *J. Phys. Paris* 9:457
71. Groma, G. I., Helgerson, S. L., Wolber, P. K., Beece, D., Dancshazy, Zs., et al. 1984. *Biophys. J.* 45:985
72. Hagler, A., Moult, J. 1978. *Nature* 272:222
73. Hall, J. E., Vodyanoy, I., Balasubramanian, T. M., Marshall, G. R. 1984. *Biophys. J.* 45:233
74. Hartshorne, R. P., Catterall, W. A. 1981. *Proc. Natl. Acad. Sci. USA* 78:4620
75. Haydon, D. A. 1964. *Recent Prog. Surf. Sci.* 1:94
76. Haydon, D. A., Hladky, S. B. 1972. *Q. Rev. Biophys.* 5:187
77. Herzberg, O., James, M. N. G. 1985. *Nature* 313:653
78. Hille, B. 1984. *Ionic Channels of Excitable Membranes.* Sunderland, Mass: Sinauer
79. Hingerty, B. E., Ritchie, R. H., Ferrell, T. L., Turner, J. E. 1985. *Biopolymers* 24:427
80. Hladky, S. B. 1979. *Curr. Top. Membr. Transp.* 12:53
81. Hodgkin, A. L., Katz, B. 1949. *J. Physiol.* 108:37
82. Hol, W. G. J., de Maeyer, M. C. H. 1985. *Biopolymers* 23:809
83. Hol, W. G. J., Halie, L. M., Sander, C. 1981. *Nature* 294:532
84. Hol, W. G. J., van Duijnen, P. T., Berendsen, H. J. C. 1978. *Nature* 273:443
85. Hollecker, M., Creighton, T. E. 1982. *Biochim. Biophys. Acta* 701:395
86. Honig, B. H., Hubbell, W. L. 1984. *Proc. Natl. Acad. Sci. USA* 81:5412
87. Hoppe, J., Sebald, W. 1981. *Chemiosmotic Proton Circuits in Biological Membranes*, p. 449. Reading, Mass: Addison-Wesley
88. Hoppe, J., Sebald, W. 1984. *Biochim. Biophys. Acta* 768:1
89. Iomoto, T. 1984. *Biophys. J.* 44:293
90. Jencks, W. P. 1969. *Catalysis in Chemistry and Biology.* New York: McGraw-Hill
91. Jordan, P. C. 1982. *Biophys. J.* 39:157
92. Jung, G., Katz, E., Schmitt, H., Voges, K.-P., Menestrina, G., Boheim, G. 1983. *Studies in Physical and Theoretical Chemistry*, Vol. 24, ed. G. Spach, p. 359. Amsterdam: Elsevier
93. Kaback, H. R. 1981. *Chemiosmotic Proton Circuits in Biological Membranes*, ed. V. P. Skulachev, P. C. Hinkle, p. 525. Reading, Mass: Addison-Wesley

94. Kakitani, H., Kakitani, T., Rodman, H., Honig, B. 1985. *Photochem. Photobiol.* 41:471
95. Kempf, C., Klausner, R. D., Weinstein, J. N., van Renswoude, J., Pincus, M., Blumenthal, R. 1982. *J. Biol. Chem.* 257:2469
96. Ketterer, B., Neumcke, B., Läuger, P. 1971. *J. Membr. Biol.* 5:225
97. Kim, S. K., Nguyen, H. L., Swaminathan, P. K., Clementi, E. 1985. *J. Phys. Chem.* 89:2870
98. Kirkwood, J. G. 1939. *J. Chem. Phys.* 7:911
99. Kollman, P. A., Wipff, G., Singh, U. C. 1985. *J. Am. Chem. Soc.* 107:2212
100. Latorre, R., Alvarez, O., Cecchi, X., Vergara, C. 1985. *Ann. Rev. Biophys. Biophys. Chem.* 14:79
101. Latorre, R., Hall, J. E. 1978. *Membrane Transport Processes*, Vol. 2, ed. D. C. Tosteson, Yu. A. Ovchinnikov, R. Latorre, p. 313. New York: Raven
102. Lau, A., McLaughlin, A., McLaughlin, S. 1981. *Biochim. Biophys. Acta* 645:279
103. Läuger, P., Benz, R., Stark, G., Bamberg, E., Jordan, P. C., Fahr, A., Brock, W. 1981. *Q. Rev. Biophys.* 14:513
104. Läuger, P., Neumcke, B. 1973. *Membranes: A Series of Advances*, Vol. 2, ed. G. Eisenman, pp. 1–59. New York: Marcel Dekker
105. Lee, C. Y., McCammon, J. A., Rossky, P. J. 1984. *J. Chem. Phys.* 80:4448
106. Lee, W. K., Jordan, P. C. 1984. *Biophys. J.* 46:805
107. Lehn, J.-M. 1978. *Acc. Chem. Res.* 11:49
108. Lehn, J.-M. 1985. *Science* 227:849
109. Lelkes, P. I., Miller, I. R. 1980. *J. Membr. Biol.* 52:1
110. Leo, A., Hansch, C., Elkins, D. 1971. *Chem. Rev.* 71:525
111. Leo, A., Hansch, C., Jow, P. Y. C. 1976. *J. Med. Chem.* 19:611
112. Levitan, E., Palti, Y. 1983. *J. Theor. Biol.* 100:107
113. Levitt, D. G. 1978. *Biophys. J.* 22:209
114. Lindau, M., Ruppel, H. 1985. *Photochem. Photobiophys.* 9:43
115. Deleted in proof
116. Mackay, D., Berens, P., Wilson, K., Hagler, A. 1984. *Biophys. J.* 46:229
117. Mahler, E. L., Eichele, G. 1984. *Biochemistry* 23:3887
118. Matthew, J. B. 1985. *Ann. Rev. Biophys. Biophys. Chem.* 14:387
119. Matthew, J. B., Hanania, G. I. H., Gurd, F. R. N. 1979. *Biochemistry* 18:1919
120. Matthew, J. B., Richards, F. M. 1982. *Biochemistry* 21:4989
121. McDaniel, R. V., McLaughlin, A., Winiski, A. P., Eisenberg, M., McLaughlin, S. 1984. *Biochemistry* 23:4618
122. McDaniel, R. V., Sharp, K., Brooks, D., McLaughlin, A. C., Winiski, A. P., et al. 1986. *Biophys. J.* In press
123. McLaughlin, S. 1977. *Curr. Top. Membr. Transp.* 9:71
124. McLaughlin, S. 1983. *Physical Chemistry of Transmembrane Ion Motions*, ed. G. Spach, p. 69. New York: Elsevier
125. McLaughlin, S., Mulrine, N., Gresalfi, T., Vaio, G., McLaughlin, A. 1981. *J. Gen. Physiol.* 77:445
126. Mezei, M., Beveridge, D. L. 1981. *J. Chem. Phys.* 74:6902
127. Mille, M., Vanderkooi, G. 1977. *J. Colloid Interface Sci.* 59:211
128. Miller, C., Bell, J. E., Garcia, A. M. 1984. *Curr. Top. Membr. Transp.* 21:99
129. Moult, J., Sussman, F., James, M. N. G. 1985. *J. Mol. Biol.* 182:555
130. Muller, R. U., Andersen, O. S. 1982. *J. Gen. Physiol.* 80:427
131. Nakanishi, K., Balogh-Nair, V., Arnaboldi, M., Tsujimoto, K., Honig, B. 1980. *J. Am. Chem. Soc.* 102:7945
132. Nakayama, H., Furihata, K., Seta, H., Otake, N. 1981. *Tetrahedron Lett.* 22:5217
133. Nelson, A. P., McQuarrie, D. A. 1975. *J. Theor. Biol.* 55:13
134. Nemethy, G., Peer, W. J., Scheraga, H. A. 1981. *Ann. Rev. Biophys. Bioeng.* 10:459
135. Neumcke, B., Läuger, P. 1969. *Biophys. J.* 9:1160
136. Noda, M., Shimizu, S., Tanabe, T., Takai, T., Kayano, T., et al. 1984. *Nature* 312:121
137. Ohshima, H., Healy, T. W., White, L. R. 1982. *J. Colloid Interface Sci.* 90:17
138. Ohshima, H., Ohki, S. 1985. *Biophys. J.* 47:673
139. Paltauf, F., Hauser, H., Phillips, M. C. 1971. *Biochim. Biophys. Acta* 249:539
140. Parsegian, V. A. 1969. *Nature* 221:844
141. Parsegian, V. A. 1975. *Ann. NY Acad. Sci.* 264:161
142. Pedersen, C. J., Frensdorff, H. K. 1978. *Angew. Chem. Int. Ed. Engl.* 11:16
143. Pethig, R. 1979. *Dielectric and Electronic Behavior of Biological Materials*, p. 63. New York: Wiley
144. Pflugrath, J. W., Quiocho, F. A. 1985. *Nature* 314:257
145. Privalov, P. 1982. *Adv. Protein Chem.* 35:1
146. Pullman, A., Etchebest, C. 1983. *FEBS Lett.* 163:199
147. Rashin, A., Honig, B. 1984. *J. Mol. Biol.* 173:515
148. Rashin, A., Honig, B. 1985. *J. Phys. Chem.* In press
149. Rees, D. C. 1980. *J. Mol. Biol.* 141:323

150. Reynolds, J. A., Gilbert, D. B., Tanford, C. 1974. *Proc. Natl. Acad. Sci. USA* 71: 2925
151. Rico, M., Gallego, E., Santoro, J., Bermejo, F. J., Nieto, J. L., Herranz, J. 1984. *Biochem. Biophys. Res. Commun.* 123: 757
152. Rogers, N. K., Moore, G. R., Sternberg, M. J. E. 1985. *J. Mol. Biol.* 182: 613
153. Rogers, N. K., Sternberg, M. J. E. 1984. *J. Mol. Biol.* 174: 527
154. Rosenberg, R. L., Tomiko, S. A., Agnew, W. S. 1984. *Proc. Natl. Acad. Sci. USA* 81: 1239
155. Rosseinsky, D. R. 1965. *Chem. Rev.* 65: 467
156. Sackmann, E., Trauble, H. 1972. *J. Am. Chem. Soc.* 94: 4482
157. Scheiner, S., Hillenbrand, E. A. 1985. *Proc. Natl. Acad. Sci. USA* 82: 2741
158. Schwarz, G. 1977. *Ann. NY Acad. Sci.* 303: 190
159. Schwarz, G. 1978. *J. Membr. Biol.* 43: 127
160. Schwarz, G. 1978. *J. Membr. Biol.* 43: 149
161. Deleted in proof
162. Shein, S. J., Kagan, B. L., Finkelstein, A. 1978. *Nature* 276: 159
163. Deleted in proof
164. Sheridan, R. P., Levy, R. M., Salemme, F. R. 1982. *Proc. Natl. Acad. Sci. USA* 79: 4545
165. Shoemaker, K. R., Kim, P. S., Brems, D. N., Marqusee, S., York, E. J., et al. 1985. *Proc. Natl. Acad. Sci. USA* 82: 2349
166. Sigrist-Nelson, K., Azzi, A. 1979. *J. Biol. Chem.* 254: 4470
167. Skulachev, V. P., Hinkle, P. C. 1981. *Chemiosmotic Proton Circuits in Biological Membranes.* Reading, Mass: Addison-Wesley
168. Stern, O. 1924. *Z. Elektrochem.* 30: 508
169. Tamkun, M. M., Talvenheimo, J. A., Catterall, W. A. 1984. *J. Biol. Chem.* 256: 7990
170. Tanford, C. 1965. *Physical Chemistry of Macromolecules*, p. 526. New York: Wiley
171. Tanford, C., Kirkwood, J. G. 1957. *J. Am. Chem. Soc.* 79: 5333
172. Terwilliger, T. C., Eisenberg, D. 1982. *J. Biol. Chem.* 257: 6016
173. Tokutomi, S., Iwasa, T., Yoshizawa, T., Ohnishi, S. 1980. *FEBS Lett.* 114: 145
174. Torrie, G. M., Valleau, J. P. 1980. *J. Chem. Phys.* 73: 5807
175. Torrie, G. M., Valleau, J. P., Patey, G. N. 1982. *J. Chem. Phys.* 76: 4615
176. Torrie, G. M., van Gunsteren, W. F., Berendsen, H. J. C., Hermanns, J., Hol, W. G. J., Postma, J. P. M. 1983. *Proc. Natl. Acad. Sci. USA* 80: 4315
177. Tosteson, M. T., Tosteson, D. C. 1981. *Biophys. J.* 36: 109
178. Trissl, H.-W. 1982. *Biophys. Struct. Mech.* 8: 213
179. Trudell, J. R. 1980. *Progress in Anesthesiology*, Vol. 2, ed. B. R. Fink, p. 261. New York: Raven
180. Tsuji, K., Neumann, E. 1983. *Biophys. Chem.* 17: 153
181. Urry, D. W. 1982. *Progress in Clinical and Biological Research*, Vol. 79, ed. B. Haler, J. Perez-Polo, J. Coulter, p. 87. New York: Liss
182. Urry, D. W., Prasad, K. U., Trapane, T. L. 1982. *Proc. Natl. Acad. Sci. USA* 79: 390
183. Valleau, J. P., Torrie, G. M. 1982. *J. Chem. Phys.* 76: 4623
184. van Gunsteren, W. F., Berendsen, H. J. C., Hermanns, J., Hol, W. G. J., Postma, J. P. M. 1983. *Proc. Natl. Acad. Sci. USA* 80: 4315
185. Wada, A. 1976. *Adv. Biophys.* 9: 1
186. Wada, A., Nakamura, H. 1981. *Nature* 293: 757
187. Warshel, A., Russell, S. T. 1985. *Q. Rev. Biophys.* 18: 283
188. Warshel, A., Russell, S. T., Churg, A. K. 1984. *Proc. Natl. Acad. Sci. USA* 81: 4785
189. Warwicker, J., Watson, H. C. 1982. *J. Mol. Biol.* 157: 671
190. Weigele, J. B., Barchi, R. L. 1982. *Proc. Natl. Acad. Sci. USA* 81: 1239
191. Wipff, G., Weiner, P., Kollman, P. 1982. *J. Am. Chem. Soc.* 104: 3249
192. Wolfenden, R. 1981. *Science* 222: 1087
193. Wright, E. M., Diamond, J. M. 1969. *Proc. R. Soc. London* B172: 227

Ann. Rev. Biophys. Biophys. Chem. 1986. 15 : 195–235

CHROMOSOME CLASSIFICATION AND PURIFICATION USING FLOW CYTOMETRY AND SORTING[1]

J. W. Gray and R. G. Langlois

Lawrence Livermore National Laboratory, Biomedical Sciences Division, University of California, P.O. Box 5507, L-452, Livermore, California 94550

CONTENTS

[1] The US Government has the right to retain a nonexclusive, royalty-free license in and to any copyright covering this paper.

PERSPECTIVES AND OVERVIEW

The field of cytogenetics has advanced during the last 20 years as a result of improvements in the procedures for classification and study of chromosomes. In conventional chromosome analysis, classification is accomplished on a cell-by-cell basis. That is, mitotic cells are collected and mounted on a microscope slide so that individual chromosomes are visible. The chromosomes are then stained to facilitate classification. Chromosomes stained for total DNA content may be classified according to total DNA content (17) and centromeric index (relative location of the centromeric constriction) (99, 100). Chromosomes stained using procedures that produce transverse bands along the length of the chromosomes [e.g. quinacrine (18, 19, 112), C-banding (1, 154)] are classified according to the banding pattern. Procedures have also been developed that allow differential identification of the chromatids within a chromosome and that permit detection of exchanges between sister chromatids (76). These cell-oriented procedures have the advantage that chromosomes within a single cell can be compared during the classification process. Thus, homologous chromosomes can be compared and the chromosomes can be arranged according to length, banding pattern, and/or DNA content. Modern cytogenetics has developed from the critical application of these techniques, especially banding analysis. Thus, classification of chromosomes in normal cells is now routine, as is identification of disease-linked chromosome aberrations such as aneuploidy (e.g. trisomy 21 associated with Down's syndrome) and translocations (e.g. the t(8 ; 14) chromosome rearrangement associated with Burkitt's lymphoma) as well as quantification of chromosomal rearrangements resulting from genetic damage (e.g. deletions, ring chromosomes, dicentrics, gaps, breaks, and sister chromatid exchanges). However, these techniques are not without limitation. The analyses are slow and labor-intensive. Thus, analysis of chromosomes for diagnostic purposes is expensive and sometimes too late to be of therapeutic value. Quantification of chromosome aberrations as a measure of induced genetic damage is sufficiently labor-intensive that it is not widely applied.

In a new approach to chromosome analysis introduced in 1975 (43), chromosomes isolated from mammalian cells are stained with DNA-specific fluorescent dyes, individually classified according to their dye contents using flow cytometry, and purified for biochemical analysis by fluorescence-activated sorting. Classification of chromosomes using flow cytometry [called flow karyotyping (42)] has the advantage of speed and precision. Thousands of chromosomes can be analyzed each second so that several hundred thousand chromosomes can be classified in a few minutes. Chromosome dye (DNA) content can be measured with an accuracy approaching 1% so that changes in DNA content as small as 10^{-15} g can be detected routinely. Flow sorting offers the possibility (12, 14, 43) of purifying microgram quantities of material from a single chromosome type. Thus, flow karyotyping is developing as a powerful adjunct to conventional chromosome classification procedures and flow sorting is developing as a powerful tool for study of the molecular biology of chromosomes.

We review here both the techniques and the applications of flow karyotyping and sorting, hereafter called flow cytogenetics. Included in our review of the techniques are the procedures for chromosome isolation and staining, the principles of flow cytometry and sorting, and the analytical procedures for classification of the chromosomes from flow cytometric data. Applications reviewed include classification of normal human chromosomes, identification of disease-linked structural and numerical aberrations, quantification of the frequency of structurally aberrant chromosomes as a measure of the extent of induced genetic damage, and purification of chromosomes of a single type to facilitate interpretation of flow karyotypes, gene mapping, and production of recombinant DNA libraries from chromosomes of a single type.

CHROMOSOME ISOLATION

The first requirement for flow cytogenetics is the preparation of an aqueous suspension of intact, dispersed chromosomes. A variety of methods have been developed for this purpose. First, cultures of established cell lines, short term primary cultures, or tissues in vivo are treated with a mitotic blocking agent such as colcemid to produce a population enriched in mitotic cells. Chromosomes are isolated from this population by treating it with agents selected to weaken the plasma membrane (e.g. hypotonic solutions or detergents) and to stabilize the mitotic chromosomes (e.g. divalent or polyvalent cations and intercalating dyes). The cells are then mechanically disrupted to liberate the isolated chromosomes. Since no isolation method works equally well for all cell types or applications, we first review some of the variables that are important for selection of a method and then describe in detail four methods that have been used extensively for chromosome isolation.

Important variables in the chromosome isolation process include the membrane characteristics of the cells, the fraction of cells that are cycling, and the total number of available mitotic cells. Most isolation methods can be employed for monolayer cultures of established cell lines or primary fibroblasts since these cells can be grown in large numbers and nearly pure mitotic cell populations can be selectively detached from monolayer cultures by shaking. These cells also can tolerate vigorous shearing methods such as homogenization or sonication (73, 122, 125). Suspension cultures of cell lines or stimulated peripheral lymphocytes contain many interphase cells and the mitotic cells are often more fragile than fibroblasts. Thus, milder isolation conditions using vortexing or needle shearing have been used to liberate chromosomes without producing excessive debris from disrupted interphase nuclei (150, 152). Analysis of chromosomes from human amniotic fluid cultures is further complicated by the small numbers of proliferating cells in these samples. However, one isolation method has been developed that is compatible with small cell numbers (135).

Hexylene Glycol Method

Mitotic cells are swollen in a hypotonic KCl buffer, then suspended in an isolation buffer containing hexylene glycol and divalent cations (Ca^{2+} or Mg^{2+}) to stabilize the chromosomes (16, 43, 44, 143). The chromosomes are released by homogenization or needle shearing. This method has been used to isolate chromosomes from primary fibroblasts, transformed cell lines, lymphocytes, and tumor cell lines (13, 46, 75, 124) and is compatible with most fluorescent stains and with subsequent quinacrine banding of the isolated chromosomes (44). Analysis of proteins from the isolated chromosomes indicates minimal alteration of histone and nonhistone proteins (144, 146–148). The DNA extracted from the isolated chromosomes has a molecular weight of greater than 3×10^7 d (152). One disadvantage of this method is that approximately 10^6 mitotic cells are required for best results, so that it cannot be used for amniotic samples.

Hypotonic PI Method

Mitotic cells are swollen in a hypotonic KCl solution containing the intercalating DNA stain propidium iodide (PI) as a chromosome stabilizing agent (2, 7, 11, 61). Triton X-100 is then added to weaken the cell membrane and the chromosomes are released from the cells by syringing. This method has been used for isolation of chromosomes from fibroblasts, lymphocytes, and cells from solid tumors (4, 61, 152). The isolated chromosomes can be banded and yield DNA fragments greater than 3×10^7 d (152). Since PI is required in the isolation buffer, the method is not easily adapted to other stains. However, a recent modification that uses a nonfluorescent psoralen derivative as the stabilizing agent appears to be compatible with other fluorescent stains (152).

Polyamine Method

Mitotic cells are suspended in a hypotonic buffer containing the polyamines spermine and spermidine, chelating agents, and reducing agents to stabilize the chromosomes (8, 120). The detergent, digitonin, is then added and the chromosomes are released by vortexing. This method has been used for isolation of chromosomes from fibroblasts, stimulated lymphocytes, and lymphoblastoid cell lines (31, 67, 150). However, it seems best suited to lymphoblasts. It has the advantage that high molecular weight DNA (2×10^8 d) can be extracted from the isolated chromosomes (8). This method is compatible with many fluorescent stains if divalent cations are added after isolation (68). Chromosomes produced by this method are highly condensed, however, making subsequent chromosome banding difficult (11, 31).

$MgSO_4$ Method

Mitotic cells are suspended in a hypotonic solution containing $MgSO_4$, dithioerythritol, and Triton X-100 and the chromosomes are released by gentle syringing (134, 135). This procedure seems well suited for many cell types, including fibroblasts, stimulated lymphocytes, human × rodent hybrid cells, human amniocytes, and chorionic villus biopsies (134, 135). This method is compatible with most DNA-specific fluorescent stains and with immunochemical staining (132), but the chromosomes are difficult to band. This method does not markedly affect the histone composition of chromosomes (132), and DNA extraction yields fragments of greater than 10^7 d (135).

Other Methods

Methods using fixed chromosomes (125), low pH buffers (6, 97, 126, 129), and high pH buffers (145) are useful for some applications. The methods for fixed chromosomes are attractive for those applications requiring storage of the chromosomes for extended periods.

CHROMOSOME STAINING

Isolated chromosomes can be stained fluorescently using DNA-specific dyes, fluorescently conjugated antibodies, or fluorescently conjugated polynucleotide probes. In this section we focus on the DNA-specific dyes that have been used extensively for both analysis and sorting. The future prospects for antibody and polynucleotide probes are discussed under Future Directions.

A variety of fluorescent dyes bind to DNA or chromosomes (for a review see 78), but only a few are suitable for flow cytogenetics. Ideally, dyes for flow cytogenetic application must provide bright, uniform staining of chromosomes with minimal fluorescence from unbound dye or from dye bound nonspecifically to other cellular components (56). For the highest uniformity, chromosomes are suspended at equilibrium with the staining solution. It is important in equilibrium staining for dyes to have enhanced fluorescence efficiency when bound to DNA.

Three general staining approaches have been used with isolated chromosomes. Each provides different information on the size and DNA base composition of individual chromosome types. Single DNA-specific stains have been used to differentiate chromosome types based primarily on differences in DNA content. Pairs of stains that preferentially bind to adenine-thymine (AT) rich or guanine-cytosine (GC) rich DNA allow measurements of both base composition and DNA content for chromosome types. Finally, stain combinations that utilize dyes that differ subtly in the DNA sequences of their binding sites have been used to highlight chromosomes with particular classes of repetitive DNA.

Staining for DNA Content

The phenanthridium dyes, ethidium bromide (EB) and propidium iodide (PI), bind to DNA by intercalation and have excitation maxima of 520–540 nm and emission maxima of 600–620 nm (39, 90, 91, 111). Both dyes undergo a 10- to 20-fold increase in fluorescence efficiency when bound to DNA (91, 106). Chromosomal fluorescence appears to be proportional to DNA content with these dyes, since their binding affinity and fluorescence efficiency is unaffected by differences in DNA base composition or sequence (91). Also, the relative fluorescence intensities measured for chromosomes stained with these dyes are highly correlated with independent measurements of relative chromosomal DNA content (42). The binding of these dyes appears to be unaffected by differences in chromosome condensation since the coefficient of variation (CV = standard deviation/mean) of the chromosome stain content in chromosomes of a single type ranges from 0.01 to 0.03, even though microscopic examination shows wide variations in condensation (73). The relative intensity of individual chromosome types also appears to be unaffected by differences in stain concentration. Thus, these dyes can be used to obtain precise measurements of the relative DNA content of individual chromosome types. One minor disadvantage of these dyes is that they also bind with high affinity to RNA. RNAase treatment of the chromosome suspension can eliminate this interference.

The bisbenzimidazole and related dyes, Hoechst 33258 (Ho) and DAPI, have also been used for flow analysis of chromosomes. These dyes excite at 360 nm and emit fluorescence at 440–470 nm (9, 59, 84, 93). Advantages of these dyes compared with EB and PI are higher extinction coefficients, higher fluorescence enhancement on binding to DNA, and minimal affinity for RNA (54, 60, 66, 84). The relative intensity of staining of individual chromosome types with these dyes, however, differs from that with EB or PI because Ho and DAPI preferentially bind to AT-rich regions in DNA.

Staining for DNA Content and Base Composition

A second strategy is to utilize stain combinations that provide measurements of both DNA content and DNA base composition. In this approach, chromosomes that cannot be resolved using one dye (44, 75) may be resolved using a second dye with different binding characteristics. Base composition should be useful for differentiating chromosome types, since studies of banding mechanisms suggest that chromosomal regions vary in average base composition (22, 34, 62, 73, 79). Two dyes, Ho and chromomycin A3 (CA3) have been particularly useful for flow cytometric measurements of DNA content and base composition. Ho preferentially binds to AT-rich regions in DNA (84, 96), while CA3 binds preferentially to GC-rich DNA (5, 57, 77, 139). These two dyes also differ in spectral properties; Ho excites maximally at 360 nm and emits maximally at 460 nm, while CA3 excites maximally at 430 nm and emits maximally at 580 nm (56). While these two dyes bind independently to chromosomes, they interact by energy transfer so that the fluorescence from the two dyes cannot be measured separately (73, 74, 77, 82). The chromosomal Ho and CA3 contents can be determined, however, by exciting these stains selectively at two different wavelengths (44, 73). With this approach, two fluorescence intensities are measured for each chromosome. The sum of the two fluorescence intensities is determined by the total DNA content and the ratio of the two fluorescence intensities is determined by the base composition.

Stains with DNA Sequence Specificity

A third staining approach exploits the fact that base composition–specific stains often require specific DNA sequences for binding (70, 96, 117, 138, 139). Thus, pairs of dyes with the same base specificity but differing sequence specificity can be used to highlight the presence of specific sequences. Most chromosomal regions have sufficient sequence complexity that such effects tend to average out, but sequence effects can be significant for regions containing large blocks of highly repetitive DNA. Sequence-specific dye combinations have been used successfully to highlight polymorphic regions of human chromosomes that contain repetitive DNA in metaphase spreads (80, 83, 116). This approach has been applied to flow cytometric analysis by staining the chromosomes with Ho and CA3 and then counterstaining them with the nonfluorescent ligands netropsin or distamycin A (68, 101). Since both of these nonfluorescent ligands have an AT binding preference, they compete with Ho for binding sites in all regions except those that are rich in sequences recognized by Ho but not by the ligands. Thus, the Ho fluorescence is decreased for all chromosomes except those having large blocks of the appropriate repetitive DNA (e.g. human chromosomes 9 and 15).

Bromodeoxyuridine Quenching

One staining method has been developed for differentiating chromosome types that replicate at different times within S-phase (27, 28). This method takes advantage of the fact that the fluorescence of Ho is quenched when Ho is bound to DNA containing the base analog bromodeoxyuridine (BrdUrd) (84), while the fluorescence of CA3 is enhanced slightly by this analog (130). Chromosomes replicating in early or late S-phase can be preferentially labeled with BrdUrd and distinguished by their extensive Ho quenching. Chromosomes are preferentially labeled by treating the cells with BrdUrd for a short time and then timing the collection of mitotic cells for chromosome isolation so that they are in early or late S-phase during BrdUrd labeling.

PRINCIPLES OF FLOW CYTOMETRY AND SORTING

Univariate Flow Karyotyping and Sorting

INSTRUMENTATION   Most flow cytometric chromosome analyses have relied on chromosome classification according to the fluorescence emitted from a single dye. The principles of flow cytometric chromosome classification are illustrated in Figure 1. During flow cytometry (98, 104, 136, 137) the stained chromosomes are injected into the center of a liquid stream (called the sheath fluid). The sheath and sample fluid then flow into a narrow channel where hydrodynamic forces compress the sample stream to a diameter of a few micrometers. There the chromosomes flow one by one through an intense beam of light, of wavelength adjusted to excite the fluorescent dye efficiently. The resulting fluorescence is recorded for each chromosome as a measure of its dye content. Since flow cytometry allows analysis of thousands of chromosomes each second, several hundred thousand chromosomes can be measured in a few minutes. The results are accumulated to show the distribution of DNA among the chromosomes of the population. A univariate fluorescence distribution measured for Chinese hamster M3-1 line chromosomes stained with Ho is shown in the insert in Figure 1. Since most types of Chinese hamster chromosomes have unique DNA contents, each type produces a distinct peak in the fluores-

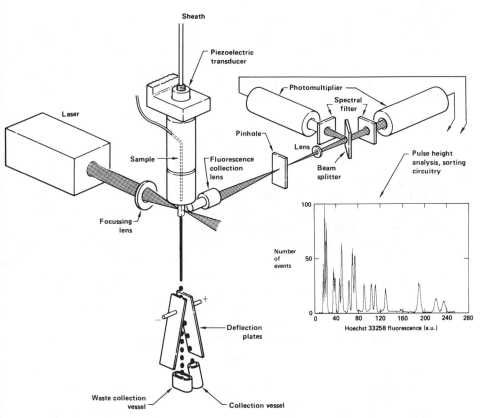

Figure 1 Schematic diagram of a single laser flow system for chromosome classification and sorting. The fluorescently stained chromosomes are injected into the center of a flow chamber and transported one by one through a laser beam. The resulting fluorescence is collected by a lens and projected onto one or more photomultipliers that produce an electrical pulse proportional in amplitude to the intensity of the fluorescence. Pulses from approximately 10^3 chromosomes/sec are amplified, digitized, and accumulated to form a fluorescence distribution as shown in the insert for Chinese hamster chromosomes stained with Hoechst 33258. After passing through the laser beam, the chromosomes continue flowing downward and eventually leave the flow chamber in a liquid jet. They flow with the jet until it breaks into droplets. The droplets containing the chromosomes to be sorted are charged and separated from the other droplets during passage through a high-voltage electric field.

cence distribution (4, 42, 75). As long as the chromosomes are stained with the same dye, the locations of the peaks are sufficiently reproducible that chromosomes can be identified from the positions of their peaks. The relative frequency of each chromosome type in the population is proportional to the area of its peak. Thus, the chromosomes in a suspension can be classified according to their fluorescence distributions, which are termed "flow karyotypes." It is important to note that the relative peak positions can vary considerably when the chromosomes are stained with dyes differing in base specificity (see Chromosome Staining). Figure 2, for example, shows the distinctly different fluorescence distributions measured for human chromosomes stained with EB (no base composition preference), Ho (AT binding preference), and CA3 (GC binding preference).

Precise measurement of chromosomal dye content is essential to the accumulation of flow karyotypes in which the maximum number of chromosomes are resolved. In modern flow cytometers, measurement precision approaching 1% can be achieved routinely (2, 4, 43, 44, 87, 119, 134–136, 150) by (a) constraining the chromosomes to flow through the measurement region in a stream of diameter of only a few micrometers, (b) maintaining constant flow velocity, (c) adjusting the wavelength of the light beam to excite the DNA-specific chromosome dye efficiently, and (d) collecting the resulting fluorescence efficiently. Measures a and b ensure that all chromosomes are excited equally so that the resulting fluorescence is proportional to the dye content. Measures c and d are necessary so that the fluorescence detected from each chromosome is maximal and the measurement precision is not reduced by statistical fluctuation in the number of photons detected for each chromosome. The flow karyotypes of the Chinese hamster and human chromosomes in Figures 1 and 2,

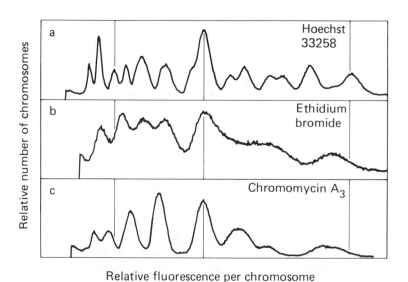

Relative fluorescence per chromosome

Figure 2 Univariate flow distributions of human chromosomes isolated from normal fibroblasts stained with Ho (*a*), EB (*b*), and CA3 (*c*) (73).

respectively, show peaks with CV = 0.01–0.05. The number of clearly resolved peaks decreases as the CV of the measurement increases.

ANALYSIS OF UNIVARIATE FLOW KARYOTYPES Flow karyotyping requires accurate estimation of the means and areas of each peak. This is commonly accomplished by matching normal distributions to each peak using a least-squares-best-fit procedure (16, 32, 43, 103, 107). That is, the experimentally measured flow karyotype, $f(x)$, is matched by the function

$$g(x) = \sum_i (0.3989 * A_i/\sigma_i) * \exp(-0.5 * [(x - \mu_i)/\sigma_i]^2) + d(x),$$

where $f(x)$ is the number of chromosomes recorded at fluorescence x; A_i, μ_i, and σ_i are the area, mean, and standard deviation of the ith peak, respectively, and $d(x)$ is a function chosen to approximate the shape of the continuum that underlies the peaks in the flow karyotype. The continuum in the flow karyotype is produced by the fragmentation of nuclei and chromosomes during the chromosome isolation process and by randomly occurring aberrant chromosomes. The continuum must be accounted for during the analysis or substantial error in the estimates for the peak areas may occur. Estimates for the relative peak means typically vary by less than 0.5% from day to day for flow karyotypes measured for chromosomes isolated from the same cell population and stained with the same dye. Thus the peak means are accurate chromosome classifiers (75). A change in the mean of any peak or the appearance of a new peak signals a chromosome alteration resulting from a change in chromosomal DNA content and/or base composition (e.g. translocation, addition, or deletion). Each peak area is proportional to the relative frequency of that chromosome type in the population of chromosomes. Thus, areas of peaks for chromosomes X and Y in males should be half the areas for the autosomal chromosomes. Monosomic and trisomic chromosomes should produce peaks with areas equal to 0.5 and 1.5 times the areas of the peaks for the autosomal chromosomes, respectively.

FLOW SORTING Flow sorting (40, 52, 55, 98, 136) allows purification of chromosomes that can be resolved in a flow karyotype as illustrated in Figure 1. For sorting, the chromosomes are entrained in a thin, electrically conductive liquid jet in air (typically 75 μm in diameter) after flow cytometric classification. They continue down the jet to the point where it is forced by the action of a vibrating piezoelectric crystal to break into droplets (at a known distance from the analysis point). At this time they become encapsulated in liquid droplets. An electric charge is induced on the droplets that contain chromosomes to be purified. The other droplets, which are either empty or contain unwanted chromosomes, are left uncharged. Two classes of chromosomes can be sorted simultaneously by

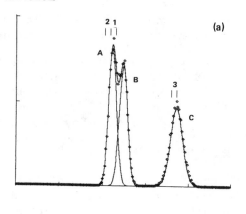

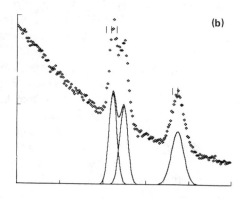

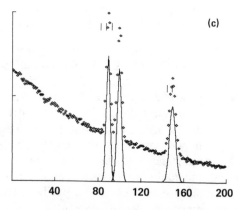

Simulated Fluorescence Intensity

charging the droplets containing one class positively and the droplets containing the other class negatively. All droplets then fall through a high-voltage electric field so that the charged droplets containing the chromosomes of interest separate from the uncharged droplets. The chromosome analysis rate (typically 1000/sec) is kept low compared to the rate at which the jet breaks into droplets (typically 20,000/sec) to reduce the probability that more than one chromosome will reside in a single droplet. Often, a purity circuit is used to abort the sorting process when a chromosome to be sorted occurs in the same droplet as another particle. The charged droplets containing the chromosomes of interest are collected for visual or biochemical analysis. The fluid used in the liquid jet for sorting is usually chosen to be the buffer used for chromosome isolation to preserve the morphology, protein composition, and DNA molecular weight of the sorted chromosomes. Several studies have shown that the sorted chromosomes have good morphology (12, 14, 43), can be banded using conventional techniques (12, 35, 44), and have near-normal chromosomal protein composition (132, 146). In addition, the molecular weight of the DNA isolated from sorted chromosomes is approximately the same as prior to sorting (10^7 to 10^8 d) (135, 152).

The purity with which a specific chromosome type can be sorted depends largely on the degree to which its peak in the flow karyotype is separated from nearby peaks, the location of the sorting window (i.e. the definition of the fluorescence intensities allowed for chromosomes to be sorted), and the presence of DNA-containing debris fragments in the suspension of isolated chromosomes. The hypothetical effects of these factors on sorting purity are illustrated in Figure 3. Three peaks produced by hypothetical chromosomes A, B, and C are illustrated in this distribution; A and B overlap significantly while C is well separated from the other two. Panel a shows only the three peaks. Panel b shows these peaks superimposed on a slowly decreasing continuum such as might be produced by the presence of DNA-containing debris. Panel c shows the same distribution as panel b except that the peak resolution is increased by a factor of two. In general, the

---

Figure 3 Hypothetical fluorescence distributions expected for a chromosome population containing only three chromosome types, A, B, and C. Chromosomes A and B differ by only 10% in total fluorescence while C is well resolved from A and B. (a) Distribution expected for these chromosomes assuming a fluorescence measurement CV of 0.04 and no debris continuum. Chromosome C can be sorted with high purity from window 3. Chromosome A can be sorted with moderate purity from window 1 and with high purity from window 2. The sorting efficiency from window 2 is low, however. (b) The same chromosomes superimposed on a debris continuum. The purity of all sorts is reduced by the presence of the debris particles. (c) The same distribution as in (b) except the measurement CV is reduced to 0.02. The purity of all sorted fractions is increased by sorting at high resolution.

sorting purity is highest for well resolved peaks well away from the debris continuum (peak *C*, panel *a*). Sorting purity is also affected by the location of the sorting windows. For example, the purity of sorting of type *A* chromosomes from window 2 would be higher than from window 1 although the efficiency of sorting would be reduced. The sorting purity is always increased by an increase in resolution (Figure 3, panel *c*).

Bivariate Flow Karyotyping and Sorting

INSTRUMENTATION   Bivariate flow karyotyping uses pairs of dyes to provide information about two chromosome properties simultaneously (44, 83, 124, 132). For example, Ho and CA3 provide information on both DNA content and base composition, thereby increasing the resolution of individual chromosome types (44, 73, 75, 89). As described under Chromosome Staining, the optimum use of Ho and CA3 requires that they be excited at different wavelengths (Ho at 360 nm and CA3 at 430 nm). This is accomplished using dual beam flow cytometry (33, 123) as illustrated in Figure 4. In this approach, isolated chromosomes stained with two fluorescent dyes (Ho and CA3 in this example) pass sequentially through two laser beams, each adjusted to preferentially excite one stain. The two fluorescence pulses resulting as a chromosome crosses the two laser beams are measured and the measurements of several hundred thousand chromosomes are accumulated to form a bivariate flow karyotype.

Figure 5 shows a bivariate flow karyotype for human chromosomes stained with Ho and CA3. All human chromosomes except 9 through 12

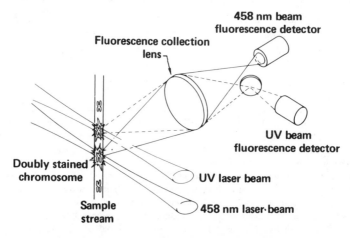

Figure 4   Schematic diagram of a dual laser sorter (33).

Hoechst 33258

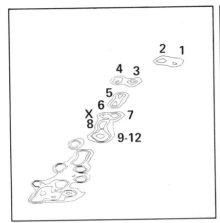 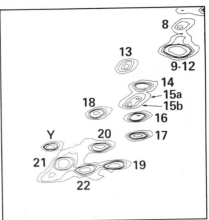

Chromomycin A3

Figure 5 Bivariate flow karyotype of human chromosomes derived from peripheral lymphocytes from a normal donor (75). *Left:* whole karyotype. *Right:* expanded view of the smaller chromosome types from the same sample. The chromosomes producing each peak are indicated.

(and sometimes 14 and 15) are better resolved in this bivariate flow karyotype than in any univariate flow karyotype. Thus, bivariate flow karyotyping has been applied extensively for chromosome classification and aberration detection. Chromosome sorting based on Ho and CA3 fluorescence has also proved useful for production of highly purified chromosome fractions.

ANALYSIS OF BIVARIATE FLOW KARYOTYPES   In bivariate flow karyotypes, chromosomes are classified according to the mean and volume of the peaks that they produce (32, 75). The most accurate analysis procedure is similar to that used for analysis of univariate flow karyotypes. That is, the peak parameters (means, volumes, and standard deviations) are estimated by fitting the sum of several bivariate normal distributions to the peaks in the bivariate flow karyotype. The number of chromosomes with CA3 and Ho fluorescence intensities of x and y, $f(x, y)$, is matched by the function

$$g(x, y) = \sum_i (0.1592 * A_i/(\sigma x_i * \sigma y_i)) * \exp\{-0.5 * [(x - \mu x_i)/\sigma x_i)^2$$
$$+ ((y - \mu y_i)/\sigma y_i)^2]\} + d(x, y),$$

where A_i is the volume, σx_i and σy_i are the standard deviations, and $(\mu x_i \mu y_i)$ is the bivariate mean of peak i; $d(x, y)$ is a function describing the debris

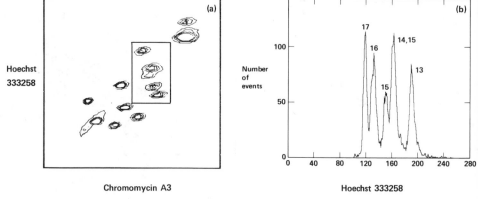

Figure 6   Simplified data analysis scheme for bivariate flow karyotypes (135). (*a*) Four peaks (produced by chromosomes 13 through 16) in a bivariate flow karyotype circumscribed by a rectangular box. (*b*) The univariate Ho distribution produced by the inscribed chromosomes.

continuum (102). The bivariate means for flow karyotypes measured for chromosomes isolated from human cells vary only slightly from day to day and from person to person. Thus, a significant change in peak mean may indicate a change in chromosomal DNA content and/or base composition. The volume of each peak is proportional to the relative frequency of occurrence of the chromosome type that produced it. The least-squares-best-fit approach to bivariate karyotype analysis has two limitations: (*a*) It is a formidable task even on a modern supercomputer and (*b*) it is difficult to select a function $d(x, y)$ to represent the debris continuum. Without a function for $d(x, y)$, estimation of the volumes of the peaks superimposed on the continuum is difficult.

A simpler method of analysis has been developed for estimation of the volumes of selected chromosome peaks in bivariate flow karyotypes for human chromosomes (135). In this approach, selected chromosomes (e.g. chromosomes 13 through 17, chromosomes 18 and 20, and chromosomes 21 and Y) are circumscribed with a rectangular box. The bivariate distribution within the box is then collapsed to form a univariate Ho distribution. This process is illustrated in Figure 6 for human chromosomes. The areas of the univariate peaks can be estimated as described above and can be used as estimates for the volumes of the peaks in the bivariate distribution.

High Speed Sorting

Many studies requiring pure fractions of chromosomes of a single type can be accomplished most easily if microgram quantities of chromosomal material are available. Unfortunately, collection of this amount may take hours using conventional sorters (29). The number 1 human chromosome contains only 4.8×10^{-13} g of DNA and the Y contains only 9.8×10^{-14} g. Further, in a suspension of chromosomes isolated from human cells, only 4% of the chromosomes are number 1 and only 2% are Y. Thus, if 1000 chromosomes can be processed each second, 13.4 hr of sorting will be required to produce 1 μg of DNA from chromosome 1 and 131 hours will be required to produce 1 μg of DNA from the Y chromosome. As a consequence, many scientifically attractive experiments are logistically impossible using conventional sorters. High speed sorting (HiSS) (113) has been developed to facilitate production of microgram quantities of chromosomal material. The major differences between HiSS and conventional sorters are the operating pressure, liquid jet velocity, and droplet production rate. Conventional sorters operate with jet velocities of about 10 m/sec and droplet production rates of about 20,000/sec. This is achieved by pressurizing the system to about 10 psi. HiSS operates with a jet velocity of about 50 m/sec and a droplet production rate of 215,000/sec. This is achieved by pressurizing the system to 200 psi. Chromosomes can be analyzed and sorted 10 to 20 times faster in HiSS than in conventional sorters. The chromosome resolution achieved during HiSS approaches that obtained in conventional dual-beam sorters so that the purity of the sorted chromosomes is high. With HiSS, production of microgram quantities of DNA from a single type of chromosome has become routine.

Slit-Scan Flow Cytometry

Slit-scan flow cytometry has been developed to measure total chromosomal stain content and the distribution of the stain along the chromosomes (24, 45–47, 94). The principles of slit-scan chromosome analysis are illustrated in Figure 7. During slit-scan analysis, hydrodynamic forces in the nozzle orient the chromosomes so they flow lengthwise through a thin scanning region. The variation of the intensity of emitted fluorescence with time is recorded for each chromosome. Chromosomes can be scanned by measuring the intensity of fluorescence as the chromosomes flow across a thin laser beam or by measuring the intensity of fluorescence passing through a thin slit onto which the image of the chromosome is projected. The spatial resolution achieved using these techniques ranges from 1–2 μm so that chromosome features of this size can be resolved. The profile shown in the insert to Figure 7 was recorded for a human number 1 chromosome. The most prominent morphological feature in this profile is the centrally located dip produced when the chromosome centromere, with its reduced DNA content, passed through the scanning laser beam. Chromosomes

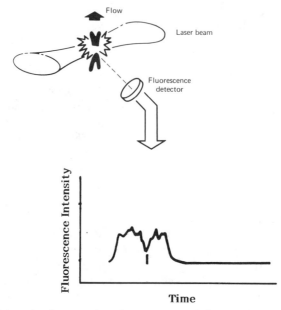

Figure 7 Schematic diagram of a slit-scan flow cytometer. Each chromosome flows lengthwise through a thin laser beam. The resulting fluorescence is recorded as a measure of the distribution of dye along the chromosome. The profile shown in the insert was recorded for a human number 1 chromosome. The centromere location in the profile is indicated by a vertical bar.

have been classified according to the location of the centromeric dip (45, 47, 94) or according to the number of centromeric dips (46). The total fluorescence of a chromosome can be calculated by integrating the area under its fluorescence profile. These applications are discussed in the following section.

APPLICATIONS OF FLOW KARYOTYPING

Classification of Chromosomes from Different Species

Flow karyotypes have been measured for several mammalian species using chromosomes derived from a number of different cell types. Figure 8 shows flow karyotypes for chromosomes derived from three different species. The female Indian muntjac (*Muntiacus muntjak*) has one of the simplest karyotypes of all mammals; the female has only three different chromosome types ($2n = 6$) (14, 92). These chromosome types can be easily resolved even with univariate analysis (Figure 8a). Because of their karyotypic simplicity, muntjac cells have been used as a model system for study of the chromosomal sites of viral integration (13). The near-diploid Chinese hamster cell line, M3-1, is karyotypically more complex with 14 different chromosome types (43). Still, most or all of these chromosomes can be resolved by univariate analysis (Figure 8*b*). Bivariate analysis (Figure 8*c*) of hamster chromosomes improves the separation of some of the smaller chromosome types, but most hamster chromosomes have similar fluorescence with both stains so that the resolution gain from bivariate analysis is modest. Flow karyotypes of embryonic hamster cells, which spontaneously transform in culture, have been used extensively to detect the karyotypic changes that may be important to cell transformation (4). The sensitivity of the flow karyotype to DNA content changes as small as 10^{-15} g has proved

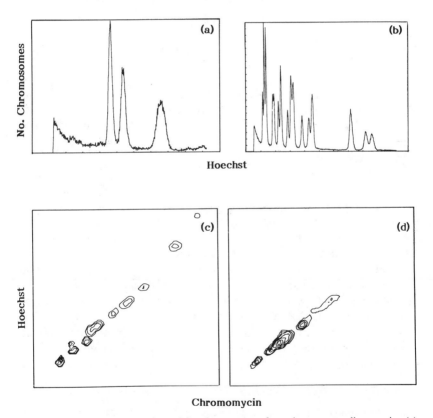

Figure 8   Flow karyotypes measured for chromosomes from three mammalian species. (*a*) and (*b*): Univariate Ho distributions for chromosomes from muntjac and Chinese hamster M3-1 cells, respectively. (*c*) and (*d*): Bivariate (Ho vs CA3) distributions for chromosomes from Chinese hamster M3-1 and mouse cells, respectively.

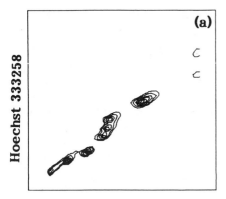 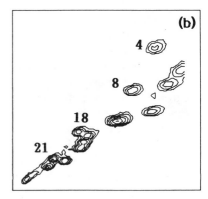

Chromomycin A3

Figure 9   Bivariate flow karyotypes of a Chinese hamster ovary (CHO) parental cell line (*a*) and a CHO × human hybrid cell line that contains human chromosomes 4, 8, 18, and 21 (*b*). The peaks produced by the human chromosomes are labeled.

especially useful in these studies. Flow karyotypic analysis of mouse chromosomes is much more difficult because of the similarity in DNA content, base composition, and centromeric heterochromatin of the 21 different types of mouse chromosomes (53, 120). Figure 8*d* shows that separate resolution of the individual mouse chromosomes is difficult, even in bivariate flow karyotypes. However, the large derivative chromosomes in cell lines containing Robertsonian fusions can be clearly separated from the remaining chromosome types (36).

Flow karyotyping has also been applied to the analysis of inter-species somatic cell hybrids. Figure 9 shows bivariate Ho vs CA3 flow karyotypes of a Chinese hamster parental cell line and a human × hamster hybrid cell line containing human chromosomes 4, 8, 18, and 21. The human chromosomes are separately resolved from the hamster chromosomes due to differences in DNA content and base composition. Flow karyotypic analysis of hybrids has been used to study the karyotypic stability of the hybrid as well as for flow sorting of specific human chromosomes (30; see Chromosome Sorting). Other mammalian species that have been studied by flow karyotyping include rat (61), chicken (127), and rat kangaroo (124).

Characterization of Normal Human Chromosomes

While fibroblast cell strains were utilized for the development of techniques for preparation of isolated chromosomes from human cells (44, 73), advances in culture and isolation methods for peripheral lymphocytes (150, 152), amniocytes (135), and chorionic villus cells have greatly extended the utility of flow karyotyping for population cytogenetic studies. High resolution univariate (51, 97, 150) and bivariate (75, 135) flow karyotypes from lymphocyte chromosomes are especially useful in this regard. Figure 5 shows a bivariate Ho vs CA3 flow karyotype of lymphocyte chromosomes. The chromosome types producing the peaks in the flow distribution were identified using quinacrine banding of chromosomes sorted from each peak (44). All human chromosome types except chromosomes 9–12 and chromosomes 14 and 15 can be separately resolved in the bivariate flow karyotype (44, 75).

The variability in peak position of individual chromosome types from replicate measurements is sufficiently small (CV = 0.005) that subtle differences among individuals can be quantitated. The variability in peak positions among individuals is significantly larger than this measurement error, with the largest variability seen for chromosomes that are known to contain polymorphic regions (75). Figure 10 shows that polymorphic variability is sufficiently large for some chromosomes that individual homologs can often be separately resolved. In some cases homolog differences in the flow karyotype correlate with G- or Q-band polymorphisms seen in metaphase spreads. Other homolog differences probably correlate with polymorphisms that are not highlighted by quinacrine, but can be visualized by other banding methods (80, 83, 116, 118). Thus, bivariate karyotyping may be a useful tool for population studies of the distribution of normal polymorphisms in humans.

Figure 11 is a statistical summary of the results of flow karyotypes measured for ten normal individuals, with each ellipse representing a 95% tolerance interval for the peak location of each chromosome type. Peaks corresponding to normal chromosome types can be unambiguously identified because these ellipses do not overlap. For most chromosome types, a structural abnormality corresponding to the gain or loss of one chromosome band (1/600 of the mitotic genome) should move any peak position outside of a normal 95% tolerance interval. Recent studies indicate that comparable precision can be obtained for chromosomes derived from amniotic fluids (J. W. Gray, B. Trask, R. Langlois, G. van den Engh, et al, manuscript in preparation). In these studies the amniotic cultures were obtained on the same day as the conventional karyotypes were completed, and only two to three additional days were required for the flow karyotypes. This interval is compatible with the time constraints of prenatal diagnosis.

Slit-scan flow cytometry provides another approach to flow karyotyping. Fluorescence profiles collected during scanning can be processed to determine both chromosome DNA content and centromeric index (CI = DNA content of the long arm of the chromosomes/total chromosome DNA

Chromomycin A3

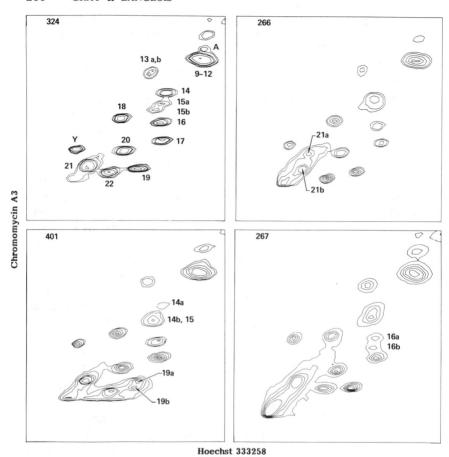

Hoechst 333258

Figure 10 Bivariate flow karyotypes measured for the smaller chromosome types from human lymphocytes from four normal donors (75). Individual homologs that are separately resolved owing to normal polymorphisms are labeled *a* and *b*.

content). Figure 12 shows a bivariate DNA vs CI distribution measured for the larger human chromosomes (J. N. Lucas, J. Mullikin, J. Gray, manuscript in preparation). Chromosomes 1 and 2, 3 and 4, and 9–12 are much better resolved with this technique than in bivariate Ho vs CA3 distributions. The principal limitation to slit-scan analysis at present is the relatively low rate (approximately 100/sec) at which chromosomes can be analyzed.

Detection of Homogeneous Aberrations

Flow karyotyping has been applied to the detection of chromosome abnormalities that occur in most or all of the cells in a population [e.g.

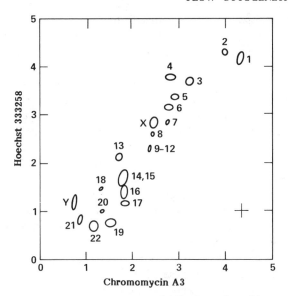

Figure 11 Statistical summary of the total variability in peak positions among normal individuals (75). The *ellipses* are 95% tolerance regions for each human chromosome type. The *cross* shows the expected change in peak position that might result from the addition or deletion of fluorescence corresponding to one chromosome band.

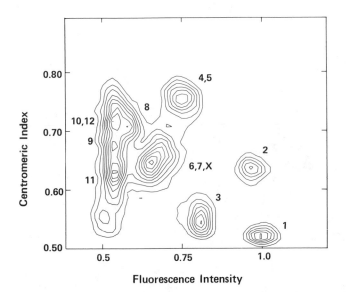

Figure 12 Bivariate DNA vs centromeric index distribution measured for the larger human chromosomes using slit-scan flow cytometry (J. Lucas, J. Mullikin, J. Gray, manuscript in preparation).

inherited chromosome defects or specific marker chromosomes that occur at high frequency in tumor cell populations (46, 48, 142; L.-C. Yu, personal communication)]. Both numerical and structural aberrations can be detected.

Numerical aberrations (aneuploidy) cause a change in the volume (or area) of the peak corresponding to the aneuploid chromosome type. Figure 13 shows the increase in volume of the peak corresponding to chromosome 21 in an amniotic sample with trisomy 21 (panel *b*) compared to the volume of the peak for chromosome 21 in a normal amniotic sample (panel *a*). While trisomies have been detected in most amniotic samples known to be aneuploid, estimation of peak volume is not possible for some samples with flow karyotypes that show elevated debris continua.

Structural aberrations such as deletions, duplications, or translocations that result in derivative chromosomes with altered DNA content or base composition cause the appearance of new peaks in the flow karyotype. Figure 14 shows the effect of a translocation on the bivariate flow karyotype of an amniotic sample. In this case there has been a translocation between chromosomes 21 and 22 and the two derivative chromosomes both produce new peaks in the flow distribution. While the presence of a new peak in the flow karyotype falling outside of a 95% confidence ellipse (see Figure 11) demonstrates the presence of a structural aberration, it is not

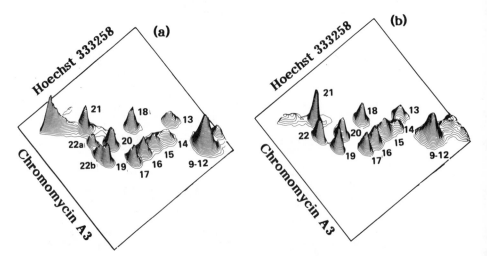

Figure 13 Isometric view of bivariate flow karyotypes measured for amniotic samples from a normal female (*a*) and a female carrying three number 21 chromosomes (*b*). Note the increased volume of the peak corresponding to chromosome 21 in the sample with trisomy 21 (J. W. Gray, B. Trask, R. Langlois, G. van den Engh, et al, manuscript in preparation).

always possible to determine which chromosome regions are involved in the rearrangement. In addition, derivative chromosomes that have the same staining characteristics as a normal chromosome type (i.e. paracentric inversions) will not be detected.

Bivariate karyotypes have several properties that make them particularly useful for characterizing human tumor cell populations. There is growing evidence that individual tumors often are associated with the presence of specific chromosome rearrangements (153). Identification of specific marker chromosomes that are characteristic of the tumor is often complicated in conventional cytogenetic analyses by high levels of karyotypic instability in tumor cell populations. Thus, many metaphase spreads must be analyzed to distinguish between specific markers present in most cells and random aberrations that differ from cell to cell. With bivariate karyotyping, characteristic markers that occur at high frequency produce peaks in the distribution, while low-frequency random aberrations contribute to a low-level continuum and will not produce peaks (46).

Bivariate karyotypes of two human tumor cell lines are shown in

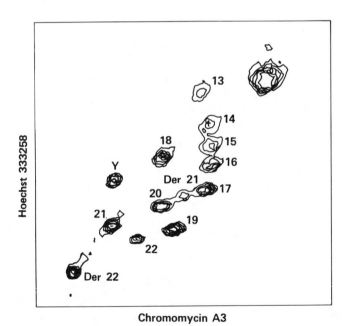

Figure 14   Bivariate flow karyotype measured for the smaller chromosomes from amniotic cells carrying the translocation t(21;22). The two derivative chromosomes from this translocation produce new peaks Der 21 and Der 22.

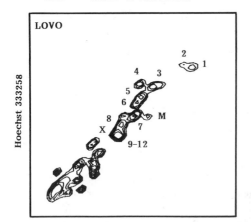 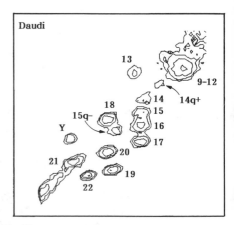

Chromomycin A3

Figure 15 Bivariate flow karyotypes measured for two two human tumor cell lines. The flow karyotype for the human colon carcinoma line (LOVO) contains an abnormal marker chromosome peak, *M* (46). The flow karyotype for the Burkitt's lymphoma line (Daudi) shows abnormal peaks labeled 14q + [resulting from a t(8;14) translocation] and 15q − (resulting from a deletion of a portion of chromosome 15).

Figure 15. The colon carcinoma cell line, LOVO (37), is karyotypically unstable with the number of chromosomes per cell varying from 37–55. Most of this variability is not apparent in the bivariate karyotype since random aberrations contribute only to the continuum. Thus, the flow karyotype shows only normal chromosomes and a high-frequency marker chromosome (*M*, Figure 15, *left*). The Burkitt's lymphoma line, Daudi, is also reported to have considerable variability in chromosome number from cell to cell, but most cells have the characteristic t(8;14) translocation as well as a deletion on one chromosome 15 (115). The bivariate karyotype (Figure 15, *right*) shows new peaks for two of these markers (14q + and 15q −), while the third marker (8q −) is obscured by the peak corresponding to the normal 9–12 group. Bivariate flow sorting can be used not only to identify characteristic marker chromosomes, but also to sort these markers for mapping of specific break points at the DNA level.

Detection of Heterogeneous Aberrations

Exposure of cells to genotoxic or clastogenic agents can result in the production of a large variety of different chromosome alterations. Detection of these heterogeneous aberrations is complicated by the fact that only rare cells have chromosome abnormalities, and the type of abnormality differs from cell to cell. As described above, these chromosome aberrations may cause increases in peak widths in flow karyotypes owing to the presence of damaged chromosomes, and can lead to increases in the continuum underlying the peaks due to the accumulation of chromosome fragments and randomly rearranged chromosomes. Several approaches have been described for estimating the magnitude of the continuum and/or peak-width increases in univariate karyotypes as a measure of induced genetic damage (3, 25, 26, 38, 49, 107, 109, 110, 140). The limitations of this approach are that relatively large amounts of damage are required to alter the flow distribution and that any variability in the chromosome isolation procedure may produce debris which is indistinguishable from induced chromosome damage.

The technique of slit-scan flow cytometry is particularly well suited for quantification of some classes of heterogeneous chromosome aberrations (46). Slit-scan provides information on the morphology of individual chromosomes so that the frequency of chromosomes with distinctive aberrant morphology can be used as a measure of total chromosome damage. This approach has been applied to detection of dicentric chromosomes as a measure of X-ray–induced chromosome damage. The slit-scan profile of a normal chromosome shows a dip in intensity at the position of the centromere because of the reduced DNA content there (23, 45–47, 94). Thus, profiles for dicentric chromosomes differ from those for normal chromosomes because the dicentric profiles have two centromeric dips (46; Figure 16). The frequency of dicentric chromosomes can be directly determined from the fraction of profiles with two centromeric dips. Figure 17 shows the frequencies of dicentric chromosomes measured visually and by slit-scan flow cytometry for Chinese hamster M3-1 cells

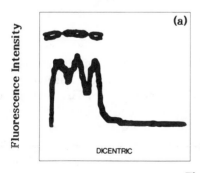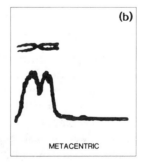

Figure 16   Slit-scan fluorescence profiles of a normal metacentric chromosome (*b*) and an abnormal dicentric chromosome (*a*). The chromosome silhouettes show the chromosomes expected to produce the two slit-scan profiles (46).

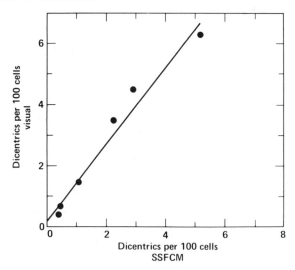

Figure 17   Dicentric chromosome frequencies measured for Chinese hamster cells exposed to 0–400 rad X-rays (46). Frequencies were estimated both by slit-scan flow cytometry and visual scoring of metaphase cells from the same samples.

exposed in vitro to various doses of X-rays. The close correlation between these methods confirms that the frequency of dicentric chromosomes can be estimated by slit-scanning. Since large numbers of chromosomes can be analyzed in a short period of time with slit-scan techniques, this approach should be particularly useful for low-dose exposures where several hundred thousand chromosomes must be scored to obtain precise estimates for the dicentric frequencies.

CHROMOSOME SORTING

Peak Identification in Flow Karyotypes

Flow sorting has proved essential for identification of the chromosomes that produce the various peaks in a flow karyotype. Chromosomes are sorted from each peak, mounted on microscope slides, banded using quinacrine, Giemsa, or other procedures, and identified visually. This procedure has been applied for Chinese hamster (43), human (12), mouse (35), and muntjac (14) cell populations. The ease with which sorted chromosomes can be banded depends in part on the procedure used for chromosome isolation. Our experience suggests that chromosomes isolated using the hexylene glycol method are most easily banded and those isolated using the polyamine procedure are most difficult to band. No matter what method is used, however, only a small fraction of the chromosomes will band. Thus, banding analyses of sorted chromosomes give information only about the major chromosome fraction in the purified population. Little information is available about the frequency and nature of the minor contaminants.

Gene Mapping

Flow sorting is a powerful tool for gene mapping. Chromosomes from each of the peaks of interest in a flow karyotype are purified by sorting. Hybridization with a labeled nucleic acid probe coding for the gene under study is then used to determine which chromosome fraction contains sequences that are homologous with the probe.

Two procedures have been applied for the detection of DNA hybridization. In the first, DNA extracted from the sorted chromosomes is digested using a restriction enzyme and the resulting DNA fragments are separated by gel electrophoresis (15, 88, 128). The separated fragments are transferred to nitrocellulose paper, hybridized to ^{32}P-labeled probe DNA, washed, and autoradiographed to reveal the presence of the hybridized probe. Figure 18 shows a univariate flow karyotype and three regions [produced by chromosomes $16+18$ (region 1), 9–12 (region 2), and 3–6 (region 3)] from which chromosomes were sorted in an effort to map the location of the globin genes (88). The insert to Figure 18 shows the result of hybridizing with cDNA prepared by reverse transcription of human globin mRNA. This autoradiograph shows that the α-globin gene hybridized to chromosomes from region 1 and the β-, γ-, and δ-globin fragments hybridized to chromosomes from region 2. These genes were mapped more precisely by hybridizing a labeled probe to DNA from chromosomes sorted from populations containing translocations that carried portions of selected chromosomes into a new location of the flow karyotype. This approach was used, for example, to map the β-, γ-, and δ-globin genes to the distal portion of the short arm of chromosome 11 (88).

In the second procedure, the DNA hybridization step has been simplified considerably by hybridizing the labeled probe DNA to chromosomes sorted directly onto nitrocellulose filters (20, 86, 89). In this approach, approximately 10,000–30,000 chromosomes are sorted onto the filter, denatured, and hybridized to ^{32}P-labeled probe DNA. The filters are then autoradiographed to determine the chromosome fractions to which the probe binds. The advantage of this approach is its simplicity and the relatively small number of chromosomes required for an analysis.

Recombinant DNA Library Production

Another important application of chromosome sorting is the production of recombinant DNA libraries containing DNA only from one or a few chromosome types (30, 31, 36, 63–65, 67, 68, 149–151). To date mainly complete digest libraries have been constructed from 10^5 to 5×10^6 sorted

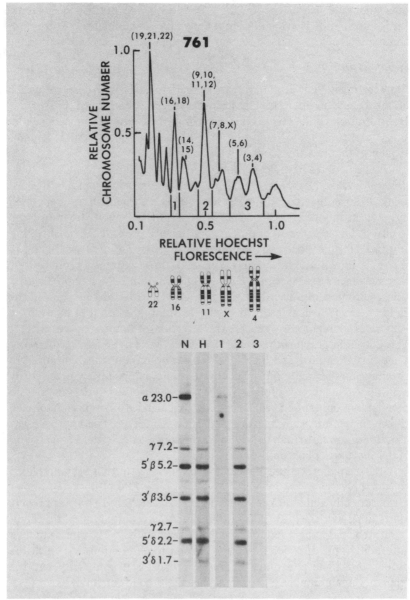

Figure 18   Flow karyotype measured for human chromosomes (*top*). Chromosomes were sorted from regions 1, 2, and 3. *Below*, the results of an EcoRI restriction enzyme analysis of DNA from a normal individual (*N*), from an individual lacking the α-globin genes (*H*), and from chromosomes sorted from regions 1, 2, and 3 following hybridization with cDNA produced from globin mRNA (88).

chromosomes. Typically, DNA extracted from the sorted chromosomes is digested to completion using restriction enzymes such as EcoRI or HindIII and cloned into vectors such as the lambda phage for which the packaging efficiency (efficiency with which chromosomal DNA can be incorporated into the cloning vector) is high. The first recombinant DNA library from sorted chromosomes consisted of 50,000 recombinant λgtWES.λB phages produced from 2×10^6 sorted human X chromosomes (31). Numerous libraries have now been constructed from sorted chromosomes. Indeed, the National Laboratory Gene Library Project at the Lawrence Livermore and Los Alamos National Laboratories has produced and begun to distribute recombinant DNA libraries produced from DNA of each of the human chromosomes. Table 1 lists libraries constructed from sorted chromosomes.

High quality library production requires high sorting purity. Since the purity depends on the degree to which the chromosomes can be resolved during flow karyotyping, dual beam chromosome sorting is often used since it allows good resolution of human chromosomes (especially chromosomes 13–22 and Y). Several groups have also found it advantageous to sort human chromosomes from human × hamster hybrid cells that contain only a few human chromosomes (29) or from human cell lines in which chromosome polymorphisms (51, 150) or rearrangements (35, 63, 67, 68, 150) leave one chromosome type particularly well resolved. Figure 9 shows an example of a bivariate Ho vs CA3 flow karyotype measured for line UV20HL21-7 containing human chromosomes 4, 8, 18, and 21. Human chromosomes 4 and 8 are especially well resolved and can be sorted with high purity. The human × hamster hybrid cells also grow well in culture and the fraction of selected human chromosomes in suspensions of chromosomes from hybrid cells is higher than in suspensions of chromosomes isolated from human cells so that sorting efficiency is increased. However, the hybrids may be karyotypically unstable. It is also important that the molecular weight of the DNA isolated from the sorted chromosomes be sufficiently high to allow efficient cloning. DNA fragment sizes suitable for production of complete digest libraries (typically 50 kb) may be obtained from sorted chromosomes isolated using most techniques. However, the polyamine isolation procedure seems to yield DNA with the highest molecular weight. Unfortunately this procedure is difficult to apply to fibroblasts.

FUTURE DIRECTIONS

The future of flow cytogenetics depends on improvements in procedures for the isolation and staining of chromosomes and on the development of improved instrumentation.

Table 1 Recombinant DNA libraries produced from sorted chromosomes

Chromosome	Number of recombinations	Chromosome	Number of recombinations	Chromosome	Number of recombinations
Human:					
1+2	6.1×10^4 (e, c, n)[a]	9	3.0×10^5 (h, c, n)	17	1.7×10^4 (h, c, n)
1+HSR	6.0×10^4 $(h, c, 58)$	10	3.9×10^5 (e, c, n)	18	8.9×10^5 (h, c, n)
3	NA[b]				2.0×10^4 (e, c, n)
4	2.3×10^4 (h, c, n)	11	1.1×10^5 (h, c, n)	19	3.0×10^5 (h, c, n)
	1.2×10^5 (e, c, n)		1.6×10^4 (e, c, n)		6.6×10^3 (e, c, n)
5	1.3×10^6 (e, c, n)	12	6.0×10^5 (e, c, n)	20	3.9×10^6 (h, c, n)
6	7.6×10^5 (h, c, n)	13	2.2×10^4 (h, c, n)	21	4.7×10^5 (h, c, n)
	4.8×10^4 (e, c, n)		1.1×10^4 (e, c, n)		3.0×10^5 $(e, l, 63)$
7	2.4×10^5 (e, c, n)	14+15	2.6×10^6 (h, c, n)	22	3.0×10^5 $(e, l, 63)$
		15	1.6×10^4 (e, c, n)		1.1×10^5 (h, c, n)
					2.5×10^4 (e, c, n)
8	7.4×10^5 (h, c, n)	16	2.5×10^4 (e, c, n)	X	5.0×10^4 $(e, l, 31)$
			1.5×10^6 (h, c, n)		6.0×10^4 $(e, c, 65)$
				dic(x)-(q24)	2.1×10^5 (e, c, n)
				Y	NR[c] $(h, c, 67)$
Mouse:					NR $(e, l, 105)$
t(X;7)1ct	1.0×10^5 $(e, l, 36)$				2.5×10^5 (h, c, n)
Hamster:					
1	6.0×10^4 $(e, a, 50)$				
2	9.0×10^4 $(e, a, 50)$				

[a] (enzyme, vector, reference): e, EcoRI complete digest library; h, HindIII complete digest library; c, Charon 21A library; l, λgtWES.λB library; a, Charon 4A library; n, National Laboratory Gene Library Project Newsletter #9, 4 September 1985. Information and libraries are available from Dr. M. A. Van Dilla, Lawrence Livermore National Laboratory, P.O. Box 5507 L-452, Livermore, CA 94550 or Dr. L. Deaven, Los Alamos National Laboratory, LS-4 Mailstop 888, Los Alamos, NM 87545.
[b] NA: not available.
[c] NR: not reported.

Chromosome Isolation

FLOW KARYOTYPING   Two problems remain in the production of chromosomes for flow karyotyping: the occurrence of DNA-containing debris and
the inability to produce high quality chromosome suspensions from cells
and tissues in which the mitotic index is low (e.g. solid tissues and cultures of
human tumor cells).

The presence of fluorescent debris in a chromosome preparation
complicates estimation of the frequency of occurrence of the chromosomes
that occur as peaks superimposed on the debris continuum. Accurate
estimation of peak area (for univariate analysis) or volume (for bivariate
analysis) is difficult without exact knowledge of the functional form of the
continuum, especially for the smaller chromosomes. The recent work of
Aten et al (3) and van den Engh et al (134, 135) suggests that continued work
on chromosome isolation procedures that require minimal manipulation
of the chromosome suspension may lead to isolation procedures that
routinely yield representative, high-quality chromosome preparations containing minimal debris.

Isolation of chromosomes from cells in which the mitotic index is low
remains difficult. It is often the case in such cultures that many nonviable,
partially degenerated cells are present. The nuclei from such cells may
disintegrate during isolation so that substantial amounts of debris-
containing DNA are released into the suspension. Improvements in this
area are likely to come from improved cell culture techniques and from
elimination of the damaged cells (e.g. by density centrifugation or enzymatic degradation prior to chromosome isolation).

CHROMOSOME SORTING   Continued development of techniques for production of high quality preparations of isolated, stained chromosome suspensions will also facilitate chromosome sorting. Efficient recovery of isolated
chromosomes from mitotic cells is important, especially since the advent of
the high speed sorter (113), in order to reduce the cell culture effort required
to support the sorting operation. Suspensions containing minimal debris
are desirable since debris may mimic chromosomes or pass unnoticed
through the sorter, contaminating the sorted chromosome fraction.

The speed of sorting may be increased by increasing the frequency of the
chromosome(s) to be sorted in the suspension since the speed of sorting is
usually limited by the rate at which chromosomes can be analyzed. Recent
studies have demonstrated that the frequency of selected chromosomes can
be increased substantially by fractionating the chromosomes according to
size using velocity sedimentation (21). The frequency of selected human and
hamster chromosomes has been increased fivefold using this technique.
Thus, the combination of preenrichment using velocity sedimentation and

high speed sorting may allow purification of chromosomes as much as 100 times faster than is possible using conventional sorting alone.

Cytochemistry

The success of flow karyotyping and the purity and ease with which chromosomes can be sorted depend on the degree to which chromosomes can be resolved during flow analysis. This, in turn, depends on the precision and distinctness with which chromosomes in suspension can be stained with fluorescent dyes. Flow karyotypic studies to date have been based primarily on the use of DNA-specific dyes like Ho, CA3, EB, and PI. Most chromosomes can be resolved using these dyes if sufficient resolution can be maintained during flow cytometry. However, this may require the use of high power lasers, specialized instrumentation, and relatively low analysis rates. Flow karyotyping and chromosome purification may be improved significantly by development of staining techniques with higher chromosome specificity. Improvements in this area may come from use of chromosome-specific monoclonal antibodies as immunochemical reagents for chromosome staining and/or from the use of chromosome-specific recombinant DNA probes as reagents for chromosome staining.

MONOCLONAL ANTIBODIES AS IMMUNOCHEMICAL REAGENTS   The utility of monoclonal antibodies as reagents for chromosome staining depends on the existence of structural and/or biochemical differences between chromosome types that can be recognized by the antibody and on the development of procedures for immunochemically staining chromosomes in suspension. Chromosomal proteins seem to be the most likely target for production of chromosome-specific monoclonal antibodies. The nonuniform distribution of nonhistone chromosome proteins among mammalian chromosomes has been documented for Chinese hamster chromosomes (145). Thus, these proteins may be good targets for antibody development. Antibodies have also been developed against proteins associated with chromosomal subregions [e.g. the centromere (103)]. These antibodies may prove useful during morphological classification of chromosomes using slit-scan flow cytometry.

The possibility of staining chromosomes in suspension using immunochemical reagents has been demonstrated by Trask et al (132). In this study, an antibody against human histone H2B was used to stain chromosomes isolated from human, mouse, and hamster cells. The chromosomes were then counterstained with a DNA-specific dye and analyzed by flow cytometry. Figure 19 shows the bivariate DNA vs FITC-immunofluorescence distribution measured for a mixture of Chinese hamster and human chromosomes. The immunofluorescence is clearly apparent and the precision of staining with the DNA dye is acceptable.

Interestingly, the human chromosomes stain much more intensely than the rodent chromosomes in spite of the fact that the H2B protein is highly conserved among mammals. This suggests that the antibody may be sensitive to conformational differences among the chromosomes from the various species, and raises the possibility of using chromosome-specific conformation as the basis for immunochemical discrimination among chromosomes.

FLUORESCENCE HYBRIDIZATION   Successful chromosome staining procedures take advantage of biochemical and/or structural differences among chromosomes. However, because these differences are usually small, sensitive, high-precision analysis procedures are required. Fluorescence hybridization, however, may be developed to take advantage of a chromosome feature that seems likely to be truly chromosome-specific— the DNA sequence itself. In this approach, chemically modified, chromosome-specific DNA (called probe DNA) is used as the staining reagent. The probe DNA is hybridized to the denatured DNA in cells and chromosomes on slides (71, 72, 115) or in suspension (133), and is made fluorescent by treatment with a fluorescent reagent that binds specifically to the chemically modified probe DNA. For example, probe DNA may be labeled with biotin and fluorescently stained with fluorescein-avidin (71, 72)

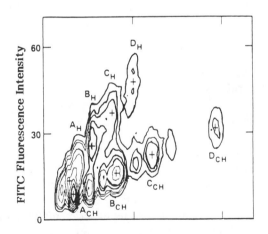

Hoechst 33258 Fluorescence Intensity

Figure 19   Bivariate DNA vs immunofluorescence distribution measured for a mixture of human and Chinese hamster chromosomes stained immunofluorescently with an antibody against human histone H2B (linked to fluorescein) and Ho. A_H, B_H, C_H, and D_H: human chromosome peaks; A_{CH}, B_{CH}, C_{CH}, and D_{CH}: Chinese hamster peaks.

or may be labeled with acetylaminofluorene (AAF) and fluorescently stained with a fluorescein-conjugated antibody against the AAF-DNA adduct (69, 131). In studies by Pinkel et al (114) where biotin-labeled whole genomic human DNA was used as a reagent to stain human chromosomes in a human × hamster hybrid cell line, the specificity and stoichiometry of probe-linked fluorescence was high. In addition, the intensity of fluorescence was proportional to the amount of human DNA in the hybrid cells. It remains to be seen whether fluorescence hybridization can be applied to chromosomes in suspension. However, recent studies by Trask et al (133) showed the feasibility of using fluorescence hybridization to stain mouse nuclei in suspension.

After development of fluorescence hybridization procedures for chromosomes in suspension, it will be necessary to develop DNA probes that are specific for a single chromosome and to apply these probes to chromosomes in suspension. One potential source for chromosome-specific probes may be the recombinant DNA libraries produced from sorted chromosomes (see Table 1). Multiple, unique sequence probes may be selected and used as a cocktail to label a selected chromosome more or less uniformly. The key in these studies will be to select a sufficiently large number of probes and to develop fluorescence hybridization procedures to allow intense, uniform chromosome staining. Another potential source for chromosome-specific probes is the increasingly large selection of newly discovered chromosome-specific repetitive DNA probes (10, 41, 85, 95, 141). One of these probes has been used successfully to stain selectively a portion of the human Y chromosome in a human metaphase spread (114).

Data Analysis

Improved procedures for quantitative analysis of flow karyotypes may permit more accurate estimation of peak areas or volumes. D. H. Moore & J. W. Gray (manuscript in preparation), for example, have developed a procedure for estimation of the peak means and areas (volumes) that does not require functional approximation of the debris continuum. This procedure takes advantage of the fact that the debris continuum usually varies slowly compared to the peaks in a flow karyotype so that the derivative of the continuum is near zero. Thus, the peak means and areas can be estimated by fitting the derivative of the sum of several normal distributions to the derivative of the experimentally measured flow karyotype. Since the derivative of the continuum is near zero, it has little effect on the analysis. This procedure has the disadvantage that it can be applied only to flow karyotypes in which the number of measurements is large. However, it seems particularly attractive as a procedure for analysis of bivariate flow karyotypes for which the debris continuum cannot be accurately approximated.

Flow Instrumentation

The utility of flow cytogenetics in the future may be affected significantly by the development of inexpensive, simple to operate instruments for routine chromosome classification and the development of a new generation of chromosome analysis instruments permitting measurements that are currently not possible.

LOW-COST INSTRUMENTATION   Current instruments for chromosome analysis and sorting utilize expensive, high-power ion lasers for chromosome excitation. These lasers are required because of the relatively low efficiency of fluorescence collection of conventional flow instruments (typically 10–15%). Thus, development of chromosome analysis flow systems with high collection efficiency (approximately 85%), similar to those developed for other applications, may facilitate use of lower power lasers (121). The laser requirements for flow cytogenetics may also be reduced by selecting fluorescent dyes that excite in the red. Potentially useful fluorescent dyes with this characteristic have been reported (81, 119). Red lasers are considerably less expensive and complex than lasers that emit in the blue, especially if the power requirements can be reduced by increasing the system efficiency (119). Thus inexpensive, reliable, and simply operated instruments may be developed specifically for chromosome analysis.

ADVANCED INSTRUMENTATION   It is now clear that chromosome shape, DNA content, and base composition are important features for flow cytometric chromosome discrimination. Thus, the development of instruments that allow simultaneous classification of chromosomes according to, for example, total Ho and CA3 fluorescence and centromeric index should allow substantially improved chromosome discrimination, especially for human chromosomes. The development of chromosome scanning instruments with increased spatial resolution may also allow improved chromosome classification by allowing measurement of chromosome features such as the banding pattern, centromere location, or distribution of fluorescence resulting from fluorescence hybridization. Scanning resolution may be increased by increasing the spatial resolution in present slit-scan flow cytometers or by developing new scanning techniques (23). One new scanning technique now under development involves scanning chromosomes through a series of fringes produced by the constructive and destructive interference of two intersecting laser beams. Information about the morphology of chromosomes is derived mathematically from the shape of the fluorescence profiles recorded as the chromosomes pass lengthwise through the fringe field and from knowledge of the shape of the fringe field itself (J. Mullikin, R. Norgren, J. Lucas, J. Gray, manuscript in preparation). Spatial resolution approaching 1 μm may be achieved from improvements in slit-scan flow cytometers and resolution approaching 0.3 μm is theoretically possible from fringe-scanning.

CONCLUSION

Flow cytogenetics has developed considerably since Gray et al (43) reported in 1975 that chromosomes could be isolated, fluorescently stained, classified, and purified using flow cytometry and sorting (43). Substantial work has already demonstrated the utility of flow karyotyping in quantitative chromosome classification. With continued development of the techniques for chromosome isolation, staining, and analysis, it seems likely that flow karyotyping will become an accepted adjunct to banding analysis for chromosome classification. Flow sorting allows purification of DNA and proteins from chromosomes of a single type to a degree that cannot be achieved by any other means. Continued development of the chromosome preparation and sorting technology should make purification of microgram (and possibly milligram) quantities of chromosomes of one type routine.

ACKNOWLEDGMENTS

The authors appreciate the expert assistance of Mrs. Angela Riggs in the preparation of this manuscript. This work was performed under the auspices of the United States Department of Energy by the Lawrence Livermore National Laboratory under contract number W-7405-ENG-48 with support from USPHS Grants HD 17665 and GM 25076.

Literature Cited

1. Arrighi, F., Hsu, T. C. 1971. *Cytogenetics* 10:81–86
2. Aten, J., Barendsen, J. B. A. 1980. In *Flow Cytometry IV*, ed. O. D. Laerum, T. Lindmo, E. Thorud, pp. 287–92. Oslo: Universitetsforlaget
3. Aten, J. A., Kooi, M. W., Bijman, J. T., Kipp, J. B. A., Barendsen, G. M. 1984. In *Biological Dosimetry—Cytometric Approaches to Mammalian Systems*, ed. W. G. Eisert, M. L. Mendelsohn, pp. 51–59. Berlin: Springer-Verlag
4. Bartholdi, M. F., Ray, F. A., Cram, L. S., Kraemer, P. M. 1984. *Cytometry* 5:534–38
5. Behr, W., Honikel, K., Hartmann, G. 1969. *Eur. J. Biochem.* 9:82–92
6. Benz, R. D., Burki, H. J. 1978. *Exp. Cell Res.* 112:143–53
7. Bijman, J. T. 1983. *Cytometry* 3:354–458
8. Blumenthal, A. B., Dieden, J. D., Kapp, L. N., Sedat, J. W. 1979. *J. Cell Biol.* 81:255–59
9. Bontemps, J., Houssler, C., Fredericq, E. 1975. *Nucleic Acids Res.* 2:971–84
10. Burk, R. D., Szabo, P., O'Brien, S., Nash, W. G., Yu, L.-C., Smith, K. D. 1985. *Chromosoma* 92:225–33
11. Buys, C. H. C. M., Koerts, T., Aten, J. A. 1982. *Hum. Genet.* 61:157–59
12. Carrano, A. V., Gray, J. W., Langlois, R. G., Burkhart-Schultz, K., Van Dilla, M. A. 1979. *Proc. Natl. Acad. Sci. USA* 76:1382–84
13. Carrano, A. V., Gray, J. W., Langlois, R. G., Yu, L.-C. 1983. In *Chromosomes and Cancer*, ed. J. D. Rowley, J. E. Ultmann, pp. 195–209. New York: Academic
14. Carrano, A. V., Gray, J. W., Moore, D. H. Jr., Minkler, J. L., Mayall, B. H., et al.

1976. *J. Histochem. Cytochem.* 24:348–54

15. Carrano, A. V., Lebo, R. V., Yu, L.-C., Kan, Y. W. 1981. In *Haematology and Blood Transfusion*, Vol. 26, *Modern Trends in Human Leukemia IV*, ed. R. Neth, et al, pp. 156–59. Berlin/Heidelberg: Springer-Verlag

16. Carrano, A. V., Van Dilla, M. A., Gray, J. W. 1979. See Ref. 98, pp. 421–51

17. Caspersson, T. 1950. *Cell Growth and Cell Function, A Cytochemical Study*. New York: Norton. 185 pp. 1st ed.

18. Caspersson, T., Lomakka, G., Zech, L. 1971. *Hereditas* 67:89–102

19. Caspersson, T., Zech, L., Johansson, C., Modest, E. F. 1970. *Chromosoma* 30:215–27

20. Collard, J. G., de Boer, P. A. J., Janssen, J. W. G., Schijven, J. F., de Jong, B. 1985. *Cytometry* 6:179–85

21. Collard, J. G., Tulp, A., Hollander, J., Bauer, F., Boezeman, J. 1980. *Exp. Cell Res.* 126:191–97

22. Comings, D. E., Kovacs, B. W., Avelino, E., Harris, D. C. 1975. *Chromosoma* 50:111–45

23. Cram, L. S., Arndt-Jovin, D., Grimwade, B., Jovin, T. 1979. *J. Histochem. Cytochem.* 27:445–53

24. Cram, S., Bartholdi, M., Wheeless, L., Gray, J. W. 1985. See Ref. 136, pp. 163–94

25. Cremer, C., Cremer, T., Gray, J. W. 1982. *Cytometry* 2:287–90

26. Cremer, C., Gray, J. W. 1982. *Mutat. Res.* 94:133–42

27. Cremer, C., Gray, J. W. 1982. *Somatic Cell Genet.* 8:319–27

28. Cremer, C., Gray, J. W. 1983. *Cytometry* 3:282–86

29. Cremer, C., Gray, J. W., Ropers, H. 1982. *Hum. Genet.* 60:262–66

30. Cremer, C., Rappold, G., Gray, J. W., Muller, C. R., Ropers, H. H. 1984. *Cytometry* 5:572–79

31. Davies, K. E., Young, B. D., Elles, R. G., Hill, M. E., Williamson, R. 1981. *Nature* 293:374–76

32. Dean, P. N. 1985. See Ref. 136, pp. 195–221

33. Dean, P. N., Pinkel, D. 1978. *J. Histochem. Cytochem.* 26:622–27

34. Disteche, C., Bontemps, J., Houssier, C., Frederic, J., Fredericq, E. 1980. *Exp. Cell Res.* 125:251–64

35. Disteche, C. M., Carrano, A. V., Ashworth, L. K., Burkhart-Schultz, K., Latt, S. A. 1981. *Cytogenet. Cell Genet.* 29:189–97

36. Disteche, C. M., Kunkel, L. M., Lojewski, A., Orkin, S. H., Eisenhard, M., et al. 1982. *Cytometry* 2:282–86

37. Drewinko, B., Romsdahl, M. M., Yang, L. Y., Ahearn, M. J., Trujillo, J. M. 1976. *Cancer Res.* 36:467–75

38. Fantes, J., Green, D., Elder, J., Malloy, P., Evans, H. J. 1983. *Mutat. Res.* 119:161–68

39. Fried, J., Perez, A. G., Clarkson, B. D. 1976. *J. Cell Biol.* 71:172–81

40. Fulwyler, M. J. 1969. *Science* 166:747–49

41. Graham, G. J., Hall, T. J., Cummings, M. R. 1984. *Am. J. Hum. Genet.* 36:25–35

42. Gray, J. W., Carrano, A. V., Moore, D. H. Jr., Steinmetz, L. L., Minkler, J., et al. 1975. *Clin. Chem.* 21:1258–62

43. Gray, J. W., Carrano, A. V., Steinmetz, L., Moore, D., Mayall, B., Mendelsohn, M. 1975. *Proc. Natl. Acad. Sci. USA* 72:1231–34

44. Gray, J. W., Langlois, R. G., Carrano, A. V., Burkhart-Schultz, K., Van Dilla, M. A. 1979. *Chromosoma* 73:9–27

45. Gray, J. W., Lucas, J., Pinkel, D., Peters, D., Ashworth, L., Van Dilla, M. A. 1980. See Ref. 2, pp. 249–55

46. Gray, J. W., Lucas, J., Yu, L.-C., Langlois, R. 1984. In *Biological Dosimetry—Cytometric Approaches to Mammalian Systems*, ed. W. G. Eisert, M. L. Mendelsohn, pp. 25–35. Berlin: Springer-Verlag

47. Gray, J. W., Peters, D., Merrill, J. T., Martin, R., Van Dilla, M. A. 1979. *J. Histochem. Cytochem.* 27:441–44

48. Green, D. K., Fantes, J. A., Buckton, K. E., Elder, J. K., Malloy, P., et al. 1984. *Hum. Genet.* 66:143–46

49. Green, D. K., Fantes, J. A., Spowart, G. 1984. See Ref. 46, pp. 67–76

50. Griffin, J., Cram, L. S., Crawford, B., Jackson, P., Schilling, R., et al. 1984. *Nucleic Acids Res.* 121:4019–34

51. Harris, P., Boyd, E., Ferguson-Smith, M. A. 1985. *Hum. Genet.* 70:59–65

52. Herzenberg, L. A., Sweet, R. G., Herzenberg, L. A. 1976. *Sci. Am.* 234:108–17

53. Hilwig, I., Gropp, A. 1972. *Exp. Cell Res.* 75:122–26

54. Hilwig, I., Gropp, A. 1975. *Exp. Cell Res.* 91:457–60

55. Horan, P., Wheeless, L. 1977. *Science* 198:149–57

56. Jensen, R. H., Langlois, R. G., Mayall, B. H. 1977. *J. Histochem. Cytochem.* 25:954–64

57. Kamiyama, M. 1968. *J. Biochem.* 63:566–72

58. Kanda, N., Schreck, R., Alt, F., Bruns, G., Baltimore, D., Latt, S. 1983. *Proc. Natl. Acad. Sci. USA* 80:4069–73

59. Kapuscinski, J., Skoczylas, B. 1977. *Anal. Biochem.* 83:252–57

234   GRAY & LANGLOIS

60. Kapuscinski, J., Skoczylas, B. 1978. *Nucleic Acids Res.* 5:3775–99
61. Kooi, M. W., Aten, J. A., Stap, J., Kipp, J. B. A., Barendsen, G. W. 1984. *Cytometry* 5:547–49
62. Korenberg, J. R., Engels, W. R. 1978. *Proc. Natl. Acad. Sci. USA* 75:3382–86
63. Krumlauf, R., Jeanpierre, M., Young, B. D. 1982. *Proc. Natl. Acad. Sci. USA* 79:2971–75
64. Kunkel, L. M., Lalande, M., Monaco, A. P., Flint, A., Middlesworth, W., Latt, S. A. 1985. *Gene* 33:251–58
65. Kunkel, L. M., Tantravahi, U., Eisenhard, M., Latt, S. A. 1982. *Nucleic Acids Res.* 10:1557–79
66. Labarca, C., Paigen, K. 1980. *Anal. Biochem.* 102:344–52
67. Lalande, M., Kunkel, L. M., Flint, A., Latt, S. A. 1984. *Cytometry* 5:101–7
68. Lalande, M., Schreck, R. R., Hoffman, R., Latt, S. A. 1985. *Cytometry* 6:1–6
69. Landegent, J., Jansen in de Wal, N., Baan, R., Hoeijmakers, J., van der Ploeg, M. 1984. *Exp. Cell Res.* 153:61–72
70. Lane, M. J., Dabrowiak, J. C., Vournakis, J. N. 1983. *Proc. Natl. Acad. Sci. USA* 80:3260–64
71. Langer, P., Waldrop, A., Ward, D. 1981. *Proc. Natl. Acad. Sci. USA* 78:6633–37
72. Langer-Safer, P., Levine, M., Ward, D. 1982. *Proc. Natl. Acad. Sci. USA* 79:4381–85
73. Langlois, R. G., Carrano, A. V., Gray, J. W., Van Dilla, M. A. 1980. *Chromosoma* 77:229–51
74. Langlois, R. G., Jensen, R. H. 1979. *J. Histochem. Cytochem.* 27:72–79, 1559
75. Langlois, R. G., Yu, L.-C., Gray, J. W., Carrano, A. V. 1982. *Proc. Natl. Acad. Sci. USA* 79:7876–80
76. Latt, S. A. 1974. *Science* 185:74–76
77. Latt, S. A. 1977. *Can. J. Genet. Cytol.* 19:603–23
78. Latt, S. A. 1979. See Ref. 98, pp. 263–84
79. Latt, S. A., Brodie, S., Munroe, S. H. 1974. *Chromosoma* 49:17–40
80. Latt, S. A., Juergens, L. A., Matthews, D. J., Gustashaw, K. M., Sahar, E. 1980. *Cancer Genet. Cytogenet.* 1:187–96
81. Latt, S. A., Marino, M., Lalande, M. 1984. *Cytometry* 5:339–47
82. Latt, S. A., Sahar, E., Eisenhard, M. E. 1979. *J. Histochem. Cytochem.* 27:65–71
83. Latt, S. A., Sahar, E., Eisenhard, M. E., Juergens, L. A. 1980. *Cytometry* 1:2–12
84. Latt, S. A., Wohlleb, J. C. 1975. *Chromosoma* 63:297–316
85. Lau, L.-F., Ying, K. L., Donnell, G. N. 1985. *Hum. Genet.* 69:102–5
86. Lebo, R. 1982. *Cytometry* 3:145–54
87. Lebo, R. V., Bastian, A. M. 1982. *Cytometry* 3:213–19
88. Lebo, R., Carrano, A. V., Burkhart-Schultz, K., Dozy, A. M., Yu, L.-C., Kan, Y. W. 1979. *Proc. Natl. Acad. Sci. USA* 76:5804–8
89. Lebo, R., Gorin, F., Fletterick, R., Kao, F.-T., Cheung, M.-C., et al. 1984. *Science* 225:57–59
90. Le Pecq, J. B. 1971. *Methods Biochem. Anal.* 20:41–86
91. Le Pecq, J. B., Paoletti, C. 1967. *J. Mol. Biol.* 27:87–106
92. Shi, L. M., Ye, Y. Y., Duan, X. S. 1980. *Cytogenet. Cell Genet.* 26:22–27
93. Lin, M. S., Comings, D. E., Alfi, O. S. 1977. *Chromosoma* 60:15–25
94. Lucas, J. N., Gray, J. W., Peters, D. C., Van Dilla, M. A. 1983. *Cytometry* 4:109–16
95. Manuelidis, L., Ward, D. C. 1984. *Chromosoma* 91:28–38
96. Martin, R. F., Holmes, N. 1983. *Nature* 302:452–54
97. Matsson, P., Rydberg, B. 1981. *Cytometry* 1:369–72
98. Melamed, M., Mullaney, P., Mendelsohn, M. L., eds. 1979. *Flow Cytometry and Sorting.* New York: Wiley. 716 pp.
99. Mendelsohn, M. L., Mayall, B. H. 1974. In *Human Chromosome Methodology*, ed. J. Yunis, pp. 311–46. New York: Academic
100. Mendelsohn, M. L., Mayall, B. H., Bogart, E., Moore, D. H., Perry, B. 1973. *Science* 179:1126–29
101. Meyne, J., Bartholdi, M. F., Travis, G., Cram, L. S. 1984. *Cytometry* 5:580–83
102. Moore, D. H. II. 1975. *UCRL Report 76507, TID*, Lawrence Livermore Natl. Lab., Livermore, Calif.
103. Moroi, Y., Peebles, C., Fitzler, M., Steigerwald, F., Tan, E. 1980. *Proc. Natl. Acad. Sci. USA* 77:1627–31
104. Muirhead, K. A., Horan, P. K., Poste, G. 1985. *Biotechnology* 3:337–56
105. Muller, C., Davies, K., Cremer, C., Rappold, G., Gray, J. W. 1983. *Hum. Genet.* 64:110–15
106. Olmsted, J. III, Kearns, D. R. 1977. *Biochemistry* 1:3647–54
107. Otto, F., Oldiges, H. 1980. *Cytometry* 1:13–17
108. Deleted in proof
109. Otto, F. J., Oldiges, H., Gohde, W., Barlogie, B., Schumann, J. 1980. *Cytogenet. Cell Genet.* 27:52–56
110. Otto, F., Oldiges, H., Gohde, W., Dertinger, H. 1980. See Ref. 2, pp. 284–86
111. Paoletti, J., Le Pecq, J. B. 1971. *J. Mol. Biol.* 59:43–62
112. Pearson, P. L. 1973. In *New Techniques in Biophysics and Cell Biology*, ed. R. H.

Pain, B. J. Smith, pp. 183–208. New York: Wiley
113. Peters, D., Branscomb, E., Dean, P., Merrill, J., Pinkel, D., et al. 1985. *Cytometry* 6:290–301
114. Pinkel, D., Straume, T., Gray, J. W. 1986. *Proc. Natl. Acad. Sci. USA.* In press
115. Rappold, G. A., Cremer, T., Hager, H. D., Davies, K. E., Muller, C. R., Yang, T. 1984. *Hum. Genet.* 67:317–25
116. Sahar, E., Latt, S. A. 1978. *Proc. Natl. Acad. Sci. USA* 75:5650–54
117. Scamrov, A. V., Beabealashvilli, R. S. 1983. *FEBS Lett.* 164:97–101
118. Schnedl, W. 1978. *Hum. Genet.* 41:1–9
119. Shapiro, H., Glazer, A., Christenson, L., Williams, J., Strom, T. 1983. *Cytometry* 4:276–79
120. Sillar, R., Young, B. D. 1981. *J. Histochem. Cytochem.* 29:74–78
121. Skogen-Hagenson, M. J., Salzman, G., Mullaney, P., Brockman, W. 1977. *J. Histochem. Cytochem.* 25:784–95
122. Stewart, G., Harris, P., Galt, J., Ferguson-Smith, M. 1985. *Nucleic Acids Res.* 13:4125–32
123. Stohr, M. 1976. In *Pulse-Cytophotometry II*, ed. W. Göhde, J. Schumann, Th. Büchner, pp. 39–45. Ghent: European Press
124. Stohr, M., Hutter, K. J., Frank, M., Futterman, G., Goerttler, K. 1980. *Histochemistry* 67:179–90
125. Stohr, M., Hutter, K. J., Frank, M., Goerttler, K. 1982. *Histochemistry* 74:57–61
126. Stubblefield, E., Cram, S., Deaven, L. 1975. *Exp. Cell Res.* 94:464–68
127. Stubblefield, E., Linde, S., Franolich, F. K., Lee, L. Y. 1978. In *Methods in Cell Biology—Chromatin and Chromosomal Protein Research II*, ed. G. Stein, J. Stein, L. J. Kleinsmith, pp. 101–13. New York: Academic
128. Stubblefield, E., Oro, J. 1982. *Cytometry* 2:273–81
129. Stubblefield, E., Wray, W. 1978. *Biochem. Biophys. Res. Commun.* 83:1404–14
130. Swartzendruber, D. E. 1977. *J. Cell. Physiol.* 90:445–54
131. Tchen, P., Fuchs, R., Sage, E., Leng, M. 1984. *Proc. Natl. Acad. Sci. USA* 81:3466–70
132. Trask, B., van den Engh, G., Gray, J., Vanderlaan, M., Turner, B. 1984. *Chromosoma* 90:295–302
133. Trask, B., van den Engh, G., Landegent, J., Jansen in de Wal, N., van der Ploeg, M. 1985. *Science* 230:1401–3
134. van den Engh, G., Trask, B., Cram, S., Bartholdi, M. 1984. *Cytometry* 5:108–17
135. van den Engh, G. J., Trask, B. J., Gray, J. W., Langlois, R. G., Yu, L.-C. 1985. *Cytometry* 6:92–100
136. Van Dilla, M. A., Dean, P. N., Laerum, O., Melamed, M., eds. 1985. *Flow Cytometry: Instrumentation and Data Analysis.* New York: Academic. 288 pp.
137. Van Dilla, M. A., Trujillo, T., Mullaney, P., Coulter, J. 1969. *Science* 163:1213–14
138. Van Dyke, M. W., Dervan, P. B. 1983. *Nucleic Acids Res.* 11:5555–67
139. Van Dyke, M. W., Dervan, P. B. 1983. *Biochemistry* 22:2373–77
140. Welleweerd, J., Wilder, M., Carpenter, S., Raju, M. 1984. *Radiat. Res.* 99:44–51
141. Willard, H. F. 1985. *Am. J. Hum. Genet.* 37:524–32
142. Wirschubsky, Z., Perlmann, C., Lindsteen, J., Klein, G. 1983. *Int. J. Cancer* 32:147–53
143. Wray, W., Stubblefield, E. 1970. *Exp. Cell Res.* 59:469–78
144. Wray, W., Stubblefield, E. 1982. In *Techniques in Somatic Cell Genetics*, ed. J. W. Shay, pp. 349–73. New York: Plenum
145. Wray, W., Stubblefield, E., Humphrey, R. 1972. *Nature New Biol.* 238:237–38
146. Wray, W., Wray, V. P. 1979. *J. Histochem. Cytochem.* 27:454–57
147. Wray, W., Wray, V. P. 1980. *Cytometry* 1:18–20
148. Wray, V. P., Wray, W. 1984. In *Chromosomal Nonhistone Proteins—Structural Associations*, ed. L. S. Hnilica, pp. 181–212. Boca Raton: CRC Press
149. Young, B. D. 1985. *Ann. NY Acad. Sci.* 450:11–23
150. Young, B. D., Ferguson-Smith, M. A., Sillar, R., Boyd, E. 1981. *Proc. Natl. Acad. Sci. USA* 78:7727–31
151. Young, B. D., Jeanpierre, M., Goyns, M. Stewart, G., Elliot, T., Krumlauf, R. 1983. *Hematol. Blood Transfus.* 28:301–10
152. Yu, L.-C., Aten, J., Gray, J., Carrano, A. V. 1981. *Nature* 293:154–55
153. Yunis, J. J. 1983. *Science* 221:227–36
154. Yunis, J. J., Roldan, L., Yasmineh, W., Lee, J. C. 1971. *Nature* 231:532–33
155. Zech, Z., Haglund, U., Nilsson, K., Klein, G. 1976. *Int. J. Cancer* 17:47–56

Ann. Rev. Biophys. Biophys. Chem. 1986. 15 : 237–57

ELECTRON MICROSCOPY OF FROZEN, HYDRATED BIOLOGICAL SPECIMENS

Wah Chiu

University of Arizona, Department of Biochemistry and Department of Molecular and Cellular Biology, Tucson, Arizona 85721

CONTENTS

PERSPECTIVES AND OVERVIEW

Transmission electron microscopy is capable of resolving object detail as small as 3 Å (18). This technique differs from other diffraction methods such as X-ray or neutron diffraction because electrons can be focused by electromagnetic lenses to form high-resolution images. Such an image has information about the relative positions of atoms or groups of atoms in an object, which is the equivalent of having the relative phases of the Fourier coefficients in the Fourier transform of the object (44, 57, 65, 93). An object can be tilted at angles up to 80° in an electron microscope (94). Images of tilted specimens can be combined by computer processing methods to reconstruct the three-dimensional configuration of the object (2).

When electron microscopy techniques have been used to study a

237

0883–9182/86/0610–0237$02.00

biological macromolecule or a macromolecular assembly, the results have frequently been challenged with two questions: (a) How reliable is the image of a biological specimen that has been fixed, stained, and dehydrated prior to the microscopic examination? (b) To what extent is the biological specimen altered by the flux of high energy electrons needed for focusing and imaging? Recent advances in the technology of electron microscopy for three-dimensional structural analysis of macromolecules and macromolecular assemblies eliminate both these concerns (16, 55). Biological specimens can be prepared in an unstained, hydrated state and their images can be recorded with little radiation damage. Image processing techniques can enhance the contrast of the images (68) and combine data from various tilt angles to determine the three-dimensional configuration of a macromolecule or a macromolecular assembly (2, 55, 95). Structural features can also be studied in different chemical states to probe for conformational changes in relationship to different functional states (45, 92, 94). A number of review papers have been published that address various aspects of three-dimensional electron microscopy of macromolecules (2, 15, 16, 41, 46). The following introductory section reviews recent advances in preparative methods for unstained, unfixed specimens of interest to molecular and cellular biologists.

Water is an essential component in all biological macromolecules and macromolecular assemblies. In its absence, they can be denatured and/or become nonfunctional. The need for a high vacuum (10^{-6}–10^{-7} torr) in the column of the electron microscope poses a technical problem for examining biological specimens in the presence of water. Three independent techniques were developed to circumvent the hydration problems. First, a hydration chamber in which water vapor pressure could be regulated was built around the specimen area of an electron microscope (73). Second, the specimen was embedded in glucose by evaporation from 1% solution (93). The rationale of this approach is that the hydroxyl groups in the glucose molecule may be used in place of water to form hydrogen bonds with the protein in order to preserve high-resolution structure. However, it is not known how much bound water is still maintained around the protein in the preparation. Third, the specimen was rapidly frozen at a temperature close to that of liquid nitrogen and then examined in a frozen, hydrated state with a microscope equipped with a low temperature stage (88).

It is well known from protein X-ray crystallography that 40–60% of the volume of most protein crystals is solvent (74, 75). When a protein crystal is dehydrated, it no longer diffracts either X-ray or electron radiation to high resolution (16, 75). The effectiveness of a method for preserving the structural integrity of a protein molecule for electron microscopy can be evaluated from the quality of a low-dose electron diffraction pattern of a thin protein crystal. Catalase crystals have been used as test specimens, and their electron diffraction patterns showed the feasibility of attaining reflections beyond 3 Å resolution with any of the above methods (73, 88, 93).

The glucose embedding method is the easiest to use because the procedure is identical to the conventional negative stain method. No alteration in an electron microscope is required. The inherent disadvantage of this method is the similarity in scattering density between protein and glucose, which results in poor contrast at low resolution (24, 93). However, the glucose embedding method has been applied for high-resolution structural studies of a number of thin crystals of proteins including purple membrane (93), crotoxin complex (20, 61), T4 DNA helix destabilizing protein (19, 23), Fc protein (49), tRNA (40), and fungal lysozyme (80).

In contrast to the glucose embedding method, both hydration chamber and frozen hydration methods require modification of the microscope. The hydration chamber technique has been used to localize the phase-separated lipid domains in cell membranes and to observe their movements (58, 59). An important drawback to this technique is the poor contrast in the image, presumably because of scattering from the large amount of water vapor required in the chamber.

In contrast to glucose embedding and hydration chamber techniques, the frozen hydration technique has yielded relatively good contrast in the low-resolution images of an ice-embedded catalase crystal (90). This result has stimulated active research that in the past few years has resulted in improvements of both the preparative procedures and the hardware needed to observe the ice-embedded specimens in an electron microscope. Applications of this technique to a broad spectrum of biological materials have been developed rapidly. This paper focuses on the recent developments in the microscopic technique of frozen, hydrated specimens and examples of its biological applications.

PREPARATIVE PROCEDURES FOR FROZEN, HYDRATED SPECIMENS

In an ideal preparation, the frozen, hydrated specimen is embedded in a layer of ice that is slightly thicker than the specimen. Table 1 shows a number of different procedures for preparing a frozen, hydrated specimen. All of these methods require a hydrophilic support film. The support film can be made hydrophilic by a vacuum deposit of a thin film of silicon monoxide (87) or by glow discharge in air or in an alkylamine vapor (36). The latter technique is the most commonly used. The support film is

Table 1   Methods of preparing frozen, hydrated biological specimens

Method	Support film	Film hydrophilicity	Water thinning	Freezing	Coolant	Reference
Sandwich between folding grids	C-film	Coating with a thin layer of silicon monoxide	Evaporation	Manual	LiN_2	(87)
Sandwich between 2 carbon films	C-film	No treatment	Evaporation	Manual	LiN_2	(60)
Sandwich between behenic acid film and polylysine-coated film	C-film	Glow discharge in air	Automatically formed	Manual	LiN_2	(14)
Sandwich between grids of different meshes	Polyimide film	No treatment necessary	Blotting	Manual or guillotine	LiN_2	(85a)
Thin film freezing	C-film or holey grid	Glow discharge in alkylamine	Blotting	Guillotine	Li ethane	(1, 36)
Spray freezing	C-film	Glow discharge in air	Spray	Mechanical device	Li ethane	(36)

generally made of carbon, but in one case was made of polyimides (85a). Alternatively, one can use a holey grid on which the frozen, hydrated specimens are suspended across the holes (1). The absence of a support film enhances the contrast of the specimen but makes it difficult to detect the contrast transfer function in the optical diffractogram of the image. The contrast transfer function is used to determine defocus and astigmatism values as part of the image reconstruction procedure (38). A recent study suggests that this may not be a problem for low-resolution work, particularly when a relatively large defocus value is used (70).

The first critical step in the frozen hydration procedure is to obtain a thin film of water by removing the excess liquid on the grid prior to its freezing. This can be done by draining off most of the bulk liquid on the grid slowly and gently with filter paper and then waiting for evaporation to leave a thin layer of water. The amount of water left on the grid can be judged by direct visual inspection or from observing an interference pattern of the water-air interface at the edge of the grid in a phase contrast light microscope. To avoid the complete drying of the prepared specimen before it is actually frozen, the evaporation can be performed in a humidity box (90) or the sample can be sandwiched between two support films (14, 60, 83, 85a, 87).

A potential drawback of reducing the water on the sample by evaporation is the resultant change in the ionic strength and the pH of the sample, which may perturb the structural integrity of the biological macromolecule. For example, a protein crystal grown in low salt can be partially disordered or even completely dissolved by the increase in the salt concentration induced by drying. To avoid the alteration of the chemical environment in the samples that results from evaporation, Glaeser's group has proposed that the sample be sandwiched between a monolayer fatty acid (e.g. behenic acid) film and a polylysine-coated carbon film (14). A thin layer of water is automatically formed between the films as excess liquid is squeezed out when the fatty acid film is adhered to the specimen grid. Therefore, the step of thinning the water by evaporation is not needed in this procedure. Since this method has been applied only to two biological specimens (13, 14), its general applicability remains to be determined. A more commonly used procedure for forming a thin layer of water is merely to squeeze out the water by blotting rapidly with filter paper (36, 69). This method generally requires that a more concentrated suspension be applied to the grid because most of the suspension is removed during the blotting. The blotting method is generally practiced in conjunction with a fast freezing procedure described later. It is likely that significant evaporation might occur during the time between blotting and freezing, and hence the original chemical composition of the sample may not be well preserved. However, this method has been shown to be satisfactory for most of the recent biological applications and would be the choice for the first trial of a new specimen. It is conceivable that more elaborate procedures will be needed for some other biological specimens.

With enough practice, some grids with an acceptably thin layer of water can be obtained using any of these procedures. However, one seldom obtains a preparation with more than 50% of the grid squares filled with ideal ice thickness with any of these methods. Therefore, development of a more reproducible method of obtaining a suitable water layer deserves further research effort.

The second critical step in the preparation of frozen, hydrated specimens is the freezing of the thin water film prepared in the first step. The sample generally contains water, salt, and the biological molecule or macromolecular assembly of interest. The freezing of these three components should be achieved without introducing significant structural changes in the biological molecules. It is well known that water can be frozen into various forms. At low temperature and low pressure the predominant forms are vitreous, cubic, and hexagonal (7, 36, 53, 76, 96). Vitreous ice forms when the water is frozen very rapidly (11, 33, 37, 53, 79). The two crystalline types form at lower freezing rates. During the formation of crystalline ice in a typical biological sample, solutes may be segregated from the water and suspended particles because only pure water crystallizes. This segregation and the resulting concentration of solutes at the grain boundaries may severely distort the structure of interest (69, 77). These problems do not occur when a sample is frozen and maintained in vitreous ice. Vitreous ice is therefore the preferred embedding matrix, especially for the examination of isolated particles.

Theoretical estimates suggest that a freezing rate of 10^7 K/sec is required to obtain vitreous ice from pure water (33, 36). The actual freezing rate needed to vitrify ice in the presence of biological molecules must be high, but is difficult to determine accurately. The choice of coolant affects the freezing rate because of the differences in the thermal properties of different coolants. Liquid nitrogen and ethane cooled by liquid nitrogen have been used as coolants for preparing frozen, hydrated biological specimens. Ethane should be used in a well-ventilated area such as a fume hood. The distance between the surfaces of ethane and liquid nitrogen should be adjusted properly to minimize the precooling of the grid in the vapor phase produced by the liquid nitrogen (53). Another factor that would affect the freezing rate is the way the grid is plunged into the coolant. Table 1 shows that manual immersion or a guillotine device are the predominant methods of delivering the grid into the coolant. The guillotine device has become preferred because it allows a higher freezing rate and is more reproducible.

In addition to the freezing rate, the surrounding biological molecules also influence the formation of vitreous ice (77). It has been suggested that biological molecules may serve as a cryoprotectant (22, 32, 77). In other words, the surfaces of biological molecules may influence the organization of water at a radial distance of 20–30 Å. For example, the ice in the aqueous channels of a protein or DNA crystal could remain vitreous under conditions at which the embedding ice away from the biomolecular crystal surface is already crystalline. This may explain the excellent structural preservation of some thin crystals even though they were frozen relatively slowly in a less effective coolant (24, 35, 50, 88).

SPECIFICATIONS FOR THE CRYO SYSTEM OF AN ELECTRON MICROSCOPE

Frozen, hydrated specimens are prepared outside the microscope and can be stored in a grid box kept under liquid nitrogen for weeks prior to observation in an electron microscope (77, 94). Each microscope has its own device for transferring the frozen specimen into the microscope column (66, 89). The delivery of the frozen specimen grid to the cryo specimen holder is either done in liquid nitrogen or in a nitrogen gas atmosphere cooled by liquid nitrogen. A frost protector surrounding the specimen grid is essential during the transfer step in order to prevent condensation of atmospheric water onto the specimen and the specimen grid holder. In the laboratory of the Medical Research Council, Cambridge, a dehumidifier was installed in the electron microscope room in order to minimize such water condensation during transfer. Some investigators have enclosed the specimen holder and airlock in a plastic bag flushed with a continuous flow of dry nitrogen gas in order to provide additional protection from condensation of water vapor during insertion (77). In any case, it is recommended that this step be performed as quickly as possible to reduce the atmospheric exposure of the cryo specimen holder.

In order to allow observation of a frozen, hydrated specimen, an electron microscope must be equipped with a cold specimen holder and an efficient anticontaminator. An ideal cold specimen holder should have the same resolution performance at low temperature as the standard specimen holder at room temperature, and should be able to tilt from 60° to −60°. The optimal temperature of the specimen holder is below the devitrification temperature of the ice (between −125° and −150°C, depending on the chemical content of the sample). The devitrification temperature for some crystalline samples may be higher than that of isolated particles because of the cryoprotectant effect of the biological molecules. For most biological applications, it is sufficient to maintain the specimen holder at a single fixed temperature. All of the commercially available cold specimen holders can be operated below $-160°$. Their guaranteed practical resolution is no better than 8 Å (9), although some may achieve better resolution under the most optimal conditions. A 5 Å resolution image of a frozen, hydrated crotoxin-complex protein crystal has been taken by T. W. Jeng (unpublished data) with a cold stage that was built in our laboratory after the design of Hayward & Glaeser (52) for a JEOL 100 electron microscope with the top entry configuration. This is the highest resolution image data yet recorded from a frozen, hydrated specimen. Our cold stage (currently operated around $-150°C$) routinely resolves lattice images of graphitized carbon at 3.5 Å and of crystalline ice at 3.66 Å when operated at higher temperature (24).

During the transfer of the frozen, hydrated specimen to the microscope column, atmospheric water that is condensed on the cryo specimen holder is degassed once in the high-vacuum environment of the microscope. The anticontaminator in the JEOL as well as those in most other electron microscopes cannot handle this additional water vapor efficiently. The consequence is rapid condensation of the water vapor or other residual gases on the cold surface of the specimen. A good anticontaminator needs to operate at a temperature lower than that of the specimen and needs to have the proper geometry for trapping the contaminants efficiently. Most laboratories have had to modify their anticontaminators before practicing the frozen hydration technique successfully and routinely. For example, we had to replace the copper braids connecting the copper rod in the liquid nitrogen dewar to the upper anticontaminator of the JEOL 100 CX electron microscope, and to improve the heat transfer at all the junctions from the dewar to the anticontaminator (24). These modifications reduced the temperature of the upper anticontaminator from $-90°$ to $-160°C$ and increased the length of usable time for each grid before excessive contamination occurred. However, our lower anticontaminator is still cooled to only $-100°C$. We recently introduced an additional anticontaminator so that our cold stage can be operated around $-150°C$ without excessive contamination of the specimens. As another example, research groups in other laboratories had to design new anticontaminators for their Philips microscopes in order to cope with the ice condensation problem (8, 56, 77, 92).

Another problem in some of the commercially available cold stages is the limitation of tilt angle to less than 30° in a eucentric goniometer. In the case of the Philips microscope, this limitation is caused by contact between the frost protector of the cryospecimen holder and the objective aperture holder. By removing the objective aperture holder from the microscope, the specimen can be tilted up to 50° (R. Henderson, private communication). To achieve a higher degree of tilt with a large field of view, one has to further modify the specimen holder. The latest model of cryo specimen holder designed by Gatan Inc. for the Philips microscope allows the specimen to tilt to $\pm 60°$ without hitting the objective aperture or pole pieces and to maintain a large field of view (P. N. T. Unwin, private communication). In a top-entry system as in the JEOL microscope, we have built a set of fixed tilt angle holders that include angles up to 60°. This is less convenient than the continuous tilting possible with a side-entry goniometer, but remains the best way of achieving high-resolution imaging given the current status of the technology. For a tilt angle beyond 60°, the grids can be bent (29, 94). The tilt angle can be determined subsequently from the electron micrographs (2, 94). Further improvements in the stability and tilting capability of the cold stage and/or the efficiency of the anticontaminator supplied by the manufacturers of electron microscopes will be required in order to bring the frozen hydration technique into routine application in the laboratories of molecular and cellular biologists.

OBSERVATION OF FROZEN, HYDRATED SPECIMENS

The presence of ice in the specimen can be detected from the characteristic diffraction intensities of ice in an electron diffraction pattern (36, 67, 96) or from the bend contours of ice in the image (39, 90). As noted above, there are different forms of ice, depending upon the preparative method and the temperature of the specimen during observation. The diffraction pattern of vitreous ice is quite easily distinguished from that of crystalline ice (36, 39, 67, 79, 96). Under optimal freezing conditions and with a proper cryo system in an electron microscope, diffraction patterns corresponding to vitreous ice can be observed. However, if one cannot observe the broad and diffuse diffraction rings of vitreous ice, one or more of the following reasons may be considered: (a) The specimen was not frozen properly. (b) The specimen holder is operated at a temperature above the devitrification temperature. For example, at $-120°C$ it takes less than a few minutes for the pure ice to transform from vitreous into cubic form (33, 36). (c) The presence of cubic or hexagonal ice resulting from water condensation during the transfer step or from the column can mask out the diffraction intensities from the vitreous ice in the specimen. After extensive electron irradiation, ice can transform into different forms dependent on the temperature. A thorough study on ice transformation as a function of cold-stage temperature and electron exposure has been carried out by Heide & Zeitler (53).

All frozen, hydrated specimens are extremely sensitive to radiation. Data from these specimens must be recorded using a minimal exposure technique

(97). Because the specimen is observed at low temperature, radiation damage becomes significant after an accumulated electron exposure 4–6 times higher than the critical exposure at room temperature (48, 51, 63, 69). No significant reduction in radiation damage of biological specimens is observed when the specimen temperature is lowered below $-130°C$ (17, 21). A typical image of a frozen, hydrated specimen can be recorded at a magnification of 40,000 × with a specimen exposure of 10 electron/$Å^2$. After the specimen is irradiated with a few tens of electrons per $Å^2$, bubbling of ice and/or disappearance of the specimen suspended across a hole is observed (6, 24, 53, 84, 85). At a lower magnification (e.g. 20,000 ×), it is conceivable that a tilt series of images from the same specimen could be recorded.

The image contrast of a frozen, hydrated specimen depends on the inherent scattering contrast of the specimen, the defocus value of the objective lens, and the ice thickness. The inherent scattering contrast of the specimen varies in different specimens according to their structure, and cannot be altered by any microscopic technique (43, 69). Setting the defocus value to several microns from optimal defocus value enhances the visibility of the low-resolution structural detail (1, 8, 70, 71), but may complicate the image processing step if the reconstruction contains data beyond the first zero of the phase contrast transfer function (38). The effect of a thick layer of ice is analogous to having a thick support film, which reduces the visibility of the specimen. Since the scattering density of ice is smaller than that of carbon film, the effect on the image contrast is less for ice than for carbon film with an equivalent thickness.

Another parameter that affects the contrast of the specimen is the extent to which the specimen is embedded in the layer of ice. The contrast of a partially or completely dried specimen is better than that of a fully hydrated specimen (69). While interpreting images of a frozen, hydrated specimen, it is necessary to question whether the specimen is fully hydrated. If the specimen is suspended across a hole, it is likely that ice is present around the specimen. However, there is no guarantee that the specimen is fully embedded in the ice. In other words, the specimen may protrude partially or entirely from the surface of the ice layer (69). Prior to deriving the structure from images of frozen, hydrated specimens, it is advisable to compare a large number of images of the specimen embedded in ice of varying thicknesses and with different defocus values in order to determine which images represent the proper specimen preparation.

BIOLOGICAL APPLICATIONS

Since the introduction of the frozen hydration technique, Glaeser, Taylor, and co-workers have demonstrated its utility, initially with catalase crystals

(90) and subsequently in the periodic array of the surface-layer protein of the outer membrane of *Spirillum serpens* (47, 86). It is remarkable that the lipid bilayer was visible in some of the side or end-on views of the outer membrane attached to this bacterial surface-layer protein. These results stimulated much subsequent work, which led to the refinement of the technique and its application to a number of biological problems. The following section presents the more recent applications of this technique in four classes of biological specimens.

Two-Dimensional Periodic Arrays of Membrane Proteins

Many cell membrane proteins form two-dimensional periodic arrays in vivo that are uniquely suitable for electron-microscopic analysis (46). Except for purple membrane, two-dimensional membrane protein crystals have not been found that diffract to better than 10 Å resolution (41, 93). At this level of resolution, one can expect to determine the quaternary structure of the protein. In negatively stained preparations, the high scattering contrast of the stain masks out the structural detail of the membrane protein domain that is surrounded by lipid. Since the scattering densities of ice and lipid are comparable, structural features of the entire length of the membrane proteins that extend in both the lipid bilayer and the aqueous spaces on either side can be revealed. The gap junction and acetylcholine receptor are excellent examples of membrane proteins studied by the frozen, hydrated specimen method in Unwin's laboratory (10, 92). Other applications include the lumenal plasma membrane of the mammalian urinary bladder (A. Brisson, P. N. T. Unwin, unpublished data) and the pore-forming OmpC protein from *Escherichia coli* outer membrane (13).

The technique also facilitates the detection of structural changes in membrane proteins under different chemical environments. For example, a gross structural change in the gap junction protein due to the presence of Ca^{2+} was speculated on the basis of the appearance of the membrane surfaces in negative stain. This conclusion was subsequently supported by the three-dimensional density maps of the frozen, hydrated membrane proteins (Figure 1). At the present time, the detectable level of structural alteration is most likely limited by the crystallinity of the membrane protein. In some cases, it may be limited by the mechanical instability of the cold specimen holder.

Thin Crystals of Proteins, Nucleic Acids, and Protein-Nucleic Acid Complexes

Electron diffraction is different from X-ray diffraction in that it requires a thin crystal for structural analysis (16). A number of macromolecules have been found to form crystals with an area of 1 μm^2 and a thickness of a few hundred angstroms (one or a few unit cells). The frozen hydration technique is an ideal method for preparing these specimens for electron microscopic studies because of the high volume of solvent in these crystals (23, 24, 34, 35, 77, 78).

The two-dimensional crystals of ribosomes extracted from the cells of early chicken embryos have been studied by negative staining and frozen hydration techniques. The three-dimensional density map of the negatively stained ribosome crystals provides a description of the gross surface morphology of the ribosome particle. The interpretation of its internal structure is complicated by the high content of RNA which may be positively stained by the uranium ions (78). Since the relative scattering density decreases from RNA to protein to ice, it is hoped that the region of the RNA can be identified as the area of highest density in the reconstructed map of the frozen, hydrated ribosomes. The projected density map of the

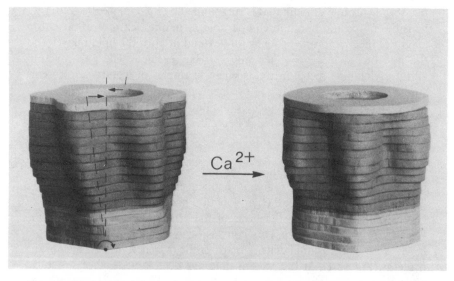

Figure 1   Models of protein configurations with and without Ca^{2+} constructed from three-dimensional density maps of frozen, hydrated gap junction membrane protein. The lighter shaded portions correspond to parts of the protein contacting fluid inside (*top*) and outside (*bottom*) the cell; the darker shaded portion would be in contact with lipid. The configurations are related to each other by tilt axes passing radially through the bases of the subunits (as depicted on the *left-hand* model). Subunits on either side of the channel tilt in opposite directions about these axes, and the resulting tangential displacements (*arrows*) are greatest at the upper, cytoplasmic face. (Kindly provided by P. N. T. Unwin. Reproduced with permission from Reference 92.)

frozen, hydrated ribosome crystals appears to have a better resolution than the map of the negatively stained crystals, particularly in the region where individual ribosomes partly overlap (77). Work on the structural determination of frozen, hydrated ribosome crystals is being continued in Unwin's laboratory.

The frozen hydration technique has also been applied to the study of thin protein crystals of crotoxin complex (62), T4 DNA helix destabilizing protein (50), and Fc fragment (49). Electron diffraction patterns of these protein crystals embedded in ice have been obtained beyond 3 Å resolution. High-resolution (better than 9 Å) images of frozen, hydrated T4 DNA helix destabilizing protein and crotoxin complex crystals have been recorded with a custom-built cold stage fitted to our JEOL 100 CX electron microscope in top entry configuration (50, 62). Since some of the T4 DNA helix destabilizing crystals are not coherent throughout the entire crystal patch, we have developed a computer processing algorithm that aligns many small image areas by minimization of their phase residuals and merges them by vectorial summation of the Fourier coefficients. Figure 2 is a projected density map with a nominal resolution of 8.4 Å in which the protein dimer is well resolved in the unit cell and the protein is clearly distinguished from the solvent space (50). Henderson has suggested an alternative image processing scheme in which the image distortion is defined by a correlation method (25, 26, 81, 95) and the distortion is subsequently removed by a real space interpolation procedure (54). This method was used in processing the images of frozen, hydrated crotoxin complex crystals. Some of the Fourier coefficients calculated from the undistorted images were detectable at 5 Å resolution with a signal-to-noise ratio better than 2 (T. W. Jeng, M. F. Schmid, unpublished data).

When the crotoxin complex protein crystal is embedded in glucose, high-resolution data of similar quality are obtained (61). Quantitative analysis of the electron diffraction patterns from these crystals embedded in ice and in glucose shows that the amplitudes of their structure factors are similar for reflections at Bragg spacings smaller than 10 Å and different for reflections at Bragg spacings larger than 10 Å (24). This makes sense because the solvent contributes strongly to the low-resolution reflections, whereas the high-resolution data are determined primarily by the protein. Since the difference in the scattering densities of protein and glucose is small, it is preferable to determine the low-resolution (< one eighth per angstrom) structure of the protein using the ice embedding method in order to avoid the problem of interpretation in the reconstructed density map. In practice, until the cryo systems of commercially available electron microscopes are improved the glucose embedding method will be much easier to use and will remain the preferred approach for tackling problems at the 3 Å resolution level. A low temperature stage is still essential for reducing radiation damage in imaging glucose-embedded protein crystals at high resolution (64). However, since the glucose-embedded sample can be cooled in the microscope column and hence no cryo transfer step is involved, the custom-built cold stage and modified anticontaminator in our JEOL microscope as well as in a number of cryo microscopes elsewhere are well suited for high-resolution (3–5 Å) imaging at low temperature (34, 52, 98).

The structural model of the double helix in DNA was based on deduction

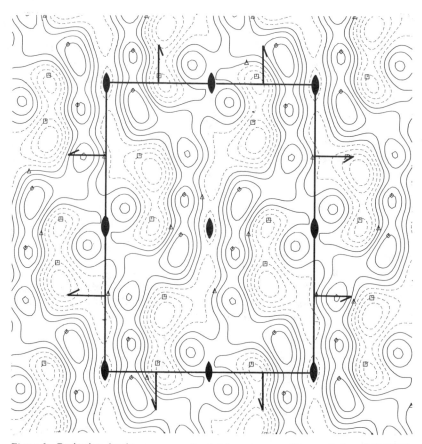

Figure 2 Projection density map reconstructed from images of frozen, hydrated T4 DNA helix destabilizing protein crystals with a nominal resolution of 8.4 Å. The phases for this reconstruction were obtained by the vector addition of 13 aligned Fourier transforms of images corresponding to about 40,000 unit cells. The average figure of merit is 0.7. The high-density (protein) contours are *solid* and the low-density (ice) contours are *dashed*. The protein dimer is well resolved in the unit cell of 47 Å × 63 Å as marked in the rectangle. (Reproduced with permission from Reference 50.)

from an X-ray fiber diffraction diagram. A similar electron diffraction pattern has also been obtained from a thin, frozen, hydrated DNA fiber (69). In that diffractogram, diffraction spacings equivalent to those of the A form of DNA were observed. In order to obtain an atomic structure of DNA, a single crystal is needed. Until recently it has been difficult to make a single crystal of oligonucleotides suitable for X-ray diffraction analysis. A thin crystal of DNA made by precipitation in ethanolic solution (42) has been found suitable for electron diffraction analysis. This crystal was sensitive to the change in ethanol concentration and hydration. A procedure has been worked out to prepare this crystal in the frozen, hydrated state (35) such that an electron diffraction pattern with reflections extending to 3.5 Å and a lattice image with resolution to 6.5 Å could be obtained (34). Before a structural model of the DNA can be deduced, higher resolution image data are required.

Viruses

Viruses are usually crystallized into large unit cells and lengthy effort would be required to determine their structure by X-ray diffraction analysis. Electron microscopy could be useful for providing a low-resolution structural model of the virus particle (4, 5) that could be used by X-ray crystallographers to determine the structure at higher resolution and at a more rapid pace (12). The opportunity to visualize frozen, hydrated virus particles that are free of stain and fixative should make this joint approach more attractive and reliable. The frozen, hydrated specimen technique can also be used to study the structures of viruses in different chemical environments that represent different assembly intermediate states. For example, the size and shape of a virus particle may depend on ionic strength and pH. The molecular envelope determination requires relatively low-resolution data which is well within the capability of the present technology (24).

The frozen hydration technique has been employed to study structures of adenovirus (1), Semliki Forest virus (1), influenza virus (8), SV 40 virus (6), and polio virus (T. W. Jeng, B. V. V. Prasad, J. Hogle, unpublished data). Figure 3 is an electron image of frozen, hydrated SV 40 virus particles that shows good contrast (6). In order to reconstruct the three-dimensional structure of a virus particle, it is necessary to identify the orientation view of the particle (27, 28). The contrast in images of frozen, hydrated viruses taken at a defocus close to the optimal value (e.g. Scherzer defocus) can be quite low, making the determination of the particle orientation difficult. The defocus can be purposely set further away from the optimal value to enhance the low-resolution contrast, but oscillation of the contrast transfer function may complicate the image processing procedure. This problem

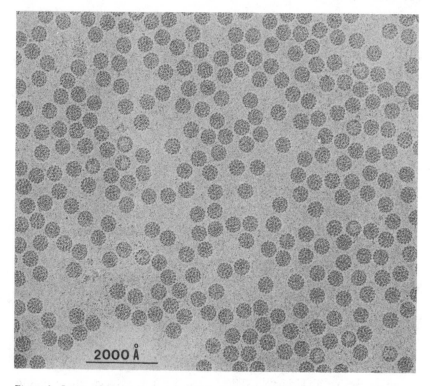

2000 Å

Figure 3 Image of frozen, hydrated SV 40 virus particles. A single virus particle has a diameter of 490 Å. The orientations of some of these virus particles can be identified in the image as close to the icosahedral symmetry axis. (Kindly provided by T. S. Baker, J. Drak, and M. Bina. See Reference 6.)

may be circumvented by recording two successive images at different defocus values. The particle orientation can be determined from images far from focus (several microns), and the three-dimensional reconstruction can be made from images taken closer to the optimal defocus value. This approach is being tested in several laboratories.

Filamentous Proteins

Many proteins assemble in filamentous form in functional or nonfunctional states. DeRosier and co-workers introduced the concept of reconstructing the three-dimensional structure of filamentous proteins from electron micrographs. They demonstrated this idea originally with images of negatively stained T4 bacteriophage tails (30, 31). A frequent observation in the optical or computer-calculated diffraction pattern of images of the negatively stained filamentous protein has been the lack of mirror symmetry across the meridian. The difference between the far and near sides of the specimen has been interpreted as resulting from a preparative artifact due to the absorptive distortion of the support film and/or uneven staining of the specimen. A recent study of frozen, hydrated T4 phage tails shows that this lack of symmetry is not found in the image of a properly prepared specimen embedded in ice suspended across a hole (70). However, some micrographs have still shown asymmetry as seen in negatively stained preparations, particularly when the thickness of the ice layer embedding the phage has been smaller than the largest dimension of the phage. The reconstruction of the frozen, hydrated T4 phage tail shows structural variation that supports the previous conclusion from negative stain studies that phage particles exhibit a polymorphic property (3, 82).

The frozen hydration technique has also been applied to the study of filaments of actin (91), filaments of actin and actin-binding protein complex (P. Flicker, R. Milligan, unpublished data), filaments of RecA and RecA-DNA complex (D. Rankert, unpublished data), and microtubules (72). Some of these filaments have diameters of 100–200 Å. Images of these frozen, hydrated filamentous proteins have sufficiently good contrast for visibility of structural features. Figure 4 is an example of an image of frozen, hydrated actin filaments (91). The resolution as measured from the highest order of layer lines in optical or computed Fourier transforms from images of the frozen, hydrated actin filaments is comparable to that of negatively stained specimens. This suggests that the intrinsic helical structure of these filaments is the factor that limits the attainable resolution in the image reconstruction. Preliminary observation of the variation in the helical parameters within and among filaments of RecA and RecA-DNA prepared by frozen hydration also suggests that the specimens have a flexible nature (M. F. Schmid, D. Rankert, unpublished data). However, it may be worthwhile to develop new computer algorithms similar to those used for two-dimensional crystalline arrays (54) in order to extend the resolution for the reconstruction of frozen, hydrated filamentous proteins.

Because the frozen hydration technique allows preservation of structure at the time of freezing, it has opened up the possibility of studying dynamic transitions of structures in macromolecular complexes free from preparative artifacts. This approach was nicely demonstrated in a recent work of Mandelkow & Mandelkow, who visualized frozen, hydrated microtubules in the course of time-dependent reactions during the disassembly process (72). This study has added new evidence to support the proposed mechanism of microtubule breakdown as occurring not only at both ends but also from inside. Short protofilaments were found to be among the early breakdown products.

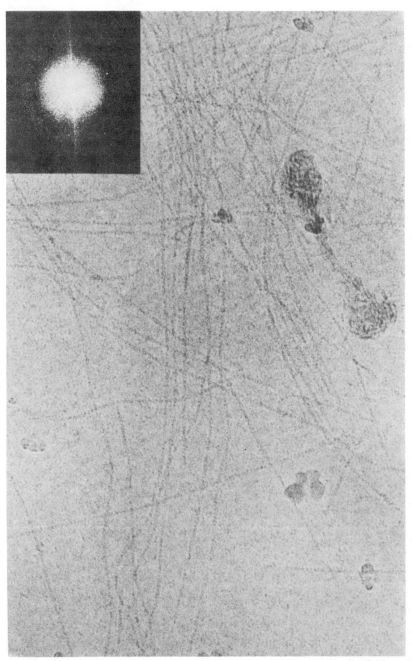

Figure 4   Image of frozen, hydrated actin filaments. The optical diffractogram of this image (*insert*) is shown with the first zero in the contrast transfer function at 1/35 Å^{-1} and the defocus calculated to be 2.9 μm. (Kindly provided by J. Trinick & co-workers.)

CONCLUSION

Direct phasing from images has made electron microscopy an attractive alternative and/or complement to X-ray diffraction analysis. Electron microscopy of frozen, hydrated specimens has emerged as a unique technique for three-dimensional structural analysis of macromolecules and macromolecular assemblies in their physiological states. Most of the biological systems have conformations that are subject to changes in response to external stimuli as well as changes in their assembly or disassembly processes. The freezing technique provides an opportunity to capture these specimens in various chemical and physiologically relevant states so structural dynamics may be studied. This technique can be extended to the study of other systems in addition to the cited examples, such as the individual enzyme complex, the antigen-antibody complex, specialized membrane vesicles, isolated nuclear envelopes, and cellular organelles.

The application of this technique requires a cryo system fitted to the electron microscope. Most of the cryo systems that are successfully used for biological applications have been partly or entirely developed in the users' laboratories. Further improvement in the preparative procedure and in the cryo system of an electron microscope is needed to make this practice routine and standard in the laboratories of molecular and cellular biologists. Because of the increased interest in the technique, it is conceivable that the industry will take a more active role in improving the current cryo systems for all electron microscopes.

Since the image contrast of frozen, hydrated specimens is relatively lower than that of negatively stained specimens, a new computer processing scheme has to be developed for treating the image of frozen, hydrated specimens. The correlation averaging technique (54, 81, 95) is expected to be quite useful as part of the image processing procedure for images of frozen, hydrated crystalline and noncrystalline specimens.

ACKNOWLEDGMENTS

I thank Robert Grant, Tzyy-Wen Jeng, and Debby Rankert for many helpful technical discussions. The assistance of Laura Degn, David Morgan, Joe Schrag, and Eva Wilson for preparing this manuscript is deeply appreciated. I thank R. M. Glaeser, Y. Talmon, and P. N. T. Unwin for their comments on this manuscript. This work has been supported by the United States Public Health Services grants numbers RR02250 and GM27061 and the Department of Defense contract number DAMD1784C4110.

256   CHIU

Literature Cited

1. Adrina, M., Dubochet, J., Lepault, J., McDowall, A. W. 1984. *Nature* 308:32
2. Amos, L. A., Henderson, R., Unwin, P. N. T. 1982. *Prog. Biophys. Mol. Biol.* 39:183
3. Amos, L. A., Klug, A. 1985. *J. Mol. Biol.* 99:51
4. Baker, T. S., Caspar, D. L. D. 1984. *Ultramicroscopy* 13:137
5. Baker, T. S., Caspar, D. L. D., Murakami, W. T. 1983. *Nature* 303:446
6. Baker, T., Drak, J., Bina, M. 1985. *Proc. 43rd Annu. Meet. Electron Microsc. Soc. Am.*, pp. 316–17, San Francisco: San Francisco Press
7. Blackman, M., Lisgarten, N. D. 1958. *Adv. Phys.* 7:189
8. Booy, F. P., Ruigrok, R. W. H., Van Bruggen, E. F. T. 1985. *J. Mol. Biol.* 184:667
9. Booy, F. P., Van Bruggen, E. F. J. 1984. *Ultramicroscopy* 12:337
10. Brisson, A., Unwin, P. N. T. 1985. *Nature* 315:474
11. Bruggeller, P., Mayer, E. 1980. *Nature* 288:569
12. Caspar, D. L. D. 1983. *Proc. 41st Annu. Meet. Electron Microsc. Soc. Am.*, pp. 416–17, San Francisco: San Francisco Press
13. Chang, C.-F., Mizushima, S., Glaeser, R. M. 1985. *Biophys. J.* 47:629
14. Chang, C.-F., Ohno, T., Glaeser, R. M. 1985. *J. Electron Microsc. Tech.* 2:59
15. Chiu, W. 1978. *Scanning Electron Microsc.* 1:569
16. Chiu, W. 1982. In *Electron Microscopy of Proteins*, ed. J. R. Harris, 2:233–59. London: Academic
17. Chiu, W., Downing, K. H., Dubochet, J., Glaeser, R. M., Heide, H. G., et al. 1985. *J. Microsc.* In press
18. Chiu, W., Glaeser, R. M. 1977. *Ultramicroscopy* 2:207
19. Chiu, W., Hosoda, J. 1978. *J. Mol. Biol.* 122:103
20. Chiu, W., Jeng, T. W. 1980. In *Electron Microscopy of Molecular Dimensions*, ed. W. Baumeister, W. Vogell, pp. 137–42. Berlin: Springer-Verlag
21. Chiu, W., Knapek, E., Jeng, T. W., Dietrich, I. 1981. *Ultramicroscopy* 6:291
22. Clegg, J. S. 1981. *Collect. Phenom.* 3:289
23. Cohen, H. A., Chiu, W., Hosoda, J. 1983. *J. Mol. Biol.* 169:235
24. Cohen, H. A., Jeng, T. W., Grant, R., Chiu, W. 1984. *Ultramicroscopy* 13:19
25. Crepeau, R. H., Edelstein, S. J. 1984. *Ultramicroscopy* 13:11
26. Crepeau, R. H., Fram, E. K. 1981. *Ultramicroscopy* 6:7
27. Crowther, R. A., Amos, L. A. 1971. *J.*
   Mol. Biol. 60:123
28. Crowther, R. A., Amos, L. A., Finch, J. T., DeRosier, D. J., Klug, A. 1970. *Nature* 226:421
29. Deatherage, J. F., Taylor, R. A., Amos, L. A. 1983. *J. Mol. Biol.* 167:823
30. DeRosier, D. J., Klug, A. 1968. *Nature* 217:130
31. DeRosier, D. J., Moore, P. B. 1970. *J. Mol. Biol.* 52:355
32. Dowell, L. G., Moline, S. W., Rinfret, A. P. 1962. *Biochim. Biophys. Acta* 59:458
33. Dowell, L. G., Rinfret, A. P. 1960. *Nature* 188:1144
34. Downing, K. H. 1984. *Ultramicroscopy* 13:35
35. Downing, K. H., Glaeser, R. M. 1980. *Biophys. J.* 32:851
36. Dubochet, J., Lepault, J., Freeman, R., Berriman, J. A., Homo, J. C. 1982. *J. Microsc.* 128:219
37. Dubochet, J., McDowall, A. W. 1981. *J. Microsc.* 124:RP3
38. Erickson, H. P., Klug, A. 1970. *Ber. Bunsenges. Phys. Chem.* 74:1129
39. Falls, A. H., Wellinghoff, S. T., Talmon, Y., Thomas, E. L. 1983. *J. Mater. Sci.* 18:2752
40. Fugiyoshi, Y., Uyeda, N., Morikawa, K., Yamagishi, H. 1984. *J. Mol. Biol.* 172:347
41. Fuller, S. D. 1981. In *Methods in Cell Biology*, ed. J. N. Turner, 22:251. New York: Academic
42. Giannoni, G., Padden, F. J., Keith, H. D. 1969. *Proc. Natl. Acad. Sci. USA* 62:964
43. Glaeser, R. M. 1971. *J. Ultrastruct. Res.* 36:466
44. Glaeser, R. M. 1982. In *Methods of Experimental Physics*, ed. G. Ehrenstein, H. Lecar, 20:391–444. New York: Academic
45. Glaeser, R. M. 1985. *Biophys. J.* 47:322a
46. Glaeser, R. M. 1985. *Ann. Rev. Phys. Chem.* 36:243
47. Glaeser, R. M., Chiu, W., Grano, D. 1979. *J. Ultrastruct. Res.* 66:235
48. Glaeser, R. M., Taylor, K. A. 1978. *J. Microsc.* 112:127
49. Grant, R. A., Degn, L. L., Chiu, W., Robinson, J. P. 1983. *Proc. 41st Annu. Meet. Electron Microsc. Soc. Am.*, pp. 730–31. San Francisco: San Francisco Press
50. Grant, R., Schmid, M. F., Chiu, W., Deatherage, J., Hosoda, J. 1986. *Biophys. J.* 49:251
51. Hayward, S. B., Glaeser, R. M. 1979. *Ultramicroscopy* 4:201
52. Hayward, S. B., Glaeser, R. M. 1980. *Ultramicroscopy* 5:3

53. Heide, H.-G., Zeitler, E. 1985. *Ultramicroscopy* 16:151
54. Henderson, R., Baldwin, J. M., Downing, K. H., Lepault, J., Zemlin, F. 1986. *Ultramicroscopy.* In press
55. Henderson, R., Unwin, P. N. T. 1975. *Nature* 257:28
56. Homo, J.-C., Booy, F., Labouesse, P., Lepault, J., Dubochet, J. 1984. *J. Microsc.* 136:337
57. Hoppe, W. 1983. *Angew. Chem. Int. Ed. Engl.* 22:456
58. Hui, S. W. 1976. *Nature* 262:303
59. Hui, S. W., Parsons, D. F. 1975. *Science* 190:383
60. Jaffe, J. S., Glaeser, R. M. 1984. *Ultramicroscopy* 13:373
61. Jeng, T. W., Chiu, W. 1983. *J. Mol. Biol.* 164:329
62. Jeng, T. W., Chiu, W. 1983. *Proc. 41st Annu. Meet. Electron Microsc. Soc. Am.,* pp. 430–31. San Francisco: San Francisco Press
63. Jeng, T. W., Chiu, W. 1984. *J. Microsc.* 136:35
64. Jeng, T. W., Chiu, W., Zemlin, F., Zeitler, E. 1984. *J. Mol. Biol.* 175:93
65. Klug, A. 1979. *Chem. Scr.* 14:245
66. Kraus, B., Kodge, E., Swann, P. 1983. *Norelco Rep.* 30:9
67. Kumai, M. 1968. *J. Glaciol.* 7:95
68. Kuo, I. A. M., Glaeser, R. M. 1975. *Ultramicroscopy* 1:56
69. Lepault, J., Booy, F. P., Dubochet, J. 1983. *J. Microsc.* 129:89
70. Lepault, J., Leonard, K. 1985. *J. Mol. Biol.* 182:431
71. Lepault, J., Pitt, T. 1984. *EMBO J.* 3:101
72. Mandelkow, E.-M., Mandelkow, E. 1985. *J. Mol. Biol.* 181:123
73. Matricardi, V. R., Moretz, R. C., Parsons, D. F. 1972. *Science* 177:268
74. Matthews, B. W. 1974. *J. Mol. Biol.* 82:513
75. Matthews, B. W. 1977. In *The Proteins,* ed. H. Neurath, R. L. Hill, C. Boeder, 3:404–590. New York: Academic
76. Mayer, E., Bruggeller, P. 1982. *Nature* 298:715
77. Milligan, R. A., Brisson, A., Unwin, P. N. T. 1983. *Ultramicroscopy* 13:1

78. Milligan, R., Unwin, P. N. T. 1983. *Proc. 41st Annu. Meet. Electron Microsc. Soc. Am.,* pp. 432–33. San Francisco: San Francisco Press
79. Narten, A. H., Venkatesh, C. G., Rice, S. A. 1976. *J. Chem. Phys.* 64:1106
80. Robinson, J. P., Jeng, T. W., Chiu, W., Hash, J. H. 1983. *Proc. 41st Annu. Meet. Electron Microsc. Soc. Am.,* pp. 732–33. San Francisco: San Francisco Press
81. Saxton, W. O., Baumeister, W. 1984. *Ultramicroscopy* 13:57
82. Smith, P. R., Aebi, U., Joseph, R., Kessel, M. 1976. *J. Mol. Biol.* 106:243
83. Talmon, Y. 1982. *10th Int. Congr. Electron Microsc., Hamburg,* 1:25. Frankfurt: Dtsch. Ges. Elektronenmikroskopie
84. Talmon, Y. 1984. *Ultramicroscopy* 14:305
85. Talmon, Y., Adrian, W., Dubochet, J. 1985. *J. Microsc.* In press
85a. Talmon, Y., Davis, H. T., Scriven, L. E., Thomas, E. L. 1979. *Rev. Sci. Instrum.* 50:698
86. Taylor, K. A. 1978. *J. Microsc.* 112:259
87. Taylor, K. A., Glaeser, R. M. 1973. *Rev. Sci. Instrum.* 44:1596
88. Taylor, K. A., Glaeser, R. M. 1974. *Science* 186:1036
89. Taylor, K. A., Glaeser, R. M. 1975. *Rev. Sci. Instrum.* 46:985
90. Taylor, K. A., Glaeser, R. M. 1976. *J. Ultrastruct. Res.* 55:448
91. Trinick, J., Cooper, J., Seymour, J., Egelman, E. 1985. *J. Microsc.* In press
92. Unwin, P. N. T., Ennis, P. D. 1984. *Nature* 307:609
93. Unwin, P. N. T., Henderson, R. 1975. *J. Mol. Biol.* 94:425
94. Unwin, P. N. T., Zampighi, G. 1980. *Nature* 283:545
95. Verschoor, A., Frank, J., Radermacher, M., Wagenknecht, T., Boublik, M. 1984. *J. Mol. Biol.* 178:677
96. Vertsner, V. N., Zhdanov, Gl. S. 1966. *Sov. Phys. Crystallogr.* 10:597
97. Williams, R. C., Fisher, H. W. 1970. *J. Mol. Biol.* 52:121
98. Zemlin, F., Reuber, E., Beckmann, E., Zeitler, E., Dorset, D. L. 1985. *Science* 229:461

Ann. Rev. Biophys. Biophys. Chem. 1986. 15 : 259–77
Copyright © 1986 by Annual Reviews Inc. All rights reserved

RECENT ADVANCES IN PLANAR PHOSPHOLIPID BILAYER TECHNIQUES FOR MONITORING ION CHANNELS

Roberto Coronado[1]

Department of Pharmacology, University of North Carolina At Chapel Hill, Chapel Hill, North Carolina 27514

CONTENTS

PERSPECTIVES AND OVERVIEW

As common as they may appear today, biochemical manipulations of ion channels are a major endeavor. A critical step was to recognize that channels only have an ion transport function when integrated in a membrane across two aqueous compartments. That function is necessarily lost when the cell membrane is disrupted. A biochemical assay of channel activity must therefore be a strategy to put together lipids and channel proteins so that flux between the two compartments is restored. In the past five years, artificial membranes in the form of planar bilayers and liposomes

[1]Present address: Department of Physiology and Molecular Biophysics, Baylor College of Medicine, Houston, Texas 77030.

259

have been used in most efforts to recover or reconstitute, as presently known, channel activity from purified cell membrane fragments and detergent extracts (for reviews see 40, 41, 50, 51). This combination of biochemical fractionation and reconstitution has lately resulted in the purification of ion channel proteins that play key roles in excitation (28, 35, 73, 91), cell-cell communication (46), and synaptic transmission (7, 10, 11). In the present article, I review current planar bilayer methods and give a summary of work from three areas in which planar bilayers are presently making a significant and enduring contribution: discovery of new natural toxins specifically targeted against ion channels, effects of phospholipid surface charge, and the identification of channels from membrane sources not directly accessible by conventional electrophysiological methods.

Basic observations of the self-assembly properties of monolayers paved the way to the present bilayer technology. Langmuir & Waugh (39) observed in 1938 that monolayers of protein (albumin) could be adsorbed to a solid support such as a platinum wire loop or a plate perforated by a small hole if the support was passed through the air-water interface in which the monolayers were suspended. By lowering a plate through an interface that contained, in addition to protein, a solution of lecithin in hexane, they showed that films formed at the hole lasted several minutes without rupturing. They concluded that the artificial membrane probably consisted of two layers of protein separated by one or two monolayers of "lipoid" material. The authors went on to suggest: "The technique we have just described for building protein membranes across holes in plates may be useful to the biologist in the study of the properties of membranes. Such a plate having a hole covered by a membrane could form a partition between two separate aqueous solutions, so that permeabilities and conductivities, etc., could be studied" (39).

The two-chamber system with a septum to support the planar bilayer was actually introduced much later by Mueller and collaborators in 1962 (62). This work established that the electrical capacitance of the artificial lipid film was consistent with an organization of the phospholipid as in a bimolecular bilayer. In the most common Mueller-type setup, each chamber holds 1–5 ml of solution and the septum separating the two chambers has an aperture of 300 μm. Planar bilayers are spontaneously assembled under water by covering the aperture with a solution of phospholipid in a lipid solvent (62). The final result is a membrane with an enormous area sealed across a hole separating two aqueous solutions, the ideal system envisioned by Langmuir for measuring electrical conductivities.

The ability to form planar lipid bilayers was not sufficient for reconstituting channels, which requires a way to insert protein molecules into the bilayer. The fusion technique (53), among many other methods, proved to be simple, reproducible, and of general application for this purpose. Using sealed native membrane fragments or vesicles derived from the sarcoplasmic reticulum of muscle, Miller and collaborators (8, 50–53, 57, 58) were able to identify electrical events associated with the fusion of single vesicles to the planar film. This work established the minimal conditions that enhance and control fusion—the rudiments of a methodology (for a review see 50). Fusion procedures define the aqueous chambers in contact with a planar bilayer as the *cis* side (side of addition of biological material) and the *trans* protein-free side. Incubation of vesicles on one side only ensures a remarkable sidedness of fused channels, either *cis*-cytoplasmic or *cis*-extracellular but seldom both simultaneously.

Fusion procedures have now been used to reconstitute channel activity from many sources including terminal cisternae of muscle sarcoplasmic reticulum (79, 80), transverse tubular membranes of skeletal muscle (1, 2, 13–16, 26, 42, 52, 55–60, 89), cilia (21), nerve axons (19, 86), human and other cardiac sarcolemma (17, 33, 69), cardiac sarcoplasmic reticulum (85), synaptic terminals (67, 68), smooth muscle (92), rat corpus striatum (63), and brain plasma membrane (5, 24, 36, 65, 66). In all of these cases, channel functions, some of them unknown in vivo, seem to be adequately preserved.

ASSEMBLY OF LIPIDS AND CHANNEL PROTEINS

Reconstitution of purified channels exploits two stable conformations that phospholipid molecules adopt in water: sealed liposomes and monolayers at air-water or hydrocarbon-water interfaces. These conformations arise as a consequence of the amphipathic structure of phospholipids and are governed in general by (*a*) the nature of the hydrocarbon or detergent used as lipid solubilizer (30, 47), (*b*) the molar fraction of components at equilibrium (best understood in terms of multicomponent phase diagrams) (30), and (*c*) the exact experimental protocol used. Na^+ channels (29, 73, 91) or acetylcholine receptors (AchR) (7, 10, 83) can be solubilized and purified in ternary mixtures of phospholipid, detergent, and water. Reassembly of the channel and the bilayer occurs when the detergent concentration is rapidly lowered below the critical micelle concentration.

Assembly of channels and lipids into a liposome solves the problem of measuring flux activity, but it creates a new problem. In natural membranes ion channels are vectorially oriented all in the same way. In reconstituted liposomes, orientation is frequently random. Randomness can be tested—i.e. using probes that bind to one side of the channel only—but it cannot be corrected. For example, Barchi and collaborators (35, 91) have shown that when a preparation of purified sodium channels (a predominant polypeptide of MW 260,000) is incorporated into phosphatidylcholine (PC) liposomes, the flux of Na^+ elicited by batrachotoxin (BTX) is blocked partially by saxitoxin (STX) added externally to the liposome and partially by STX added internally during liposome assembly. Hartshorne et al (28, 34) recently confirmed this randomness using a purified Na^+ channel preparation from rat brain, incorporated into Mueller-type planar bilayers. Tetrodotoxin (TTX) blocking sites, which in nerve and muscle are external, are found in planar bilayer experiments on both the *cis* side (the side where liposome-bilayer fusion occurs) and on the protein-free *trans* side. It was also shown that any given channel has only one TTX site, either *cis* or *trans* but not both (29). Thus the randomness of orientation present in the reconstituted vesicles is reflected when they are fused into the planar bilayer. Hartshorne et al's results (29, 34) contrast with Krueger et al's original observation (36) that when sodium channels were fused into the planar bilayer from native rat brain vesicles, sidedness was strictly preserved: TTX blockade was elicited by *cis* toxin only. One explanation in this case may be that in native vesicles fused to planar bilayers the sidedness of incorporated channels reflects a selective fusion of either inside-out or right-side-out vesicles. However, the results of Hartshorne et al (29) indicate that the orientation of the channel protein per se does not influence this selection. In this context, one interesting system is the vesicular preparation of skeletal muscle transverse tubules in which, under almost identical fusion conditions, calcium-activated potassium channels are inserted with the cytoplasmic end on the *cis* chamber (the side where vesicles are fused) (42) while sodium channels have exactly the opposite orientation (57, 58). This suggests that the two channels may indeed belong to vesicle populations with opposite orientations.

A study of lipid effects and channel orientation has been conducted with purified AchR from *Torpedo californica*. After reconstitution of AchR into liposomes of soybean lipid, up to 85% of the bound alpha-bungarotoxin (α-BGT) (a marker for the extracellular end) is accessible from the external solution (7). The apparent number of α-BGT sites can be decreased to approximately 50% by increasing the mole fraction of cholesterol in the liposome and by subjecting the preparation to one or more cycles of freezing and thawing (7). Thus, inclusion of cholesterol or increase of internal volume per liposome by freeze-thaw cycles tends to alleviate restrictions on the preferentially right-side-out orientation of AchR. The origin of the effect of cholesterol is not known, nor do we know why the mean size of liposomes loaded with AchR is considerably larger than that of control liposomes without protein (7). The latter, however, may not be altogether unaccountable considering that AchR, a large multisubunit glycoprotein of MW 280,000, makes up about 7% of the reconstituted membrane weight (10). This large mass of protein is likely to impose restrictions on the packing of lipid and detergent (54).

AchR channel activity can at present be measured either as fluxes in liposomes (7, 11) or as single-channel currents in planar bilayers (37). Montal, Schindler, and collaborators (37, 76, 77) have suggested that these two techniques could be made interchangeable by transforming liposomes into monolayers and then reassembling the monolayers into planar bilayers (61, 82). This would enable the experimenter to combine affinity ligand binding and evoked electrical response in the same preparation of purified channels. By modifying the monolayer spreading method of Trurnit (88, 90), Montal & Schindler have shown (37, 76, 77) that proteolipid monolayers containing active AchR can be formed at an air-water interface by diffusion of membrane vesicles from the bulk solution to the surface. A planar membrane may be formed across a hole wetted with hydrocarbon by apposing a proteolipid monolayer containing channels and a monolayer containing only phospholipids (37, 77). The assembly of planar bilayers can occur in this situation only if the surface pressure of monolayers is in excess of 25 dyne cm^{-1}. Under conditions optimized for recovery of AchR, Labarca et al (37) have shown that the surface density of BGT sites is about 80% of that present in liposomes; the latter has been estimated at 1.2×10^{10} AchR molecules cm^{-2}. Control experiments show that activity of AchR in monolayers is not likely to be contaminated by activity of AchR present in liposomes adsorbed to the monolayer. Thus, in at least one case (the AchR) conditions have been found to transfer stoichiometric quantities of channels from liposomes to planar bilayers.

In the most recent technique, patch-clamp electrodes are used to assemble a bilayer film at the tip entry of a pipette filled with saline (18, 27, 78, 81). This method has its origin in the well-known observations of Blodgett and Langmuir (9, 38) that phospholipid monolayers can be stacked on a flat glass surface by carefully dipping the glass several times into a solution containing a lipid monolayer on its surface. To form a bilayer over the tip of a patch electrode, the pipette tip is lifted from and then immediately resubmerged in a solution with a surface monolayer. This results in the formation of a high-resistance giga-seal. Recordings of ion channels of cellular origin and peptide ionophores only active in bilayers have established that the bulk of the film formed at the tip is a lipid bilayer (18, 22, 27, 44, 45, 71, 78, 81, 86). The nature of the phospholipid-glass interactions seems to be critical for the assembly of stable phospholipid films in patch electrodes. Many phospholipids (e.g. phosphatidylethanolamines from several sources) form good seals. Seal resistance for these lipids, like that of patch electrodes on cell surfaces, is in the range of 1–20 gΩ. Many other lipids [e.g. phosphatidylcholines (PC) (except diphytanoyl-PC)] form extremely poor seals. Knowledge of the chemical reactivity and electrostatic phenomena at the surface of glass (43) has contributed clues about the bonding of lipid to glass. Coronado (12) recently described the effect of divalent cations on the assembly of negatively charged phosphatidylserine (PS) and neutral dioleoyl-PE (DOPE) on patch pipettes. PS bilayers with seals in the gigaohm range formed only when divalent cations were present in the pipette and bath solutions. In contrast, DOPE seals were independent of divalent ions at the pH where DOPE is neutral (pH 6.5) or predominantly positively charged (pH 1.5). At pH 10, when most PE molecules possess a net negative charge, DOPE seals, like PS seals, became divalent cation dependent. These results suggest that the conductivity of the lipid-glass junction is largely dominated by the charge of lipid polar head groups (12).

The advantage of patch electrodes when applied to recordings in vitro [as in the monolayer-dipping (18, 27, 78, 81), the liposome-patch (74, 83, 84), or the bilayer-patch (6) techniques] clearly resides in the improved time resolution and low electrical noise that can be obtained from bilayers of small area. This has been recently demonstrated by Rosenberg et al (74) in liposome-patch recordings of purified Na^+ channels where channel opening was elicited by pulse protocols on the millisecond time scale; the results were identical to those obtained in vivo. By the same token, however, the main disadvantage of patch electrodes is that the probability of finding reconstituted channels in a given patch is low owing to the low density of channels usually achieved by reconstitution.

MUSCLE CALCIUM CHANNELS

Recording of calcium channels in planar bilayers has proved a challenge. In tissues such as heart, calcium channels are present in minute quantities. Their intrinsic low density and the overwhelming presence of K^+ and Cl^- channels of large unitary conductance, which are easily detected in planar bilayers containing heart sarcolemma (17, 33, 69), have made progress slow (but see 22). A second challenge is more fundamental. Calcium channels in vivo display fast and complex kinetics (31, 32). Activation kinetics of tens of milliseconds is followed by an equally fast turning-off or inactivation process. A second, slow inactivation, on the time scale of minutes, irreversibly closes or "washes-off" the calcium channel once a cell is ruptured by a microelectrode. Thus until recently there has been little progress in recovering calcium channel activity from fractionated membranes. Nelson et al (66) and Ehrlich et al (21) nevertheless reported calcium channels in planar bilayers from rat brain microsomes and cilia of *Paramecium* with conduction and permeability properties that match to some extent those found in vivo. These channels may be distant relatives of those recorded in neurons and muscle, inasmuch as they are spontaneously open in the absence of voltage steps or agonists, can be recorded in symmetrical *cis* and *trans* 0.1 M Ca^{2+}, and apparently are insensitive to calcium antagonists.

In skeletal muscle, two important advances have overcome the inherent difficulties of recording calcium channels in vitro. One is the preparation of vesicles from the T-tubule and the terminal cisternae of twitch muscle (49, 72). These vesicles are highly "fusogenic" to bilayers, and channel activity present in them is remarkably well preserved (1, 42, 52, 57, 80). Soon after the discovery of calcium currents in skeletal muscle (75), Almers et al (3) showed that these currents originated predominantly in the T-system. The T-system comprises a network of invaginated plasma membrane that relays action potentials, generated at the muscle surface, deep into the muscle fiber. The T-system makes extensive anatomical contacts with the terminal cisternae of sarcoplasmic reticulum (SR), the intracellular compartment that stores and releases Ca^{2+} during the excitation-contraction cycle. A different set of calcium channels, pharmacologically distinct from those in T-tubules, is thought to mediate release from the SR (49, 64). Owing to the subcellular dimensions of the T-system and the intracellular location of sarcoplasmic reticulum, to date none of these channels have been accessible by patch electrodes in situ. This complicating factor has provided a strong rationale for a biochemical approach to the problem of calcium release and the use of suitable vesicular preparations.

The second advance is the discovery of drugs that stabilize calcium channels in the open conformation. In the case of plasma membrane calcium channels, agonists such as Bay K-8644 (23, 87) induce calcium channels to remain open for hundreds of milliseconds. Likewise in terminal cisternae vesicles of SR, Ogawa & Ebashi (70) and others (49, 64) described the chemical gating of calcium release by adenine nucleotides (ATP, AMP-PCP, and adenosine). Hence, the expectation was soon raised that these compounds might chemically gate fluxes through calcium channels that otherwise were normally closed (1, 23, 58, 80). The use of calcium agonists in the search for calcium channels in T-tubules (1) and SR terminal cisternae (80) has been analogous to the use of batrachotoxin for the Na$^+$ channel (36). The calcium agonist permits a pharmacological separation of the divalent current through the calcium channel from the agonist-insensitive residual current, which could in principle be carried by other channels present in these vesicles. Also, by increasing the lifetime of channels, these agonists solve in part the inherent bandwidth limitations associated with recordings in planar bilayers.

Affolter & Coronado (1, 2) have shown that T-tubule calcium channels in planar bilayers have a slope conductance of 20 pS under saturating 0.1 M *cis* Ba^{2+}. Recording of activity over long periods under constant depolarized 0 mV conditions requires micromolar concentrations of agonists Bay K-8644 or CGP-28392 in the *cis* vesicle-containing chamber (Figure 1). The agonist induces a population of stable open channels that tend to cluster into bursts lasting 1–5 s. The number of bursts per unit time and the probability of observing an open channel increase with *cis* positive potentials. The use of quaternary derivatives of verapamil such as D890, which is impermeant through the lipid phase, has confirmed that channels orient in the planar bilayer with the cytoplasmic face on the *cis* chamber (2);

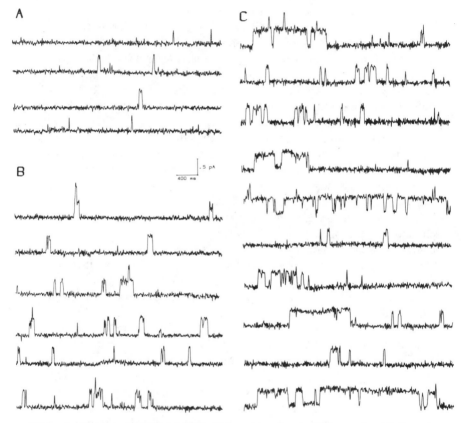

Figure 1   Barium current through single calcium channels activated by agonists. (*A*): Records at HP = 0 mV in the absence of agonists; bilayer was formed in symmetrical *cis-trans* solution containing 0.05 M NaCl, 0.1 mM EGTA, 10 mM Hepes-tris pH 7.0. Afterwards, *cis* $BaCl_2$ was raised to 0.1 M and 100 μg of vesicular T-tubule protein were added to the *cis* side. (*B*), (*C*): Records at HP = 0 mV in 3 μM CGP-28392 (*B*) or 3 μM Bay-K8644 (*C*). Using the solution described in (*A*), the bilayer was incubated for 5 min with agonist just prior to addition of T-tubule protein. Agonist was added from a concentrated solution (3 mM) in ethanol. Time mark (400 ms) and current mark (0.5 pA) is the same for all records (1).

this was also suggested by the voltage dependence of burst formation (1). Dihydropyridine antagonists such as nitrendipine block the channel, surprisingly, from either side of the bilayer. To my knowledge, this property was not recognized in earlier experiments with whole cells. It could indicate that dihydropyridine, like tertiary verapamil derivates (31), gains access to the channel mainly by partitioning first into the bilayer lipid. When Ba^{2+} is replaced by a divalent cation from the rest of group IIa, the slope conductance follows the sequence $Ba^{2+} = 20$ pS $> Sr^{2+} = 12$ pS $> Ca^{2+} = 10$ pS $\gg Mg^{2+} < 2$ pS. The high barium conductance and the lack of significant Mg^{2+} permeability indicate that the channel is endowed with permeation properties that are characteristic of muscle calcium channels recorded in whole cells (4, 32). Another permeation property unequivocally identifies this channel as the carrier of the T-tubule calcium current: In the absence of divalent cations to support current, the channel transports Na^+ or K^+ ions at saturating rates higher than for divalent cations alone (16). Almers & McCleskey (4) have shown in frog skeletal muscle fibers that this conduction feature is intimately linked to the mechanism of ion permeation.

The properties of SR calcium channels in planar bilayers closely parallel those of calcium channels studied by isotopic fluxes in vesicles. Adenine nucleotides can stimulate calcium release from vesicles derived from the terminal cisternae of the SR (49, 64, 70), and release rates can be as high as 100 s^{-1} when stimulated by ATP or AMP-PCP (49, 80). Efflux is blocked by micromolar concentrations of the polycationic dye ruthenium red and millimolar Mg^{2+} (49). Using the planar bilayer fusion technique, Smith et al (80) identified the calcium release channel in terminal cisternae SR on the basis of *cis* activation by nucleotides, blockade by ruthenium red, and selectivity for divalent cations (Figure 2). Unlike the T-tubule channel, the SR calcium channel exhibits an unusually large unitary conductance of 170 pS in 50 mM Ba^{2+} or 125 pS in Ca^{2+}. The channel discriminates against monovalents by a permeability ratio, $P(Ba^{2+})/P(Cs^+) = 11.4$.

A major experimental problem in the detection of single SR Ca^{2+} channels was the fact that in SR, the bulk of the membrane conductance is by monovalent cations and anions (50, 51, 53). To eliminate the contribution of monovalent channels, vesicles of the terminal cisternae were fused in choline chloride buffers. The myoplasmic *cis* side was then perfused with chloride-free buffer containing large organic cations, and the intra-SR *trans* side was perfused with high Ba^{2+} or Ca^{2+} to support current. Sidedness was demonstrated by the agonist effect of nucleotide on the *cis* chamber only (80). In the absence of nucleotide, open channel lifetimes are biexponential with values $t_1 = 8.4$ ms and $t_2 = 24.2$ ms. After 1 mM *cis* ATP, the long lifetime is increased almost fourfold ($t_2 = 95.2$ ms). ATP does not seem to exert its activating effect through a protein kinase or other ATP

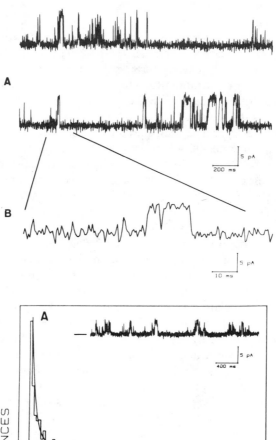

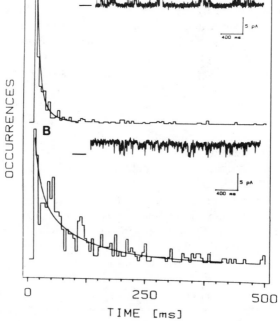

hydrolyzing reaction since nonhydrolyzable analogs may substitute. The significance of a calcium channel activation by nucleotide, which in muscle is present at millimolar levels, is not known. ATP is unlikely to function as a second messenger during excitation-contraction coupling because nucleotide levels probably remain constant throughout the cycle. Results with vesicles indicate that free Mg^{2+} is a strong regulator of calcium release and an efficient blocker of the channel in planar bilayers (J. S. Smith, personal communication). This interaction might be one in vivo mechanism for preventing constant channel activation by cytoplasmic ATP.

CONTRIBUTION OF LIPID SURFACE CHARGE

The activity of ion channels may be modulated in various ways by the composition of the phospholipid bilayer. The most direct and physically tractable way is through interactions with membrane surface charges. Cellular plasma membranes contain approximately 15% negatively charged lipid, mostly phosphatidylserine (PS) and phosphatidylinositol (PI). The immediate effect of these charges at the surface of a bilayer is a negative electrostatic surface potential which is balanced by a local accumulation of cations and depletion of anions (48).

Recent work with planar bilayers on K^+ (8, 56), Ca^{2+} (16), and Na^+ (J. F. Worley, R. J. French & B. K. Krueger, in preparation) channels of nerve and muscle (Figure 3) has significantly clarified the effect of surface charge on ion conduction. In planar bilayers the phospholipid composition is under experimental control. The presence or absence of surface charge effects is thus easily demonstrated by imposing on the channel various mole fractions of the acidic lipid PS. Likewise, the possible contribution of local fixed charges present in the channel protein itself (or tightly bound lipid) may be unmasked when channels are inserted in planar bilayers in which the bulk of the bilayer lipid is composed of the neutral zwitterionic phosphatidylethanolamine (PE). As shown in Figure 3, a common finding is

---

Figure 2 Incorporation of SR Ca^{2+} channels into planar bilayers. *Top records (A), (B)*: Single-channel current fluctuations in *cis* 2.5 μM free Ca^{2+}, 125 mM Tris, 250 mM Hepes, pH 7.4; *trans* 50 mM Ba^{2+}, 250 mM Hepes, pH 7.4. At the holding potential of records, HP = −20 mV; the mean open channel current was 7.5 pA. (*B*) corresponds to an expanded time scale. *Bottom histograms (A), (B)*: Nucleotide stimulation of the SR Ca^{2+} channel. (*A*): open time histogram of data recorded at HP = 0 mV using 50 mM *trans* Ca^{2+} as current carrier. Insets are representative current traces. The horizontal bar to the left of each record denotes the closed channel current. The fitted exponentials correspond to $t_1 = 8.4$ ms (y-intercept = 459), $t_2 = 24.2$ ms (y-intercept = 60). (*B*): open time histogram after 1 mM *cis* ATP: $t_1 = 13.3$ msec (y-intercept = 29), $t_2 = 95.3$ (y-intercept = 16) (80).

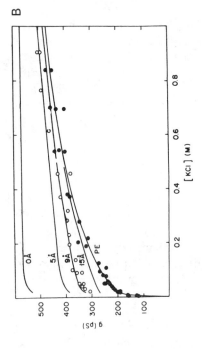

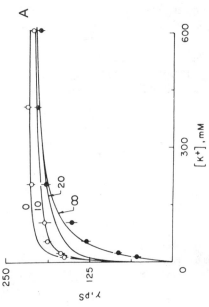

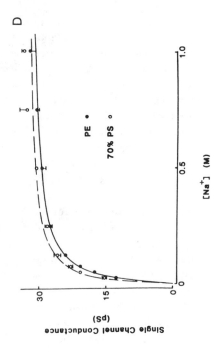

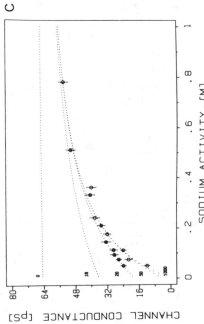

that surface charge effects, even in 100% PS bilayers, are remarkably small; these effects are significant only at low ionic strength. For the SR K^+ channel (8) and the calcium-activated K^+ channel (56), single channel conductances are 2-fold to 2.5-fold higher in PS than in PE in the limit of 10–20 mM potassium ion concentration. For sodium current through batrachotoxin-activated Na^+ channels (J. F. Worley et al, in preparation) or sodium current through Bay K-8644–activated calcium channels (16), this difference is much less. These results show that the entry of ions into these channels is not affected by the full value of the surface potential of PS, so the ions are heavily insulated from the bilayer lipid (see 8).

Bell & Miller (8) have interpreted the insulation of the SR K^+ channel assuming that for a channel embedded into a membrane of negative charges the ion concentration relevant to conduction is not that given by the bulk solution but is that existing at the entryway or mouth of the channel, separated by some distance from the bulk lipid. This distance, albeit unknown in the absence of structural information, may be estimated as an equivalent vertical distance, x, separating the center of the channel's mouth

---

Figure 3 Saturation of single channel conductance in neutral and charged phospholipid bilayers. *A*: SR K^+ channel. Single channel conductance vs K^+ concentration at various distances from a negatively charged bilayer. Curves calculated according to Equations 1 and 2 of text represent the expected conductance of a channel protruding the indicated distance (Å) from a bilayer containing the measured surface charge density of 0.93 charges/nm². Data plotted are from PE/PC membranes (*filled circles*) and PE/PS (*open circles*) (8). *B*: C^{2+}-activated K^+ channel. Single channel conductance as a function of KCl concentration. Unit conductance was measured for channels incorporated into PE (*filled circles*) or PS (*open circles*). The solid lines labeled "0 Å" to "15 Å" are computer fits (Equations 1 and 2 of text) that allow bulk K^+ concentration to be expressed as the local K^+ concentration at various distances from the surface of a PS bilayer, according to Gouy-Chapman-Stern theory. The curve labeled "PE" assumes that there is no surface charge effect in this lipid (56). *C*: Ca^{2+} channel. Sodium current through single calcium channels was measured in the absence of *trans*-extracellular ($< 1 \mu M$) divalent ions. *Trans* Na^+ was held constant at 50 mM and *cis* Na^+ was varied as indicated on the x-axis. Conductance in PE (*open circles*) and PS (*filled circles*) bilayers corresponds to the slope conductance at 0 mV. Data in PE is described by a rectangular hyperbola with a midpoint of 0.2 M and a saturating conductance of 68 pS. *Dotted lines* (Å) correspond to a rectangular hyperbola in which bulk Na^+ was replaced by $Na(x)$, the local concentration of Na^+ ions at a distance x, for a 100% PS bilayer (Equations 1 and 2 of text). In each curve, conductance was solved as a function of local Na^+ concentration at a constant x (16). *D*: Batrachotoxin-activated Na^+ channel. Single channel conductance as a function of the sodium ion concentration for neutral (*filled circles*) and negatively charged (*open circles*) membranes. Each point represents the slope of linear current-voltage relationships for sodium channels at the indicated sodium concentration on both sides of the membrane ($< 10 nM$ divalent cation). In neutral and negatively charged membranes the conductance can be described by a rectangular hyperbola. The data at each point were determined from 3–30 membranes (J. F. Worley, R. J. French, B. K. Krueger, personal communication).

from the plane of the lipid charges, i.e. as if all the insulation were given by protein mass that protrudes from the bilayer into the bulk solutions. It follows from the Gouy-Chapman theory (8, 48) that if the surface potential and surface charge density are known (given by the Grahame equation and a calibrated density of PS in the bilayer), then $\psi(x)$, the electrostatic potential at distance x, and $C^+(x)$, the monovalent ion concentration at x, may be solved from

$$\psi(x) = [2RT/F] \ln [\{1 + A \exp(-Dx)\}/\{1 - A \exp(-Dx)\}] \qquad \text{1a.}$$

$$A = [\exp(F\psi(0)/2RT) - 1]/[\exp(F\psi(0)/2RT) + 1] \qquad \text{1b.}$$

where D is the reciprocal of the Debye length and $\psi(0)$ is the potential at $x = 0$, i.e. the surface potential. The concentration of ions at a distance x follows from a Boltzmann distribution in the electrostatic potential field,

$$C^+(x) = C^+(b) \exp[-F\psi(x)/RT], \qquad 2.$$

where $C^+(b)$ is the monovalent bulk solution. This set of equations contains x as the only free parameter and thus has provided a simple estimate of the equivalent distance separating the pore entry from the bulk lipid (8, 16, 56). It is clear from the x values fitted in Figure 3, all in the range of 10–20 Å, that such large insulating distances may only be physically realized by proteins of large dimensions. This is in perfect agreement with available data on AchR channels (11) and on the molecular weights of Na^+ channel polypeptides (91) and the nitrendipine-binding protein linked to calcium channels (20). Figure 3 also shows that conductance-activity relationships in neutral lipid alone, when no lipid charges are imposed on channels, are described by the bulk activity and a simple Langmuir isotherm (except the Ca^{2+}-activated K^+ channel). This implies that the entryways per se do not provide extra fixed charges, as these charges, if present, would distort a simple Langmuir-type saturation. In the case of SR K^+ and Ca^{2+} channels this conclusion is rigorously justified by linear Scatchard plots of the data in neutral PE membranes (8, 16). The Ca^{2+}-activated K^+ channel seems to be a special case in which saturation cannot be fitted by a single-site model (56). The possibilities of a charge at the mouth of the channel and of multiple occupancy through a "charge-free" pore have been discussed (56).

CHARYBDOTOXIN: A K^+ CHANNEL TOXIN

Little is known about the biochemistry of Ca^{2+}-activated potassium channels, largely owing to the lack of a suitable ligand for use as a biochemical probe for the channel. Miller et al (52) recently used planar bilayers to describe a protein inhibitor of single Ca^{2+}-activated K^+ channels of mammalian skeletal muscle. The inhibitor is a minor component of the venom of the Israeli scorpion, *Leiurus quinquestraitus* (LQV). The partially purified active component is a basic protein of MW 7000. Addition of LQV to the external *trans* solution of a control channel induces long-lived nonconductive quiescent periods separated by bursts of activity (Figure 4). Within such bursts, the rapid opening and closing of the channel and its unitary conductance are identical to those seen in control conditions. LQV is active (and reversible upon perfusion) only in the *trans* chamber, i.e. opposite to the side on which Ca^{2+} activates the channel. Increase in toxin concentration shortens the duration of active bursts but has no effect on the average duration of the nonconducting periods, as if active toxin in the venom associates reversibly with one active channel. The finding that burst duration is inversely proportional to toxin concentration was used to develop a functional assay to fractionate the venom by a two-step chromatography. This is a novel example of planar bilayer reconstitution at its best. The mechanism of action of charybdotoxin (CTX), the active peptide in LQV, appears to be by direct occlusion of the extracellular conduction pathway. This hypothesis is supported by competition experiments with TEA, a low-affinity blocker of K^+ channels, and by the release of CTX blockade by TEA. CTX has an affinity in the nanomolar range; this affinity, combined with the long blocked times, makes CTX a suitable candidate to be used in ligand binding, as a photoaffinity ligand or possibly as a basis for affinity chromatography. The specificity of CTX, which is currently under investigation (52), shows that it is not effective against batrachotoxin-activated sodium channels or K^+ channels of squid axons.

Charybdotoxin is only one of a large group of toxins currently being examined in planar bilayers (for a review see 60). Moczydlowski et al (55, 60) described the U-conotoxins, a series of 22–amino acid peptides purified from the venom of the marine snail *Conus geographus*. Electrophysiological studies have shown that U-conotoxin GIIIA blocks action potentials of mammalian muscle without affecting nerve action potential. In planar bilayers, GIIIA induces discrete blocking events of the batrachotoxin-activated Na^+ channel from rat muscle but not from rat brain under similar conditions. A statistical analysis of the blocked-unblocked events indicated that GIIIA in muscle displays blockade parameters similar to those of tetrodotoxin (TTX) or saxitoxin (STX) (58). However, GIIIA apparently recognizes structural differences between the nerve and muscle Na^+ channels.

CONCLUSIONS

The high diffusion rate of ions through single open pores is the sweet secret of the phenomenal success of single channel reconstitution methods. After all, it is only necessary to insert and monitor one molecule in a planar

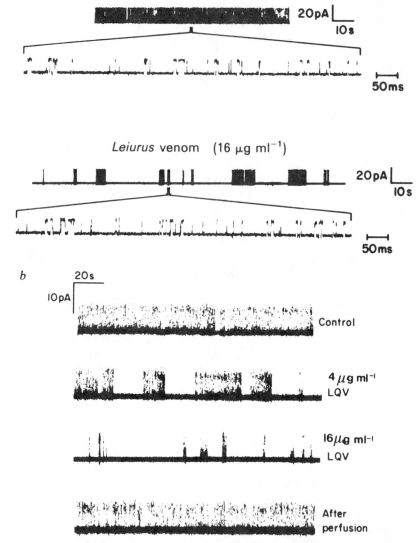

Figure 4 Effect of *Leiurus* venom on the Ca^{2+}-activated K^+ channel. (*a*): Control record taken for 2 min, after which LQV (16 μg ml^{-1}) was added to the external solution and records were collected for a further 2 min. All traces were recorded with continuous stirring. A 200-fold expanded time trace is displayed below each record to illustrate the channel's gating. (*b*): Concentration-dependence and reversibility of *Leiurus* venom. The traces show channel behavior before addition of crude venom (*top trace*), in the presence of venom at the indicated concentrations (*two middle traces*), and after extensive perfusion of the external chamber with venom-free medium (52).

bilayer, reproducibly of course, to claim such success. The fusion method (53) seems to meet such criteria of reproducibility. As they stand however, the present methods exclude the possibility of monitoring ion pumps and electrogenic carriers, along with a variety of other cellular transporters of low turnover rate. It is a problem of yields: One pA of pump current in a planar bilayer would probably require the concerted action of over 10,000 pump molecules. At present such a quantity of molecules cannot be inserted into planar films. At least in the case of one protein, the AchR, crucial progress is being made in the handling of large numbers of molecules (37). The most promising approach appears to be the transfer of molecules from liposomes to planar bilayers by the use of monolayers. Spreading of protein monolayers is not a new trick. In fact, the vast literature immediately applicable to channel reconstitution invites revision (25).

ACKNOWLEDGMENTS

I am grateful to Dr. Jeffrey S. Smith and to the editors of the *Annual Review of Biophysics and Biophysical Chemistry* for comments and suggestions on the manuscript. Work from this laboratory is supported by NIH grant R01 GM-32824.

Literature Cited

1. Affolter, H., Coronado, R. 1985. *Biophys. J.* 48:341
2. Affolter, H., Coronado, R. 1985. *Biophys. J.* In press
3. Almers, W., Fink, R., Palade, P. T. 1981. *J. Physiol.* 312:177
4. Almers, W., McCleskey, E. W. 1984. *J. Physiol.* 353:585
5. Andersen, O. S. 1985. In *Reconstitution of Ion Channel Proteins*, ed. C. Miller. New York: Plenum. In press
6. Andersen, O. S., Mueller, R. U. 1982. *J. Gen. Physiol.* 80:403
7. Anholt, R., Fredkin, D. R., Deerinck, T., Ellisman, M., Montao, M., Lindstrom, J. 1982. *J. Biol. Chem.* 275:7122
8. Bell, J. E., Miller, C. 1984. *Biophys. J.* 45:279
9. Blodgett, K. B. 1934. *J. Am. Chem. Soc.* 36:495
10. Changeux, J. P., Devillers-Thery, A., Chemouilli, P. 1984. *Science* 225:1335
11. Criado, M., Eibl, H., Barrantes, F. J. 1984. *J. Biol. Chem.* 259:9188
12. Coronado, R. 1985. *Biophys. J.* 47:851
13. Coronado, R., Affolter, H. 1985. *Biophys. J.* 47:434a
14. Coronado, R., Affolter, H. 1985. Presented at Ann. Meet. Am. Heart Assoc., 58th Sci. Sess., Washington, DC, Nov. 10–14. *Circ. Res.* (Abstr.) In press
15. Coronado, R., Affolter, H. 1985. See Ref. 5
16. Coronado, R., Affolter, H. 1985. *J. Gen. Physiol.* In press
17. Coronado, R., Latorre, R. 1982. *Nature* 298:849
18. Coronado, R., Latorre, R. 1983. *Biophys. J.* 34:231
19. Coronado, R., Latorre, R., Mautner, H. 1984. *Biophys. J.* 45:289
20. Curtis, B. M., Catterall, W. A. 1984. *Biochemistry* 23:2113
21. Ehrlich, B. E., Finkelstein, A., Forte, M., Kung, C. 1984. *Science* 225:427
22. Ehrlich, B. E., Schen, C. R., Garcia, M. L., Kaczorowski, G. J. 1985. *Proc. Natl. Acad. Sci. USA*. In press
23. Freedman, S. B., Miller, R. J. 1984. *Proc. Natl. Acad. Sci. USA* 81:5580
24. French, R. J., Blaustein, M. B., Romine, W. O., Tam, K., Worley, J. F., Krueger, B. K. 1985. *Biophys. J.* 47:191a
25. Gorter, E. 1941. *Ann. Rev. Biochem.* 10:619
26. Guo, X., Moczydlowski, E. 1985. *Biophys. J.* 47:438a
27. Hanke, W., Methfessel, C., Wilmsen, H.,

Katz, E., Boheim, G. 1983. *Biochim. Biophys. Acta* 727:108

28. Hartshorne, R. P., Catterall, W. A. 1984. *J. Biol. Chem.* 259:1667
29. Hartshorne, R. P., Keller, B. U., Talvenheimo, J. A., Catterall, W. A., Montal, M. 1985. *Proc. Natl. Acad. Sci. USA* 82:240
30. Helenius, A., Simons, K. 1975. *Biochim. Biophys. Acta* 415:29
31. Hescheler, J., Pelzer, D., Trube, G., Trautwein, W. 1982. *Pflügers Arch.* 393:287
32. Hess, P., Tsien, R. W. 1984. *Nature* 309:453
33. Hill, J. A., Coronado, R., Strauss, H. C. 1985. See Ref. 14
34. Keller, B., Hartshorne, R., Talvenheimo, J., Catterall, W., Montal, M. 1985. *Biophys. J.* 47:439a
35. Kraner, S. D., Tanaka, J. C., Roberts, R. M., Barchi, R. L. 1985. *Biophys. J.* 47:440a
36. Krueger, B. K., Worley, J. F., French, R. J. 1983. *Nature* 303:172
37. Labarca, P., Lindstrom, J., Montal, M. 1984. *J. Gen. Physiol.* 83:473
38. Langmuir, I. 1920. *Trans. Faraday Soc.* 15:62
39. Langmuir, I., Waugh, D. F. 1938. *J. Gen. Physiol.* 21:745
40. Latorre, R., Alvarez, O., Cecchi, X., Vergara, C. 1985. *Ann. Rev. Biophys. Biophys. Chem.* 14:79
41. Latorre, R., Miller, C. 1983. *J. Membr. Biol.* 71:11
42. Latorre, R., Vergara, C., Hidalgo, C. 1982. *Proc. Natl. Acad. Sci. USA* 79:805
43. Lee, M. L., Wright, B. W. 1980. *J. Chromatogr.* 184:235
44. Levitan, I. 1985. See Ref. 5
45. Lievano, A., Sanchez, J., Darzon, A. 1985. *Biophys. J.* 47:145a
46. Makowski, L., Caspar, D. L. D., Phillips, W. C., Baker, T. S., Goodenough, D. A. 1984. *Biophys. J.* 45:208
47. McIntosh, T. J., Simon, S. A., MacDonald, R. C. 1980. *Biochim. Biophys. Acta* 597:445
48. McLaughlin, S. 1977. *Top. Membr. Transp.* 9:71
49. Meissner, G. 1984. *J. Biol. Chem.* 259:2365
50. Miller, C. 1983. *Physiol. Rev.* 63:1209
51. Miller, C. 1984. *Ann. Rev. Physiol.* 46:549
52. Miller, C., Moczydlowski, E., Latorre, R., Phillips, M. 1985. *Nature* 313:316
53. Miller, C., Racker, E. 1976. *J. Membr. Biol.* 30:283
54. Mimms, L., Zampighi, G., Nozaki, Y., Tanford, C., Reynolds, J. A. 1981. *Bio-*

chemistry 20:833
55. Moczydlowski, E. 1985. *Biophys. J.* 47:190a
56. Moczydlowski, E., Alvarez, O., Vergara, C., Latorre, R. 1985. *J. Membr. Biol.* 83:273
57. Moczydlowski, E., Garber, S. S., Miller, C. 1984. *J. Gen. Physiol.* 84:665
58. Moczydlowski, E., Hall, S., Garber, S. S., Strichartz, G. S., Miller, C. 1984. *J. Gen. Physiol.* 84:687
59. Moczydlowski, E., Latorre, R. 1983. *J. Gen. Physiol.* 82:511
60. Moczydlowski, E., Uehara, A., Hall, S. 1985. See Ref. 5
61. Montal, M., Mueller, P. 1972. *Proc. Natl. Acad. Sci. USA* 69:3561
62. Mueller, P., Rudin, D. O., Tien, H. T., Wescott, W. C. 1962. *Circulation* 26:1167
63. Murphy, R. B., Vodyanoy, V. 1984. *Biophys. J.* 45:22
64. Nagasaki, K., Kasai, M. 1983. *J. Biochem. Tokyo* 84:1101
65. Nelson, M. T. 1985. *Biophys. J.* 47:67a
66. Nelson, M. T., French, R. J., Krueger, B. K. 1984. *Nature* 308:77
67. Nelson, M. T., Reinhart, R. 1984. *Biophys. J.* 45:60
68. Nelson, M. T., Roudna, M., Bamberg, E. 1983. *Am. J. Physiol.* 245:C151
69. Nelson, M. T., Worley, J. F., Rogers, T. B., Lederer, W. J. 1985. *Biophys. J.* 47:143a
70. Ogawa, Y., Ebashi, S. 1976. *J. Biochem.* 80:1149
71. Orozco, C. B., Suarez-Isla, B. A., Froehlich, J. P., Heller, P. F. 1985. *Biophys. J.* 47:57a
72. Rosemblatt, M., Hidalgo, C., Vergara, C., Ikemoto, N. 1981. *J. Biol. Chem.* 256:8140
73. Rosenberg, R. L., Tomiko, S. A., Agnew, W. S. 1984. *Proc. Natl. Acad. Sci. USA* 81:1239
74. Rosenberg, R. L., Tomiko, S. A., Agnew, W. S. 1984. *Proc. Natl. Acad. Sci. USA* 81:5594
75. Sanchez, J. A., Stefani, E. 1978. *J. Physiol.* 238:197
76. Schindler, H. 1979. *Biochim. Biophys. Acta* 555:316
77. Schindler, H. 1980. *FEBS Lett.* 122:77
78. Schuerholz, T., Schindler, H. 1983. *FEBS Lett.* 152:187
79. Smith, J. S., Coronado, R., Meissner, G. 1985. *Biophys. J.* 47:451a
80. Smith, J. S., Coronado, R., Meissner, G. 1985. *Nature* 316:446
81. Suarez-Isla, B. A., Wang, K., Lindstrom, J., Montal, M. 1983. *Biochemistry* 22:2319

82. Takagi, M., Azuma, K., Kishimoto, U. 1965. *Annu. Rep. Biol. Works Fac. Sci. Osaka Univ.* 13:107
83. Tank, D. W., Huganir, R. L., Greengard, P., Webb, W. W. 1983. *Proc. Natl. Acad. Sci. USA* 80:5129
84. Tank, D. W., Miller, C., Webb, W. W. 1982. *Proc. Natl. Acad. Sci. USA* 79:7749
85. Tomlins, B., Williams, A. J., Montgomery, R. A. P. 1984. *J. Membr. Biol.* 80:191
86. Torres, R. M., Coronado, R., Bezanilla,
F. 1984. *Biophys. J.* 45:38a
87. Towart, R., Schramm, M. 1984. *Trends Pharmacol.* 5:111
88. Trurnit, H. J. 1960. *J. Colloid Sci.* 15:1
89. Vergara, C., Latorre, R. 1983. *J. Gen. Physiol.* 82:543
90. Verger, R., Pattus, F. 1976. *Chem. Phys. Lipids* 16:285
91. Weigele, J., Barchi, R. 1982. *Proc. Natl. Acad. Sci. USA* 79:3651
92. Wolf, D., Cecchi, X., Naranjo, D., Alvarez, O., Latorre, R. 1985. *Biophys. J.* 47:386a

Ann. Rev. Biophys. Biophys. Chem. 1986. 15 : 279–319
Copyright © 1986 by Annual Reviews Inc. All rights reserved

ACTIVE TRANSPORT IN *ESCHERICHIA COLI*: Passage to Permease[1]

H. Ronald Kaback

Roche Institute of Molecular Biology, Roche Research Center, Nutley, New Jersey 07110

CONTENTS

[1] Abbreviations: $\Delta\bar{\mu}_{H^+}$, the proton electrochemical gradient across the membrane; RSO, right-side-out; ISO, inside-out; ΔpH, pH gradient; $\Delta\Psi$, membrane potential; FAD, flavin adenine dinucleotide; D-LDH, D-lactate dehydrogenase; PMS, phenazine methosulfate; TMPD, N,N,N',N'-tetramethylphenylenediamine; DAD, diaminodurene; CCCP, carbonylcyanide-*m*-chlorophenylhydrazone; DCCD, N,N'-dicyclohexylcarbodiimide; TPMP$^+$, triphenylmethylphosphonium; TPP$^+$, tetraphenylphosphonium; DMO, 5,5'-dimethyloxazolidine-2,4-dione; octylglucoside, octyl-β-D-glucopyranoside; NPG, *p*-nitrophenyl-α-D-galactopyranoside; TDG, β-D-galactosyl 1-thio-β-D-galactopyranoside; TMG, methyl 1-thio-β-D-galactopyranoside; SDS-PAGE, sodium dodecylsulfate polyacrylamide gel electrophoresis; Q_1H_2, ubiquinol-1; *p*-CMBS, *p*-chloromercuribenzene-sulfonate; DEPC, diethylpyrocarbonate.

279

0883–9182/86/0610–0279$02.00

PERSPECTIVES AND OVERVIEW

In the same way that the DNA double helix revolutionized molecular genetics, the chemiosmotic hypothesis, formulated and refined by Peter Mitchell during the 1960s (111–115), revolutionized bioenergetics. It is now the conceptual framework for a wide array of bioenergetic phenomena. Curiously, however, the elegance and importance of this hypothesis have received relatively little attention because the chemiosmotic hypothesis was formulated initially to explain oxidative phosphorylation and is still identified with this traditionally controversial field; because biochemists are generally uncomfortable with ephemeral entities such as electrochemical ion gradients; and because various disciplines within the area of bioenergetics use different terminology to describe similar phenomena.

Although the phenomenon of active transport (i.e. concentration of solute against a gradient at the expense of metabolic energy) has been recognized for some time, insight into the biochemistry involved has begun only recently. The reasons for the changing emphasis from the phenomenological to the molecular level are basically threefold: (a) The formulation of the chemiosmotic hypothesis (111–115) stipulates that the immediate driving force for many processes in energy-coupling membranes is an electrochemical gradient of hydrogen ion ($\Delta\bar{\mu}_{H^+}$). (b) Techniques have been developed that enable detection and quantitation of electrochemical ion gradients in microscopic systems. (c) Isolated membrane vesicles and proteoliposomes that have been reconstituted with purified components are available and can be manipulated biochemically and genetically so as to yield information on a molecular level.

This contribution is not a general review; its purpose is to discuss active transport in *Escherichia coli* as a representative example of current developments. However, the work discussed is highly relevant to other systems, as evidenced by the profusion of similar experimental systems that have been developed from other cells, organelles, and epithelia. Furthermore, the discussion is pertinent not only to active transport but, in a broad sense, to the general problem of energy transduction in biological membranes. For example, the primary function of mitochondrial or chloroplast membranes is to convert respiratory energy or light, respectively, into chemical energy (i.e. ATP). However, in *E. coli* respiratory energy is converted into work in the form of electrochemical or osmotic gradients.

MEMBRANE VESICLES AND ACTIVE TRANSPORT: PHENOMENOLOGY

Preliminary evidence garnered in the 1960s (65, 73) suggested that cytoplasmic membrane vesicles from *E. coli* might provide a useful model system for studying active transport. This early promise has been more than fulfilled. Numerous studies demonstrate that vesicles prepared from *E. coli* and other bacteria, eukaryotic cells, intracellular organelles, and epithelia catalyze the accumulation of many different solutes under appropriate experimental conditions. In some instances initial rates of transport are comparable to those of the intact cell (102, 166) and the vesicles accumulate many solutes to concentrations markedly higher than in the external medium (140). Moreover, it has been demonstrated directly with *E. coli* that each vesicle in the preparations is probably functional (164).

Right-side-out (RSO) bacterial membrane vesicles are prepared by lysis of osmotically-sensitized cells (i.e. protoplasts or spheroplasts). RSO vesicles consist of osmotically-intact, unit-membrane–bound sacs that are 0.5–1.0 μm in diameter (67). They are probably more accurately described as "ghosts" (73), as a single structure is obtained from each cell if mechanical stress is avoided. The vesicles are devoid of internal structure, their metabolic activities are restricted to those provided by the enzymes of the membrane itself, and numerous observations demonstrate that the vesicle membrane retains the same polarity and configuration as the membrane in the intact cell (127–129; in addition, cf 172).

Alternatively, by subjecting cells to low shear forces in a French pressure cell, inside-out (ISO) vesicles can be prepared (53, 149). ISO vesicles are about ten times smaller than RSO vesicles and the yield from the preparation is low, but it is apparent that they have a polarity opposite to that of the intact cell (cf 144 for a review).

Transport by membrane vesicles per se is practically nil, and the energy source for uptake of any given substrate can be determined by studying which compounds or experimental manipulations drive accumulation. In addition, metabolic conversion of transport substrates and energy sources is minimal. These properties constitute an advantage over intact cells for the study of active transport, as the system provides clear definition of the phenomena.

The transport systems elucidated in RSO vesicles from *E. coli* fall into four main categories: (*a*) group translocation in which a covalent change is exerted upon the transported molecule so that the reaction itself results in passage of the solute through the diffusion barrier; (*b*) primary active transport in which a hydrogen ion is extruded, thereby generating $\Delta \bar{\mu}_{H^+}$ (interior negative and/or alkaline); (*c*) secondary active transport in which solute is either accumulated or extruded against a concentration gradient in response to $\Delta \bar{\mu}_{H^+}$; and (*d*) passive diffusion of certain weak acids or weak bases and lipophilic ions, followed by equilibration with the pH gradient (ΔpH) and the electrical potential ($\Delta\Psi$), respectively, across the membrane.

In *E. coli* membrane vesicles uptake of D-glucose, D-fructose, D-mannose, and certain other carbohydrates occurs by vectorial phosphorylation via the phosphoenolpyruvate-phosphotransferase system (PTS) [a group translocation mechanism (66)]. The PTS, described originally in 1964 by Roseman and his colleagues (92) and subsequently studied in great detail (25), catalyzes transfer of phosphate from P-enolpyruvate through a sequence of specific proteins to the appropriate carbohydrates in such a manner that the carbohydrate is translocated across the membrane and accumulated as a result of phosphorylation. Although the system is critical for the bacteria that employ it and the PTS plays a central role in regulating carbohydrate metabolism (153), it is not ubiquitous among bacteria and has not been found thus far in organisms phylogenetically higher than bacteria.

As opposed to group translocation, transport of a variety of other solutes by *E. coli* membrane vesicles occurs by secondary active transport (69). Accumulation of these solutes is mediated by highly specific permeases or transporters and is coupled to the oxidation of D-lactate to pyruvate; oxidation is catalyzed by a flavin adenine dinucleotide (FAD)–linked, membrane-bound D-lactate dehydrogenase (D-LDH) that has been purified to homogeneity (35, 84) and synthesized in vitro (157) from a hybrid plasmid containing the *dld* gene (200). Electrons derived from D-lactate are passed to oxygen via a membrane-bound respiratory chain, and in this sequence of reactions respiratory energy is converted into work in the form of active transport. Although other oxidizable substrates such as L-lactate, succinate, α-glycerol-P, and NADH may stimulate transport, they are not as effective as D-lactate unless ubiquinone is added to the vesicles (172, 173) or the vesicles contain *o*- and/or *d*-type terminal oxidases (K. Matsushita, L. Patel, H. R. Kaback, unpublished data). Active transport in the vesicle system is also driven by nonphysiological electron carriers such as reduced phenazine methosulfate (PMS) (87), pyocyanine (172), N,N,N',N'-tetramethylphenylenediamine (TMPD), or diaminodurene (DAD). [TMPD and DAD are effective only when the *o*- and/or *d*-type terminal oxidases are present (K. Matsushita, L. Patel, H. R. Kaback, unpublished data).] Some of these electron carriers, particularly reduced PMS, drive transport much more effectively than physiological electron donors such as D-lactate.

Solute accumulation in RSO membrane vesicles also occurs in the absence of oxygen when appropriate anaerobic electron transfer systems are present (88). Lactose and amino acid transport under anaerobic conditions can be coupled to α-glycerol-P oxidation with fumarate as electron acceptor or to oxidation of formate utilizing nitrate as electron acceptor. Both of these anaerobic electron transfer systems are induced by appropriate growth conditions, and components of both systems are bound loosely to the membrane, necessitating the use of a modified procedure for vesicle preparation (89).

Although a convincing body of evidence demonstrates that high-energy phosphate bond energy is not directly involved in secondary active transport (69, 70), internally generated ATP drives solute accumulation in RSO membrane vesicles from *Salmonella typhimurium* and, under certain conditions, *E. coli* (62). In these experiments, *S. typhimurium* induced for P-glycerate transport (154) are loaded with pyruvate kinase and ADP by lysing spheroplasts under appropriate conditions. Vesicles so prepared catalyze proline or serine accumulation in the presence of P-enolpyruvate, and the activity is blocked by the protonophore carbonylcyanide-*m*-chlorophenylhydrazone (CCCP) and by the H^+-ATPase inhibitor N,N'-dicyclohexylcardobiimide (DCCD), but not by anoxia or cyanide. In contrast, respiration-driven solute accumulation is abolished by CCCP and by anoxia or cyanide, but not by DCCD. Moreover, P-enolpyruvate does not drive accumulation effectively in vesicles that lack the P-glycerate transport system. The results support an overall mechanism in which P-enolpyruvate gains access to the interior of the vesicles by means of the P-glycerate transporter and is then acted on by pyruvate kinase to phosphorylate ADP. ATP formed inside the vesicles is hydrolyzed by the H^+-ATPase to drive transport. Using the plasmid pBR322 as vector and *E. coli* as host, a fragment of *S. typhimurium* DNA encoding the P-glycerate transporter was cloned, and RSO vesicles prepared from the host in the presence of pyruvate kinase and ADP also catalyzed ATP-dependent solute accumulation. Parenthetically, more recent studies (202) demonstrate that the P-glycerate utilization system consists of four genes: *pgtP*, which encodes the P-glycerate transporter, a membrane-bound protein with an apparent M_r of 37 kd; *pgtA*, an activator gene; and *pgtB* and *pgtC*, two regulatory genes.

Finally, solute accumulation in RSO vesicles is driven by artificially imposed K^+ gradients ($K^+_{in} \rightarrow K^+_{out}$) in the presence of valinomycin, a K^+-specific ionophore (57, 58, 76, 160). When membrane vesicles prepared in K^+-containing buffers are diluted into media lacking the cation and valinomycin is added, efflux of K^+ creates a diffusion potential ($\Delta\Psi$, interior negative) across the membrane that drives substrate accumulation. Alternatively, artificially imposed ΔpH (interior alkaline) produces similar effects (76). The finding that accumulation occurs under these conditions has important implications for the mechanism of respiration- and ATP-driven secondary active transport.

THE CHEMIOSMOTIC HYPOTHESIS

Since it is beyond the scope of this contribution to provide an extensive general review of the chemiosmotic hypothesis and its ramifications, references 41, 48, 49, 56, 113, 114, 123, and 168 are recommended for additional discussion.

In the most general sense (Figure 1), the concept of the hypothesis is that the immediate driving force for many processes in energy-coupling membranes is a proton electrochemical gradient ($\Delta\bar{\mu}_{H^+}$) composed of electrical and chemical parameters according to the following relationship:

$$\Delta\bar{\mu}_{H^+}/F = \Delta\Psi - 2.3\,RT/F\,\Delta pH. \qquad 1.$$

Here $\Delta\Psi$ represents the electrical potential across the membrane and ΔpH is the chemical difference in proton concentration across the membrane. R is the gas constant, T is absolute temperature, and F is the Faraday constant; $2.3\,RT/F$ is equal to 58.8 at room temperature.

Accordingly, the basic energy-yielding processes of the cell, respiration or absorption of light, generate $\Delta\bar{\mu}_{H^+}$, and the energy stored therein is utilized to drive a number of seemingly unrelated phenomena such as formation of ATP from ADP and inorganic phosphate, secondary active transport, and transhydrogenation of NADP by NADH. Recently it has been shown that $\Delta\bar{\mu}_{H^+}$ or one of its components is also involved in bacterial motility (26), nitrogen fixation (94), transfer of genetic information (42, 43, 77, 95, 156, 189), sensitivity and resistance to certain antibiotics (103, 106), cellulose synthesis (24), and processing of secreted proteins (21, 23, 29). Importantly, many processes driven by $\Delta\bar{\mu}_{H^+}$ are reversible. Thus, hydrolysis of ATP via the H^+-ATPase leads to the generation of $\Delta\bar{\mu}_{H^+}$. Similarly, transport of solutes down a concentration gradient (i.e. the reverse of active transport) can also generate $\Delta\bar{\mu}_{H^+}$ (74). Clearly, therefore, the "common currency of energy exchange," particularly in bacteria, is not ATP, but $\Delta\bar{\mu}_{H^+}$.

The molecular mechanism of H^+ translocation is unsolved for any of the

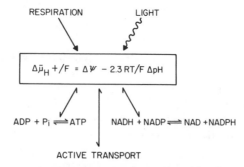

Figure 1 Generalized chemiosmotic hypothesis. Respiration or absorption of light leads to the generation of a proton electrochemical gradient ($\Delta\bar{\mu}_{H^+}$) that provides the immediate driving force for oxidative phosphorylation, photophosphorylation, secondary active transport, and transhydrogenation.

processes described. However, since respiration-driven $\Delta\bar{\mu}_{H^+}$ generation is particularly relevant to secondary active transport in bacteria and because it is not immediately obvious why respiration should generate $\Delta\bar{\mu}_{H^+}$, a cursory discussion is important to put the problem into perspective. Although the concept is controversial, Mitchell (113, 114) has postulated that "loops" in the respiratory chain may be an important means of H^+ translocation across the membrane during electron flow (Figure 2). According to this elegantly simple notion, the electron and H^+ carriers that comprise the respiratory chain are disposed alternately and asymmetrically across the membrane in such a manner that the first component of a "coupling site" on the inner surface (e.g. a reduced flavoprotein that transfers both electrons and H^+) passes two H^+ ions and two electrons to a second component on the outer surface. However, the second intermediate accepts electrons only (e.g. a nonheme-iron protein), and the two H^+ ions are lost to the medium on the side of the membrane opposite their origin. Electrons from the second component on the outer surface are then passed to a third component on the inner surface, and this component (e.g.

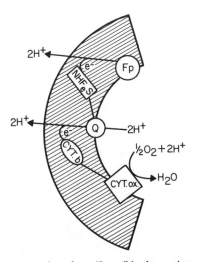

Figure 2 Conceptual representation of two "loops" in the respiratory chain. Electrons and protons are transferred from flavoprotein (*Fp*) (e.g. a dehydrogenase) on the inner surface of the membrane to a nonheme iron-sulfur center (*NHFeS*) oriented toward the outer surface. Since the NHFeS center accepts electrons (e^-) but not H^+, $2H^+$ are released into the medium on the external surface of the membrane. Electrons are then transferred to quinone (*Q*) on the inner surface where Q abstracts $2H^+$ from the medium bathing the inner surface of the membrane. Electrons are then transferred to cytochrome b (*CYT.b*) with release of $2H^+$ into the external medium and finally to a terminal oxidase that reduces oxygen to water on the inner surface.

quinone) can accept H^+ in addition to electrons. Thus, as the third component accepts electrons from the second component, two H^+ ions are taken up from the medium bathing the inner surface of the membrane. The process is then repeated with different redox intermediates. The overall result is that H^+ ions move from one side of the membrane to the other with a net flux of electrons in the opposite direction. In addition, Mitchell (116) has postulated the existence of a "protonmotive ubiquinone cycle" to explain the absence of appropriate H^+ carriers following cytochrome b in the mitochondrial respiratory chain.

Given this type of scheme, it is apparent that the "protonmotive stoichiometry" (i.e. the stoichiometry between H^+ extrusion and electron transfer) in the respiratory chain cannot exceed 2, since the known H^+ carriers do not react with more than two H^+ ions during a single turnover. Although Mitchell's laboratory and others (cf 117) reported stoichiometries of 2, the findings were challenged by a number of other workers who have reported values of 3 or 4 (5, 98, 134, 145, 146, 167, 194, 196). With *E. coli* membrane vesicles, inhibition of electron transfer at the end of the respiratory chain leads to rapid collapse of $\Delta\bar{\mu}_{H^+}$, while inhibition at the dehydrogenase level causes only slow collapse (72, 102, 160). In addition, oxidation of either D-lactate (165) or D-alanine (46, 125) by the appropriate dehydrogenase leads to generation of $\Delta\bar{\mu}_{H^+}$ even when the enzymes are bound to the wrong surface of the membrane (i.e. the outside). Thus, it seems unlikely that either of these dehydrogenases comprises part of a loop. Finally, purified mitochondrial cytochrome oxidase (cf 195) and terminal oxidases purified from *Paracoccus denitrificans* (169) and the thermophile PS-3 (170) appear to catalyze vectorial H^+ translocation as well as vectorial electron transfer in the absence of H^+ carriers. In contrast, the experimental evidence with the o- (47, 83, 104, 105) and d-type (85, 110) terminal oxidases from *E. coli* is consistent with the notion that these enzymes function as "half-loops." Therefore, although the loop mechanism remains attractive in its simplicity, it is not likely to be the only mechanism for respiration-dependent $\Delta\bar{\mu}_{H^+}$ generation.

The salient features of the chemiosmotic hypothesis with respect to secondary active transport are presented schematically in Figure 3. As shown, $\Delta\bar{\mu}_{H^+}$ generated either by respiration or through the action of the H^+-ATPase drives transport via different substrate-specific permeases (i.e. porter or carrier proteins) that extend through the membrane, binding a substrate on the external surface of the membrane and releasing it on the internal surface. In addition, it is postulated that $\Delta\bar{\mu}_{H^+}$ drives transport by several different mechanisms depending on the nature of the substrate as follows.

1. Cationic substrates such as lysine or K^+ are transported by "uniport,"

a mechanism driven specifically by $\Delta\Psi$. Accordingly, a transmembrane uniporter facilitates electrophoresis of external cationic substrate in response to the internally negative $\Delta\Psi$.

2. Transport of acids such as lactate or succinate, in contrast, is driven by the interior alkaline ΔpH across the membrane. The protonated acid is translocated across the membrane, and since the internal space is alkaline relative to the external medium, the acid dissociates. H^+ is pumped out and the anion accumulates. Formally, this is classified as "symport" because H^+ is translocated with substrate. It is important, however, to make the distinction between this type of mechanism and that postulated for neutral substrates (e.g. sugars and neutral amino acids), in which H^+ is cotransported on the permease rather than on the substrate. In this case, the permease is presumed to bind substrate and H^+ independently and to couple the energy released from the downhill translocation of H^+ in response to $\Delta\bar{\mu}_{H^+}$ (interior negative and/or alkaline) to the uphill transport of substrate against a concentration gradient.

3. Finally, solutes such as Na^+ or Ca^{2+} that are pumped out of the cell are transported by "antiport." A transmembrane, substrate-specific antiporter couples downhill translocation of H^+ to efflux of substrate against a concentration gradient.

Although these considerations are simple and lead to straightforward predictions, the predictions become more complicated if the stoichiometry

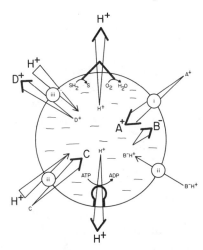

Figure 3 The chemiosmotic hypothesis and secondary active transport. The proton electrochemical gradient (interior negative and alkaline) generated by respiration or ATP hydrolysis drives influx of cationic substrates (A^+) by uniport (*i*), acidic (B^-H^+) and neutral substrates (C) by symport (*ii*), and efflux of cations like Na^+ or Ca^{++} (D^+) by antiport (*iii*). See text for further discussion.

between H^+ and substrate is greater than unity. For example, if $H^+ : Na^+$ antiport has 1 : 1 stoichiometry the process will be electrically neutral and respond to the ΔpH across the membrane. On the other hand, if the stoichiometry is $2 H^+ : 1 Na^+$ the mechanism becomes electrogenic and will respond to ΔpH and/or $\Delta \Psi$. Such considerations are discussed in more detail below.

Another important aspect of the proposed mechanisms is reversibility. Efflux of cations under nonenergized conditions should lead to generation of $\Delta \Psi$ (interior negative), while efflux of acids should generate ΔpH (interior alkaline) and efflux of sugars or neutral amino acids should give rise to $\Delta \bar{\mu}_{H^+}$. In contrast, Na^+ or Ca^{2+} influx should generate ΔpH and/or $\Delta \Psi$ depending on the stoichiometry of the antiport reaction. As shown by Michels et al (108) and Otto et al (126) the reverse reactions may have more than academic importance. Thus, carrier-mediated efflux of glycolytically generated lactic acid can provide a significant increase in growth yield under certain conditions owing to generation of $\Delta \bar{\mu}_{H^+}$.

Clearly, the chemiosmotic formulation predicts that solute transport is an indirect process whereby respiration or ATP hydrolysis drives substrate accumulation with $\Delta \bar{\mu}_{H^+}$ as an intermediate. As such, the experimental approach to the problem at this level of resolution depends upon techniques that allow determination of $\Delta \bar{\mu}_{H^+}$ in systems that are not readily amenable to electrophysiology and upon the demonstration that changes in $\Delta \bar{\mu}_{H^+}$ correlate with changes in the accumulation of appropriate substrates.

Determination of $\Delta \Psi$

In 1971 Grinius et al (44) initiated the use of lipophilic ions to measure the polarity of $\Delta \Psi$ in mitochondria and chloroplasts. The ions are sufficiently lipophilic to enter the hydrophobic core of the membrane and are able to delocalize their charges, thereby allowing passive equilibration with $\Delta \Psi$ (50).

Studies with radioactively labeled lipophilic cations extended this approach considerably and provided convincing evidence that $\Delta \Psi$ can be measured quantitatively in intact *E. coli* and RSO membrane vesicles. Thus tetraphenylphosphonium (TPP$^+$), tetraphenylarsonium, and Rb$^+$ (in the presence of valinomycin) equilibrate fastest with $\Delta \Psi$. However, these cations and triphenylmethylphosphonium (TPMP$^+$), triphenylmethyl-arsonium, triphenylmethylammonium, and dimethyldibenzylammonium (in the presence of tetraphenylboron) all accumulate to the same steady-state level over submicromolar to millimolar ranges of concentration. Moreover, the magnitude of $\Delta \Psi$ determined from steady-state levels of accumulation of these cations is virtually identical to that obtained from fluorescence quenching studies with 3,3'-diisopropylthiodicarbocyanine (188).

Most importantly, using giant cells of *E. coli* grown in the presence of 6-amidinopenicillanic acid, Felle et al (32) measured $\Delta\Psi$ by two independent techniques: directly with intracellular microelectrodes and indirectly from the steady-state distribution of [³H]TPP⁺. Under various conditions the two methods yielded values that agree very closely. With both techniques, $\Delta\Psi$ (interior negative) approximated -85 mV at pH 5.0 and -142 mV at pH 8.0 with an average slope of -22 mV/pH unit over the range of pH 5.0–7.0. In a parallel study of membrane vesicles using TPP⁺ distribution alone to measure of $\Delta\Psi$, values of about -90 mV at pH 5.0 and -110 mV at pH 7.5–8.0 with an average slope of -6 mV/pH unit were obtained. Although the difference between intact cells and RSO vesicles is yet to be understood, the results lend firm support to the conclusion that distribution studies with lipophilic cations in *E. coli* provide an excellent quantitative measure of $\Delta\Psi$. It is also noteworthy that distribution studies with [³H]TPP⁺ yield $\Delta\Psi$ values similar to those obtained electrophysiologically in a number of eukaryotic systems (100, 101, 201).

Since ISO vesicles have a polarity opposite to that of intact cells and RSO vesicles, $\Delta\Psi$ generated by respiration or ATP hydrolysis is interior positive, necessitating the use of lipophilic anions. Although thiocyanate distribution studies with ISO vesicles (144) are less detailed than studies with lipophilic cations in *E. coli*, they demonstrate that respiration or ATP hydrolysis generates a $\Delta\Psi$ of opposite polarity but similar magnitude to that observed in RSO vesicles. These results, which are interesting in their own right, provide additional support for the validity of the distribution measurements with lipophilic cations.

Determination of ΔpH

The basic principle behind measurement of ΔpH is clear-cut. If the internal compartment under consideration is more alkaline than the external medium, as in intact *E. coli* or RSO vesicles, a permeant weak acid is used; the most popular is 5,5'-dimethyloxazolidine-2,4-dione (DMO) (187). The protonated form of the acid penetrates the membrane passively, but once the acid reaches the internal space it dissociates and the anion, which is impermeant, accumulates. Conversely, if the inside is more acid than the outside, as in ISO vesicles, permeant weak bases such as methylamine are used. In this case the unprotonated species is passively permeant and is protonated internally to form a positively charged, impermeant species. In general, if the pK of the probe is 2 pH units or more removed from internal pH, ΔpH can be calculated by substitution of the steady-state distribution ratio into the Nernst equation [i.e. $\Delta pH = 2.3\ RT/F$ log (distribution ratio)]. On the other hand, if the pK of the probe is within 2 pH units of internal pH a more complicated calculation is used to obtain accurate values (cf 142). Although the principles described are straightforward, determination of ΔpH is very dependent on the method used to measure uptake of the probes. Thus for many years it was impossible to demonstrate that *E. coli* membrane vesicles generate a significant ΔpH (cf 68).

In 1976 Padan et al (133) made the critical observation that *E. coli* rigidly maintains an internal pH of 7.5–7.8 and they demonstrated that the magnitude of ΔpH depends upon external pH. At pH 5.5–6.0 ΔpH has a maximal value of about 2 pH units (i.e. -120 mV, interior alkaline). This decreases to zero at about pH 7.6 and then reverses at higher pH such that the interior of the cell becomes more acid than the exterior. In addition, Rottenberg (150) demonstrated that acetate may be utilized to determine ΔpH in mitochondria and suggested that this weak acid might be more useful than DMO in vesicles because it is less permeant. Initial experiments with RSO vesicles using filtration assays revealed a small amount of acetate uptake that was sensitive to external pH, but the acetate concentration gradient was so small that the putative ΔpH appeared to be thermodynamically insignificant. It was not until flow dialysis (17) was utilized that the problem was resolved (139–141, 143, 178). By this means, the concentration of solute in the medium bathing the vesicles is monitored continuously without experimental manipulations that cause artifactual loss of accumulated probe.

By use of flow dialysis and various permeant weak acids (i.e. acetate, propionate, butyrate, DMO, benzoate, or acetylsalicylate) it is easily demonstrated that RSO vesicles from *E. coli* generate a ΔpH (interior alkaline) of about -120 mV (i.e. 2 pH units) at pH 5.5 in the presence of reduced PMS or D-lactate (139, 140, 142, 143). This value is remarkably similar to that reported by Padan et al (133) with intact cells. Furthermore, similar experiments with ISO vesicles using methylamine showed that these vesicles establish a ΔpH of similar magnitude but opposite polarity (i.e. interior acid) (144). Finally, Navon et al (118) and Ogawa et al (124) provided confirmation for the quantitative nature of the measurements. They utilized high resolution ^{31}P nuclear magnetic resonance spectroscopy to measure ΔpH in intact *E. coli*, and their results are very similar to those obtained from distribution studies with permeant weak acids in intact cells and RSO membrane vesicles.

Effect of External pH on $\Delta\bar{\mu}_{H^+}$, $\Delta\Psi$, and ΔpH

As reported originally with intact *E. coli* (133, 204), ΔpH (interior alkaline) varies dramatically with external pH in RSO membrane vesicles (139, 140, 142, 143). The value of ΔpH is almost constant from pH 5.0 to 5.5 (-115 to

-120 mV), decreases above pH 5.5, and is negligible at about pH 7.5–7.8. Furthermore, preliminary evidence suggests that RSO vesicles, like intact cells, may also acidify the internal compartment at pH 8.0–8.5 (L. Patel & H. R. Kaback, unpublished). In contrast, $\Delta\Psi$ (interior negative) is about -90 mV at pH 5.0 and increases to about -110 mV at pH 7.5–8.0. As a result, $\Delta\bar{\mu}_{H^+}$ exhibits a maximum of about -220 mV at pH 5.5 and a minimum of about -110 mV at pH 7.5–8.0. With the exception that $\Delta\Psi$ increases with external pH to a smaller extent, the results with RSO membranes are both qualitatively and quantitatively similar to those obtained with intact cells.

Clearly, the variation in ΔpH with external pH results from the capacity of the system to maintain internal pH at pH 7.5–7.8. This observation is conceptually important. Since the internal space is small relative to the external medium, the system does not have to extrude many protons in order to establish ΔpH.

Although the reason for collapse of ΔpH from pH 5.5 to 7.5 is unclear (143, 144), a simple possibility is the operation of an antiport mechanism with an alkaline pH optimum. According to this explanation, as external pH increases beyond pH 5.5 exchange of internal cation for external H^+ would occur at an increasingly rapid rate, leading to collapse of ΔpH with a compensatory increase in $\Delta\Psi$ (cf below). In an effort to investigate this possibility, Reenstra et al (144) undertook a series of experiments with ISO vesicles. Unlike RSO vesicles, which extrude Na^+ (159, 178), ISO vesicles catalyzed Na^+ accumulation when a $\Delta\bar{\mu}_{H^+}$ (interior positive and/or acid) was present across the membrane, and under no circumstances was K^+ or Rb^+ accumulation observed. However, the properties of the $H^+:Na^+$ antiport activity make it difficult to conclude that this reaction accounts for the phenomenon. The concentration gradient of Na^+ established by ISO vesicles was constant with pH from 5.7–8.0, Na^+ accumulation was driven by either ΔpH (interior acid) or $\Delta\Psi$ (interior positive), and Na^+ accumulation did not lead to dissipation of ΔpH over the pH range studied. Finally and remarkably, the pH profile for ΔpH in ISO vesicles was essentially the mirror image of that observed in RSO vesicles. The ΔpH was maximal at pH 7.0–7.5 and decreased at acid pH. This makes it even more difficult to envisage that $H^+:Na^+$ antiport, by functioning at alkaline pH specifically, could be responsible for the collapse of pH from pH 5.5 to 7.5. In summary, the mechanism responsible for the maintenance of internal pH at acid pH remains a fascinating unsolved problem.

On the other hand, evidence has been obtained that indicates that the ability of cells to establish an internally acid ΔpH above pH 8.0–8.5 is due to electrogenic $H^+:Na^+$ antiport. Mutants of *E. coli* (161, 203) and *Bacillus alkalophilus* (90, 91) have been isolated that do not grow at alkaline pH, do not establish ΔpH (interior acid) at alkaline pH, and do not exhibit $H^+ : Na^+$ antiport activity.

Effect of Ionophores

As demonstrated with RSO (139, 140, 142, 143) and ISO vesicles (144), $\Delta\Psi$ and ΔpH can be altered reciprocally with little or no change in $\Delta\bar{\mu}_{H^+}$ and no change in respiration. In RSO vesicles, for example, $\Delta\Psi$ (interior negative) decreases as increasing concentrations of valinomycin are added in the presence of K^+ at pH 5.5 or 7.5 because valinomycin-mediated K^+ influx occurs at a rate that approximates electrogenic H^+ efflux. With the decrease in $\Delta\Psi$ at pH 5.5, ΔpH (interior alkaline) increases for reasons discussed below. As a result of the reciprocal nature of the alterations, $\Delta\bar{\mu}_{H^+}$ is either unaffected or decreases slightly.

In contrast, nigericin, an ionophore with $H^+ : K^+$ exchange activity, induces effects that are opposite to those of valinomycin at pH 5.5. With increasing concentrations of nigericin, ΔpH decreases with increase in $\Delta\Psi$ and $\Delta\bar{\mu}_{H^+}$ remains constant or decreases slightly. Importantly, at pH 7.5, where there is no ΔpH, the ionophore has no effect on $\Delta\Psi$.

Finally, protonophores like CCCP collapse both $\Delta\Psi$ and ΔpH, thereby leading to dissipation of $\Delta\bar{\mu}_{H^+}$ (143). The observation that CCCP abolishes both ΔpH and $\Delta\Psi$ indicates that protons are the electrogenic species.

As opposed to the increased respiration induced by ionophores in mitochondria (i.e. the system exhibits respiratory control), no such effect is observed with *E. coli* membrane vesicles. This can be accounted for by considering the interrelationship between the components of $\Delta\bar{\mu}_{H^+}$ (144). At steady state under conditions where both $\Delta\Psi$ and ΔpH are present across the membrane, the magnitude of $\Delta\bar{\mu}_{H^+}$ is determined by the activity of the H^+ pump and the back-leak of H^+ through the membrane, and each parameter of $\Delta\bar{\mu}_{H^+}$ limits the magnitude of the other. With RSO vesicles, $\Delta\Psi$ (interior negative) draws H^+ towards the interior of the vesicles and thus limits the pH gradient. Similarly, ΔpH (interior alkaline) limits $\Delta\Psi$ (interior negative) because H^+ diffuses into the vesicles down the concentration gradient, thus decreasing net extrusion of positive charge and decreasing $\Delta\Psi$. Therefore, dissipation of either $\Delta\Psi$ or ΔpH removes a force that is limiting for the other parameter, allowing it to increase without a corresponding increase in the rate of H^+ extrusion. The same explanation in reverse applies to ISO vesicles.

$\Delta\bar{\mu}_{H^+}$ and Solute Accumulation

The studies described above not only provide direct confirmation of one of the major contentions of the chemiosmotic hypothesis, that respiration or ATP hydrolysis leads to generation of a transmembrane $\Delta\bar{\mu}_{H^+}$; they also establish an experimental framework within which to test specific predictions.

EFFECT OF SOLUTE ACCUMULATION ON ΔpH AND $\Delta\Psi$ If solute accumulation is driven by $\Delta\bar{\mu}_{H^+}$, and if passive accumulation of weak acids and lipophilic cations reflects the individual components of $\Delta\bar{\mu}_{H^+}$, it follows that weak acid and/or lipophilic cation accumulation should be diminished in the presence of solutes such as lactose or glucose-6-P that are accumulated in relatively large amounts. Both predictions are confirmed experimentally. When RSO vesicles containing *lac* permease or the glucose-6-P transporter are allowed to accumulate acetate in the presence of reduced PMS and valinomycin, addition of lactose or glucose-6-P causes release of about 50% of the accumulated acid (140). In contrast, in vesicles that contain neither of these porters acetate accumulation is unaffected by either substrate. Similarly, lactose causes a decrease in TPMP$^+$ accumulation, and no effect is observed with vesicles devoid of *lac* permease (160). A good explanation for the effects is that both solutes are transported in symport with H$^+$ (112). It is also notable that downhill transport of lactose and certain other sugars in de-energized cells (51, 78, 96, 190–193) or RSO vesicles (22, 135) occurs with alkalinization of the external medium.

EFFECT OF IONOPHORES ON SOLUTE ACCUMULATION   Before proceeding, we must reemphasize a few important points. (*a*) $\Delta\bar{\mu}_{H^+}$ is maximal at pH 5.5 where approximately half of the total driving force is ΔpH and the other half is $\Delta\Psi$. (*b*) At pH 7.5, $\Delta\bar{\mu}_{H^+}$ is reduced by about half and consists solely of $\Delta\Psi$. (*c*) At pH 5.5, ΔpH and $\Delta\Psi$ can be manipulated reciprocally with little or no effect on $\Delta\bar{\mu}_{H^+}$. (*d*) Nigericin has no effect on $\Delta\Psi$ at pH 7.5 and therefore no effect on $\Delta\bar{\mu}_{H^+}$.

With these considerations as a framework, Ramos & Kaback (140) studied the effects of valinomycin and nigericin on steady-state levels of accumulation of 14 different substrates at pH 5.5 and 7.5. Direct correlation between variations in accumulation of a particular solute and variations in $\Delta\bar{\mu}_{H^+}$, $\Delta\Psi$, or ΔpH are observed in only a few instances at pH 5.5. Nevertheless, certain qualitative statements are justified: (*a*) Accumulation of lactose, proline, tyrosine, serine, glycine, leucine, and surprisingly lysine, glutamate, and succinate at pH 5.5 responds to increasing concentrations of valinomycin or nigericin in a manner that correlates reasonably well with the effect of these ionophores on $\Delta\bar{\mu}_{H^+}$. That is, accumulation of each substrate is progressively and mildly inhibited or relatively unaffected by increasing concentrations of either ionophore. (*b*) Accumulation of glucose-6-P, lactate, gluconate, and glucuronate at pH 5.5 is stimulated by valinomycin and inhibited by nigericin; these effects are clearly analogous to the effects of the ionophores on ΔpH. (*c*) Regardless of whether accumulation of a particular substrate is stimulated, inhibited, or un-affected by valinomycin or nigericin at pH 5.5, in each case valinomycin markedly inhibits accumulation at pH 7.5 and nigericin has no effect whatsoever. Generally, therefore, at pH 5.5 the systems fall into two categories: those driven preferentially by $\Delta\bar{\mu}_{H^+}$ and those driven preferentially by ΔpH. Furthermore, all of the systems, including those driven by ΔpH at pH 5.5, are driven by $\Delta\Psi$ at pH 7.5.

Since coupling between accumulation of a particular solute and $\Delta\bar{\mu}_{H^+}$, ΔpH, or $\Delta\Psi$ varies with the external pH, it is not surprising that quantitative correlations are observed at pH 5.5 in only a few instances. With tyrosine, leucine, lysine, and succinate, for example, there is reason-able correlation between the effects of valinomycin and nigericin on $\Delta\bar{\mu}_{H^+}$ and the effects of the ionophores on accumulation. Lactose and glycine accumulation are also coupled preferentially to $\Delta\bar{\mu}_{H^+}$ at pH 5.5, but there is a bias towards $\Delta\Psi$ because valinomycin inhibits accumulation more effectively than it dissipates $\Delta\mu_{H^+}$. Similarly, although accumulation of glucose-6-P, lactate, gluconate, and glucuronate is coupled to ΔpH at pH 5.5, in only one case (i.e. lactate) is accumulation in complete equilibrium with ΔpH; nigericin does not inhibit accumulation of these acids as effectively as it dissipates ΔpH at pH 5.5. Finally, the reader is referred to Robertson et al (148), who investigated the kinetics of many of these transport systems under similar conditions. Basically, the results support the conclusions drawn from studying steady-state levels of accumulation (e.g. that initial rates of glutamate and lysine transport are driven by $\Delta\bar{\mu}_{H^+}$); however, certain subtleties are also revealed (i.e. that succinate transport is biased kinetically towards ΔpH).

H$^+$: SOLUTE STOICHIOMETRY   The overall import of the results described above is that $\Delta\bar{\mu}_{H^+}$ is the immediate driving force for secondary active transport. However, the experiments also reveal certain details that are not explained. For instance, as discussed above, accumulation of organic acids presumably depends on the relative alkalinity of the internal pH, and should be unaffected by $\Delta\Psi$ (interior negative). Since there is no ΔpH across the membrane at pH 7.5–7.8, however, this notion cannot account for acid accumulation at high external pH. Furthermore, LeBlanc et al (99), using artificially imposed pH gradients (interior alkaline) and diffusion potentials (interior negative), have provided support for the argument that glucose-6-P and glucuronate transport are electrically neutral at acid pH and electrogenic at alkaline pH. In addition, when steady-state levels of accumulation of certain substrates (i.e. lactose, proline, lysine, and suc-cinate) are examined as a function of pH, it appears that $\Delta\bar{\mu}_{H^+}$ is insufficient to account for the magnitude of the concentration gradients at alkaline pH if the stoichiometry between H$^+$ and substrate is 1:1 (140, 141).

There is a simple explanation for these observations within the bounds of the chemiosmotic hypothesis (141, 151). Perhaps the stoichiometry between H^+ and substrate varies as a function of external pH in such a manner that it is 1 : 1 at pH 5.5 but increases to higher values as external pH increases. If, for example, the stoichiometry between H^+ and proline or lactose were 2 : 1 at pH 7.5 rather than 1 : 1, the concentration gradients would be compatible thermodynamically with $\Delta\bar{\mu}_{H^+}$ at pH 7.5 [i.e. if the stoichiometry is 2 : 1, the concentration gradient varies as the square of the charge gradient (113, 192)]. Similarly, at pH 5.5 transport of certain organic acids might occur by a formal chemiosmotic mechanism, i.e. one H^+ ion (or two if the acid is glucose-6-P) is taken up per mole of undissociated acid, while at pH 7.5 two or more H^+ ions might be taken up per mole of acid, one (two with glucose-6-P) in association with the substrate itself and one in association with the permease. By this means, transport of glucose-6-P, lactate, glucuronate, and gluconate at pH 7.5 would become electrogenic, having become symport mechanisms.

Subsequent to the appearance of papers of Ramos & Kaback (140, 141) supporting these ideas, studies with intact cells (4, 32, 204) cast doubt on the suggestion that there might be a discrepancy between steady-state lactose accumulation and $\Delta\bar{\mu}_{H^+}$ at alkaline pH. Specifically, it was shown that in intact cells, as opposed to RSO vesicles, $\Delta\Psi$ increases with pH so as to compensate almost completely for the decrease in ΔpH. In intact cells, therefore, $\Delta\bar{\mu}_{H^+}$ does not decrease as drastically with increasing pH as it does in vesicles, and the steady-state level of lactose accumulation at high pH can be accommodated without a change in H^+ : lactose stoichiometry. In addition, more direct studies of H^+ : lactose symport in de-energized cells do not indicate a change in stoichiometry at high pH (4, 204).

On the other hand, many studies with intact cells and RSO vesicles demonstrate that ΔpH is zero at pH 7.5–7.8. Thus, it is difficult to account for the transport of certain organic acids without postulating a pH-dependent increase in H^+ : substrate stoichiometry (99, 140, 141, 143). Direct measurements in intact cells supporting this notion have been reported (175).

Na$^+$-Dependent Transport

Although most bacterial transport systems probably catalyze H^+ symport, several instances have been reported in which solute transport depends upon Na^+ or Li^+ (cf 90, 91, 178 for reviews). Moreover, the studies of Stock & Roseman (171), Lanyi et al (97), and Krulwich (90, 91), in particular, indicate that symport mechanisms are operative.

RSO membrane vesicles from *S. typhimurium* grown under appropriate conditions catalyze methyl 1-thio-β-D-galactopyranoside (TMG) transport in the presence of Na^+ or Li^+ (15, 178), and TMG-dependent Na^+ uptake is also observed when a K^+ diffusion potential (interior negative) is imposed. Cation-dependent TMG accumulation varies with $\Delta\bar{\mu}_{H^+}$, and the vesicles catalyze Na^+ efflux in a manner that is consistent with the operation of a $H^+ : Na^+(Li^+)$ antiport mechanism. The results are consistent with a model in which TMG : $Na^+(Li^+)$ symport is driven by $\Delta\bar{\mu}_{H^+}$, which functions to maintain low intravesicular $Na^+(Li^+)$ concentrations through H^+: $Na^+(Li^+)$ antiport. Similar mechanisms have been suggested for light- and respiration-dependent amino acid transport in vesicles from *Halobacterium halobium* (97) and *B. alkalophilus* (90, 91).

Studies with $H^+ : Na^+$ antiport mutants in *B. alkalophilus* (45, 90, 91) and *E. coli* (161, 203) demonstrate that such cells exhibit, in addition to defective growth at alkaline pH, pleiotropic defects in Na^+-dependent substrate translocation. It has been suggested that a number of Na^+:substrate symport systems may share a common Na^+-translocating subunit with the $H^+ : Na^+$ antiporter. Mapping studies with the *E. coli* mutant (152) show, however, that the mutation is coincident with the gene for sigma factor of RNA polymerase. Although the relationship between sigma factor, $H^+ : Na^+$ antiport, and Na^+:substrate symport is unclear, it is possible that the pleiotropic defect exhibited by the mutant is related to a transcriptional alteration rather than a defect in a common Na^+-translocating subunit.

PASSAGE TO PERMEASE: THE β-GALACTOSIDE TRANSPORT SYSTEM

The β-galactoside or *lac* transport system in *E. coli* has been studied more extensively than any other bacterial transport system. It was described originally in 1955 by Cohen, Kepes, Rickenberg and their co-workers (13, 14, 81, 82, 147) and is part of the well-known *lac* operon that allows the organism to utilize the disaccharide lactose. In addition to regulatory loci, the *lac* operon contains three structural genes: (*a*) the Z gene encoding β-galactosidase, a cytosolic enzyme that cleaves lactose upon entry into the cell; (*b*) the Y gene encoding *lac* permease or *lac* carrier protein which catalyzes transport of lactose through the plasma membrane; and (*c*) the A gene encoding thiogalactoside transacetylase, an enzyme that catalyzes the acetylation of thiogalactosides with acetyl-CoA as the acetyl donor. (The enzyme has no known physiological function save detoxification of thiogalactosides, which are not found commonly in nature.)

In 1963, Mitchell (112) postulated explicitly that β-galactoside transport occurs in symport with H^+ (Figure 3) and that $\Delta\bar{\mu}_{H^+}$ is the immediate driving force for accumulation. West (190) and West & Mitchell (191, 192) then demonstrated that addition of lactose to de-energized cells causes alkalinization of the external medium, thus providing the first experimental support for the hypothesis. As discussed above, during the following five or six years evidence accumulated demonstrating virtually unequivocally that the *lac* permease catalyzes H$^+$: β-galactoside symport. More recently focus has shifted to a molecular level.

Purification of Functional Lac Permease

Although the kinetics (38, 74, 76, 148), substrate specificity (155), and genetics (59) of β-galactoside transport were studied intensively and the *lac Y* gene product was known to be a membrane protein (80), relatively little progress was made toward purification until 1980 because all attempts to solubilize the protein in a functional state were unsuccessful (cf 132). In 1978 (177) the *lac Y* gene was cloned into a recombinant plasmid, allowing amplification of the permease (176), elucidation of its nucleotide sequence and the amino acid sequence of the permease (6), and synthesis of the permease in vitro (28). Shortly thereafter, Newman & Wilson (122) solubilized the permease in octyl-β-D-glucopyranoside (octylglucoside) and successfully reconstituted transport activity in proteoliposomes using octylglucoside dilution (138). At about the same time, Kaczorowski et al showed that *p*-nitrophenyl-α-D-galactopyranoside (NPG) is a highly specific photoaffinity label for the permease (75). By using a strain of *E. coli* with amplified levels of *lac Y*, [^{3}H]NPG to photolabel the permease specifically, and transport activity of reconstituted proteoliposomes, the product of the *lac Y* gene was purified to homogeneity in a completely functional state (34, 121). Subsequently, another similar method was developed for the same purpose (198).

In the initial steps of the purification (183), *E. coli* membranes are extracted sequentially with high concentrations of urea and cholate to effect about a threefold purification of the carrier in situ. Both of these operations are based upon earlier studies (132, 136) that demonstrated that the treatment of RSO membrane vesicles with these reagents removes significant amounts of protein with little or no effect on permease activity. Subsequent extraction with octylglucoside in the presence of *E. coli* phospholipids solubilizes most of the permease but only about 15% of the remaining protein, leading to additional fourfold enrichment and twelvefold purification relative to the original membrane. The octylglucoside extract is then subjected to DEAE-Sepharose column chromatography under isocratic conditions at pH 6.0. Transport activity and most of the protein-associated photolabel is eluted in a symmetrical peak slightly behind the void volume of the column. Overall, the procedure results in 35-fold purification from the crude membrane fractions with a yield of about 50% based on the recovery of the photolabel.

Since photolabeling studies with [^{3}H]NPG indicate that 3% of the protein in the membrane of the amplified strain is *lac* permease, 35-fold enrichment of the photolabeled material suggests that a high degree of purification is achieved. This is confirmed by sodium dodecylsulfate polyacrylamide gel electrophoresis (SDS-PAGE). The purified material yields a single broad band[2] with M_r of about 33 kd, which is in close agreement with published values (64, 80). When membranes are prepared from cells that were not induced, the band corresponding to purified permease is only a minor constitutent of the octylglucoside extract of urea/cholate-treated membranes; thus the purified protein exhibits an important property expected of the product of *lac Y* in the recombinant plasmid (176, 177). The amino-acid composition of the purified protein closely matches the composition predicted from the DNA sequence of *lac Y* (121), indicating that purified permease has a molecular weight similar to that predicted (6) from the DNA sequence. Furthermore, N-terminal sequencing of the first 13 amino acids of purified permease is in complete agreement with the DNA sequence, providing additional evidence that the preparation has a high degree of purity.

It is not known why permease yields a spuriously low M_r on SDS-PAGE, although a high content of hydrophobic amino acids in the protein suggests that the phenomenon may be due to unusually high binding of SDS. Furthermore, it is not known why the permease migrates as a broad band. However, it is noteworthy that the protein maintains a high degree of secondary structure in SDS, as judged by circular dichroic measurements. Regardless, when the permease is subjected to SDS-PAGE at increasing concentrations of polyacrylamide and the data are treated quantitatively (1, 120), an extrapolated M_r of about 46 kd is obtained. A similar value is obtained by gel permeation chromatography on Sephacryl S-300 in hexamethylphosphoric triamide (86).

Properties of Reconstituted Proteoliposomes

Proteoliposomes prepared by octylglucoside dilution followed by freeze-thaw/sonication are unilamellar vesicles about 100 nm in diameter with no internal structure (19, 37). Relatively low magnification of platinum/carbon replicas of freeze-fractured proteoliposomes containing purified *lac* permease confirm the unilamellar nature of the preparation (Figure 4*A*). Higher magnification reveals that both the convex and concave fracture surfaces exhibit a relatively uniform distribution of particles that are about 70 Å in diameter (Figure 4*B*) and that the surfaces are smooth when

---

[2] At higher protein concentrations a less intense band is also observed at about M_r 60 kd. Since this band is observed after photoaffinity labeling with NPG and reacts with antibody prepared against purified *lac* carrier, it is probably an aggregate of the *lac* carrier protein.

liposomes are formed in the absence of permease (Figure 4*C*). Since particles, but no pits, are observed on both surfaces of the membranes, it seems likely that the permease has equal affinity for the phospholipids in each leaflet of the bilayer. Given the mass of the permease (6, 121, 176), a size of 70 Å suggests that the particles contain one or two polypeptides, depending on the degree to which the shadowing increases the particle diameter (19).

The micrographs in Figure 5 compare selected proteoliposomes at high magnification. On the etched proteoliposome in Figure 5*C* a boundary is indicated between the true outer surface of the proteoliposomes and the surrounding ice at the etch line in the lower right corner. The true outer surface of the membrane is smooth, suggesting that the permease is largely embedded within the bilayer. The unidirectionally replicated intramembrane particles (Figures 5*A*, *C*) appear identical in size and shape. The white shadows are not always conical, as would be expected from a smooth-surfaced globular protrusion from the hydrophobic fracture plane, but there does not appear to be a consistent pattern within the shadow that would suggest a defined substructure. A few of the intramembrane particles in the rotary-replicated proteoliposome (circled in Figure 5*B*) display a groove or cleft, but other patterns of metal deposition are also observed in the same fracture surface.

More recent studies (18) have utilized tantalum rather than platinum/carbon for shadowing, as tantalum yields a higher level of resolution. With micrographs shadowed in this way the intramembrane particles appear to be closer to 50 Å in diameter and more particles exhibit a cleft.

When proteoliposomes containing purified permease are equilibrated with $^{86}Rb^+$, treated with valinomycin, and diluted into sodium phosphate, efflux occurs very slowly; at least 80% of the label is retained at 20 min (37). On addition of CCCP, which specifically increases permeability to H^+, a marked increase in the rate of Rb^+ efflux is evident. If the same experiments are performed in the absence of valinomycin, Rb^+ efflux is negligible and addition of CCCP has no significant effect. The observations demonstrate, albeit indirectly, that the proteoliposomes are highly impermeable to the ions present in the reaction mixture (i.e. H^+, Rb^+, Na^+, Cl^-, and P_i). Thus, the slow rate of Rb^+ efflux observed in the presence of valinomycin is caused by generation of $\Delta\Psi$ (interior negative) that is maintained for long periods because the proteoliposomes are impermeant to counterions. Addition of CCCP, on the other hand, provides a pathway for H^+, resulting in dissipation of $\Delta\Psi$ with rapid downhill movement of Rb^+.

Proteoliposomes with the properties described are ideally suited for H^+: solute symport studies. Morphologically, the preparation consists of a population of unilamellar, closed, unit membrane–bound sacs that are

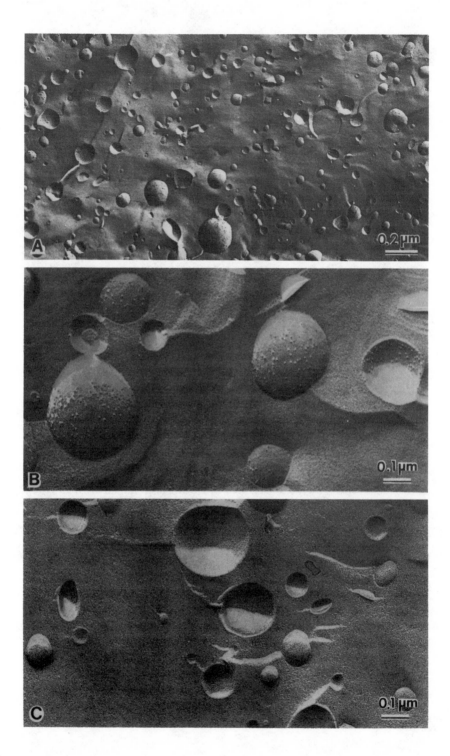

relatively uniform in diameter and contain no internal structure. These properties correlate with the pseudo–first-order efflux and exchange kinetics observed for Rb^+ and lactose (37). Furthermore, the proteoliposomes are passively impermeable to many ions. Thus, certain aspects of H^+: lactose symport that were impossible to document with RSO vesicles (i.e. stimulation of efflux by ionophores; cf below) are demonstrable in the reconstituted system. Generally, proteoliposomes reconstituted with *lac* permease exhibit all of the phenomena described in RSO membrane vesicles, but the results are significantly more clear-cut and thereby provide firmer support for certain ideas concerning reaction mechanisms.

A Single Polypeptide is Required for Lactose Transport

Although *lac* permease purified to apparent homogeneity catalyzes all of the translocation reactions characteristic of the β-galactoside transport system in cells and RSO vesicles, evidence has been presented that suggests that the transport system may require more than a single gene product (60, 137, 184, 185). For this reason, kinetic experiments were performed carefully on proteoliposomes reconstituted with purified permease. Turnover numbers were determined for various modes of translocation and were compared to those calculated from published V_{max} for RSO membrane vesicles (181; Table 1). Both the turnover number of the permease and its apparent K_m for lactose are virtually identical in proteoliposomes and membrane vesicles with respect to $\Delta\Psi$-driven lactose accumulation, counterflow, facilitated diffusion (i.e. lactose influx under nonenergized conditions), and efflux.

Furthermore, Matsushita et al (104) have demonstrated that proteoliposomes reconstituted with both purified *o*-type terminal oxidase and *lac* permease mimic one of the basic physiological functions of the bacterial membrane, respiration-driven active transport. The *o*-type oxidase was extracted from the membrane of an *E. coli* mutant deficient in the *d*-type terminal oxidase with octylglucoside after treatment of membranes with urea and cholate. Oxidase was then purified to homogeneity by DEAE-Sepharose chromatography. The enzyme contains four polypeptides (M_r 66, 35, 22, and 17 kd) and two b-type cytochromes (b558 and b563), and catalyzes oxidation of ubiquinol-1 (Q_1H_2) and other electron donors with specific activities 20- to 30-fold higher than crude membranes. Activity is

---

Figure 4   Freeze-fracture electron microscopy of proteoliposomes reconstituted with purified *lac* permease (*A*, *B*) and liposomes prepared in the absence of protein (*C*). Platinum/carbon replicas of freeze-fractured preparations formed by octylglucoside dilution followed by one cycle of freeze-thaw/sonication are shown. Study performed by Joseph Costello, Department of Anatomy, Duke University Medical Center.

highly dependent upon exogenous phospholipids, and certain quinone analogs are potent inhibitors. Proteoliposomes reconstituted with purified o-type oxidase contain randomly distributed 85–90 Å intramembrane particles on the convex and concave fracture surfaces. During oxidase turnover the system generates a $\Delta\bar{\mu}_{H^+}$ (interior negative and alkaline) of −115 to −140 mV (104, 105).

Changes in external and internal pH measured with a glass electrode and entrapped pyranine, respectively, demonstrated that during ubiquinol oxidation H^+ is released on the external surface of the membrane and consumed on the internal surface (47, 105). In contrast, with TMPD, an electron donor that is essentially unprotonated at neutral pH, little change in external pH is observed until CCCP is added, at which point the medium becomes alkaline. The results are consistent with the proposal that oxidase turnover generates a $\Delta\Psi$ (interior negative) due to vectorial electron flow from the outer to the inner surface of the membrane. The pH gradient (interior alkaline), on the other hand, appears to result from scalar (i.e. nonvectorial) reactions that consume H^+ at the inner surface and/or release H^+ at the outer surface of the membrane.

In the presence of Q_1H_2, proteoliposomes reconstituted with both the o-type terminal oxidase and *lac* permease accumulate lactose against a concentration gradient. This accumulation is completely abolished by valinomycin and nigericin (104). Since uptake in the absence of Q_1H_2 or in the presence of valinomycin and nigericin represents equilibrium with the medium, it is apparent that the steady-state level of lactose accumulation observed during oxidase turnover represents a concentration difference of at least 10- to 20-fold across the membrane. Moreover, from comparison of

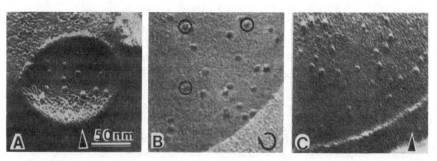

Figure 5 Comparison of intramembrane particles at high magnification. *A*, freeze-fracture of a concave vesicle showing well-defined globular intramembrane particles. *B*, freeze-fracture rotary-shadowed convex surface showing globular particles that display substructure as a notch or cleft (*encircled*). *C*, freeze-fracture etch of a convex surface. The etch line is in the *lower right* under the *arrowhead*. No particles are seen on the hydrophilic etch surface. From Costello et al (19).

Table 1   Comparison of turnover numbers for the *lac* carrier protein : ML 308-225 membrane vesicles versus proteoliposomes reconstituted with purified carrier

	Turnover numbers (sec)	
Reaction	Membrane vesicles[a] $[K_m \text{ (mM)}]$	Proteoliposomes $[K_m \text{ (mM)}]$
$\Delta\Psi$-driven influx ($\Delta\Psi = 100$ mV)	16 [0.2]	18 [0.5]
Counterflow	16–39 [0.45]	28 [0.65]
Facilitated diffusion	8–15.5 [20]	13 [7]
Efflux	8 [2.1]	6 [2.0]

[a] Determination of the amount of *lac* carrier protein in ML 308-225 membrane vesicles is based on photolabeling experiments with [³H]NPG that indicate that the carrier represents about 0.5% of the membrane protein. From Viitanen et al (181).

lactose transport induced by Q_1H_2 to that induced by valinomycin-mediated K^+ diffusion potentials ($K_{in}^+ \rightarrow K_{out}^+$), and from quantitation of the magnitude of the $\Delta\Psi$ generated under each condition, it is clear that lactose transport activity is commensurate with the magnitude of the bulk phase $\Delta\bar{\mu}_{H^+}$. Finally, freeze/fracture studies on the doubly reconstituted proteoliposomes demonstrate that intramembrane particles corresponding to the permease and the *o*-type oxidase can be differentiated morphologically and that the proteins do not appear to interact in the plane of the membrane.

Given these results it seems likely that only a single polypeptide species, the product of the *lac Y* gene, is necessary for each of the reactions catalyzed by the *lac* transport system *E. coli*. In addition, the double reconstitution experiment provides yet another strong line of evidence, this time on a molecular level, for the concept that secondary active transport is driven by a transmembrane $\Delta\bar{\mu}_{H^+}$.

Mechanistic Studies

Lactose efflux down a concentration gradient in RSO vesicles was observed to occur in symport with H^+. Translocation appears to be limited either by deprotonation of the carrier on the outer surface of the membrane or by a step corresponding to the return of the unloaded carrier to the inner surface of the membrane (74, 76). In addition, the observations are consistent with the following conclusions (Figure 6): (*a*) Efflux occurs by an ordered mechanism in which lactose is released first from the permease, followed by loss of the symported H^+. (*b*) The permease recycles in the protonated form during exchange and counterflow. (*c*) Reactions catalyzed by the unloaded permease involve net movement of negative charge. Experiments with proteoliposomes containing purified permease confirm and extend many of these ideas (37, 180).

Transient accumulation of Rb$^+$ during lactose efflux in the presence of valinomycin argues that lactose translocation is coupled to movement of a charged species such that $\Delta\Psi$ (interior negative) is generated. In addition the process is abolished by CCCP, suggesting that H$^+$ is the electrogenic species, and efflux-induced Rb$^+$ uptake is blocked by p-chloromercuribenzenesulfonate (pCMBS), a sulfhydryl reagent that inactivates the permease. The rate of lactose efflux is also enhanced by ionophores that collapse $\Delta\bar{\mu}_{H^+}$, and artificial imposition of $\Delta\Psi$ (interior negative) dramatically slows efflux with no effect on apparent K_m.

Efflux is pH-dependent, increasing more than 100-fold from pH 5.5 to pH 9.5 in a sigmoidal fashion with a midpoint at about pH 8.3 (180). In contrast, exchange is insensitive to pH and much faster than efflux, particularly at relatively acid pH values (below pH 7.5). Therefore, the rate-determining step for efflux must involve either deprotonation of the permease on the external surface of the membrane or the reaction corresponding to return of the unloaded permease to the inner surface of the membrane, as these are the only steps by which efflux and exchange differ (Figure 6).

Assuming that loss of lactose and H$^+$ is necessary for reinitiation of an efflux cycle, external pH might influence the rate of turnover in either of two

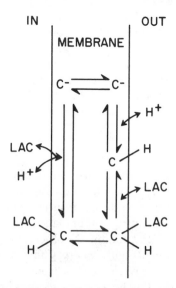

Figure 6   Schematic representation of reactions involved in lactose efflux. C represents the *lac* permease. The order of substrate binding at the inner surface of the membrane is not implied. From Kaczorowski & Kaback (74).

ways. First, deprotonation could be slow thereby limiting the overall rate of efflux in a pH-dependent manner. Although H^+ transfer between accessible amino acid residues and water in soluble enzymes is usually fast, little is known about such reactions with hydrophobic membrane proteins. Second, pH could alter the equilibrium between protonated and unprotonated forms of the permease, favoring the unprotonated form at more alkaline pH. Since it is assumed that the protonated form cannot recycle without substrate, efflux would be at least partially controlled by external pH. The rate-determining step might then involve "movement" of the unloaded permease to the inner surface of the membrane. The observation that efflux increases with pH is consistent with either possibility. In contrast, if deprotonation is not obligatory for exchange, H^+ might remain bound during this mode of translocation, rendering exchange insensitive to pH. If efflux is an ordered mechanism in which release of lactose by the permease is followed by loss of H^+ (Figure 6), deprotonation and/or return of the unloaded permease could be slow and could appear as the limiting step for efflux.

Counterflow experiments conducted at various pH values reveal that external lactose affects H^+ loss from the permease, and therefore support the ordered efflux mechanism shown in Figure 6. When external lactose was saturating, counterflow was unaffected by pH; moreover, transient formation of $\Delta\Psi$ observed during efflux was abolished under these conditions. The results can be interpreted as follows: On initiation of efflux, lactose and H^+ bind to the permease on the inner surface of the membrane (in unspecified order) and are translocated to the outer surface. Lactose is released, but in the presence of excess labeled substrate, rebinding and influx occur rapidly before deprotonation occurs. Under these conditions, therefore, H^+ release is infrequent and pH has no effect on the overall phenomenon. When external [^{14}C]lactose is limiting, however, rebinding of labeled substrate is less frequent, allowing deprotonation and return of the unloaded permease. Moreover, as pH is increased, deprotonation and return of the unloaded permease are enhanced, resulting in further diminution of counterflow. Inhibition of efflux-generated $\Delta\Psi$ formation by external lactose is also readily explained by this scheme. When lactose is present externally at saturating concentrations, release of lactose and rebinding of substrate occur rapidly before deprotonation can occur, and the ability of the system to generate $\Delta\Psi$ is negated.

Given the suggestion that H^+ loss could be limiting for efflux, one means used to investigate the putative mechanism further was to search for a solvent deuterium isotope effect (76, 180). At equivalent pH and pD (i.e. pD = pH + 0.4) (63) the rate of lactose-facilitated diffusion (influx or efflux) was about three to four times slower in deuterium than in protium, while the rate of exchange was identical. Furthermore, during counterflow with external lactose below K_m, the magnitude of the overshoot was two to three times greater in deuterium than in protium. With respect to the kinetic model (cf Figure 6), high external lactose prevented deprotonation of the permease (the C-H form) and it recycled across the membrane in the fully loaded state, catalyzing 1:1 exchange of internal unlabeled lactose with external [^{14}C]lactose. When external lactose was below the K_m, however, D_2O increased the efficiency of counterflow, particularly at higher pH (pD) values. Under these conditions, the C-H or C-D form of the permease partitioned between two pathways; one involved loss of protium or deuterium which resulted in net efflux and the other involved rebinding of lactose prior to loss of protium or deuterium which resulted in exchange (i.e. counterflow). The former pathway was favored at high pH (pD) values, but the C-D form was deprotonated slower than the C-H form, favoring binding of external [^{14}C]lactose. Consequently, the frequency with which the permease returned to the inner surface in the loaded form was enhanced relative to the unloaded form in the presence of D_2O, and the effect was more pronounced at alkaline pH.

Interestingly, $\Delta\Psi$-driven lactose accumulation exhibited essentially no solvent deuterium isotope effect (180). The observations as a whole suggest that under conditions where turnover is driven by a lactose concentration gradient, the rate of translocation is determined by a step or steps involving protonation or deprotonation, or by the subsequent step, i.e. the reaction corresponding to the return of the unloaded permease. In contrast, when there is a driving force on H^+ (i.e. in the presence of $\Delta\bar{\mu}_{H^+}$), these steps are no longer rate-determining. It is cautioned that the solvent deuterium isotope effects described cannot be attributed definitively to a true kinetic isotope effect as opposed to a pK_a effect (deuterium increases the pK_a values of various functional groups from 0.4 to 0.7 pH units) because the isotope effect on efflux disappears at pH 9.0 and above. However, at such alkaline pH the rate of efflux approaches the rate of exchange, suggesting that the rate-determining step for efflux may change at high pH. In any event, it seems that different steps are limiting when turnover is driven by $\Delta\bar{\mu}_{H^+}$ or by a solute concentration gradient.

Secondary Structure of the Permease

Circular dichroic measurements with purified *lac* permease have indicated that $85 \pm 5\%$ of the amino acid residues are arranged in helical secondary structures whether the protein is solubilized in octylglucoside or reconstituted into proteoliposomes (33). This finding led to a systematic examination of primary structure as determined from the DNA sequence of *lac Y* (6). When sequential hydrophilicity and hydrophobicity (i.e. hydropathy) were evaluated according to the method of Kyte & Doolittle (93), it became apparent that the permease has a number of relatively long hydrophobic regions punctuated by shorter hydrophilic regions. In light of the circular dichroism data, this treatment suggests strongly that most if not all of the hydrophobic segments are α-helical; moreover, it is likely that they are embedded in the bilayer. About 12 of the longest hydrophobic regions exhibit a mean length of 24 ± 4 amino acid residues, and they comprise approximately 70% of the length of the polypeptide. The mean length of these segments correlates remarkably well with the mean lengths calculated for similar domains found in four other integral membrane proteins involved in H^+ translocation (bacteriorhodopsin and the three subunits of the F_0 portion of the H^+-ATPase). A 24-residue peptide in α-helical conformation is expected to be a maximum of 36 Å long, corresponding roughly to the thickness of the hydrophobic core of the membrane.

According to the predictions of Chou & Fasman (12), eighteen regions of the permease contain reverse turns (180° reversals). Fifteen of the putative turns (83%) fall within hydrophilic regions between the hydrophobic segments postulated to traverse the bilayer.

Based on these considerations, the model shown in Figure 7 is proposed. The permease is postulated to consist of 12 α-helical segments that traverse the membrane in zig-zag fashion as shown for bacteriorhodopsin (30, 52). Similar secondary-structure models containing 13 (3) and 14 (186) transmembrane α-helical segments have also been proposed, the first based on hydropathic profiling and the second based on Laser-Raman spectroscopy plus a method for structural prediction that accounts for amphipathic helices.

Experiments with proteolytic enzymes and site-directed polyclonal antibodies (antibodies directed against synthetic peptides corresponding in sequence to specified regions of the permease) provide support for certain general aspects of the models. Thus, photoaffinity-labeled permease in RSO or ISO vesicles is accessible to proteases, demonstrating that the permease extends through the bilayer (39). Moreover, antibodies directed against peptides corresponding to the C-terminus (8, 162, 163) and hydrophilic segments 5 and 7 (Figure 7) bind preferentially to ISO vesicles relative to RSO vesicles, indicating that each of these sites in the permease is on the same side of the membrane, the cytoplasmic surface. In addition, lactoperoxidase-catalyzed iodination (3) suggests that the N-terminus is on the cytoplasmic surface. Antibodies directed against each of the other putative hydrophilic regions (Figure 7) do not bind to vesicles of either orientation, although the antibodies react with the permease on immunoblotting (20). Presumably, these parts of the protein are either buried in the membrane or are inaccessible within the tertiary structure of the

308

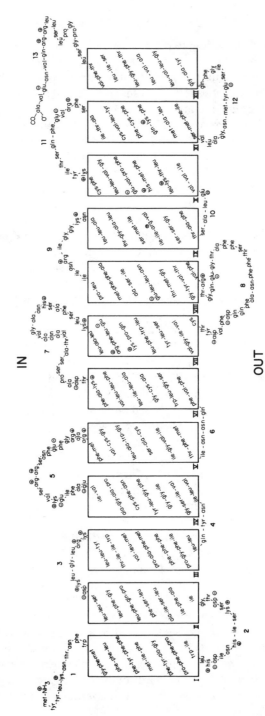

Figure 7 Secondary structure model of *lac* permease based on the hydropathic profile of the protein as determined by Foster et al (33). Hydrophobic segments are shown in boxes as transmembrane, α-helical domains connected by hydrophilic segments. As discussed in the text, the C-terminus and hydrophilic segments 5 and 7 (with the N-terminus as hydrophilic segment 1) are on the cytoplasmic surface of the membrane.

native polypeptide. It is evident that in order to obtain more definitive structural information, it will be necessary to crystallize the permease.

Immunological Reagents

Monoclonal antibodies (Mabs) against purified *lac* permease were studied to determine their effects on lactose transport in RSO vesicles and in proteoliposomes reconstituted with purified permease (9, 10). Of more than 60 Mabs tested, only one (4B1) inhibits transport. Furthermore, the nature of the inhibition is highly specific; 4B1 inhibits only those translocation reactions that involve net H^+ movement (active transport, influx, and efflux under nonenergized conditions and lactose-induced H^+ influx). In contrast, 4B1 has little effect on exchange and no effect on generation of $\Delta\bar{\mu}_{H^+}$ or on the ability of the permease to bind ligand. Thus, 4B1 alters the relationship between lactose and H^+ translocation at the level of the permease. Study of counterflow with external lactose at saturating and subsaturating concentrations reveals that 4B1 mimics the effects of deuterium oxide. The results suggest that the Mab either inhibits the rate of deprotonation or alters the equilibrium between protonated and unprotonated forms of the permease. Monovalent Fab fragments prepared from 4B1 inhibit transport much as intact IgG does, in a qualitative sense. However, 4B1 IgG is approximately twice as effective as the Fab fragments on a molar basis, suggesting that the intact IgG binds bivalently while the Fab fragments bind 1 : 1. Support for this conclusion has been provided by binding experiments with radiolabeled 4B1 and its Fab fragments.

Radiolabeled 4B1 and 5F7 (another Mab that does not inhibit transport) were found to bind to distinct, nonoverlapping epitopes in the permease (55). Immunofluorescence microscopy and radioiodinated IgG and Fab fragments were used to show that both Mabs bind to spheroplasts and to RSO membrane vesicles, but only to a small extent to ISO vesicles. Thus, unlike the C-terminus and hydrophilic regions 5 and 7, which are on the cytoplasmic surface of the membrane, the 4B1 and 5F7 epitopes are on the external (i.e. periplasmic) surface of the membrane. In RSO vesicles 4B1 binds with a stoichiometry of 1 mol IgG per 2 mol permease, while 4B1 Fab fragments bind 1 : 1. Importantly, these binding stoichiometries are preserved in proteoliposomes reconstituted with purified permease. With respect to the 4B1 epitope, therefore, the orientation of the permease in the reconstituted system appears to be similar to that in the native membrane.

In contradistinction, experiments with site-directed polyclonal antibodies against the C-terminus of the permease indicate that the reconstituted system is not entirely representative of the native membrane (8, 162, 163). Radiolabeled anti–C-terminal Fab fragments, which bind to ISO vesicles relatively exclusively, were found to bind to proteoliposomes containing the permease; this indicates that a significant percentage of the C-terminus is on the outside of the membrane in the reconstituted system (i.e. on the wrong side of the membrane). In addition, treatment of reconstituted proteoliposomes with carboxypeptidase partially degraded the ultimate C-terminus with no effect on the 4B1 epitope or on transport activity. Finally, anti-hydrophilic segment 7, which also binds to ISO vesicles, did not bind to proteoliposomes; this suggests that segment 7, like the 4B1 epitope, maintains the proper orientation after reconstitution. Therefore, although the obvious possibility that reconstituted permease molecules are scrambled has been considered, the data are more consistent with the idea that a portion of the molecules undergoes intramolecular dislocation of the C-terminus during reconstitution with no effect on catalytic activity. In this context, it is also significant that 4B1 enhanced binding of Mab 4A10R in the reconstituted system. Mab 4A10R is directed against a cytoplasmically disposed epitope that is partially related to the C-terminus of the permease (i.e. in the native membrane, the epitopes for 4A10R and 4B1 are on opposite sides of the membrane) (54). Since it is unlikely that this effect of 4B1 is intermolecular, a more reasonable interpretation is that these conformationally coupled epitopes, which are normally on opposite sides of the membrane, are present on the external surface of the proteoliposomes within the same permease molecules. Finally, it is notable that C-terminal dislocation has been reported with reconstituted cytochrome b5 (31, 174) and H-2K histocompatibility antigen (7).

RSO vesicles from a mutant "uncoupled" for β-galactoside:H^+ symport (197) were specifically defective in the ability to catalyze accumulation of TMG in the presence of $\Delta\bar{\mu}_{H^+}$ (54). Moreover, efflux under nonenergized conditions was slow and unaffected by ambient pH of 5.5–7.5, and TMG-induced H^+ influx was only about 15% of that in vesicles containing wild-type permease. Alternatively, mutant vesicles bound NPG and Mab 4B1 to the same extent as wild-type vesicles and catalyzed facilitated diffusion and equilibrium exchange as well as wild-type vesicles. When counterflow was studied with external substrate at saturating and subsaturating concentrations, the mutation, like Mab 4B1, also simulated the effects of deuterium oxide. That is, the mutation had no effect on the rate or extent of counterflow when external substrate was saturating, but stimulated counterflow when external substrate was below the apparent K_m. Also, replacement of protium with deuterium stimulated counterflow in wild-type vesicles when external substrate was limiting, but the isotope had no effect on mutant vesicles under the same conditions. It is suggested that the mutation results in a lac permease with a higher pK_a, thereby either limiting the rate of deprotonation or altering the equilibrium between protonated and deprotonated forms of the permease. Although Mab 4B1 binds similarly to wild-type and mutant RSO vesicles, Mab 4A10R binds to ISO vesicles from the mutant only 30% as well as it binds to the same preparation from the wild type. Furthermore, antibodies against hydrophilic domains 5 and 7 bind threefold better and one-fifth as well, respectively, to ISO vesicles from the mutant than to ISO vesicles from the wild type. Thus, these immunological probes are able to discriminate between wild-type permease and the uncoupled variant. In addition, the results suggest that the mutation causes a significant alteration in the conformation of the permease, particularly on the cytoplasmic surface of the membrane.

Structure/Function Considerations

In addition to acting thermodynamically as the driving force for active lactose transport, $\Delta\bar{\mu}_{H^+}$ alters the distribution of the permease between two very different kinetic pathways (76, 148, 181). In the absence of $\Delta\bar{\mu}_{H^+}$, transport exhibits an apparent K_m of about 20 mM for lactose; when $\Delta\bar{\mu}_{H^+}$ is applied, the apparent K_m decreases by a factor of about 100 to 0.2 mM. Moreover, the distribution of the permease between the two pathways varies with the square of ΔpH or $\Delta\Psi$. For these reasons, it was suggested very tentatively that the permease might exist in two forms, monomer and dimer; the monomer would catalyze facilitated diffusion (high apparent K_m) and the dimer would catalyze active transport (low apparent K_m). Finally, it was suggested that $\Delta\bar{\mu}_{H^+}$ might cause an alteration in subunit interactions (i.e. dimerization). In addition, genetic studies that demonstrated that certain *lac Y* mutations are dominant (109) also lend credence to the idea that oligomeric structure may be important for *lac* carrier function.

Other studies demonstrate that inactivation of the permease by various maleimides (16) is enhanced two- to three-fold in the presence of $\Delta\bar{\mu}_{H^+}$, and a similar effect was observed with the histidine reagent diethylpyrocarbonate (DEPC) (130). Thus, the shift between the two kinetic pathways may involve structural or conformational changes that alter the reactivity of certain functional groups in the permease. Furthermore, treatment of vesicles with DEPC or photooxidation in the presence of rose bengal, an operation that photooxidizes histidine residues, inactivated $\Delta\bar{\mu}_{H^+}$-driven lactose transport with no effect on the ability of the permease to bind NPG or catalyze facilitated diffusion of lactose (36, 130). In addition, treatment with DEPC prevented the permease from responding to $\Delta\bar{\mu}_{H^+}$. That is, after treatment with DEPC, the permease no longer exhibited a decrease in K_m in response to $\Delta\bar{\mu}_{H^+}$, and only the high-K_m, facilitated-diffusion pathway is observed.[3] In all likelihood, therefore, one or more histidine residues in the permease are involved in the response of the protein to $\Delta\bar{\mu}_{H^+}$.

Although the permease appears to be monomeric in dodecylmaltoside (199) and hexamethylphosphoric triamide (86), and dimeric in dodecyl octaethylene glycol monoether (61), the oligomeric state of the protein in solution may have little relationship to its oligomeric state in the membrane. Electron inactivation analysis (79) indicated that if the permease is an oligomer in the membrane, it is no bigger than a dimer (40). In these experiments, vesicles containing *lac* permease were frozen in liquid nitrogen before or after generation of $\Delta\bar{\mu}_{H^+}$ and then subjected to a high-intensity electron beam for various time periods. Since the vesicles become very permeable after low doses of irradiation, it was necessary to extract and reconstitute the permease in order to assay activity. Thus, after irradiation the samples were extracted with octylglucoside, reconstituted into proteoliposomes, and tested for transport activity. Under all conditions the decrease in activity exhibited pseudo–first-order kinetics as a function of radiation dosage, allowing straightforward application of target theory. When permease activity solubilized from nonenergized vesicles was assayed, the results were consistent with a functional molecular weight of 45–50 kd, a value similar to that determined by other means (cf above). Importantly, moreover, comparable values were obtained when the octylglucoside extract was irradiated, and target masses observed for D-LDH and DCCD-sensitive ATPase activity in the same vesicle preparations are comparable to the known molecular weights of D-LDH and the F_1 portion of the H^+-ATPase. Remarkably, when the same procedure was carried out with vesicles energized prior to freezing and irradiation, a functional molecular weight of 85–90 kd was obtained with no change in the target mass of D-LDH. Moreover, when the vesicles were energized in the presence of CCCP, the target mass of the permease returned to 45–50 kd.

Recent studies by Dornmair et al (27), in which the size of the permease was derived from rotational diffusion measurements, are also consistent with the idea that the permease is monomeric in the membrane, but the monomeric state appears to be maintained in the presence of $\Delta\Psi$. It should be noted, however, that the studies were carried out with permease that was labeled with eosinyl-maleimide, which reacts at Cys148 and thereby inactivates the protein. In addition, the labeled permease was reconstituted

---

[3] It is noteworthy, however, that after the vesicles are inactivated with DEPC, H^+ influx in response to an inwardly-directed lactose gradient is no longer observed (135). Thus, modification of the carrier in this manner appears to uncouple lactose and H^+ movements at the level of the carrier.

into dimyristoylphosphatidylcholine, a phospholipid that supports minimal activity (11).

Oligonucleotide-Directed, Site-Specific Mutagenesis

Although chemical modification of specific amino-acid residues in a protein can provide important information, there are obvious drawbacks to this approach. Site-directed mutagenesis has been used to introduce single amino-acid changes into proteins (205), and this strategy has recently been applied to the *lac* permease.

Based on substrate protection against N-ethylmaleimide (NEM) inactivation, Fox & Kennedy (80) postulated that there is an essential sulfhydryl group in the permease located at or near the active site, and Cys148 was later shown to be the critical residue (2). Trumble, Viitanen, et al (179, 182) cloned *lac Y* into single-stranded M13 phage DNA and, using a synthetic deoxyoligonucleotide primer 21 bases long that is complementary to the *lac Y* template with the exception of a single mismatch, converted Cys148 in the permease into a glycine residue. Cells bearing mutated *lac Y* exhibited initial rates of lactose transport that were about one fourth that of cells bearing the wild-type gene on the same recombinant plasmid. Transport activity was less sensitive to inactivation by NEM, and galactosyl 1-thio-β-D-galactopyranoside (TDG) afforded no protection whatsoever against inactivation. Furthermore, permease with serine in place of Cys148 (71, 119, 182) catalyzed transport at least as well as wild-type permease and exhibited the same properties with regard to NEM inactivation and TDG protection. The findings indicate that although Cys148 is important for substrate protection against sulfhydryl inactivation it is not obligatory for lactose: H^+ symport, and that another sulfhydryl group elsewhere within the permease may be required for full activity. Strikingly, site-directed mutagenesis of Cys154 indicates that a sulfhydryl group at this position may be critical for permease activity (71, 107). Thus, permease with glycine in place of Cys154 exhibited essentially no transport activity, while substitution of Cys154 with serine also caused marked, though less complete, loss of activity. In addition, R. Brooker & T. H. Wilson have replaced Cys176 or Cys234 with serine residues with less than 50% loss of activity (personal communication). Experiments designed to evaluate the role of the remaining four cystine residues in the permease are currently in progress.

As discussed above, experiments with DEPC and rose bengal in RSO vesicles suggest that histidine residues may play an important role in coupling H^+ and lactose translocation. For this reason, each of the four histidine residues in the *lac* permease were replaced with arginine residues by site-directed mutagenesis (71, 131). Replacement of His35 and His39 with arginine had no apparent effect on activity. In contrast, replacement of either His205 or His322 with arginine caused dramatic loss of transport activity, although the cells contained a normal complement of permease molecules as determined by immunoadsorption assays. Although substitution of His205 or His322 by arginine resulted in the loss of ability to catalyze active lactose transport, permease molecules with arginine at residue 322 facilitated downhill lactose movements at high concentrations of the disaccharide. Furthermore, permease with Arg322 that has been purified and reconstituted into proteoliposomes has been shown to catalyze facilitated diffusion of lactose at a significant rate without concomitant translocation of H^+ (i.e. the permease is uncoupled); purified permease with Arg205 had no activity.

Although site-directed mutagenesis is a powerful technique that allows highly specific changes in primary structure to be made at will, problems concerning the interpretation of a given alteration on catalytic activity remain. Thus, the change of a given amino acid residue might lead to a change in catalytic activity for at least two reasons: (a) The mutagenized residue might play a primary role in the reaction mechanism or (b) the replacement might lead to an alteration in activity via a secondary structural change. With regard to the His205 and His322 replacements, therefore, the question is whether these residues play a direct role in lactose: H^+ symport or whether their substitution with arginine leads to a structural change with secondary, long-range effects.

Although it is impossible to answer the question satisfactorily without detailed structural information, the following considerations are noteworthy in this context: Arginine was chosen specifically to replace histidine to maintain positive charge in the mutagenized residues, and replacement of His35 and His39 had no effect on permease activity. More importantly, it should be possible to resolve the question at least to some extent by further mutational analyses. It should be possible to obtain revertants of Arg205 and Arg 322 that encode catalytically active permease molecules; by sequencing lac Y in these reversion mutants, the nature of the reversions can be determined. If all of the back mutations involve arginine-to-histidine codons at the appropriate site, the conclusion that histidine plays a direct, primary role in lactose: H^+ symport would seem justified. On the other hand, if a significant number of revertants involve nucleotides encoding other residues in the permease, it would seem unlikely that histidine residues per se are directly involved in the symport mechanism. With this idea in mind, a number of spontaneous revertants of the Arg205 mutant have been isolated, and partial sequencing of lac Y in all of the plasmids reveals back mutations of arginine to histidine in codon 205.

Finally, site-directed mutagenesis has been used to introduce a few other alterations, in addition to the mutations described, into the *lac* permease: (*a*) Gln60 was replaced with a glutamic acid residue, and thereby a negative charge was introduced into the second putative α-helix of the permease (71, 158). Although permease molecules altered in this manner catalyzed active transport in a fashion indistinguishable from that of wild-type permease, the Glu60 mutation was more susceptible to heat inactivation. (*b*) A segment of the polypeptide from Met372 to Pro405 was deleted, and thus the last putative transmembrane α-helix from the permease was excised (71). Remarkably, permease altered in this manner was either not inserted into the membrane or was proteolyzed subsequent to insertion. That is, as judged by immunoadsorption assays, the cytoplasmic membrane is grossly deficient in permease after excision of putative helix 12.

Literature Cited

1. Banker, G. A., Cotman, C. W. 1972. *J. Biol. Chem.* 247 : 5856
2. Beyreuther, K., Bieseler, B., Ehring, R., Müller-Hill, B. 1981. *Methods in Protein Sequence Analysis*, ed. M. Elzine, p. 139. Clifton, NJ : Humana
3. Bieseler, B., Prinz, H., Beyreuther, K. 1985. *Proc. NY Acad. Sci.* In press
4. Booth, I. R., Mitchell, W. J., Hamilton, W. A. 1979. *Biochem. J.* 182 : 687
5. Brand, M. D., Reynafarje, B., Lehninger, A. L. 1976. *J. Biol. Chem.* 251 : 5670
6. Büchel, D. E., Gronenborn, B., Müller-Hill, B. 1980. *Nature* 283 : 541
7. Cardoza, J. D., Kleinfeld, A. M., Stallcup, K. C., Mescher, M. F. 1984. *Biochemistry* 23 : 4401
8. Carrasco, N., Herzlinger, D., Mitchell, R., DeChiara, S., Danho, W., et al. 1984. *Proc. Natl. Acad. Sci. USA* 81 : 4672
9. Carrasco, N., Tahara, S. M., Patel, L., Goldkorn, T., Kaback, H. R. 1982. *Proc. Natl. Acad. Sci. USA* 78 : 7652
10. Carrasco, N., Viitanen, P., Herzlinger, D., Kaback, H. R. 1984. *Biochemistry* 23 : 3681
11. Chen, C.-C., Wilson, T. H. 1984. *J. Biol. Chem.* 259 : 10150
12. Chou, P. Y., Fasman, G. D. 1974. *Biochemistry* 13 : 222
13. Cohen, G. N., Monod, J. 1957. *Bacterial. Rev.* 21 : 169
14. Cohen, G. N., Rickenberg, H. V. 1955. *C. R. Acad. Sci.* 240 : 466
15. Cohn, D., Kaback, H. R. 1980. *Biochemistry* 19 : 4237
16. Cohn, D. E., Kaczorowski, G. J., Kaback, H. R. 1981. *Biochemistry* 20 : 3308
17. Colowick, S. P., Womack, F. C. 1969. *J. Biol. Chem.* 244 : 774
18. Costello, M. J., Escaig, J., Matsushita, K., Carrasco, N., Menick, D. R., Kaback, H. R. 1985. *Biophys. J.* In press
19. Costello, M. J., Viitanen, P., Carrasco, N., Foster, D. L., Kaback, H. R. 1984. *J. Biol. Chem.* 259 : 15579
20. Danho, W., Makofske, R., Humiec, F., Gabriel, T., Carrasco, N., Kaback, H. R. 1985. *Proc. 9th Am. Peptide Symp. Toronto*, ed. C. M. Deber, K. D. Kopple. Rockford, Ill : Pierce Chemical. In press
21. Daniels, C. J., Bole, D. G., Quay, S. C., Oxender, D. L. 1981. *Proc. Natl. Acad. Sci. USA* 78 : 5396
22. Daruwalla, K. R., Paxton, A. T., Henderson, P. J. F. 1981. *Biochem. J.* 200 : 611
23. Date, T., Zwizniski, C., Ludmerer, S., Wickner, W. 1980. *Proc. Natl. Acad. Sci. USA* 77 : 827
24. Delmer, D. P., Benziman, M., Padan, E. 1982. *Proc. Natl. Acad. Sci. USA* 79 : 5282
25. Dills, S. S., Apperson, A., Schmidt, M. R., Saier, M. H. Jr. 1980. *Microbiol. Rev.* 44 : 385
26. Doetsch, R. N., Sjoblad, R. D. 1980. *Ann. Rev. Microbiol.* 34 : 69
27. Dornmair, K., Corin, A. F., Wright, J. K., Jähnig, F. 1985. *EMBO J.* 4 : 3633
28. Ehring, R., Beyreuther, K., Wright, J. K., Overath, P. 1980. *Nature* 283 : 537
29. Enequist, H. G., Hirst, T. R., Hardy, S. J. S., Harayama, S., Randall, L. L. 1981. *Eur. J. Biochem.* 116 : 227
30. Engelman, D. M., Henderson, P., McLachlan, A. D., Wallace, B. A. 1980. *Proc. Natl. Acad. Sci. USA* 77 : 2023

31. Enoch, H. G., Fleming, P. J., Stritt-matter, P. 1979. *J. Biol. Chem.* 254:6483
32. Felle, H., Porter, J. S., Slayman, C. L., Kaback, H. R. 1980. *Biochemistry* 19:3585
33. Foster, D. L., Boublik, M., Kaback, H. R. 1983. *J. Biol. Chem.* 258:31
34. Foster, D. L., Garcia, M. L., Newman, M. J., Patel, L., Kaback, H. R. 1982. *Biochemistry* 21:5634
35. Futai, M. 1973. *Biochemistry* 12:2468
36. Garcia, M. L., Patel, L., Padan, E., Kaback, H. R. 1982. *Biochemistry* 21:5800
37. Garcia, M. L., Viitanen, P., Foster, D. L., Kaback, H. R. 1983. *Biochemistry* 22:2524
38. Ghazi, A., Shechter, E. 1981. *Biochim. Biophys. Acta* 645:305
39. Goldkorn, T., Rimon, G., Kaback, H. R. 1983. *Proc. Natl. Acad. Sci. USA* 80:3322
40. Goldkorn, T., Rimon, G., Kempner, E. S., Kaback, H. R. 1982. *Proc. Natl. Acad. Sci. USA* 81:1021
41. Greville, G. D. 1969. *Curr. Top. Bioenerg.* 3:1
42. Grinius, L. 1980. *FEBS Lett.* 113:1
43. Grinius, L., Bervinskiene, J. 1976. *FEBS Lett.* 72:151
44. Grinius, L. L., Jasaitis, A. A., Kadziauskas, Yu. P., Liberman, E. A., Skulachev, V. P., et al. 1971. *Biochim. Biophys. Acta* 216:1
45. Guffanti, A. A., Cohn, D. E., Kaback, H. R., Krulwich, T. A. 1981. *Proc. Natl. Acad. Sci. USA* 78:1481
46. Haldar, K., Olsiewski, P. J., Walsh, C., Kaczorosski, G. J., Bhaduri, A., Kaback, H. R. 1982. *Biochemistry* 21:4590
47. Hamamoto, T., Carrasco, N., Matsushita, K., Kaback, H. R., Montal, M., 1985. *Proc. Natl. Acad. Sci. USA* 82:2570
48. Harold, F. M. 1972. *Bacteriol. Rev.* 36:172
49. Harold, F. M. 1978. In *The Bacteria*, Vol. 6, ed. I. C. Gunsalus, L. N. Ornston, T. R. Sokatch, p. 463. New York: Academic
50. Haydon, D. A., Hladky, S. B. 1972. *Q. Rev. Biophys.* 5:187
51. Henderson, P. J. F., Giddens, R. A., Jones-Mortimer, M. C. 1977. *Biochem. J.* 162:309
52. Henderson, R., Unwin, P. N. T. 1975. *Nature* 257:31
53. Hertzberg, E., Hinkle, P. C. 1974. *Biochem. Biophys. Res. Commun.* 58:178
54. Herzlinger, D., Carrasco, N., Kaback, H. R. 1985. *Biochemistry* 24:221
55. Herzlinger, D., Viitanen, P., Carrasco, N., Kaback, H. R. 1984. *Biochemistry* 23:3688
56. Hinkle, P. C., McCarty, R. E. 1978. *Sci. Am.* 238:104
57. Hirata, H., Altendorf, K. H., Harold, F. M. 1973. *Proc. Natl. Acad. Sci. USA* 70:1804
58. Hirata, H., Altendorf, K. H., Harold, F. M. 1974. *J. Biol. Chem.* 249:2939
59. Hobson, A. C., Gho, D., Müller-Hill, B. 1977. *J. Bacteriol.* 131:830
60. Hong, J.-S. 1977. *J. Biol. Chem.* 252:8582
61. Houssin, C., le Mair, M., Aggerbeck, L. P., Shechter, E. 1985. *Arch. Biochem. Biophys.* In press
62. Hugenholtz, J., Hong, J.-S., Kaback, H. R. 1981. *Proc. Natl. Acad. Sci. USA* 78:3446
63. Jencks, W. P. 1969. *Catalysis in Chemistry and Enzymology.* New York: McGraw-Hill
64. Jones, T. H. D., Kennedy, E. P. 1969. *J. Biol. Chem.* 244:5981
65. Kaback, H. R. 1960. *Fed. Proc.* 19:130
66. Kaback, H. R. 1970. *Ann. Rev. Biochem.* 39:561
67. Kaback, H. R. 1971. *Methods Enzymol.* 22:99
68. Kaback, H. R. 1972. *Biochim. Biophys. Acta* 265:367
69. Kaback, H. R. 1974. *Science* 186:882
70. Kaback, H. R. 1976. *J. Cell Physiol.* 89:575
71. Kaback, H. R. 1986. *Ann. NY Acad. Sci.* 456:291
72. Kaback, H. R., Barnes, E. M. Jr. 1971. *J. Biol. Chem.* 246:5523
73. Kaback, H. R., Stadtman, E. R. 1966. *Proc. Natl. Acad. Sci. USA* 55:920
74. Kaczorowski, G. J., Kaback, H. R. 1979. *Biochemistry* 18:3691
75. Kaczorowski, G. J., LeBlanc, G., Kaback, H. R. 1980. *Proc. Natl. Acad. Sci. USA* 77:6319
76. Kaczorowski, G. J., Robertson, D. E., Kaback, H. R. 1979. *Biochemistry* 18:3697
77. Kalasauskaite, E., Grinius, L. 1979. *FEBS Lett.* 99:287
78. Kashket, E. R., Wilson, T. H. 1973. *Proc. Natl. Acad. Sci. USA* 70:2866
79. Kempner, E. A., Schlegel, W. 1979. *Anal. Biochem.* 92:2
80. Kennedy, E. P. 1970. In *The Lactose Operon*, ed. J. R. Bethwith, D. Zipser, p. 49. New York: Cold Spring Harbor Lab.
81. Kepes, A. 1971. *J. Membr. Biol.* 4:87
82. Kepes, A., Cohen, G. N. 1962. In *The Bacteria*, Vol. 4, ed. I. C. Gunsalus, R. Stanier, p. 179. New York: Academic

83. Kita, K., Kasahara, M., Anraku, Y. 1982. *J. Biol. Chem.* 257:7933
84. Kohn, L. D., Kaback, H. R. 1973. *J. Biol. Chem.* 248:7012
85. Koland, J. G., Miller, M. J., Gennis, R. B. 1984. *Biochemistry* 23:445
86. König, B., Sandermann, H. Jr. 1982. *FEBS Lett.* 147:31
87. Konings, W. N., Barnes, E. M. Jr., Kaback, H. R. 1971. *J. Biol. Chem.* 246:5857
88. Konings, W. N., Boonstra, J. 1976. *Curr. Top. Membr. Transp.* 9:177
89. Konings, W. N., Kaback, H. R. 1976. *Proc. Natl. Acad. Sci. USA* 70:3376
90. Krulwich, T. A. 1982. In *Membranes and Transport*, Vol. 2, ed. A. Martonosi, p. 75. New York: Plenum
91. Krulwich, T. A. 1983. *Biochim. Biophys. Acta* 726:245
92. Kundig, W., Ghosh, S., Roseman, S. 1964. *Proc. Natl. Acad. Sci. USA* 52:1067
93. Kyte, J., Doolittle, R. F. 1982. *J. Mol. Biol.* 157:105
94. Laane, C., Krone, W., Konings, W., Haaker, H., Veeger, C. 1980. *Eur. J. Biochem.* 103:39
95. Labedan, G., Goldberg, E. B. 1979. *Proc. Natl. Acad. Sci. USA* 76:4669
96. Lam, V. M. S., Daruwalla, K. R., Henderson, P. J. F., Jones-Mortimer, M. C. 1980. *J. Bacteriol.* 143:396
97. Lanyi, J. K., Renthal, R., MacDonald, R. E. 1976. *Biochemistry* 15:1603
98. Lawford, H. G. 1977. *Can. J. Biochem.* 56:13
99. LeBlanc, G., Rimon, G., Kaback, H. R. 1980. *Biochemistry* 19:2522
100. Lichtshtein, D., Dunlop, K., Kaback, H. R., Blume, A. J. 1979. *Proc. Natl. Acad. Sci. USA* 76:2580
101. Lichtshtein, D., Kaback, H. R., Blume, A. J. 1978. *Proc. Natl. Acad. Sci. USA* 76:650
102. Lombardi, F. J., Kaback, H. R. 1972. *J. Biol. Chem.* 247:7844
103. Mates, S., Eisenberg, E. S., Mandel, L. J., Patel, L., Kaback, H. R., Miller, M. H. 1982. *Proc. Natl. Acad. Sci. USA* 79:6693
104. Matsushita, K., Patel, L., Gennis, R. B., Kaback, H. R. 1983. *Proc. Natl. Acad. Sci. USA* 80:4889
105. Matsushita, K., Patel, L., Kaback, H. R. 1984. *Biochemistry* 23:4703
106. McMurry, L., Petrucci, R. E., Levy, S. B. 1980. *Proc. Natl. Acad. Sci. USA* 77:3974
107. Menick, D. R., Sarkar, H. K., Poonian, M. S., Kaback, H. R. 1985. *Biochem. Biophys. Res. Commun.* 132:162
108. Michels, P. A. M., Michels, J. P. J., Boonstra, J., Konings, W. N. 1979. *FEMS Microbiol. Lett.* 5:357
109. Mieschendahl, M., Büchel, D., Bocklage, H., Müller-Hill, B. 1981. *Proc. Natl. Acad. Sci. USA* 78:7652
110. Miller, M. J., Gennis, R. B. 1985. *J. Biol. Chem.* In press
111. Mitchell, P. 1961. *Nature* 191:144
112. Mitchell, P. 1963. *Biochem. Soc. Symp.* 22:142
113. Mitchell, P. 1966. *Chemiosmotic Coupling and Energy Transduction*, Bodmin, Cornwall, England: Glynn Research. 111 pp.
114. Mitchell, P. 1966. *Chemiosmotic Coupling in Oxidative and Photophosphorylation*, Bodmin, Cornwall, England: Glynn Research. 192 pp.
115. Mitchell, P. 1973. *J. Bioenerg.* 4:63
116. Mitchell, P. 1976. *J. Theor. Biol.* 62:327
117. Mitchell, P., Moyle, J. 1979. *Biochem. Soc. Trans.* 7:887
118. Navon, G., Ogawa, S., Shulman, R. G., Yamane, T. 1977. *Proc. Natl. Acad. Sci. USA* 74:888
119. Neuhaus, J., Soppa, J., Wright, J. K., Riede, I., Blocker, H., et al. 1985. *FEBS Lett.* 185:83
120. Neville, D. M. Jr. 1971. *J. Biol. Chem.* 246:6328
121. Newman, M. J., Foster, D., Wilson, T. H., Kaback, H. R. 1981. *J. Biol. Chem.* 256:11804
122. Newman, M. J., Wilson, T. H. 1980. *J. Biol. Chem.* 255:10583
123. Nicholls, D. G. 1982. *Bioenergetics.* New York: Academic. 490 pp.
124. Ogawa, S., Shulman, R. G., Glynn, P., Yamane, T., Navon, G. 1970. *Biochim. Biophys. Acta* 502:45
125. Olsiewski, P. J., Kaczorowski, G., Walsh, C. T., Kaback, H. R. 1981. *Biochemistry* 20:6272
126. Otto, R., Sonenberg, A. S. M., Veldkamp, H., Konings, W. N. 1980. *Proc. Natl. Acad. Sci. USA* 77:5502
127. Owen, P., Kaback, H. R. 1979. *Proc. Natl. Acad. Sci. USA* 75:3148
128. Owen, P., Kaback, H. R. 1979. *Biochemistry* 18:1413
129. Owen, P., Kaback, H. R. 1979. *Biochemistry* 18:1422
130. Padan, E., Patel, L., Kaback, H. R. 1979. *Proc. Natl. Acad. Sci. USA* 80:6221
131. Padan, E., Sarkar, H. K., Viitanen, P. V., Poonian, M. S., Kaback, H. R. 1985. *Proc. Natl. Acad. Sci. USA* 82 : 6765
132. Padan, E., Schuldiner, S., Kaback, H. R. 1979. *Biochem. Biophys. Res. Commun.* 91:854
133. Padan, E., Zilberstein, D., Rottenberg, H. 1976. *Eur. J. Biochem.* 63:533

134. Papa, S., Guerrieri, F., Lorusso, M., Izzo, G., Boffoli, D., Stefanelli, R. 1970. *FEBS Symp.* 45:37

135. Patel, L., Garcia, M. L., Kaback, H. R. 1982. *Biochemistry* 21:5805

136. Patel, L., Schuldiner, S., Kaback, H. R. 1975. *Proc. Natl. Acad. Sci. USA* 72:3387

137. Plate, C. A., Suit, J. L. 1981. *J. Biol. Chem.* 256:12974

138. Racker, E., Violand, B., O'Neal, S., Alfonzo, M., Telford, J. 1979. *Arch. Biochem. Biophys.* 198:470

139. Ramos, S., Kaback, H. R. 1977. *Biochemistry* 16:848

140. Ramos, S., Kaback, H. R. 1977. *Biochemistry* 16:854

141. Ramos, S., Kaback, H. R. 1977. *Biochemistry* 16:4271

142. Ramos, S., Schuldiner, S., Kaback, H. R. 1979. *Methods Enzymol.* 55:680

143. Ramos, S., Schuldiner, S., Kaback, H. R. 1976. *Proc. Natl. Acad. Sci. USA* 73:1892

144. Reenstra, W. W., Patel, L., Rottenberg, H., Kaback, H. R. 1980. *Biochemistry* 19:1

145. Reynafarje, B., Brand, M. D., Lehninger, A. L. 1976. *J. Biol. Chem.* 251:7442

146. Reynefarje, B., Lehninger, A. L. 1978. *J. Biol. Chem.* 253:6331

147. Rickenberg, H. V., Cohen, G. N., Buttin, G., Monod, J. 1956. *Ann. Inst. Pasteur Paris* 91:829

148. Robertson, D. E., Kaczorowski, G. J., Garcia, M. L., Kaback, H. R. 1980. *Biochemistry* 19:5692

149. Rosen, B. P., McClees, J. S. 1974. *Proc. Natl. Acad. Sci. USA* 71:5042

150. Rottenberg, H. 1975. *J. Bioenerg.* 7:61

151. Rottenberg, H. 1976. *FEBS Lett.* 66:159

152. Rowland, G. C., Giffard, P. M., Booth, I. R. 1984. *FEBS Lett.* 173:295

153. Saier, M. H. 1982. See Ref. 90, p. 27

154. Saier, M. H. Jr., Wentzel, D. L., Feucht, B. U., Justice, J. J. 1975. *J. Biol. Chem.* 250:5089

155. Sandermann, H. Jr. 1977. *Eur. J. Biochem.* 80:507

156. Santos, E., Kaback, H. R. 1981. *Biochem. Biophys. Res. Commun.* 99:1153

157. Santos, E., Kung, H.-F., Young, I. G., Kaback, H. R. 1982. *Biochemistry* 21:2085

158. Sarkar, H. K., Viitanen, P. V., Poonian, M. S., Kaback, H. R. 1985. *Biochemistry.* In press

159. Schuldiner, S., Fishkes, H. 1978. *Biochemistry* 17:706

160. Schuldiner, S., Kaback, H. R. 1975. *Biochemistry* 14:5451

161. Schuldiner, S., Padan, E. 1982. See Ref. 90, p. 65

162. Seckler, R., Wright, J. K. 1984. *Eur. J. Biochem.* 142:269

163. Seckler, R., Wright, J. K., Overath, P. 1983. *J. Biol. Chem.* 258:10817

164. Short, S. A., Kaback, H. R., Kohn, L. D. 1974. *Proc. Natl. Acad. Sci. USA* 71:1461

165. Short, S. A., Kaback, H. R., Kohn, L. D. 1975. *J. Biol. Chem.* 250:4291

166. Short, S. A., White, D. C., Kaback, H. R. 1972. *J. Biol. Chem.* 247:7452

167. Sigel, E., Carafoli, E. 1978. *Eur. J. Biochem.* 89:119

168. Skulachev, V. P., Hinkle, P. C., eds. 1981. *Chemiosmotic Proton Circuits in Biological Membranes.* Reading, Mass: Addison-Wesley

169. Solioz, M., Carafoli, E., Ludwig, B. 1982. *J. Biol. Chem.* 257:1579

170. Sone, N., Hinkle, P. C. 1982. *J. Biol. Chem.* 257:12600

171. Stock, J., Roseman, S. 1971. *Biochem. Biophys. Res. Commun.* 44:132

172. Stroobant, P., Kaback, H. R. 1975. *Proc. Natl. Acad. Sci. USA* 72:3970

173. Stroobant, P., Kaback, H. R. 1979. *Biochemistry* 18:226

174. Takagaki, Y., Radhakrishnan, R., Gupta, C. M., Khorana, H. G. 1983. *J. Biol. Chem.* 258:9128

175. Taylor, D. J., Essenberg, R. C. 1979. *Proc. Intl. Cong. Biochem., 11th, Toronto*, p. 460

176. Teather, R. M., Bramhall, J., Riede, I., Wright, J. K., Fürst, M., et al. 1980. *Eur. J. Biochem.* 108:223

177. Teather, R. M., Müller-Hill, B., Abrutsch, V., Aichele, G., Overath, P. 1978. *Mol. Gen. Genet.* 159:239

178. Tokuda, H., Kaback, H. R. 1977. *Biochemistry* 16:2130

179. Trumble, W. R., Viitanen, P. V., Sarkar, H. K., Poonian, M. S., Kaback, H. R. 1984. *Biochem. Biophys. Res. Commun.* 119:860

180. Viitanen, P., Garcia, M. L., Foster, D. L., Kaczorowski, G. J., Kaback, H. R. 1983. *Biochemistry* 22:2531

181. Viitanen, P., Garcia, M. L., Kaback, H. R. 1984. *Proc. Natl. Acad. Sci. USA* 81:1629

182. Viitanen, P. V., Menick, D. R., Sarkar, H. K., Trumble, W. R., Kaback, H. R. 1985. *Biochemistry.* In press

183. Viitanen, P. V., Newman, M. J., Foster, D., Wilson, T. H., Kaback, H. R. 1985. *Methods Enzymol.* In press

184. Villarejo, M. 1980. *Biochem. Biophys. Res. Commun.* 93:16

185. Villarejo, M., Ping, C. 1978. *Biochem. Biophys. Res. Commun.* 82:935

186. Vogel, H., Wright, J. K., Jähnig, F. 1985. *EMBO J.* 4:3625
187. Waddel, W. J., Butler, T. C. 1959. *J. Clin. Invest.* 38:720
188. Waggoner, A. J. 1979. *Methods Enzymol.* 55:689
189. Wagner, E. F., Ponta, H., Schweiger, M. 1980. *J. Biol. Chem.* 255:534
190. West, I. C. 1970. *Biochem. Biophys. Res. Commun.* 41:655
191. West, I. C., Mitchell, P. 1972. *J. Bioenerg.* 3:445
192. West, I. C., Mitchell, P. 1973. *Biochem. J.* 132:587
193. West, I. C., Wilson, T. H. 1973. *Biochem. Res. Commun.* 50:551
194. Wikström, M., Krab, K. 1978. *FEBS Lett.* 91:8
195. Wikström, M., Krab, K. 1979. *Biochim. Biophys. Acta* 549:177
196. Wikström, M., Saari, H. T. 1977. *Biochim. Biophys. Acta* 462:347
197. Wong, P. T. S., Kashket, E. R., Wilson, T. H. 1970. *Proc. Natl. Acad. Sci. USA* 65:63
198. Wright, J. K., Overath, P. 1984. *Eur. J. Biochem.* 138:497
199. Wright, J. K., Weigel, U., Lustig, A., Bocklage, H., Mieschendahl, M., et al. 1983. *FEBS Lett.* 162:11
200. Young, I. G., Jaworowski, A., Poulis, M. 1982. *Biochemistry* 21:2092
201. Young, J. D.-E., Unkleless, J. C., Kaback, H. R., Cohn, Z. A. 1983. *Proc. Natl. Acad. Sci. USA* 80:1357
202. Yu, G.-Q., Goldrich, D., Kaback, H. R., Hong, J.-S. 1985. Unpublished experiment
203. Zilberstein, D., Padan, E., Schuldiner, S. 1980. *FEBS Lett.* 116:177
204. Zilberstein, D., Schuldiner, S., Padan, E. 1979. *Biochemistry* 18:669
205. Zoller, M. J., Smith, M. 1983. *Methods Enzymol.* 100:468

Ann. Rev. Biophys. Biophys. Chem. 1986. 15 : 321–53

IDENTIFYING NONPOLAR TRANSBILAYER HELICES IN AMINO ACID SEQUENCES OF MEMBRANE PROTEINS

D. M. Engelman, T. A. Steitz, and A. Goldman

Department of Molecular Biophysics and Biochemistry, Yale University, New Haven, Connecticut 06511

CONTENTS

PERSPECTIVES AND OVERVIEW

In recent years amino acid sequences for many integral membrane proteins have been determined. At the same time theoretical arguments and experimental evidence have accumulated to indicate that transbilayer

321

0883–9182/86/0610–0321$02.00

helices are a major motif in integral membrane protein structure. It may be possible to determine the location of such transmembrane helices directly from amino acid sequences using scales of polarity and sequential search protocols. The purpose of this review is to examine the approaches that have been used and to assess their utility. It is an opportune time to consider the issues, since the structure of three integral membrane proteins that are contained in the photosynthetic reaction center of *Rhodopseudomonas viridis* have been determined at high resolution. This new structural information, combined with increasing evidence concerning the structure of bacteriorhodopsin, permits a critical test of the main ideas involved in searching sequences for transbilayer structural elements.

We find that search procedures based on a moving window that scans a sequence twenty residues at a time are suitable for finding the transbilayer helices that are known to exist. For very nonpolar helices separated by polar polypeptides, many of the proposed polarity scales succeed equally well; where the helices contain more polar groups, however, the choice of scales becomes critically important.

In our discussion we consider the arguments that support the notion that helical structure will be a dominant motif in integral membrane protein organization. We introduce and discuss the problem of suitable scaling of amino acids in terms of their polar and nonpolar characteristics, and discuss further the use of such scales in prediction of protein structure. Finally, we examine the cases in which the validity of predictions can be assessed. It is our contention that a suitable scale and protocol can lead to the successful identification of transmembrane helical structures in integral membrane proteins.

INTRODUCTION

Since the determination of the first protein structures, it has been generally observed that the interiors of proteins tend to contain fewer charged and polar residues and more nonpolar residues than the surfaces in contact with water (3, 44, 45, 67, 103). The role of the hydrophobic effect in protein folding has received constant and detailed attention since Kautzmann's influential discussion (43). The notion that hydrophobicity is an energetic determinant in protein folding has led to attempts to characterize the surfaces in contact with water (9, 10, 51), to document the hydrophobic components of interior regions of proteins (10, 12, 28, 39, 79), and to develop quantitative scales of the relative polarity of each amino acid in a polypeptide (12, 19, 21, 22, 29, 40, 49, 53, 57, 60, 62, 63, 74, 77, 88, 89, 91, 93–95, 99–101, 105). The nature of polarity scales and their formulation are the subject of an ongoing discussion (e.g. 12, 29, 77, 78). The main theme of this discussion is how amino acids partition from water into the interiors of globular proteins. In our treatment a different focus is taken: We examine segments of amino acid sequences as they interact with the nonpolar region of a lipid bilayer. Clearly such interactions are dominated by the hydrophobic effect and the set of interactions involved is different from that involved in the complex interior of a globular protein. Our discussion of scales focuses on the peculiarities of the lipid-protein interface.

HELICES IN LIPID BILAYER ENVIRONMENTS

Theoretical Considerations

Helical structure is known to be induced in polypeptides in nonaqueous environments (85, 86). The large free energy cost of transferring an unsatisfied hydrogen bond donor or acceptor from an aqueous to a nonpolar environment or of breaking such a bond in a nonpolar environment suggests that hydrogen bonds must be systematically satisfied as proteins are inserted into a membrane environment. In the nonaqueous interior of a lipid bilayer where the alternative of hydrogen bonding to water is absent, the energy of each hydrogen bonded pair compared to the unpaired state is approximately 6 kcal/mol (1a), so the lipid environment is extremely unfavorable for unfolding a polypeptide.

Typical energetics for the formation of helices in nonaqueous and aqueous environments and the transfer of a polypeptide between them are summarized in Figure 1. In order to evaluate the relative stability of a transmembrane helix and an unfolded polypeptide in the lipid environment, one must consider at least three factors: hydrogen bonding, conformation entropy, and van der Waals interactions. Although the conformational entropy term favors the unfolded state, it is considerably smaller than the energy term owing to hydrogen bonding in the lipid environment as mentioned above. Since an unfolded polypeptide chain can have many conformations and a folded helix has a well-defined structure, the entropy of the folded structure is much lower. The approximate magnitude of this term is about 1.25 kcal/mol per peptide bond (69). For a 20–amino acid helix about 24 kcal/mol favoring the unfolded state would result from entropy. In a 20–amino acid transmembrane helix, 16 hydrogen bonds would form in the nonaqueous region, contributing -96 kcal/mol favoring the helical conformation. The energy changes due to van der Waals interactions, while important in dictating some details of the final structure, would be small on the scale of energies being considered here since the unfolded chain would have interactions with solvent that would be replaced by interactions with itself as it folded. We conclude that a 20–amino acid α-helix would have a total difference free energy of stabilization in the lipid bilayer of approximately 70 kcal/mol compared with the unfolded state (Figure 1).

If we assume that the 20-residue peptide forms a helix at relatively low energy cost in an aqueous environment (7, 47) and that the side chains of the helix are nonpolar, then the chain would be more stable by tens of kcal/mol as a transbilayer helix traversing the nonpolar region of the lipids than it would be either as a helix or as an unfolded chain in the aqueous environment (by about 30 kcal/mol in the example shown in Figure 1). Spontaneous helix formation in water results from a balance of favorable hydrophobic and hydrogen-bonding energies with unfavorable entropy contributions (7, 47).

If the alternative, insertion of a hydrophobic chain of amino acids into a bilayer followed by folding, is considered, the role of hydrogen bonds becomes immediately apparent. Completion of the thermodynamic cycle in Figure 1 results in the conclusion that the insertion of the unfolded chain is extremely unfavored (+42 kcal/mol). The total free energy includes unfavorable hydrogen bond contributions and favorable hydrophobic

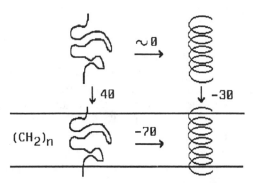

Figure 1 The formation and insertion of a polyalanine helix 20 residues long (21). We assume that the formation of the helix in solution will be at least marginally stable and thus require about 0 kcal/mol. The equilibrium free energy for transferring the helix to a position spanning the nonpolar region of a bilayer includes −32 kcal/mol from the hydrophobic effect; +5 kcal/mol have been included as the entropic term.

The alternative pathway from the unfolded state may be considered, in which the chain first inserts and then forms a helix. To obtain free energies for the process, we consider the combination of chain entropy effects and hydrogen bonding, and obtain an approximate value of −70 kcal/mol for the folding of the chain in the nonaqueous environment. Completing the cycle then gives a value +40 kcal/mol for the process of moving the chain from the aqueous to the nonaqueous environment without folding it. It is clear that the process for insertion of a random polypeptide chain is highly unfavorable, and that some folding that results in the formation of hydrogen bonds must occur prior to the entry of the polypeptide into the nonaqueous environment.

energies, so the unfavorable hydrogen bond energies are even larger than the total. Thus, we conclude that a polypeptide coil cannot be inserted into the bilayer and then fold, but rather a secondary structure must form before insertion into the bilayer (21, 22). This is the major argument favoring the existence of helical structures in membranes: Partially assembled, hydrophobic helices are energetically favored to insert into the bilayer whereas random coils or partial β-sheets (e.g. a beta hairpin) are not.

Although in the nonpolar interior of soluble proteins peptide backbone hydrogen bonds are satisfied by the formation of either α-helix or β-sheet structures, we expect that the helix will be found to be the dominant secondary structure in lipid bilayers (21, 31, 85, 86). Obviously, a single crossing of the lipid bilayer can only be achieved by a helix if all H-bonds are to be satisfied. We have previously argued (21) that the requirement of cotranslational insertion (70, 81) and folding of globular membrane proteins into the lipid bilayer limits the possible secondary structures that can be inserted to a helical hairpin in most cases. One can imagine that pairs of amphipathic but hydrophobic helices might be stable both in an aqueous environment where they are synthesized and in the bilayer where they are assembled into protein. In aqueous solution, the more polar faces of the helix pairs can face water, whereas in the bilayer the helices can rotate to face the polar groups inward, away from lipid (22).

If the alternative of β-sheet structure is considered, it is clear that progressive insertion during protein synthesis would be problematical. The beta strands would have to be inserted as hairpins or single strands in which many hydrogen bond donors and acceptors would be left unsatisfied. While a β-barrel can be imagined as a structural alternative to helices (e.g. 31), the entire barrel would have to form in solution prior to its insertion into the bilayer. Folding in solution requires that many hydrophilic residues be outside the barrel whereas stability in the membrane environment requires the reverse. It may be that different conditions of polarity, such as the creation of large aqueous channels, permit alternative structures of this kind (80, 83). Nonetheless, the use of helical structures as an efficient strategy for progressively satisfying hydrogen bond requirements in nonaqueous environments leads to the expectation that helices are major constituents of membrane protein organization.

Experimentally Observed Membrane Protein Secondary Structures

Present structural data on four polypeptides support the existence of transbilayer helices. At moderate resolution the bacteriorhodopsin structure shows the presence of seven transmembrane rods that have the appropriate dimensions and packing to be α-helices (32, 52). Spectroscopic studies suggest the presence of a large amount of helix (58), and the prevalent interpretation is that the structure contains seven transbilayer helices. The other three polypeptides are subunits of the photosynthetic reaction center. The recent determination of the reaction center structure has led to the conclusion that two of the four proteins contain five transmembrane helices each and a third subunit contains a single transmembrane helix (14, 15). These appear to be the only structures traversing the lipid bilayer, although the position of the bilayer is inferred from the structure in the crystal. In the case of the photosynthetic reaction center the structure is known at high resolution and is unambiguous. Thus, it can be argued that structural data support the presence of 18 helical segments in three globular and one anchored membrane protein. These helices can usefully serve as tests of procedures for defining transmembrane segments (see below).

It is known that other kinds of transmembrane structures exist. Studies of matrix porin from *Escherichia coli* outer membranes suggest very strongly that β-sheet is the dominant secondary structural feature (18, 80, 83). An assembly of porin molecules forms an aqueous channel through the lipid bilayer. As the channel is large and can accommodate many polar groups, there are additional possibilities for a suitable structure that can be assembled into the bilayer. These possibilities do not exist in a globular membrane protein that is surrounded by lipid and that does not contain an aqueous channel (or an anchored protein). We confine our discussion to the prediction of helical structures that can be tested using the set of globular and anchored membrane protein structures.

DETERMINATION OF HELIX LOCATIONS FROM SEQUENCE DATA

If one assumes that transmembrane α-helices are present in the structure of an integral membrane protein, methods to locate them in the protein sequence on theoretical grounds would be very useful. Many methods for evaluating polarity in amino acid sequences have been developed and applied (2, 19, 20, 22, 50, 78, 88, 93, 94) following the original approach of Rose & Roy (76, 79). Each uses a progressive analysis in which successive regions of the polypeptide are evaluated with respect to some scale of polarity as shown in Figure 2. In the following sections we examine the polarity scales to determine which are appropriate, consider the choice of window length for the moving analysis, and present tests of the significance and success of the predictions. All the methods have as their goal the identification of amino acid sequences that are sufficiently hydrophobic

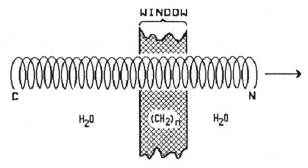

Figure 2 Schematic representation of the free energy computation for locating helices that are stable as transmembrane structures. The amino acid sequence of a protein is arranged as a continuous α-helix and is moved through a nonaqueous window. For each segment of the polypeptide chain the free energy for transferring the segment from the aqueous to the nonaqueous environment is calculated. The free energy transfer is plotted versus the N-terminal amino acid in the segment under consideration. In order to represent favorable insertions as peaks in the graph, the sign for the free energy is reversed, representing the transfer from the nonaqueous to the aqueous environment.

and sufficiently long (≥ 20 amino acids) to imply the existence of a transmembrane helix.

To frame the discussion of polarity scales we discuss in some detail the scale that we have developed during the past several years (19, 21, 22, 88), partly because we believe it is the most appropriate scale and partly so we can use it as a point of reference in our discussion of other scales and approaches. We assume that various side-chain components may be considered separately, that the details of helical structure are important in establishing an appropriate scale, and that the bilayer interior is a region of dielectric constant 2 containing no hydrogen bond donors or acceptors.

Appropriate Scales of Hydrophobicity for Bilayers and Protein Interiors

There is considerable diversity of opinion concerning the appropriate choice of polarity scale. Scales have been developed on the basis of solubility measurements (13, 29, 48, 63, 90), vapor pressures of side-chain analogs (33, 100–102), and analysis of side-chain distributions with soluble proteins (11, 29, 39, 40, 74, 79, 99). The use of side-chain distributions is complicated by the fact that hydrophobic residues are frequently found on protein aqueous surfaces (72) and by the fact that side chains span regions of different polarity (29, 77). The partition and vapor pressure measurements differ in that different assumptions are made concerning an appropriate analog of the protein dielectric interior. These issues have been extensively discussed in recent articles (12, 29, 77, 78). It is not surprising that the interior of a protein presents difficulties for modeling. The dielectric environment is extremely nonuniform, being influenced by the presence of many polar groups and hydrogen bond networks; the use of a bulk dielectric constant cannot represent its detailed fluctuations. It may prove necessary, as Guy (29) suggests, to consider a more detailed view of protein structure involving a distinction between the deep interior and the surface regions of a protein.

The hydrophobic environment in a membrane interior is simple compared with a protein interior. The hydrocarbon chains create a comparatively uniform, nonpolar environment. As the environment presents no hydrogen bond donors or acceptors, and as its dielectric constant is lower than that of a protein interior (59), scales developed from the examination of soluble, globular proteins would seem inappropriate for investigating amino acid side chains exposed to the lipid environment. Transfer free energy experiments based on the solubility of compounds in water and a nonpolar solvent analogous to a lipid bilayer are confounded by the very low solubility of even moderately polar compounds in media having dielectric constants of 2. The small number of solved structures creates an inadequate data base for the kind of statistical treatment used in categorizing side chains in globular soluble proteins. A promising approach is examination of the partitioning of compounds between an aqueous phase and the vapor state (33, 102). An alternative approach is the use of theoretical and experimental values for components of each amino acid side chain to derive a polarity scale (19, 21, 22, 93, 95). These alternatives are discussed further below.

A Polarity Scale for Identifying Transmembrane Helices

The arguments that led to the development of the Goldman, Engelman, Steitz (GES) hydrophobicity scale (19, 21, 22, 88) are outlined below. The development is rather similar to that of Von Heijne's early work (93, 94) but differs in some important details that are discussed later.

The major energetic factors favoring the partitioning of an amino acid side chain from aqueous solution into a membrane bilayer are hydrophobic interactions; those factors favoring its solution in the aqueous phase are interactions of polar and charged side chains with water. In order to make a quantitative estimate of the relative energies involved, the free energy of transfer of both the hydrophobic and hydrophilic components of each amino acid from water into oil were assigned. In order to consider the specific case of an α-helical polypeptide in a low-dielectric environment it is important that the scale be specifically adapted to the details of such a structure. This presents a dilemma. Since experimental scales have previously been based on the properties of individual amino acids in solubility measurements (13, 29, 48, 63, 90) or in transfer to the vapor phase from water (33, 100–102), they do not specifically address the circumstance of amino acids in helices. However, no experimental scale has been developed for helical structures.

We therefore developed a mixed scale (19, 21, 22, 88) in which the nonpolar properties of the amino acids as they exist in a helix were calculated using a semitheoretical approach that combines separate experimental values for the polar and nonpolar characteristics of groups in the amino acid side chains. This procedure, in essence, divides amino acids more finely than a simple consideration of main chain versus side chain characteristics.

Initially (19, 21, 88) we assumed an average hydrophobicity for a 20-residue α-helix based on the surface area of a typical helix. To the favorable baseline hydrophobicity of − 30 kcal/mol of helix we added the unfavorable energetic contribution arising from burying the various polar and charged residues. Use of this scale on the sequence of bacteriorhodopsin showed seven plausible hydrophobic regions (88). This scale was modified (22; Goldman, unpublished, 1982) to calculate the hydrophobicities for each of the 20 amino acids as they occur in an α-helix.

The hydrophobic component (Table 1) of the free energy of water-oil transfer can be calculated from the surface area of an amino acid side chain in an α-helix (9, 51, 71). Hydrophobic interactions tend to reduce the nonpolar surface area in contact with water. Their approximate magnitude has been obtained by measuring the partitioning of compounds between water and nonpolar solvents. The hydrophobic free energy thus measured has been shown (9, 71) to correlate linearly with total surface area in contact with water (51). Thus, calculation of the total contact surface area of a polypeptide that can be removed from interaction with water leads to an approximate value for the hydrophobic transfer free energy. We have used the surface area computations of Richmond & Richards (73) to obtain the surface area for each amino acid as it would be exposed in an α-helix of polyalanine. (The solvent-accessible surface varies somewhat in actual cases, depending on the neighboring residues, but this is a second-order effect.) The surface areas could then be converted into hydrophobic free energies (Table 1). In this way the experimental free energies of transfer can be adapted to the specific case of amino acids in an α-helix.

The free energy for inserting charged groups into a bilayer can be considered as having two components: the energy required to produce an uncharged species by protonation or deprotonation, and the energy required to partition the uncharged but polar portions of side chains from water to the nonaqueous phase (21). Our calculations using the Born approximation (5) showed that the transfer of a formal charge from an aqueous to a nonaqueous phase requires very substantial energy, probably on the order of 40 kcal/mol (21). Recent calculations by Honig & Hubbell (35) have led to a similar conclusion. The alternative of producing the uncharged species and partitioning it requires 10–17 kcal/mol (21, 35). Therefore, we consider that potentially charged amino acids (glutamic acid, lysine, aspartic acid, histidine, and arginine) will be transferred as the uncharged species. If we assume that the process occurs at or near neutrality, we can calculate the energy required for protonation or deprotonation by assuming a standard pK and a requirement for 99% conversion to the uncharged species. The energies obtained are included in the hydrophilic energies listed in Table 1.

There are also energy costs associated with the transfer of uncharged polar groups. These energies arise principally from the participation of side

Table 1   Transfer free energies for amino acid side chains in α-helical polypeptides[a]

	Hydrophobic	Hydrophilic	Water-oil
Phe	−3.7		−3.7
Met	−3.4		−3.4
Ile	−3.1		−3.1
Leu	−2.8		−2.8
Val	−2.6		−2.6
Cys	−2.0		−2.0
Trp	−4.9	3.0	−1.9
Ala	−1.6		−1.6
Thr	−2.2	1.0	−1.2
Gly	−1.0		−1.0
Ser	−1.6	1.0	−0.6
Pro	−1.8	2.0	0.2
Tyr	−3.7	4.0	0.7
His	−3.0	6.0	3.0
Gln	−2.9	7.0	4.1
Asn	−2.2	7.0	4.8
Glu	−2.6	10.8	8.2
Lys	−3.7	12.5	8.8
Asp	−2.1	11.3	9.2
Arg	−4.4	16.7	12.3

[a] Values are given for the hydrophobic and hydrophilic components of the transfer of amino acid side chains from water to a nonaqueous environment of dielectric 2. The hydrophobic term is based on a treatment of the surface area of the groups involved. The hydrophilic term principally involves polar contributions arising from hydrogen bonding interaction. Also included in the hydrophilic term is the energy required to convert the charged side chains to neutral species at pH 7 (19, 21, 22, 88).

Table 2 Approximate water-oil transfer free
energies for various groups[a]

Group	G (H_2O-Oil)
—OH	4.0
—NH_2	5.0
—COOH	4.3
C=O	2.0

[a] Values are derived principally on the basis of
observations using nonpolar oils. The studies on
which they are based are summarized by Davis (7).

chain groups in hydrogen bonds with water. It is difficult to treat the
hydrogen bonding potential explicitly; one must rely to a large extent on
experimental measurements based on the solubility of various compounds.
From extensive reviews of the data (13, 91) we conclude that the energies
required for transfer of polar groups from water to oil are approximately as
shown in Table 2.

Additional important specific considerations emerge regarding α-helices.
Serine and threonine in α-helical segments of proteins are known to
participate in shared hydrogen bonds with the backbone carbonyl groups
(27). Such sharing reduces the free-energy contribution opposing transfer
from the aqueous environment to the nonpolar region of the membrane. A
further consideration is the interaction of groups along the helical axis,
which is discussed below.

The contributions from different polar interactions were combined for
the hydrophilic term in Table 1. The net transfer free energies are the sum of
hydrophobic and hydrophilic components for each amino acid. The scale
uses a finer division of properties than some other scales have employed,
treating the contributions of hydrophobic surfaces and individual side-
chain polar groups separately. Its strengths are that it specifically addresses
the issue of helical structure and that it is based on a transfer from an
aqueous to a low-dielectric hydrocarbon region. In our discussion of other
approaches below, the GES scale is taken as a point of comparison.

Comparison with Other Polarity Scales

We now concentrate our attention on other scales that have been developed
for the examination of transbilayer helices. Of greatest importance in our
discussion are the scales of Von Heijne (94, 95) and Kyte & Doolittle (50)
and a scale based on partitioning between water and the vapor phase (33,
100). Other scales have emerged from an examination of partitioning of
amino acids into protein interiors or into more polar solvents such as

alcohols, e.g. the scales of Nozaki & Tanford (63), Rose & Roy (79), Guy (29), and Janin (39); these scales are not considered in detail for the reasons presented above, the main point being that partitioning of side chains from water to a protein interior is not equivalent to partitioning of side chains from water to a lipid environment.

Figure 3 shows a comparison of the GES scale (21, 22, 88) with that of Von Heijne (the VH scale). Here we use Von Heijne's revised scale (93, 94) since the original scale (95) had several incorrect chemical assumptions. The GES and VH scales correspond rather closely, with the exceptions of threonine, serine, proline, and lysine. In the cases of serine and threonine the differences are accounted for by the consideration in the GES scale of hydrogen bonds between side chains and main chains within an α-helix.

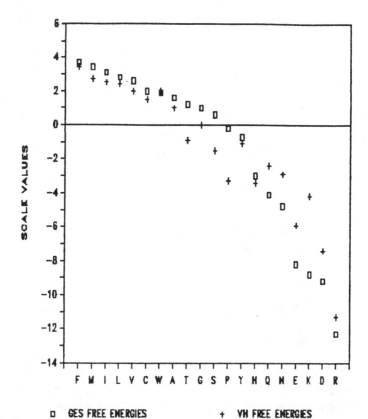

□ GES FREE ENERGIES   + VH FREE ENERGIES

Figure 3 Comparison of the GES (19, 21, 22, 88) and VH (93, 94) scales. □: GES free energies; + : VH free energies. Free energies are represented on the vertical axis in kcal/mol for the transfer of each side chain from a nonaqueous to an aqueous environment. The scales are generally similar with the exception of Ser, Thr, Pro, Lys.

Proline is given a much more polar character in the VH scale, as it is assigned two hydrogen bonds with water. This is incorrect since the only hydrogen bonding group is the main-chain carbonyl group that is not satisfied because of the closure of the imino ring. On the other hand, lysine is given a more nonpolar character by Von Heijne than aspartic or glutamic acid. Although lysine is more hydrophobic (by ~ 1 kcal/mol) than aspartic or glutamic acids, its pK is further removed from neutrality, giving rise to a hydrophilic component that eliminates the difference in our view (at pH 7.0). With these qualifications, the VH scale is, on the whole, rather similar to the GES scale.

A much used scale is that proposed by Kyte & Doolittle (the KD scale) (50). In an extensive and carefully reasoned article they examined a number of alternative polarity scales. A combination of scales based on the observed behavior on partitioning from the aqueous environment to protein interiors and on water-vapor partition gave the best agreement with known cases from soluble, globular proteins. While this scale has many virtues, it is clear that the model under consideration does not address accurately the conformational and environmental aspects we have discussed above.

In Figure 4 the scale derived by Kyte & Doolittle is compared with the GES scale. The most striking difference is that the polarities of aspartic acid, glutamic acid, lysine, and arginine are not as strong in the KD scale as in the GES scale. This is a consequence of the scaling procedures used by Kyte & Doolittle to merge different scales in their analysis. As in the case of the VH scale, the contributions of threonine and serine are considered more polar than we think appropriate. Further, tryptophan is considered a substantially or partially polar amino acid because of the ring nitrogen. While this may explain why tryptophan is predisposed to orient near interfaces, the polarity seems inappropriate in terms of the overall nonpolarity of the side chain. In general, the KD scale is in reasonable although not detailed agreement with the GES scale on the matter of the hydrophobic amino acid side chains. The differences, however, have important consequences in the prediction of transmembrane helices in cases where polar or potentially charged groups are in regions traversing the membrane (see below).

A significant contribution has been provided by measurements of water-vapor partition coefficients for model compounds containing amino acid side chain components. Since the vapor state does not provide hydrogen bonding groups, it would seem to be a good choice as an analog for a nonpolar bilayer interior. Free energies derived from these measurements (33, 100) have been merged and corrected by Kyte & Doolittle (50) to give a vapor-water transfer free energy scale (the VW scale).

Figure 5 shows a comparison of the VW scale and the GES polarity scale.

Some very large differences are evident. In developing their hydropathy scale, Kyte & Doolittle found it reasonable to adjust a number of the VW values based on chemical arguments. For example, because of the nonpolar character of its side chain, phenylalanine would be unlikely to have an equal probability of being found in an aqueous environment and in a membrane or protein interior. Similarly, on this experimental scale methionine is found to have a slightly polar character, in contradiction to its occurrence in the interior of known proteins and the apparently nonpolar character of its side chain. Also, cysteine is given a polar character, as is tryptophan. The cases of threonine and serine are interesting; the magnitudes of polarity are in agreement with the presumed value for the solvation of a hydroxyl group given in Table 2. Of course, as with the scales previously discussed the

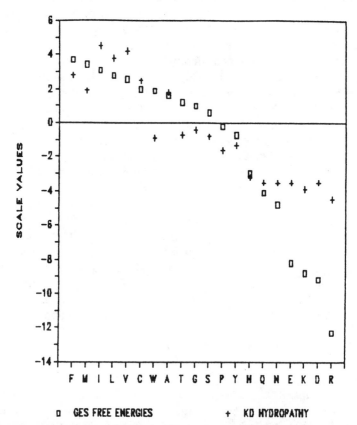

□ GES FREE ENERGIES   + KD HYDROPATHY

Figure 4   Comparison between the GES scale and the KD hydropathy scale (50). Values for the scales are on the vertical axis (free energy for the GES scale and the hydropathy for the KD scale).

structural assumptions regarding a free amino acid versus an amino acid in a helix alter the view of threonine and serine as polar amino acids. Rather striking is the extreme polar character accorded tyrosine using the transfer energy measurements. It is possible that interactions in the vapor phase, such as the dimerization of carboxylate groups, will distort estimates of the transfer energy. Furthermore, the present data do not give values for glycine, proline, or arginine.

While the VW scale appears useful a priori, the measurements that have so far been made using water-vapor transfer have resulted in somewhat perplexing conclusions concerning polarity. Kyte & Doolittle felt compelled to modify the direct conclusions from the transfer measurements and to reset the point on the scale at which zero transfer free energy is located.

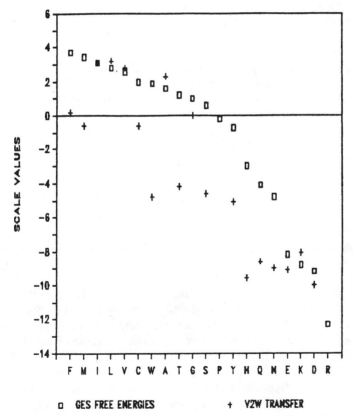

Figure 5   Comparison of GES and VW scales (29). The vertical axis represents transfer free energy in kcal/mol.

Rose et al (77) have noted that these values are not in agreement with other measures of polarity or with the observed occurrence of amino acids in protein interiors. We agree that some features of the scale derived from transfer free energies are surprising and that the scale may not be useful in efforts to predict transmembrane segments of polypeptide chains.

Polarity scales based on transfer of amino acids from water to various alcohols have been widely used (33, 63). Guy (29) has summarized the results from a number of experiments of this kind and has put them on a common scale for alcohol transfer. A striking fact is that the polar amino acids are assigned values near zero on this scale. Thus, virtually any polypeptide would be predicted to partition into a nonpolar phase based on this analysis. Clearly a restructuring would be needed for such a polarity scale to be useful in the kinds of scanning procedures under consideration in this article. Moreover, the arguments we have made concerning the suitability of alcohol partitioning measurements suggest that the scale would be inappropriate for this application.

Overview of Polarity Scales

In the foregoing discussion the GES scale was elaborated and compared with other approaches for the specific case of amino acid side chains in nonpolar environments. To apply any of these scales to the identification of transmembrane helices requires additional considerations and a computational approach. By applying the scales to prediction problems, differences in their properties become apparent and arguments concerning suitability are clarified. The application of scales is discussed below.

USE OF SCALES TO IDENTIFY TRANSMEMBRANE HELICES

The general approach to identifying transmembrane helices that has emerged in several publications is to use a scanning procedure by which an amino acid sequence can be progressively evaluated in terms of its polarity and hence its tendency to form transbilayer helices. Progressive analysis was first used in the study of globular, soluble proteins (76). In early versions, such methods invoked a smoothing algorithm applied to the detailed, residue-by-residue values of polarity or hydrophobicity. More recently, the approach taken for studies of membrane proteins has been to use a window scan of amino acid sequences (e.g. 21, 50, 88, 94). In a window scan, the sum of hydrophobicity or polarity values for a number of amino acids is taken progressively through the sequence (Figure 2). A plot is made of the position of a reference amino acid, either the first or the middle amino acid in the region being summed, versus the value of the summed polarity or average polarity in the region. The use of such scans involves several choices of approach. These include attention to factors that arise as a consequence of the detailed structural model under consideration as well as choice of the window length appropriate for the search.

Energetic Consequences of Helical Conformation

In examination of potential α-helical structures, specific considerations of structural details are important. Two of these are the polar interaction of side chains along the helical structure and the different exposure of side chains to solvent when a helix is compared to isolated amino acids or extended chains. If nearby side chains have the potential to interact in the nonpolar environment, their interaction will modify the energy calculation for a transfer from the aqueous to the nonaqueous environment. For example, if the side chains of aspartic or glutamic acid are located one turn of a helix away from the side chains of arginine or lysine, interactions are possible and expected. Whether such interactions actually involve the formation of an ion pair or the formation of a strong hydrogen bond (1a) is an issue that cannot be addressed with the present information. The reader is referred to the excellent article by Honig & Hubbell in which the issues of group interactions in a nonpolar environment are treated (35) and to the review in this volume (36). These articles conclude that the energy required to transfer polar groups as ion pairs or as strongly hydrogen-bonded structures is certainly less than that required to transfer the groups separately.

At issue is the question of how much less energy is needed to transfer ion pairs. There are examples in protein structures in which it appears that the energy needed to transfer an ion pair from the aqueous environment to a protein interior may be very small. Benzamidine binds strongly to the catalytic pocket in trypsin. In this case it appears that the cost of forming the internal ion pair between the amide and a carboxyl group in the active site is very small (4). On the other hand, the treatment of Honig & Hubbell suggests that 10–15 kcal may be required to move a carboxyl and amino group as an interacting pair from the aqueous to the nonaqueous environment. It is therefore appropriate to include a term in the scanning procedure to allow for the interaction of polar groups. We have suggested (19, 88) that the value of this term might be 5–10 kcal/mol. The exact value is not known at present. However, some reduction of the energy requirement of the groups taken separately is appropriate where the groups are located 1,4 or 1,5 in the amino acid sequence. In the calculations presented below, the GES scale includes 10 kcal/mol as the contribution from paired amino and carboxyl groups along a helix.

Entropy of Immobilization

An additional factor that must be considered in the computation of an energy profile is the entropy of immobilization involved in moving a macromolecule from a solution to a lipid bilayer. Of the six degrees of freedom a molecule has in solution, three are restricted by binding to the lipid bilayer. In the case of a loss of all six degrees of freedom in enzyme substrate interactions, the extreme value for the entropy of immobilization is thought to be about 20 kcal/mol (65a). The loss of one translational and two rotational degrees of freedom would reduce this value to about 10 kcal/mol (38). The fact that the macromolecule is not totally immobilized with the lost degrees of freedom (owing to the fluid character of the lipid bilayer) means that some further reduction is in order. We have adopted the use of 5 kcal/mol as the unfavorable free energy term, which represents the immobilization of a polypeptide chain binding to a lipid bilayer.

Choice of Window Length and Scanning Procedures

In choosing a window length for sequence analysis to locate transmembrane helical structures, two factors are important: the hydrophobic width of the bilayer itself and the orientation of a possible helix with respect to the bilayer plane. Progressive sequential analysis requires some decision as to the length of sequence that will be examined at each step. In early analyses windows as short as 7 (50) and as long as 20 amino acids (88, 19) were used. Others have adopted smoothing procedures or have used model functions to smooth the erratic behavior of small averaging windows.

The hydrophobic thickness of a lipid bilayer may vary considerably depending on the composition of the lipid fatty acyl chains and on the content of cholesterol. It has been shown, for example, that the hydrophobic thickness of the bilayer is proportional to chain length for fluid phosphatidylcholine bilayers formed from a series of phosphatidylcholines with different fatty-acid chain lengths. Thus, the thickness can vary by more than a factor of two (54). In choosing the length of the test window it would be optimal if one knew the hydrophobic thickness of the particular bilayer into which a protein was to be inserted. Nonetheless, a typical value for many lipid bilayers is of the order of 30 Å. For an α-helix to span a 30 Å distance, 21 residues are required because the interval between residues along the helix axial direction is 1.5 Å.

If the helix is tilted with respect to the bilayer plane, a longer helix can be accommodated in the hydrophobic region. It may be that a protein that consists of many helices contains helices of different tilts, and that a series of test window dimensions can reveal the presence of more extensive hydrophobic helices. A final point is that lipid bilayers in the fluid state

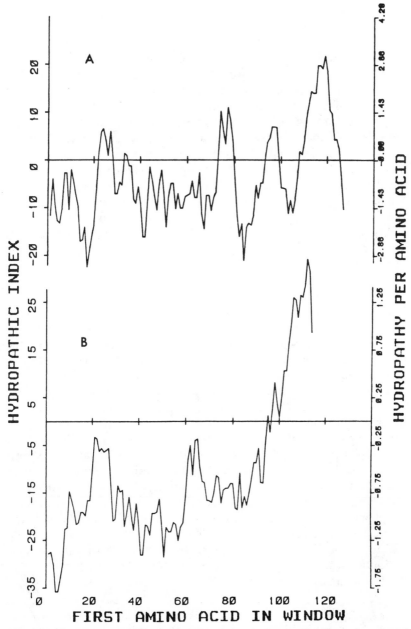

Figure 6   The sequence of cytochrome b5 (87) is analyzed using the KD hydropathy scale and windows of 7 (*A*) and 20 (*B*) amino acids. When a window of 7 is used, several important peaks occur in the profile. A lower background prediction is obtained when a 20-amino acid window is used, clearly contrasting with the experimentally known C-terminal anchoring region of the protein.

appear to be readily distorted (55, 66). Thus, the presence of a helix with a nonpolar dimension that does not match the hydrophobic thickness of the bilayer may be accommodated through distortion of the bilayer thickness.

The above discussion suggests that a reasonable choice for the test window is on the order of 20 amino acids, but that no unique number can be readily assigned. It would appear that short windows (on the order of 10 amino acids) or long windows (on the order of 30 amino acids) are unlikely to be optimal choices. It is possible that an inappropriate choice of window may give misleading results. Figure 6 shows the sequence of cytochrome b5, a protein that is anchored by its hydrophobic carboxyl terminus to the membrane of the endoplasmic reticulum (87). The KD scale was used with a window of 7 amino acids as Kyte & Doolittle originally specified (50). A peak is seen corresponding to the hydrophobic carboxy terminus, but additional peaks also appear. If the window is instead set at 20 amino acids, it becomes clear that the truly significant hydrophobic feature is the carboxy terminus and that the other peaks do not extend above zero.

Comparison of Scales

If a membrane-traversing region is extremely hydrophobic in character, virtually any scale will reveal its presence. Difficulties arise, however, in cases in which a helix contains some polar amino acids. Since the partitioning of helices is so strongly favored by the presence of exclusively nonpolar amino acids, stable structures are possible in which one or more amino acids in the middle of an otherwise nonpolar helix have strongly polar character (21). An example of this kind is provided by bacteriorhodopsin.

Figure 7 shows analyses of the bacteriorhodopsin sequence (46, 64) using the KD scale, the GES polarity scale, and the VW scale. The analyses shown in Figure 7 *B, C,* and *D* were each carried out with a window of 20 amino acids. It is evident that the analysis using the KD scale leads to a clear identification of only two helices; the other five expected helices are much less plainly revealed except as broad maxima. The VW scale gives only five peaks, and the values are radically shifted so that the free energies would lead to the prediction that the structures are unstable with the possible exceptions of one helix. Using the GES scale, seven distinct peaks separated by clear minima are observed.

Figure 7   Different analyses of the sequence of bacteriorhodopsin. *A* and *B* show the effect of using the KD scale and windows of 7 and 20 amino acids respectively. The appropriate choice of 20 does not reveal many of the helices nor does the choice of 7. *C* and *D* show the application of the VW and GES scales respectively, each with a window of 20. The VW scale, while revealing many of the helices in profile, appears far too negative in predicting stability. The GES scale, on the other hand, shows seven well-defined maxima which are thought to correspond to the seven helices present in the structure (see Figure 12).

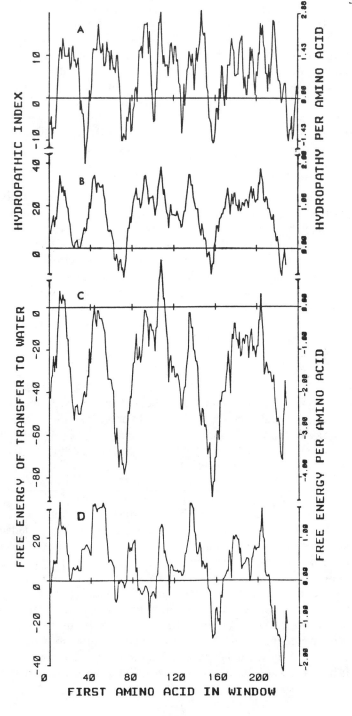

Figure 8 shows analyses of the sequence of glycophorin (92), which is known to span the red cell membrane (6), using the three scales and a window of 20 amino acids. In this case, where a very strongly nonpolar helix appears to be present, each scale gives a clear identification.

All these scales give, as expected, very similar results for the cases in which helices can be identified on the basis of inspection for nonpolar amino acids. The more difficult task of identifying helices that contain polar amino acids or that are not separated by clearly or strongly polar regions appears to be best accomplished using the GES scale.

Significance of Peaks in the Sequence Analysis

Given a choice of scale and window, the question arises of how peaks in the analysis are to be interpreted. An interesting test of the magnitude required for a peak to be biologically significant uses a series of deletion mutations in the anchoring peptide of the vesicular stomatitis viral coat protein (1). This protein appears to be anchored by a single 20–amino acid membrane-spanning sequence near its carboxy terminus (42, 75). Using genetic techniques several altered forms have been produced in which the hydrophobic region of the presumed anchor sequence has been varied in length and the cellular location of the modified protein has been determined.

Figure 9 shows the relative membrane stabilization calculated for the different modified coat proteins using a window of 20 amino acids and the GES scale. Rose and colleagues have determined the disposition of the different modified proteins (1) and have found that reduction of the anchoring sequence to 8 amino acids does not anchor the protein. However, proteins with a hydrophobic sequence of 14 or more amino acids are clearly anchored. With an anchor sequence of 12 amino acids, the protein appears to bind well in cytoplasmic membranes but only sparingly in the plasma membrane (which may be thicker). We can therefore say from inspection of Figure 9 that the peak corresponding to 20 kcal/mol appears to correlate with stable insertion and anchoring of the membrane protein.

It is not the case, however, that all proteins containing hydrophobic sequences identified in this way are membrane-spanning or, indeed, even membrane-associated proteins. An example is the sequence of trypsinogen (30), which is analyzed in Figure 10. Here a clear hydrophobic stretch is identified that, were it known to be a membrane protein, would be suspected as a transmembrane segment. Trypsinogen is, of course, a secreted, soluble protein. It is hazardous to assume that proteins that show peaks of about 20 kcal/mol must be integral membrane proteins.

We have confined our attention to prediction methods using polarity scales to identify nonpolar helices. Additional transmembrane structures may be found where the constraints are different, as in assemblies that form

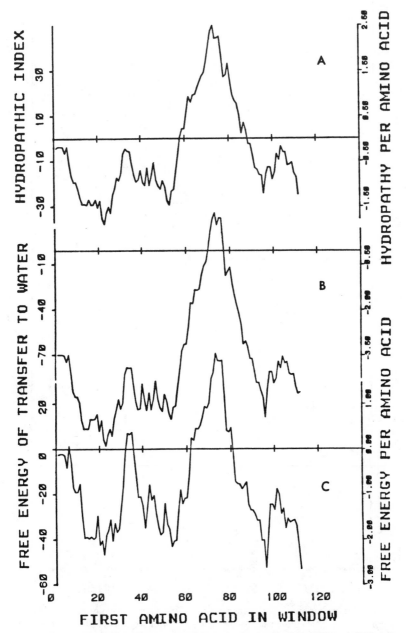

Figure 8   The sequence of glycophorin (92) is examined using the KD, VW, and GES scales. Each scale reveals the transmembrane region of the polypeptide.

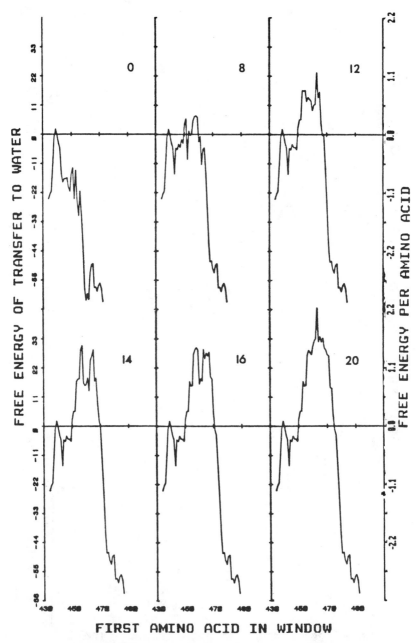

Figure 9 Profiles for the anchoring segment of the VSV G protein in progressively shortened versions (1). The VSV G protein has a membrane-anchoring sequence of 20 amino acids, which has been progressively shortened by genetic modification (1). Shown are the complete deletion of the 20 amino acid region (1), and anchoring segments of 8, 12, 14, 16, and the native 20 amino acids. For successful anchoring in cytoplasmic and plasma membranes 12–14 amino acids are required, so it appears that an anchoring energy of about 20 kcal/mol is sufficient.

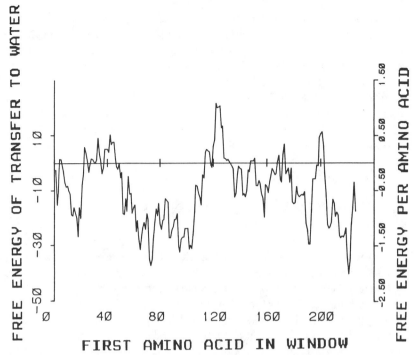

Figure 10   The free energy profile for trypsinogen (4) determined using a window of 20 amino acids and the GES scale. Note that a strongly hydrophobic peak exists near the significance level.

aqueous pores. In recent years there have been a number of efforts to examine the possible presence of amphiphilic helices that may provide polar pores (17, 24, 28). These have led, for example, to detailed models for the disposition of chains in the acetylcholine receptor (24, 26, 68). Some of the predictions have been criticized on statistical grounds (25). A major difficulty is how to distinguish whether a potential amphiphilic helix exists in the bilayer as part of a pore or in a soluble globular domain, since most helices in soluble proteins are amphiphilic (82). While these efforts may reveal additional aspects of membrane protein structure, we cannot test these aspects in the absence of well-established structural observations; consequently, our decision has been to set them aside until their veracity can be tested experimentally.

Segments of an amino acid sequence that form two closely spaced helical regions with a turn between them may not be readily identified as a helix pair by any of the procedures described above. If the region between the helices contains no amino acids of strikingly polar character, the turn may not be revealed. This does not mean that the ends of the helices are nonpolar or that the turn is unstable. It is well known that in the region in which a helix ends, a fractional charge exists as a consequence of charge separation in the peptide bonds aligned along the helix (34, 84, 96). Furthermore, the hydrogen bond donors or acceptors are not satisfied by backbone acceptors or donors. Not only does this render the end of the helix strongly polar [and suggest that helix ends will not be found in the interior of membrane bilayers (21, 31)] but the polarity produced in this way would not be revealed by the progressive analysis employed in scanning sequences for the polar characteristics of the amino acid side chains present. The structure of the inserted portion of the cytochrome b5 molecule is not known, but a hairpin of helices would not be excluded merely by the fact that two distinct peaks are not seen in the analysis shown in Figure 7.

TESTS OF PREDICTIONS USING KNOWN STRUCTURES

While there are many proteins, such as the red cell membrane glycophorin (92), for which the transmembrane structure is strongly implied by a range of data (e.g. 6), in only a few cases is helical structure established with a high level of confidence. The best-established examples are found in the structure of the photosynthetic reaction center of *Rhodopseudomonas viridis*, which has recently been determined at high resolution (14, 15). The macromolecular assembly consists of four polypeptide chains; two of these (L and M) are globular integral membrane proteins and one (H) is an anchored membrane protein. While the sequences of the *R. viridis* proteins are not yet published, they are highly homologous (61; H. Michel, personal communication) to the sequences of photosynthetic reaction centers from other organisms such as *R. capsulata* (104). The crystal structure shows a region in which bundles of helices traverse an apparently nonpolar region. Although the structure was crystallized in the presence of detergent and not in the presence of phospholipid, the distribution of polar and nonpolar groups suggests that a defined region containing a number of helices spans the membrane. Using the published sequences of the *R. capsulata* subunits it is therefore possible to construct a test of the prediction methods.

All of the putative membrane-spanning helices observed in the crystal structure are predicted from the hydrophobicity analyses of the sequences. Figure 11 shows the sequence analysis for the L, M, and H subunits made using the GES scale and a window of 20 amino acids. Four helices each are

---

Figure 11   Sequences for the H, L, and M subunits of the photosynthetic reaction center. Sequences of *R. capsulata* (104), the GES scale, and a window of 20 amino acids were used to examine the structure. One transmembrane helix is predicted for the H subunit, and five for both the L and M subunits.

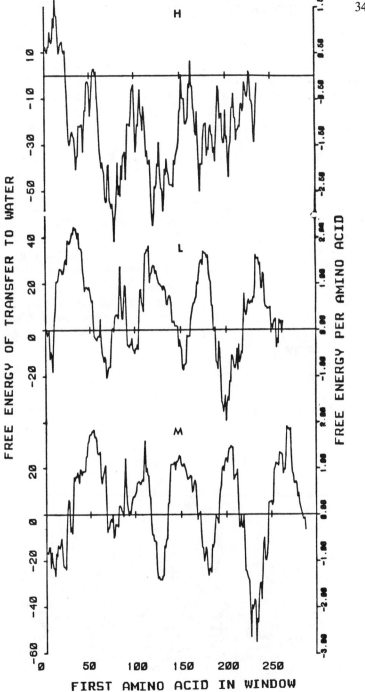

suggested by broad maxima in both the L and M cases, and a fifth helix is suggested by a relatively sharp maximum between the first two broad maxima. In the L subunit there is an additional peak that is at the margin of significance and is located near the first sharp peak. It does not correspond to a transmembrane helix in the structure. The H-subunit profile shows a single broad maximum suggesting a single transmembrane helix. The maxima from the polarity profiles were used for the predictions shown in Table 3.

Also in Table 3 are the positions of helices in *R. viridis* established from the crystal structure of Deisenhofer et al (14, 15). The agreement is striking. Of the 220 amino acids assigned by the polarity profile to 11 helices, it appears that only 2 amino acids lie outside of the helices that are actually found in the protein structures. The observed helices are actually somewhat longer than the scanning window of 20 residues. This is not surprising, since the actual helices may have hydrophilic extensions beyond the region of the nonpolar lipid bilayer. As these sequences contain very nonpolar regions with few polar amino acids, helix predictions are relatively insensitive to the choice of scale used (see above). The agreement between the predicted and established transmembrane helix location is striking and is highly encouraging for those who wish to apply prediction methods to membrane protein sequences.

Table 3 Comparison of predicted and observed membrane spanning helices in photosynthetic reaction centers[a]

Subunit	Helix	Predicted	Observed
L	A	32–51	32–55
	B	84–103	84–112
	C	116–135	115–140
	D	175–194	170–199
	E	233–252	225–251
M	A	52–71	52–78
	B	111–130	110–139
	C	148–167	142–167
	D	206–225	197–225
	E	267–186	259–285
H	A	12–31	12–37

[a] Predictions are based on the energy plots shown in Figure 11. They are based on the amino acid sequences from *R. capsulata* (104), which are known to be highly homologous to those of *R. viridis* (61). The structures of the subunits are known at high resolution for *R. viridis* (14, 15) and the transmembrane helices are located as shown. It is to be expected that actual helices may be longer than those predicted on the basis of spanning the nonpolar region of the lipid bilayer.

Although it is less well established, the structure of bacteriorhodopsin also provides a test of predictive methods. The structure is known to contain seven transmembrane helices (32, 52), and use of the current GES polarity scale on the sequence showed seven nonpolar regions (19, 22, 88), suggesting the locations of such helices in the amino acid sequence (Figure 7). The first hydrophobicity analysis of the bacteriorhodopsin sequence was performed using the earlier version of the GES scale (88). This analysis prompted a revision in the proposed positions of helices F and G in the sequence from the original model (20). While the application of the initial scale suggested locations for all seven helices, the current scale more convincingly delineates the existence and positions of helices C and G. Subsequent experiments have been consistent with and thus support the use of the computer-generated model. While the exact sequence locations of the helices in the actual structure are not known, a number of recent chemical-modification and protease-digestion studies narrow the possible locations substantially (8, 16, 26, 37, 41, 56, 65, 97, 98). There remains some ambiguity in the precise location of the short loop connecting helices F and G, but the water-accessible portions of the rest of the sequence are well defined. The regions that are predicted and those that are defined by various modification and digestion studies are compared in Figure 12. As with the case of the photosynthetic reaction centers, the agreement is excellent.

CONCLUSIONS

We conclude that the prediction of regions of integral membrane protein polypeptide sequences that span the nonpolar region of the membrane as helical structures is sound and useful.

Where membrane proteins have extremely nonpolar regions as transbilayer elements, virtually any scale of polarity can reveal their presence. On the other hand, if polar groups are present, scales that take into account the details of helical structure and transfer from water to a lipid bilayer interior are more successful in revealing important possible helices. Of the scales developed, the GES scale is most appropriate (see Table 1).

Scanning procedures should employ a window of a length approximating the hydrophobic dimension of a lipid bilayer. While this is a variable, a value of 20 amino acids is a reasonable choice.

Using studies with altered protein sequences, it is possible to establish that a peak of about 20 kcal/mol on the GES scale with a window of 20 amino acids is a significant feature. However, it is important to note that not all polypeptide sequences that contain such features have membrane-spanning helices; some soluble proteins may include features of this kind,

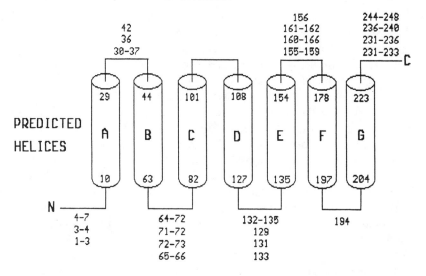

OBSERVED AQUEOUS MODIFICATIONS

Figure 12 Predicted and observed features of bacteriorhodopsin topology. Seven helices are predicted on the basis of the GES scale (19, 22, 88; Figure 7). The predicted hydrophobic regions are indicated on the presumed transmembrane helices, indicated by the amino acid sequence numbers. Modifications using reagents active in the aqueous phase should reveal the regions between predicted helices. Such reagents include enzyme, lactoperoxidase-catalyzed iodination, and antibody binding. Experimentally observed modifications are shown either as spans of amino acids in the case of enzyme cleavages or as single amino acids in the cases of modification or antigenic identification (8, 16, 26, 37, 41, 56, 65, 97, 98). The prediction of helix B has two possible extremes; that which is preferred on the basis of experimental observation is shown. Some debate concerning the location of helix F continues, and recent antibody experiments suggest that the helix may be located a few amino acids toward the amino terminus from the location shown here. The loop between helix C and D is short, and may not be accessible to the reagents used.

and extramembrane domains of membrane proteins may also include such structures. Caution is therefore recommended in the absence of confirmatory evidence.

The most striking observation, however, is that all of the known transbilayer structural elements in helical membrane proteins are accurately predicted by the polarity analysis we discuss in this article. It appears that the prediction of some secondary structural elements of membrane proteins may be, in this sense, more successful than that of proteins in the aqueous milieu.

ACKNOWLEDGMENTS

We are indebted to a number of people who have helped us with the manuscript, particularly to N. Stewart for her able assistance in its production and to H. Michel and K. Dill for useful discussions. This work was supported by the National Institutes of Health Grant GM-22778.

NOTE

The program written by A. Goldman to examine sequences as we have discussed is available upon request.

Literature Cited

1. Adams, G. A., Rose, J. K. 1985. *Cell* 41: 1007
1a. Allen, L. C. 1975. *Proc. Natl. Acad. Sci. USA* 72: 1401
2. Argos, J. K., Rao, J. K. M., Hargrave, P. A. 1982. *Eur. J. Biochem.* 128: 565
3. Blake, C. C. F., Koenig, D. F., Mair, G. A., North, A. C. T., Phillips, D. C., Sarma, V. R. 1965. *Nature* 206: 757
4. Bode, W., Schwager, P. 1975. *J. Mol. Biol.* 98: 693
5. Born, M. 1920. *Z. Phys.* 1: 45
6. Bretscher, M. S. 1971. *Nature New Biol.* 231: 229
7. Brown, J. E., Klee, W. A. 1971. *Biochemistry* 10: 470
8. Brunner, J., Franzusoff, A. J., Lüscher, B., Zugliani, C., Semenza, G. 1985. *Biochemistry* 24: 5422
9. Chothia, C. 1974. *Nature* 248: 338
10. Chothia, C. 1975. *Nature* 254: 304
11. Chothia, C. 1976. *J. Mol. Biol.* 105: 1
12. Chothia, C. 1985. *Ann. Rev. Biochem.* 53: 537
13. Davis, S. S., Higuchi, T., Rytting, J. H. 1974. *Adv. Pharm. Sci.* 4: 73
14. Deisenhofer, J., Epp, O., Miki, K., Huber, R., Michel, H. 1984. *J. Mol. Biol.* 180: 385
15. Deisenhofer, J., Epp, O., Miki, K., Huber, R., Michel, H. 1985. *Nature* In press
16. Dumont, M. E., Trewhella, J., Engelman, D. M., Richards, F. M. 1985. *J. Membr. Biol.* In press
17. Eisenberg, D., Weiss, R. M., Terwilliger, T. C. 1982. *Nature* 299: 371
18. Engel, A., Massalzki, A., Schindler, H., Dorset, D. L., Rosenbusch, J. 1985. *Nature* 317: 645

19. Engelman, D. M., Goldman, A., Steitz, T. 1982. *Methods Enzymol.* 88: 81
20. Engelman, D. M., Henderson, R., McLaughlin, A. D., Wallace, B. A. 1980. *Proc. Natl. Acad. Sci. USA* 77: 2023
21. Engelman, D. M., Steitz, T. A. 1981. *Cell* 23: 411
22. Engelman, D. M., Steitz, T. 1984. In *The Protein Folding Problem*, ed. D. Wetlauffer, p. 87. Boulder, Colo: Westview
23. Deleted in proof
24. Finer-Moore, J., Stroud, R. M. 1984. *Proc. Natl. Acad. Sci. USA* 81: 155
25. Flinta, C., Von Heijne, G., Johannson, J. 1983. *J. Mol. Biol.* 168: 193
26. Gerber, G. E., Gray, C. P., Wildenauer, D., Khorana, H. G. 1977. *Proc. Natl. Acad. Sci. USA* 74: 5426
27. Gray, T. M., Mathews, B. W. 1984. *J. Mol. Biol.* 175: 75
28. Guy, H. R. 1984. *Biophys. J.* 45: 249
29. Guy, H. R. 1985. *Biophys. J.* 47: 61
30. Hartley, B. S., Brown, J. R., Kauffman, D. L., Smillie, L. B. 1965. *Nature* 207: 1157
31. Henderson, R. 1979. *Soc. Gen. Physiol. Ser.* 33: 3
32. Henderson, R., Unwin, P. N. T. 1975. *Nature* 257: 28
33. Hine, J., Mookerjee, P. K. 1975. *J. Org. Chem.* 40: 292
34. Hol, W. G. J., van Duijnen, P. T., Berendsen, H. J. C. 1978. *Nature* 294: 532
35. Honig, B. H., Hubbell, W. L. 1984. *Proc. Natl. Acad. Sci. USA* 81: 5412
36. Honig, B. H., Hubbell, W. L., Flewelling, R. F. 1986. *Ann. Rev. Biophys. Biophys. Chem.* 15: 163
37. Huang, K. S., Bayley, H., Liao, M.-J.,

London, E., Khorana, H. G. 1981. *J. Biol. Chem.* 256:3802
38. Jahnig, E. 1983. *Proc. Natl. Acad. Sci. USA* 80:3691
39. Janin, J. 1979. *Nature* 277:491
40. Janin, J. 1979. *Bull. Inst. Pasteur* 77: 337
41. Katre, N. V., Finer-Moore, J., Hayward, S. B., Stroud, R. M. 1984. *Biophys. J.* 46:195
42. Katz, F. N., Rothman, J. E., Lingappa, V. R., Blobel, G., Lodish, H. F. 1977. *Proc. Natl. Acad. Sci. USA* 74:3278
43. Kautzmann, W. 1959. *Adv. Protein Chem.* 14:1
44. Kendrew, J. C. 1962. *Brookhaven Symp. Biol.* 15:216
45. Kendrew, J. C., Dickerson, R. E., Strandberg, B. E., Hart, R. G., Davies, D. R. 1960. *Nature* 185:422
46. Khorana, H. G., Gerber, G. E., Herlihy, W. C., Gray, C. P., Anderegg, R. J., et al. 1979. *Proc. Natl. Acad. Sci. USA* 76: 5046
47. Kim, P. S., Baldwin, R. L. 1982. *Ann. Rev. Biochem.* 51:459
48. Klein, R., Moore, M., Smith, M. 1971. *Biochim. Biophys. Acta* 233:420
49. Krigbaum, W. R., Komoriya, A. 1979. *Biochim. Biophys. Acta.* 576:204
50. Kyte, J., Doolittle, R. F. 1982. *J. Mol. Biol.* 157:105
51. Lee, B., Richards, F. M. 1971. *J. Mol. Biol.* 55:379
52. Leifer, D., Henderson, R. 1983. *J. Mol. Biol.* 163:451
53. Levitt, M. 1976. *J. Mol. Biol.* 104:59
54. Lewis, B. A., Engelman, D. M. 1983. *J. Mol. Biol.* 166:203
55. Lewis, B. A., Engelman, D. M. 1983. *J. Mol. Biol.* 166:211
56. Liao, M. J., Huang, K.-S., Khorana, H. G. 1984. *J. Biol. Chem.* 259:4194
57. Manovalan, P., Pomuswamy, P. K. 1978. *Nature* 275:673
58. Mao, D., Wallace, B. A. 1984. *Biochemistry* 23:2667
59. Matthew, J. B. 1985. *Ann. Rev. Biophys. Biophys. Chem.* 14:387
60. Meirovitch, H., Rackovsky, S., Scheraga, H. 1980. *Macromolecules* 13: 1398
61. Michel, H., Weyer, K. A., Gruenberg, H., Oesterhelt, D., Lottspeich, F. 1985. *EMBO J.* 4:1667
62. Nishikawa, K., Ogi, T. 1980. *Int. J. Pept. Protein Res.* 16:19
63. Nozaki, Y., Tanford, C. 1971. *J. Mol. Chem.* 246:2211
64. Ovchinnikov, Y., Adbulaev, N., Feigira, M., Kiselev, A., Lobonov, N. 1979. *FEBS Lett.* 100:219
65. Ovchinnikov, Y. A., Abdulaev, N. G.,

Vasilov, R. G., Vturina, I. Y., Kuryatov, A. B., Kiselev, A. V. 1985. *FEBS Lett.* 179:343
65a. Page, M. I., Jencks, W. P. 1971. *Proc. Natl. Acad. Sci. USA* 68:1678
66. Parsegian, V. A., Fuller, N., Rand, R. P. 1979. *Proc. Natl. Acad. Sci. USA* 76: 2750
67. Perutz, M. F., Muirhead, H., Cox, J. M., Goaman, L. C. G. 1968. *Nature* 219: 131
68. Popot, J.-L., Changeux, J.-P. 1984. *Physiol. Rev.* 64:1162
69. Privalov, P. L. 1979. *Adv. Protein Chem.* 33:167
70. Redman, C. M., Sabatini, D. D. 1966. *Proc. Natl. Acad. Sci. USA* 56:608
71. Reynolds, J. A., Gilbert, D. B., Tanford, C. 1974. *Proc. Natl. Acad. Sci. USA* 71: 2925
72. Richards, F. M. 1977. *Ann. Rev. Biophys. Bioeng.* 6:151
73. Richmond, T., Richards, F. 1978. *J. Mol. Biol.* 119:537
74. Robson, B., Osguthorpe, D. J. 1979. *J. Mol. Biol.* 132:19
75. Roe, J. K., Welch, W. J., Sefton, B. M., Esch, F. S., Ling, N. C. 1980. *Proc. Natl. Acad. Sci. USA* 77:3884
76. Rose, G. D. 1978. *Nature* 272:586
77. Rose, G. D., Geselowitz, A. R., Lesser, G. J., Lee, R. H., Zehfus, M. H. 1985. *Science* 229:834
78. Rose, G. D., Gierasch, L. M., Smith, J. A. 1985. *Adv. Protein Chem.* 37:1
79. Rose, G. D., Roy, S. 1980. *Proc. Natl. Acad. Sci. USA* 77:4643
80. Rosenbusch, J. P. 1974. *J. Biol. Chem.* 249:8019
81. Rothman, J., Lodish, H. 1977. *Nature* 269:775
82. Schiffer, M., Edmundson, A. B. 1967. *Biophys. J.* 7:121
83. Schindler, M., Rosenbusch, J. P. 1984. *FEBS Lett.* 173:85
84. Sheridan, R. P., Levy, R. M., Salemme, F. R. 1982. *Proc. Natl. Acad. Sci. USA* 79:4545
85. Singer, S. J. 1962. *Adv. Protein Chem.* 17:1
86. Singer, S. J. 1971. *Structure and Function of Biological Membranes.* New York: Academic. 145 pp.
87. Spatz, L., Strittmatter, P. 1971. *Proc. Natl. Acad. Sci. USA* 68:1042
88. Steitz, T., Goldman, A., Engelman, D. 1982. *Biophys. J.* 37:124
89. Sweet, R. M., Eisenberg, D. 1983. *J. Mol. Biol.* 171:479
90. Tanford, C. 1962. *J. Am. Chem. Soc.* 84: 4240
91. Tanford, C. 1980. *The Hydrophobic Effect.* New York: Wiley

92. Tomita, M., Furthmayr, H., Marchesi, V. T. 1978. *Biochemistry* 17:4756
93. Von Heijne, G. 1981. *Eur. J. Biochem.* 116:419
94. Von Heijne, G. 1981. *Eur. J. Biochem.* 120:275
95. Von Heijne, G., Blomberg, C. 1979. *Eur. J. Biochem.* 97:175
96. Wada, A. 1976. *Adv. Biophys.* 9:1
97. Walker, J. E., Carne, A. F., Schmitt, H. W. 1979. *Nature* 278:655
98. Wallace, B. A., Henderson, R. 1975. *Biophys. J.* 39:233
99. Wertz, D. H., Scheraga, H. A. 1978. *Macromolecules* 11:9

100. Wolfenden, R. 1983. *Science* 222:1087
101. Wolfenden, R., Andersson, L., Cullis, P. M., Southgate, C. C. B. 1981. *Biochemistry* 20:849
102. Wolfenden, R., Cullis, P. M., Southgate, C. C. F. 1979. *Science* 206:575
103. Wyckoff, H. W., Hardman, K. D., Allewell, N. M., Inagami, T., Johnson, L. N., Richards, F. M. 1967. *J. Biol. Chem.* 242:3984
104. Youvan, D. C., Bylina, E. J., Alberti, M., Begusch, H., Hearst, J. E. 1984. *Cell* 37:949
105. Yunger, L. M., Cramer, R. D. 1981. *Mol. Pharmacol.* 20:602

Ann. Rev. Biophys. Biophys. Chem. 1986. 15 : 355–76

MASS MAPPING WITH THE SCANNING TRANSMISSION ELECTRON MICROSCOPE[1]

J. S. Wall and J. F. Hainfeld

Biology Department, Brookhaven National Laboratory, Upton, New York 11973

CONTENTS

PERSPECTIVES AND OVERVIEW

The scanning transmission electron microscope (STEM) has established a comfortable niche in the field of biological instrumentation through use of its quantitative imaging process for mass analysis of single molecules. The more traditional methods of mass measurement (e.g. chromatography, centrifugation, light, small-angle X-ray, and neutron scattering) measure

[1] The submitted manuscript has been authored under contract DE-AC02-76CH00016 with the US Department of Energy. Accordingly, the US Government retains a nonexclusive royalty-free license to publish or reproduce the published form of this contribution or to allow others to do so for US Government purposes.

355

averages over a total population, whereas the STEM can examine one molecule at a time, giving total mass, size and shape, internal mass distribution, and degree of departure from a standard or model. Although simple mass measurements continue to be a "bread and butter" application, we are attempting to enlarge our niche in three directions: (*a*) inference of dynamic solution conformation from images of molecules prepared under various conditions, (*b*) two- and three-dimensional determination of mass distribution of complex objects, and (*c*) labeling of specific sites with heavy atom clusters.

Quantitative Electron Microscopy

The high contrast of dark field imaging and the superior signal-to-noise ratio (S/N) associated with the STEM annular detector make possible imaging of unstained biological molecules at relatively low dose (9). Without the complications of added stain, it becomes possible to relate image intensity directly to the local projected mass of the specimen. The STEM method of image acquisition (serial, one point at a time, in a TV-type raster) facilitates digital storage and analysis. Furthermore, the STEM electron detectors consist of scintillators and photomultipliers that have high linearity and quantum efficiency. Therefore the STEM is ideally suited to the type of quantitative analysis described below. This is not to assert that such analysis cannot be done with conventional electron microscopes. The pioneering work in mass measurement was carried out by Zeitler & Bahr (79) using a conventional electron microscope with film and an integrating densitometer. However, the STEM offers clear advantages for low-dose imaging of unstained molecules.

Electron Optics

The STEM obtains high resolution (2–3 Å) by use of a field emission electron source (roughly the electron optical equivalent of a laser) that can be imaged to produce adequate intensity in a diffraction-limited spot (7). Under these conditions the resolution of a STEM is identical to that of a conventional electron microscope using an equivalent objective lens and aperture (8).

The major advantage of the STEM configuration shown in Figure 1 is the separation in space of the components that affect resolution and contrast. The probe-forming components (field emission gun, condenser lens, aperture, deflection coils, and objective lens) are all above the specimen, while the space below the specimen is completely free for optimization of detectors (annular detectors for elastic scattering, energy loss spectrometer for inelastic scattering). This means that every electron emerging from the bottom of a thin specimen can be counted with an appropriate detector to convey information about the specimen volume irradiated at that instant.

In the conventional microscope the situation is quite the opposite. There is only a single imaging channel with a very limited acceptance angle, restricted to minimize lens aberrations. This means that a conventional microscope operated in the dark field mode can utilize only ~5% of the available elastically scattered electrons and none of the unscattered or inelastically scattered electrons (29, 66). To compensate for this loss, the specimen dose must be 20 times higher in a dark field conventional microscope for the same signal-to-noise ratio. In the bright field mode, collection is similar for the two types of microscope, while for phase contrast the conventional microscope is superior (80). A further complication in the conventional microscope is the complexity of the contrast transfer function; features with different spatial frequencies are imaged with differing contrast as a function of defocus. Therefore great care is required in interpreting small details in the image (or relatively complex image processing is required to correct for such effects as contrast reversals).

In the STEM, on the other hand, the large-angle annular detector signal

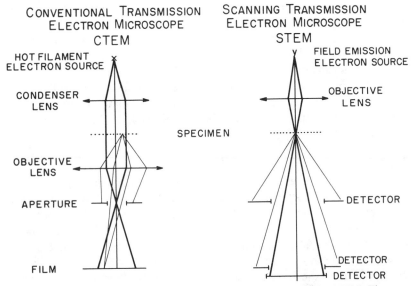

Figure 1 Schematic representation of the image-forming system in the STEM. Electrons from a field emission gun (not shown) are focused to a 2.5 Å diameter spot on the specimen by a high-quality objective lens (focal length = 1 mm, spherical aberration constant = 0.6 mm). Electrons passing through the specimen are registered on one of three concentric annular detectors (subtending 0.0–0.015, 0.015–0.04, and 0.04–0.2 rad).

(used in mass measurement) is very simple to interpret. The contrast is highest at the best focus and the image becomes blurred with defocus, but its integrated intensity does not change. Low spatial frequencies are imaged with full contrast, while higher frequencies are somewhat attenuated (54). Furthermore, when this STEM annular detector is used the signal is said to be "incoherent" in the sense that each atom in the probe makes its own contribution to the signal independently of its relationship to its neighbors (18). Thus, the intensity of the large-angle annular detector signal is directly proportional to the number of atoms in the beam at any given time (weighted by their atomic number) and is essentially independent of their spatial arrangement.

The instrument built at Brookhaven has a field emission gun that operates at 40 keV at 10^{-10}–10^{-12} A beam current. The gun is located in a separate ion-pumped vacuum chamber maintained at less than 10^{-10} torr. A condenser lens and most of the deflection coils are located outside a beam tube that connects the gun and lens chambers. The objective lens has a 1 mm focal length and is cooled to $\sim -170°C$ by helium gas circulating in a closed loop between the STEM and an external heat exchanger located in a liquid nitrogen dewar. The specimen is held at $-150°C$ during observation to eliminate contamination and to reduce mass loss due to electron irradiation; at this temperature mass loss is about four times slower than at room temperature.

Specimens are freeze-dried in a separate system, transferred to the STEM under vacuum (and at low temperature if desired), and inserted into the stage through an air lock. Scans are controlled digitally and detector signals are stored in a digital frame buffer under direction of a VAX 11-750 computer, which also supports additional terminals for image analysis. Image data are available immediately for analysis or can be stored digitally on magnetic tape.

Visualization of Heavy Atoms

The high performance possible with the STEM was demonstrated most dramatically by direct visualization of single heavy atoms on thin carbon substrates (10). This appeared to be an important development for biological structure studies and was a key motivating factor in construction of additional instruments at Johns Hopkins and at Brookhaven in the early 1970s. However, it was not appreciated at that time how severely radiation damage limited the imaging of specimens selectively stained with single heavy atoms. At the dose required for single-atom imaging ($1000 \text{ e}^-/\text{Å}^2$ or more) the biological matrix is destroyed and the atoms move significantly (6, 63). Two approaches promise to overcome this limitation: use of clusters containing many heavy atoms and image averaging of low-dose images of ordered arrays or aligned single particles. Heavy-atom labeling is emerging as a complementary technique to mass measurement; mass measurements are used to map the matrix of the molecule while the heavy-atom stains can be used to highlight features of interest.

To conclude this section we wish to mention areas not covered in this review. Carlemalm et al have recently reviewed applications of the STEM to sectioned tissue, using the ratio of elastic to inelastic scattering to produce "Z contrast" (3). There is also considerable interest in elemental mapping using characteristic electron energy losses (32, 46). A great many applications have arisen in materials science. Finally it is worth pointing out that the results discussed in this review have been obtained with STEMs designed specifically for this application, not add-on attachments for conventional electron microscopes.

MASS MEASUREMENT WITH THE STEM

Advantages

Direct measurement of particle mass is relatively straightforward, given the linearity of the STEM imaging process described above. One need only mark out the region of the image containing the object of interest, subtract the background value of the carbon substrate, and multiply the net intensity within the area by a calibration constant. The calibration factor is a constant that is checked for each specimen using tobacco mosaic virus (TMV) as an internal mass standard. The technique has been described in more detail by a number of authors (14, 15, 20, 28, 59, 61).

The major application of simple STEM mass analysis is in providing a link to biochemistry. If the molecular weight of the starting material is known and all objects in the STEM image are simple multiples of this, then particle identification is unambiguous. On the other hand, if the background is littered with small objects or the mass histogram is not sharply peaked, then caution is indicated. When such problems arise the STEM image is often the best diagnostic tool for improving the quality of the preparation.

Beyond particle identification, basic STEM mass measurements can provide the following information not readily available from other techniques: (a) heterogeneity of a preparation, (b) mass per unit length or mass per unit area, (c) mass change correlated with a biochemical treatment, (d) mass added to a standard object such as a DNA strand, and (e) internal mass distribution. In the case of DNA-binding proteins, length measurements can give the location of binding sites to within a few base pairs (23). If several proposed models for a structure are available, one can compare their projected mass distributions with that in the STEM image either point by point or by some statistical measure such as the apparent radius of gyration. Specific examples are discussed below to illustrate some of these advantages.

Accuracy

The accuracy of STEM mass measurements rivals that of most standard biochemical techniques short of sequencing and is not sensitive to the type of chemical bonding (i.e. polysaccharides are measured with the same sensitivity as proteins). The major errors are due to counting statistics of the scattered electrons and thickness variations in the thin carbon substrate. These factors are diagramed in Figure 2 for an incident dose of $10\,e^-/\text{Å}^2$ as a

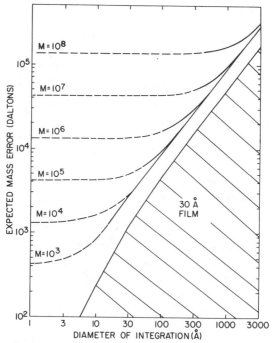

Figure 2   Expected statistical errors in mass measurement of single particles at $10\,e^-/\text{Å}^2$. The error arises from two components: thickness irregularities in the thin carbon substrate and statistical fluctuation (shot noise) in the number of electrons scattered from the molecule plus background. The carbon noise is a function of diameter of integration only and is shown by the region to the right of the figure that is *shaded* to indicate that it sets the limit to accuracy as the dose is increased (60). The curves labeled with particle mass values indicate the expected total error (counting + substrate) for a particle of that mass contained in a diameter of integration indicated (or an arbitrary contour having the same area). The leftmost portions of the curves are *dashed* to indicate that particles of the indicated mass are unlikely to be found in such a small area. Derivation of these curves is described in more detail in reference 61.

function of the diameter of the measuring circle circumscribing the particle. For example, a molecule of glutamine synthetase (GS) of 600 kd inside a measuring circle of 160 Å diameter would have counting noise of 15.7 kd and substrate noise of 7.7 kd for a net noise of 17.5 kd (factors added in quadrature), corresponding to an expected measuring error of 2.9%. The actual standard deviation (SD) observed for a good GS specimen is typically 3%.

As a second example, consider the measurement of a DNA molecule 500 Å long containing 150 base pairs of roughly 100 kd. The appropriate measuring area would be a rectangle 40 Å × 500 Å which would have the same area as a circle 160 Å in diameter. From Figure 2, the total expected error is 14 kd, with 7 kd from substrate noise and the remaining 12 kd from counting noise at $10 \text{ e}^-/\text{Å}^2$. This suggests that an added mass of 40 kd ($3 \times$ SD) should be detectable. If the added mass were localized within a smaller region, the detection limit would be lower.

Figure 2 can be used to predict measuring accuracy and signal-to-noise ratio for objects ranging in size from heavy atom clusters containing 12 gold atoms (62) to large viruses such as vesicular stomatitis virus (VSV) of 270 Md (53). For a dose (D) other than $10 \text{ e}^-/\text{Å}^2$ the counting noise component of the accuracy scales with $D^{-1/2}$. However, at high doses it is not possible to obtain accuracy below the limit set by substrate noise (shaded portion of Figure 2).

The dose of $10 \text{ e}^-/\text{Å}^2$ has been selected for a good compromise between statistical precision and structural integrity. Radiation damage limits the dose that can be used for faithful imaging of biological specimens. A rough rule of thumb for imaging in the Brookhaven STEM at $-150°C$ is $R \cong 2D$ where R is the resolution in Å and D is the dose in $\text{e}^-/\text{Å}^2$. The resolution at $D = 10 \text{ e}^-/\text{Å}^2$ would be about 20 Å, which is about the limit currently set by specimen preparation.

Mass loss due to radiation damage is also an important factor in accuracy. This loss is a linear function of dose in the initial region and is roughly 2.5% for protein at $10 \text{ e}^-/\text{Å}^2$ and $-150°C$. The TMV internal standard (95% protein) loses mass at the same rate so, to first order, the effect of mass loss can be corrected by scaling with the TMV. However, nucleic acids lose mass more slowly and lipids and polysaccharides lose mass more rapidly, so a dose-response curve for mass loss is an important component of any new study. At high dose ($100–1000 \text{ e}^-/\text{Å}^2$) most specimens reach a "stable" state with no further mass loss. The terminal mass loss factor depends on many variables of specimen preparation and microscope vacuum; typical values range from $\sim 50\%$ for protein to 100% loss for carbohydrate.

The TMV used as an internal mass standard is a rod-shaped particle

3000 Å long and 180 Å in diameter with a molecular weight of 39 Md. Note (from Figure 2) that TMV (with an area equivalent to that of an 830 Å diameter circle) should have a mass error of less than 0.5%. Therefore the TMV standard does not contribute significantly to the errors of mass measurement. In principle the TMV internal calibration is not required, but is useful in detecting any drift in microscope calibration or any problems in specimen preparation.

Specimen Preparation

The goals in specimen preparation for the STEM are twofold: (a) reproducible adsorption of the molecules of interest from solution onto the thin carbon substrate with the least possible conformational change and (b) dehydration with minimal artifact or structural perturbation. Each sample requires a unique concentration and buffer system which often must be found by trial and error. Reproducible adsorption is usually obtained by glow discharge treatment of the carbon substrate (to render it hydrophilic). However, we find that such treatment adds contaminants that show up as increased noise in background regions or as increased particle mass. Furthermore, the interaction with the substrate is so strong that it appears to disrupt some intramolecular interactions, flattening the particle against the background.

We have eliminated the need for glow discharge treatment of the carbon substrate by avoiding any exposure of the specimen side of the film to air (64). The thin carbon (evaporated onto single-crystal NaCl) is floated off the NaCl block onto a water surface. We pick up a small portion of the film from above by touching it with a titanium EM grid previously coated with holey carbon, being careful to retain a droplet of water. We invert the grid and exchange the droplet with injection buffer by wicking away ~90% of the original droplet and adding a new solution, again being careful not to let the carbon dry. We then inject an equal volume of the sample below the surface of the droplet, allow it to stand ~1 min, and then wash it several times with volatile buffer (usually ammonium acetate).

This technique gives gentle but reproducible attachment with less structural perturbation as judged by the sharpness of filament edges (see below) and reproducible fine detail. The binding, unlike that achieved through glow discharge, appears to be at least partially reversible; some molecules can be released from the substrate by subsequent addition of a higher concentration of another protein. The importance of injection cannot be overemphasized. Many proteins denature at an air-water interface, which is what touches the grid first in the usual method of applying a drop to a dry grid. This may lead to a coating of denatured molecules or a thick background deposit which may perturb the intact specimen (as in Kleinschmidt spreading). Denaturation can be reduced by treatment with fixatives such as glutaraldehyde, but this may add significantly to the mass (38).

The specimen can be dehydrated by simple air drying but this has two serious drawbacks. First, surface tension forces tend to distort features. Second, the region around the specimen tends to be the last to dry and residual salts are thereby deposited in a meniscus around the particle, possibly causing pH or ionic strength changes. We have found such air-dried samples unacceptable for mass measurement.

Freeze-drying provides an acceptable solution to the above problems if certain precautions are followed. The sample must be wicked to the thinnest layer possible (without drying), then fast frozen in liquid propane or liquid nitrogen slush to avoid ice-crystal damage. The freeze-drying must be in a clean system (with water pumped at high speed) under well-controlled conditions to prevent ice crystal growth as much as possible. Finally, the sample must be transferred to the STEM under vacuum to prevent rehydration and subsequent air-drying.

The specimen-handling hardware for the Brookhaven STEM was designed with these requirements in mind. The samples, wicked to a thin layer and fast-frozen, are transferred to a boat filled with liquid nitrogen (LN_2) in the specimen transport cartridge mounted in a stainless steel chamber. After six specimens are loaded, a LN_2-cooled metal trap is lowered to within 1 mm of the specimens, blocking any line of sight to a warm surface. As soon as the LN_2 in the boat has evaporated to a level slightly below that of the specimens the chamber is evacuated, first with a sorption pump, then with an ion pump to better than 10^{-8} torr. After completion of the freeze-drying program (usually overnight) the boat is withdrawn into the transfer cartridge, sealed, and carried under vacuum to the STEM air lock where the specimens are transferred one at a time into the STEM cold stage. The same hardware can be used for cold transfers of frozen hydrated preparations.

The major advantage of freeze-dried specimens is their ease of interpretation. There is no surrounding medium to obscure less dense portions of a structure, so edge determination is relatively unambiguous. Specimens have been prepared routinely in this manner for several years. The best measures of good freeze-drying are a clean background and particles with sharp edges. With the image analysis software described below, one can be much more precise in comparing the TMV radial mass profile with that expected from X-ray or the background histogram with that of a clean carbon film. These programs are sufficiently rapid that one can attempt to improve specimen preparation in a systematic manner; one can detect even minor advances and can track down the reasons for chance occurrences of superior specimens.

Results

The first STEM mass measurements were carried out by Wall (59) at the University of Chicago and demonstrated that filamentous fd virus had the expected mass per unit length, but that "split" particles (previously taken as evidence for a two-stranded structure) were actually dimers associated side by side in register. A subsequent study by Lamvik (27) showed that rabbit myosin had a mass per unit length consistent with 2.7 molecules per crown. Langmore & Wooley (30, 31) measured the mass of individual nucleosomes and demonstrated that they are disc-shaped with less mass in the center. Using the Basel STEM, Fngel measured the mass of protein P23 of T4 viral preheads and the mass per unit length of F-pili (14). In Heidelberg, Freeman & Leonard (19) repeated measurements of fd phage, F-pili, and rabbit muscle thick filaments and reported new studies of pfl phage, muscle thick filaments from other sources, actin filaments, glutamine synthetase, glutamate dehydrogenase, tomato bushy stunt virus, and Semliki Forest virus. At Brookhaven, Cohen et al (5) demonstrated that pyruvate carboxylase contained eight subunits rather than six as suggested by some earlier studies. Lipka et al (34) measured the mass per unit length of fd phage with no label and with one platinum or one tetra-mercury cluster per coat-protein subunit, demonstrating the radiation sensitivity of the mercury compound. Woodcock et al (73) compared the mass of native nucleosomes with reconstitutes made from DNA + histones H3 and H4 or from DNA + all four histones.

The major early study of the Brookhaven group concerned the structure of fibrinogen (38). At that time there was still some controversy as to whether the "real" molecule was compact with dinodular and trinodular forms as aggregates, or trinodular with compact forms as fragments. Our collaborators, Mosesson and Haschemeyer, were major proponents of the two opposing views. Both were proven at least partially right since we were able to demonstrate that both forms had a molecular weight of 340 kd and could be found side by side in the same field. Trinodular forms predominated in unfixed preparations while fixing increased the fraction of compact forms. Mass measurements of end vs middle domains of trinodular molecules distinguished molecules with part of one chain proteolytically removed (38). Directing of Fab antibody fragments to end or middle domains resulted in molecules with increased mass in the predicted domains (65).

At about the same time the Basel group published studies determining molecular weight, overall dimensions, and correlation with results of conventional rotary shadowing for laminin and fibronectin (17). They also measured the molecular weight of the gene 20 product from bacteriophage T4 and compared unstained images with results of conventional negative staining (12).

In a pioneering study with the Basel STEM, Engel et al studied the HPI layer from the cell wall of *Micrococcus radiodurans*, determining the mass per unit cell of this ordered array. By use of image averaging they were able to compute a mass map at ~ 30 Å resolution showing protrusions as small as 1.3 kd in the overall structure of 655 kd (16).

A similar type of analysis was applied to single myosin molecules imaged with the Basel STEM. In addition to measuring total mass of the head region, Walzthony et al formed mass maps for various conformations. Of special interest was the observation of molecules with thicker neck region and greater mass than the commonly observed "pear-shaped heads" (67).

An extensive series of studies of intermediate filaments with the Brookhaven STEM has been reviewed recently (47). These studies demonstrate that several classes of mass per unit length exist both for native and reconstituted filaments (52) and that filaments from different sources are all built on the same plan of a conserved central core surrounded by variable end domains protruding from it (48, 49, 51).

STEM analysis was essential in establishing the structural model of the dynein ATPase proposed by Johnson (25, 26). In this model the dynein of 2 Md molecular weight is like a "bouquet of flowers" with three globular domains connected to a base structure by thin strands. The two heads of 450 kd and one of 550 kd observed by STEM correlate well with the biochemical evidence of two large polypeptides with MW > 300 kd in $2:1$ stoichiometry (25, 26). Similar results were obtained for *Chlamydomonas* except that the dynein was isolated in two separate components (71). This work has also been extended to dyneins from bull sperm and sea urchin.

Multienzyme complexes are another step in structural organization amenable to STEM analysis. We have studied α-ketoglutarate dehydrogenase (T. Wagenknecht, J. Frank, N. Francis, D. J. DeRosier, personal communication) and pyruvate dehydrogenase (2). In the latter case we were able to establish the stoichiometry of the 5.7 Md intact complex and the mass, size, and shape of each of the components. Through comparison of mass distributions of partially reassembled complexes, it has also been possible to distinguish components bound to the cubic core at faces or edges (E1 or E3) (77), and thus to deduce the structural assembly of the complex.

At the next level of complexity, VSV has been characterized as to total mass (270 Md) and mass of various isolated components. From these values it has been possible to arrive at stoichiometry for the various components of the viron (53). These results are being extended to include radial mass profiles.

Other substances analyzed with the Brookhaven STEM are coated vesicles from liver and brain (50), erythrocyte ghost vesicles (42), methylenetetrahydrofolate reductase (36), clotting factor V (39), phosphorylase kinase (57), muscle thick filaments from *Limulus* (M. M. Dewey, D. Colflesh, B. Gaylinn, S. F. Fan, J. Wall, J. F. Hainfeld, personal communication), haptoglobin (68), glutamine synthetase (21), meningococcal antigens (72), ribosomes (1), nucleosome dimers (74), chromatin fibers (69), fd gene 5/DNA complexes (4), ribonucleoprotein complexes (75), bacterial flagellae (S. Trachtenberg, D. J. DeRosier, S.-I. Aizawa, R. M. Macnab, personal communication), yeast viruses (43), and acetylcholine receptors (70, 22). Other current projects include clotting factor VIII, RNA polymerase II and III, DNA-recA, DNA repair enzymes, Alzheimer's filaments, T7 virus, and polyoma virus.

To conclude this section we mention studies of DNA-binding proteins that interact specifically with DNA. STEM studies in this area have recently been reviewed by Hough et al (24). Generally speaking, the DNA and protein images can be treated as separate; the protein is analyzed as if it were a free-standing molecule, after subtraction of a DNA backbone from the image. The position along the DNA strand gives the location of a binding site to within a few base pairs (23). This method has been used to determine the number of molecules of T antigen bound at two specific sites within the control region of SV40 DNA (35) as well as to study the mode of interaction of eukaryotic RNA polymerase II with DNA (24). Beer is using similar techniques in the study of RNA polymerase III (M. Beer, personal communication). We expect this type of DNA-protein study to become increasingly attractive as technical details are worked out.

CONFORMATION OF MOLECULES IN SOLUTION

The study of conformation of molecules in solution is the first of three relatively new areas we discuss. The possibility for such a study results from the lack of any constraints on pH or ionic strength imposed by our specimen preparation protocol (in contrast with negative staining). The first study to take advantage of this flexibility in specimen preparation was the fibrinogen work described above; specimens were examined for conformational changes at pH 3, 7, and 9 (the changes were not significant) (38). A similar study on plasma fibronectin showed compact molecules at pH 7, but extended forms at pH 2.8 and 9.3 (55). In the rRNA study described below we demonstrated conformational differences between images of molecules prepared in 10^{-6} M salt and those in 10^{-5} M salt.

In our ribosomal study with Oostergetel and Boublik (40), the central question was the extent to which rRNA can fold to form a ribosome-like structure without added protein. It was expected that the rRNA would be fully extended at low ionic strength and would become more compact as increasing ionic strength shielded repulsive charges. We were somewhat surprised to observe the molecules shown in Figure 3, which were deposited from 10^{-6} M NaCl. The contour length is roughly one quarter that expected for fully stretched 28S rRNA from the large eukaryotic subunit, and both molecules show topologically equivalent branching patterns with a small fork at the right end, a major side branch one third of the length from the right end, etc. That 28S rRNA should still show secondary structure in distilled water should not be surprising, since circular dichroism measurements indicate secondary structure can be removed only by heating to 60°C under these conditions. Figures 4a–c show how increasing ionic strength renders the molecule more compact; however, the molecule never becomes as compact as the 50S subunit, which is shown in Figure 4d for comparison. Although the structure becomes more difficult to interpret, the major side arm can still be recognized in many molecules. A convenient way to assign a number to the degree of compactness is to compute the apparent radius of gyration R_g after subtracting the carbon film background. Molecules similar to those shown in Figures 4a–c have values of $R_g = 600 \pm 150$, 240 ± 30, and 160 ± 25 Å respectively, while $R_g = 87 \pm 8$ Å for intact 50S subunit images similar to those shown in Figure 4d.

In order to make a more specific statement about the significance of the structures observed at low ionic strength, we compared them with the theoretical folding predictions of Michot et al (37) based on the sequence. Histogram analysis of side arm positions was in excellent agreement with the theoretical predictions (40).

The obvious next step will be to add back various ribosomal proteins, observing their binding sites along the molecular skeleton of the rRNA as well as any perturbation of this skeleton. This project is now under way with components from *Escherichia coli* ribosomes, for which the steps in reassembly have been well documented. In a more general case, it will be of interest to examine other large RNA structures currently thought to be relatively featureless to see if they contain similar structures that might be associated with regulatory functions.

RADIAL DENSITY PROFILES

The STEM image intensity distribution represents a projection of the three-dimensional arrangement of atoms onto a plane. In general it is not possible to retrieve the three-dimensional distribution without multiple projections

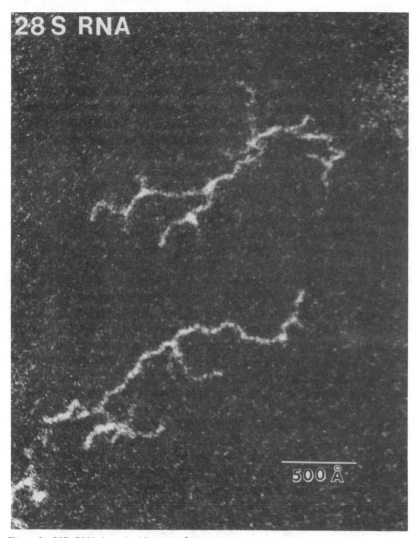

Figure 3   28S rRNA deposited from 10^{-6} M NaCl. Note the similarity of branching patterns of the two molecules. Specimen provided by M. Boublik, Roche Institute.

and tomographic analysis. Recording even a single view of an essentially undamaged unstained structure is problematic; a tilt series would incur concomitant beam damage. However, symmetry considerations can circumvent the need for multiple images. In the simplest case of spherical or cylindrical symmetry, there is only one radial density distribution that can give rise to a given projection.

(a)   (b)

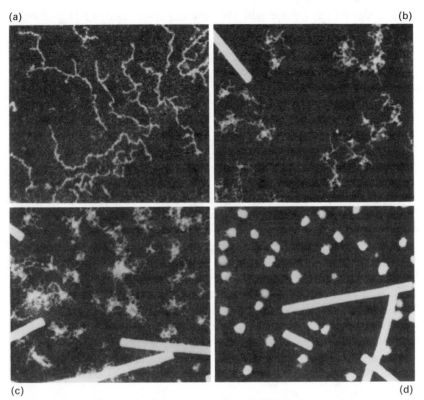

(c)   (d)

Figure 4   STEM images of (*a*) unstained freeze-dried 28S rRNA from BHK cells in water; (*b*) 60 mM ammonium acetate at pH 7.0; (*c*) 30 mM Tris/HCl, 20 mM MgCl$_2$, 360 mM KCl at pH 7.6 (reconstitution buffer). For reference, 60S ribosomal subunits in 60 mM ammonium acetate are shown in *d*. TMV (straight rods, 180 Å diameter) was included as an internal reference. Reprinted from Oostergetel et al (40).

In collaboration with Steven et al, we have worked out a practical scheme for analyzing STEM data to extract the projection data and compute the associated radial mass profile (48). The processing is done in real space owing to the coarseness of sampling inherent in STEM data where the beam diameter is usually smaller than the spacing between image points and to the ease of limiting band width to eliminate negative density values resulting from noise. The analysis is similar to mass measurement except for the use of cross-correlation analysis to find the best center or center line (for a filament). The image points are rebinned according to their distance from the best center (line) and are averaged over as many spheres or as great a length of filament as required for noise reduction. A second program transforms the projected mass into radial density.

In order to test the method, we have applied it to TMV (the internal control included in all our specimens). Figure 5 shows mass projections for RNA-free protein helix (*a*, *b*, and *c*) and intact virus (*d*, *e*, and *f*). These curves can be interpreted directly with some effort, but the radial density profile shown in Figure 6 makes the interpretation fairly obvious. Both types of filament have a central hole, relatively uniform density between 20 and 80 Å radius, and a sharp outer edge at 80 Å. The intact TMV (dashed curve) has a higher density peak at 40 Å radius corresponding to the RNA. This peak is absent in the RNA-free TMV. These peak positions also correlate well with X-ray profiles of TMV.

Radial profile analysis provides a sensitive measure of specimen quality both at low dose and as a function of increasing irradiation. The first feature to be lost is the central hole, while the edge profile remains relatively unchanged (48). We are using this type of analysis to characterize the quality of our specimen preparation as well as to develop an objective measure of resolution (currently 20–40 Å, depending on the type of

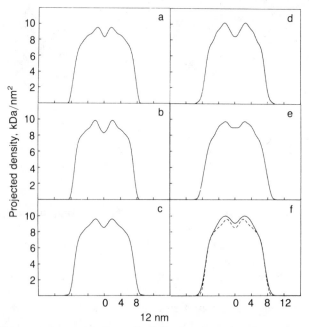

Figure 5   Transverse mass projections of individual coat-protein polymers (*a* and *b*) and of virons (*d* and *e*). The global average projections are given in *c* and *f* (*c* is also overlaid in *f* for comparison). Individual segments were obtained by averaging over lengths of ∼0.3 μm, while global averages were from lengths of ∼1.5 μm. Reprinted from Steven et al (52).

specimen). A dose-response curve for resolution would be of great help in designing experiments.

A crude early version of this type of analysis was used in the dynein study to determine the orientation of the head groups relative to microtubules (25). More recent studies have characterized the radial mass distributions of intermediate filaments (51). Tail tubes from bacteriophage T4 were found to have a central hole that might be filled with six strands of a protein (probably gene product 48) thought to be the length-determining ruler (13). A similar type of analysis has been applied to face-on views of pyruvate dehydrogenase to deduce the locations of E1 and E3 subunits relative to the E2 core (assumed to be roughly spherical) (77). Steven (personal communication) and co-workers are extending earlier mass measurements on VSV to include radial mass profiles, which allow them to assign radial positions of most of the viral components. In addition, we are actively collaborating with Safer to determine the radial profile of actin +/− tropomyosin +/− troponin +/− gold clusters (D. Safer, personal communication).

The radial profiles obtained as described above are straightforward to interpret compared to profiles of ice-embedded specimens obtained in the conventional microscope. Edges are defined with respect to vacuum instead of with respect to ice with density similar to protein, and the complications of the phase contrast transfer function are avoided (poor contrast at low spatial frequencies, especially on the equator, and oscillations at high spatial frequencies). Therefore the two methods are complementary, STEM for absolute values and boundary contours and conventional EM for internal details.

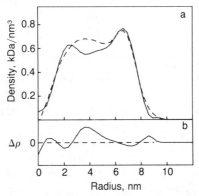

Figure 6   (*a*) Comparison between radial mass profiles of average dry density of TMV coat protein helices (*solid line*) and virons (*dashed line*), derived from the projections shown in Figures 5*c* and *f*. (*b*) The difference curve between these traces has a single significant peak of positive density at a radius of 40 Å. Reprinted from Steven et al (52).

The next step in this type of analysis will be to take into account helical symmetry, thereby producing several projections at different offsets relative to the helical repeat. These can be used with standard three-dimensional helical reconstruction programs to give a full three-dimensional density map for helical objects.

SPECIFIC LABELING WITH HEAVY-ATOM CLUSTERS

Mass analysis has been used to great advantage to map the overall shape of biological structures, but is of little help in locating specific sites of interest. The original proposal to use single heavy-atom labels requires extensive image averaging to detect such labels at low doses because of the relatively small scattering cross section (60). Since we could not take full advantage of the 2 Å resolution of the STEM without major improvements in specimen preparation, it seemed reasonable to search for an intermediate-sized label commensurate with our attainable resolution on biological samples (~ 20 Å). Conventional EM labels using ferritin and antibodies or colloidal gold can give a resolution of ~ 300 Å. A good candidate for a high-resolution label seemed to be a tetra-mercury compound developed by Lipka et al, but this was found to be unstable in the electron beam (34). Our next choices were gold and tungsten clusters that each contained 11 heavy atoms and manifested some possibility for covalent attachment to biological molecules (to permit column purification of products and to remove nonspecifically bound clusters). We showed that the gold cluster was visible in the STEM with a signal-to-noise ratio of 6 at a dose of $\sim 100 \, e^-/Å^2$ (62). This is still a relatively high dose for biological studies, but appears manageable if a low-dose image is first formed showing the biological matrix, followed by the higher-dose image that is required for cluster detection. For the thin samples studied thus far, cluster motion does not appear to be a problem. If the sample is suitable for image averaging it is possible to use the low-dose image directly, with the higher-dose image perhaps referred to to check the alignment on the fiducial marks provided by the clusters.

We have worked most extensively with the gold cluster shown in Figure 7. It is composed of 11 gold atoms surrounded by benzyl phosphines containing a total of 21 amino groups that are highly reactive. In our first experiment Safer biotinylated gold clusters which, when reacted with avidin, formed polymers with gold clusters at the biotin binding sites of avidin (45). This was the first time that specific sites on isolated molecules were visualized to ~ 10 Å resolution. Another application was labeling of carbohydrate moieties on glycoproteins. The unmodified gold cluster reacts strongly with aldehydes such as those formed by periodate oxidation of sugars. We tested this with the glycoprotein haptoglobin and demonstrated specific labeling (33).

The next step in utilization of the gold cluster was preparation of monofunctional reagents reactive toward SH, NH_2, or other groups. This could be done by blocking all but one of the 21 amino groups and converting the remaining group as described by Frey et al (41, 76, 78). Alternatively one may form the cluster from stoichiometrically measured starting materials so that most of the product is monofunctional and can be purified with reasonable yield, as described by Safer (44). Figure 8 shows a fibrinogen molecule labeled with Safer's reagent specific for SH groups. We are also studying actin filaments labeled by Safer with the same reagent.

Another possible label is the tungsten cluster $W_{12}O_{40}PO_4$. It is possible to remove one of the tungstens and insert in its place an organometallic compound with the desired function. A number of compounds have been tried with some success, but further work is needed to make the tungsten clusters usable. The main advantage of the tungsten clusters is their extreme

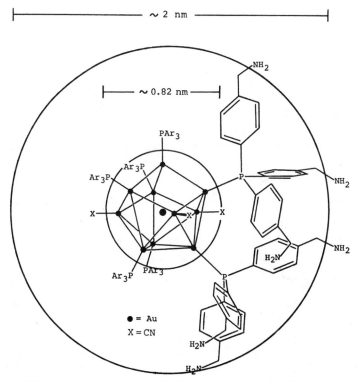

Figure 7   Schematic structure diagram for the 11-gold cluster. Reprinted from Wall et al (62).

Figure 8 Fibrinogen Metz labeled with Safer's reagent specific for SH groups (44). This variety of fibrinogen has two additional easily reduced SH groups thought to be located near the center of the molecule. The trinodular molecule shown has two gold clusters located within the central domain. Image width is 0.064 μm. Specimen provided by M. Mosesson.

resistance to radiation damage; in comparison the gold cluster decomposes at a dose of ~ 1000 e$^-$/Å^2 (62).

One apparently successful application of the tungsten cluster has been insertion of a long-chain hydrocarbon tail. This compound is very polar and forms micelles in solution or inserts into membranes (J. F. Hainfeld, unpublished data). We are exploring the use of this compound for studies of membrane cycling.

The number of possible schemes for using such labels is limitless. However, labeling is still at the trial and error stage of development and many technical problems remain to be overcome.

CONCLUSIONS

The past few years have witnessed an explosion in the productivity of the STEM. With the flood of specimens have come significant improvements in techniques of specimen preparation and image analysis. However, the full potential of the instrument for solving problems in structural biology is still far from realization. We predict that in the next few years the STEM will move gradually into the niche now occupied solely by X-ray and neutron diffraction.

ACKNOWLEDGMENTS

The Brookhaven STEM Facility is supported by the US Department of Energy and by NIH Biotechnology Resource Branch Grant No. RR 01777. Heavy-atom labeling is supported by NIGMS Grant No. GM 31975. We also wish to acknowledge other current and former members of the STEM group: Paul Furcinitti, Jim Lipka, Keith Monson, Lynn Maelia, Frank Kito, Kristin Elmore, and George Latham.

Literature Cited

1. Boublik, M., Robakis, N., Hellmann, W., Wall, J. S. 1982. *Eur. J. Cell Biol.* 27 : 177
2. CaJacob, C. A., Frey, P. A., Hainfeld, J. F., Wall, J. S., Yang, H. 1985. *Biochemistry* 24 : 2425
3. Carlemalm, E., Colliex, C., Kellenberger, E. 1982. *Adv. Electron. Electron Phys.* 63 : 269
4. Casadevall, A., Day, L. 1985. *Virology* 145 : 260
5. Cohen, N., Duc, J. A., Beegen, H., Utter, M. 1979. *J. Biol. Chem.* 254 : 9262
6. Cole, M. D., Wiggins, J. W., Beer, M. 1977. *J. Mol. Biol.* 117 : 387
7. Crewe, A. V., Eggenberger, D. N., Wall, J. S., Welter, L. M. 1968. *Rev. Sci. Instrum.* 39 : 576
8. Crewe, A. V., Wall, J. S. 1970. *Optik* 30 : 461
9. Crewe, A. V., Wall, J. S. 1970. *J. Mol. Biol.* 48 : 375
10. Crewe, A. V., Wall, J., Langmore, J. 1970. *Science* 168 : 1338
11. Deleted in proof
12. Driedonks, R. A., Engel, A., tenHeggeler, B., van Driel, R. 1981. *J. Mol. Biol.* 152 : 641
13. Duda, R. L., Wall, J. S., Hainfeld, J. F., Sweet, R. M., Eiserling, F. A. 1985. *Proc. Natl. Acad. Sci. USA* 82 : 5550
14. Engel, A. 1978. *Ultramicroscopy* 3 : 273
15. Engel, A. 1982. *Micron* 13 : 425
16. Engel, A., Baumeister, W., Saxton, W. O. 1982. *Proc. Natl. Acad. Sci. USA* 79 : 4050
17. Engel, J., Odermatt, E., Engel, A. 1981. *J. Mol. Biol.* 150 : 97
18. Fertig, J., Rose, H. 1977. *Ultramicroscopy* 2 : 269
19. Freeman, R., Leonard, K. R. 1980. *J. Microsc.* 122 : 275
20. Hainfeld, J. F., Wall, J. S., Desmond, E. J. 1982. *Ultramicroscopy* 8 : 263
21. Haschemeyer, R. H., Wall, J. S., Hainfeld, J. F., Maurizi, M. R. 1982. *J. Biol. Chem.* 257 : 7252
22. Holtzman, E., Wise, D., Wall, J., Karlin, A. 1982. *Proc. Natl. Acad. Sci. USA* 79 : 310
23. Hough, P. V. C., Mastrangelo, I. A., Wall, J. S., Hainfeld, J. F., Simon, M. N., Manley, J. L. 1982. *J. Mol. Biol.* 160 : 375
24. Hough, P. V. C., Mastrangelo, I. A., Wall, J. S., Hainfeld, J. F., Wilson, V. G., et al. 1985. *Biotechnology* 3 : 549
25. Johnson, K. A., Wall, J. S. 1983. *J. Cell Biol.* 96 : 669
26. Johnson, K. A., Wall, J. S. 1983. *J. Submicrosc. Cytol.* 15 : 181
27. Lamvik, M. K. 1978. *J. Mol. Biol.* 122 : 55
28. Lamvik, M. K., Langmore, J. P. 1977. *Proc. 10th Annu. Scanning Electron Microsc. Symp.*, p. 401. Chicago : IIT Res. Inst.
29. Langmore, J., Wall, J., Isaacson, M. 1973. *Optik* 38 : 335
30. Langmore, J. P., Wooley, J. C. 1975. *Proc. Natl. Acad. Sci. USA* 72 : 2691
31. Langmore, J. P., Wooley, J. C. 1977. *Biophys. J.* 17 : 115a
32. Leapman, R. D. 1984. *Ultramicroscopy* 12 : 281
33. Lipka, J. J., Hainfeld, J. F., Wall, J. S. 1983. *J. Ultrastruct. Res.* 84 : 120
34. Lipka, J. J., Lippard, S. J., Wall, J. S. 1979. *Science* 206 : 1419
35. Mastrangelo, I. A., Hough, P. V. C., Wilson, V. G., Wall, J. S., Hainfeld, J. F., Tegtmeyer, P. 1985. *Proc. Natl. Acad. Sci. USA* 82 : 3626
36. Matthews, R. G., Vanoni, M. A., Hainfeld, J. F., Wall, J. S. 1984. *J. Biol. Chem.* 259 : 11647
37. Michot, B., Hassouna, N., Bachellerie, J.-P. 1984. *FEBS Lett.* 167 : 263
38. Mosesson, M. W., Hainfeld, J., Haschemeyer, R. H., Wall, J. S. 1981. *J. Mol. Biol.* 153 : 695
39. Mosesson, M. W., Nesheim, M. E., DiOrio, J., Hainfeld, J. F., Wall, J. S. 1985. *Blood* 65 : 1158
40. Oostergetel, G., Wall, J. S., Hainfeld, J.

F., Boublik, M. 1985. *Proc. Natl. Acad. Sci. USA* 82:5598
41. Reardon, J. E., Frey, P. A. 1984. *Biochemistry* 23:3849
42. Regen, S. L., Shin, J., Hainfeld, J. F., Wall, J. S. 1984. *J. Am. Chem. Soc.* 106:5756
43. Reilly, J. D., Bruenn, J., Held, W. 1984. *Biochem. Biophys. Res. Commun.* 121:619
44. Safer, D., Bolinger, L., Leigh, J. S. Jr. 1986. *J. Inorg. Biochem.* 26: In press
45. Safer, D., Hainfeld, J. F., Wall, J. S., Riordan, J. 1982. *Science* 218:290
46. Somlyo, A. P. 1984. *Nature* 309:516
47. Steinert, P. M., Steven, A. C., Roop, D. R. 1985. *Cell* 42:411
48. Steven, A. C., Hainfeld, J. F., Trus, B. L., Steinert, P. M., Wall, J. S. 1984. *Proc. Natl. Acad. Sci. USA* 81:6363
49. Steven, A. C., Hainfeld, J. F., Trus, B. L., Wall, J. S., Steinert, P. M. 1983. *J. Cell Biol.* 97:1939
50. Steven, A. C., Hainfeld, J. F., Wall, J. S., Steer, C. J. 1983. *J. Cell Biol.* 97:1714
51. Steven, A. C., Trus, B. L., Hainfeld, J. F., Wall, J. S., Steinert, P. M. 1984. *Ann. NY Acad. Sci.* 455:371
52. Steven, A. C., Wall, J. S., Hainfeld, J. F., Steinert, P. M. 1982. *Proc. Natl. Acad. Sci. USA* 79:3101
53. Thomas, D., Newcomb, W. W., Brown, J. C., Wall, J. S., Hainfeld, J. F., Steven, A. C. 1985. *J. Virol.* 54:598
54. Thomson, M. G. R. 1975. In *Physical Aspects of Electron Microscopy and Microbeam Analysis*, ed. B. M. Siegel, D. R. Beaman, pp. 29–45. New York: Wiley
55. Tooney, N. M., Mosesson, M. W., Amrani, D. L., Hainfeld, J. F., Wall, J. S. 1983. *J. Cell Biol.* 97:1686
56. Deleted in proof
57. Trempe, M. R., Carlson, G. M., Hainfeld, J. F., Furcinitti, P. S., Wall, J. S. 1986. *J. Biol. Chem.* In press
58. Deleted in proof
59. Wall, J. S. 1971. *A high resolution scanning microscope for the study of single biological molecules.* PhD thesis. Univ. Chicago
60. Wall, J. S. 1978–1979. *Chem. Scr.* 14:271

61. Wall, J. S. 1979. In *Analytical Electron Microscopy*, ed. J. J. Hren, J. I. Goldstein, D. Joy, p. 333. New York/London: Plenum
62. Wall, J. S., Hainfeld, J. F., Bartlett, P. A., Singer, S. J. 1982. *Ultramicroscopy* 8:397
63. Wall, J. S., Hainfeld, J. F., Bittner, J. W. 1978. *Ultramicroscopy* 3:81
64. Wall, J. S., Hainfeld, J. F., Chung, K. D. 1985. *Proc. 43rd Annu. EMSA Meet.*, ed. G. W. Baily, p. 318. San Francisco: S. F. Press
65. Wall, J. S., Hainfeld, J. F., Haschemeyer, R. H., Mosesson, M. W. 1983. *Ann. NY Acad. Sci.* 408:164
66. Wall, J. S., Isaacson, M., Langmore, J. 1974. *Optik* 39:359
67. Walzthony, D., Bahler, M., Eppenberger, H. M., Walliman, T., Engel, A. 1984. *EMBO J.* 3:2621
68. Wejman, J. C., Hovsepian, D., Wall, J. S., Hainfeld, J. F., Greer, J. 1984. *J. Mol. Biol.* 174:319, 343
69. Williams, S. P., Athey, B. D., Muglia, L. J., Schappe, R. S., Gough, A. H., Langmore, J. P. 1986. *Biophys. J.* In press
70. Wise, D. S., Wall, J., Karlin, A. 1981. *J. Biol. Chem.* 256:12624
71. Witman, G. B., Johnson, K. A., Pfister, K. K., Wall, J. S. 1983. *J. Submicrosc. Cytol.* 15:193
72. Wolanski, B. S., McAleer, W. J., Hilleman, M. R. 1983. *J. Ultrastruct. Res.* 83:21
73. Woodcock, C. L. F., Frado, L.-L. Y., Wall, J. S. 1980. *Proc. Natl. Acad. Sci. USA* 77:4818
74. Woodcock, C. L. F., Frank, J. 1984. *J. Ultrastruct. Res.* 89:295
75. Wooley, J., Chung, S.-Y., Wall, J., LeStourgeon, W. 1986. *Biophys. J.* In press
76. Yang, H., Frey, P. A. 1984. *Biochemistry* 23:3836
77. Yang, H., Hainfeld, J. F., Wall, J. S., Frey, P. A. 1985. *J. Biol. Chem.* 30:16049
78. Yang, H., Reardon, J. E., Frey, P. A. 1984. *Biochemistry* 23:3857
79. Zeitler, E., Bahr, G. F. 1962. *J. Appl. Phys.* 33:847
80. Zeitler, E., Thomson, M. G. R. 1970. *Optik* 258:359

Ann. Rev. Biophys. Biophys. Chem. 1986. 15:377–402

APPLICATIONS OF NMR TO STUDIES OF TISSUE METABOLISM

M. J. Avison, H. P. Hetherington, and R. G. Shulman

Department of Molecular Biophysics and Biochemistry, Yale University, New Haven, Connecticut 06511

CONTENTS

PERSPECTIVES AND OVERVIEW

From its beginnings as a tool for the elucidation of biochemical pathways and bioenergetic status in unicellular organisms (76, 78, 111, 112, 169, 154), the field of NMR studies in vivo has grown to encompass not only the study of isolated perfused organs (49, 59, 75, 98), but also the study of various aspects of the biochemistry, physiology, and pathophysiology of these same organs in the intact animal (4, 24, 54, 58, 87, 104, 149, 170, 174). In recent

377

0883–9182/86/0610–0377$02.00

years several groups have begun to extend the techniques developed in animals to the study of clinically relevant conditions in humans (60, 84, 153, 185, 189).

A comprehensive review of all areas of NMR studies in vivo would be either unacceptably long or very superficial. For this reason we have restricted this review to studies published since 1980, except where an earlier study is particularly relevant to the topic under discussion. Furthermore, we have concentrated on areas that have been extending the scope of NMR in vivo. One specific omission is review of NMR studies of tumors, since a comprehensive review has recently appeared (68).

NMR spectroscopy in vivo provides a noninvasive window on the metabolic soul of the cell. It permits us to monitor a wide range of processes including cellular energetics (75, 177), carbohydrate (7, 49), amino acid (22, 161), and lipid metabolism (42, 49), and the regulation of cytoplasmic pH (3, 64, 82). Furthermore, NMR does not introduce any chemical or mechanical perturbations into the system, so multiple measurements can be made. It is this feature that holds promise for future clinical studies of metabolism in humans. NMR is a relatively insensitive technique, however, and tissue metabolites must be present in relatively high concentrations to be detectable. However, the sensitivity depends on the nucleus detected. Studies of tissue metabolism in vivo have generally made use of the ^{31}P, ^{13}C, ^{1}H, and ^{19}F nuclei. What are their relative advantages and disadvantages?

The ^{31}P nucleus has 100% natural abundance, so no labeling is required to detect ^{31}P NMR spectra. The NMR sensitivity of this nucleus is such that tissue metabolites in concentrations greater than ~ 1.0 mM can be detected in a few minutes. Unfortunately, not many phosphorus-containing metabolites are present in these concentrations, although those that are have made ^{31}P NMR a powerful probe of tissue pH and bioenergetics. The broadness of the peaks in the ^{31}P NMR spectrum in vivo impairs spectral resolution, especially in the phosphomonoester region. The resolution generally does not improve with increasing magnetic field strengths as much as in the cases of ^{1}H, ^{13}C, and ^{19}F, because the line widths increase at about the same rate as the spectral dispersion (71). The measurement of intracellular pH appears to be easier at higher field strength (21).

The ^{13}C nucleus has a fourfold lower intrinsic sensitivity than that of ^{31}P and furthermore has a natural abundance of only 1.1%. This is a disadvantage in as much as it precludes the monitoring of tissue metabolites other than lipids (42) and glycogen (4, 174) without the use of labeled precursors. This disadvantage is outweighed by the information that the use of these labels affords, namely measurements of the fluxes of metabolites through specific pathways and the control points in these pathways. The large chemical shift dispersion and relatively sharp lines are a further advantage of ^{13}C NMR. A potential problem with the use of ^{13}C in vivo is the need for proton decoupling, since at high frequencies the decoupling field may lead to sample heating.

^{1}H NMR has several significant advantages. The ^{1}H nucleus has the highest NMR sensitivity of any stable nucleus ($\sim$ 16-fold more intense than ^{31}P), has 100% natural abundance, and is found in virtually all metabolites. The ubiquity of the ^{1}H nucleus can be problematic, since (a) its chemical shift range is only 10 ppm, (b) there are many coupled spin systems that at low field do not give simple multiplet structures, and (c) the spectra contain many intense broad resonances from proteins and lipids. Last, but certainly not least, it is necessary to suppress the much larger signal from tissue water in order to observe the metabolites of interest.

The ^{19}F nucleus has high NMR sensitivity (similar to that of ^{1}H) and a large chemical shift range. Also, since there are no naturally occurring fluorinated compounds, the spectra are free of background signals. To date there have not been many applications of ^{19}F NMR to in vivo studies of metabolism, but the few actual studies indicate promise.

We now review recent studies using these four nuclei.

^{31}P NMR STUDIES

The largest number of studies in vivo to date have used ^{31}P NMR. These studies have generally sought information about the energetic status and the cytoplasmic pH of the cell or tissue, the former from the levels of intracellular high- and low-energy phosphates (primarily ATP, PCr, and P_i) and the latter from the chemical shift of P_i or some other endogenous or added titratable phosphorus-containing group (see below). Systems studied to date include isolated mitochondria (135–137), *Escherichia coli* (138, 173, 183, 184), *Chlorella* (122), yeast (5, 41), heart (28, 87, 98, 102, 104, 115, 117), liver (48, 49, 94, 97, 104, 175), brain (37, 47, 58, 60, 142, 148, 179), kidney (2, 14, 75, 104, 158), muscle (46, 56, 57, 93, 106, 120, 159, 177), and tumors (68–71, 84, 86, 110, 119, 131). In addition, the various techniques of magnetization transfer have been applied to the study of the H^+-ATPase in *E. coli* (33), yeast (5, 41), heart (114, 167), brain (168), and kidney (75, 104, 158), and the creatine kinase reaction in brain (168), heart (59, 114–116), and skeletal muscle (120). The reader is referred to several excellent reviews of work done in these areas (6, 34, 68, 78, 105, 149, 154, 169).

Compartmentation and Visibility

The ^{31}P NMR spectra of cells and tissues generally contain at least eight identifiable resonances: three arising from the α, β, and γ phosphates of ATP (and any other nucleoside triphosphates), the α and γ being coresonant with the α and β peaks, respectively, of any nucleoside diphosphates present; a phosphocreatine peak in spectra from brain and muscle; a P_i peak with a pH-dependent chemical shift; poorly resolved phosphodiester peaks (GPC, GPE, and SEP); poorly resolved phosphomonoester peaks (AMP, G6P, F6P, PCho, PEth, and IMP); and finally a large broad peak that lies beneath the whole spectrum and has been variously attributed to phospholipids (83) and bone (1).

Although the values for intracellular ATP are in good agreement with freeze-clamp studies, it has become clear that the concentrations of ADP, P_i, and in at least one case PEth are significantly lower in vivo than the freeze-clamp data indicate (75, 88, 97, 158). Thus ^{31}P NMR measurements indicate that cytoplasmic $[P_i]$ is 0.5–0.6 mM in the starved rat liver in vivo (97), 2.4 μmol/g dry wt in the perfused rat kidney (75), 1.6–2.8 μmol/g wet wt in rat leg muscle in vivo (1, 106), 0.9 ± 0.3 μmol/g wet wt in cat biceps (fast twitch), 6.1 ± 0.3 μmol/g wet wt in cat soleus (slow twitch muscle) (120), 1.2–3.6 mM in perfused rat heart (depending on the substrate provided) (117), ~ 1.3 μmol/g wet wt in the rat brain in vivo (21), and ~ 1.9 μmol/g wt in the rabbit brain in vivo (21).

In most cases direct detection of ADP in the ^{31}P NMR spectrum is impossible. The free cytoplasmic [ADP] can, however, be calculated, provided that the creatine kinase (CK) reaction is at equilibrium, from the K_{eq} for the reaction using the [ATP], [PCr], and pH obtained by NMR. We return to the question of whether the CK reaction is in equilibrium later. Of course this strategy can be used only in tissues that have high cytoplasmic CK activity, such as muscle and brain. Values for cytoplasmic [ADP] obtained in this way were 29–142 μM in perfused rat heart (depending on substrate conditions and work rate) (115, 117), less than 70 μM in rat skeletal muscle (106), ~ 20 μM in resting cat biceps (120), and 20–30 μM in rabbit brain (142, 148), all in vivo. In perfused rat kidney the cytoplasmic [ADP] was much higher, and could be detected by ^{31}P NMR from the difference in intensity of the NTP-γ and NTP-β peaks. This method was used to estimate a cytoplasmic [ADP] of 0.9 ± 0.7 μmol/g dry wt in perfused kidney (75). A later study found the renal [ADP] to be 0.27 μmol/g wet wt (158). Both these values are much higher than the concentration of less than 0.1 μmol/g wet wt estimated in vivo (158). Finally we note that in E. coli the ATP and ADP peaks are resolvable under some circumstances (183).

Various explanations of the discrepancy between the ^{31}P NMR and chemical measurements of tissue $[P_i]$ and [ADP] have been proposed, including (a) hydrolysis of high energy phosphates during the chemical extraction procedure, and (b) sequestration of P_i and ADP in an NMR-invisible form. While there is evidence that the first explanation may be true in part, several observations lead us to believe that there are indeed significant pools of NMR-invisible P_i and ADP. Freeman et al observed more than twice as much $[ATP + ADP + P_i]$ by enzymatic assay as by NMR in perfused rat kidney, although the [ATP] measured by the two methods was the same (75). Since the kidney has no PCr or CK, there could have been no buffering of the [ATP] in the course of the extraction; this suggests that the ADP and P_i measured enzymatically were not artifacts of extraction. Also, in studies of intact human muscle Taylor et al (177) found that in the initial phase of recovery following exercise the P_i peak often disappeared; however, there was no corresponding increase in PCr, even though there was no change in the ATP level. Furthermore, the rate of P_i decrease was twice the rate of PCr increase. Taylor et al interpreted these observations by proposing that P_i moves from the cytoplasm into compartments (mitochondria, sarcoplasmic reticulum) where it becomes NMR-invisible. In contrast, a recent study found no change in the [PCr $+ P_i$] during recovery (106). The reason for this difference remains unclear. In studies of perfused rat heart the [PCr + ATP] measured by freeze-clamp and that measured by NMR were the same; in this preparation, the higher ADP content found by freeze-clamp was probably not artifactual, but rather reflected a large pool of NMR-invisible ADP (98) as found in the kidney (75, 158). These results have profound consequences for the calculation of the cytoplasmic phosphorylation potential and for the identification of the factors chiefly responsible for its regulation.

Measurement of pH_i

One of the earliest and still most widespread uses of ^{31}P NMR has been the measurement of intracellular pH. Most studies have made use of the pH sensitivity of the P_i chemical shift, together with the assumption that the P_i signal originates from the cytoplasmic space. As discussed above, this is generally reasonable since the P_i in organelles such as mitochondria appears to be present in an NMR-invisible form. Roberts et al have shown that the pK_a of P_i is sensitive to the ionic strength of the solution (157). However, Gillies et al have pointed out that this dependence, within the physiological range, is very small, corresponding to less than 0.1 pH unit (81). In some cases the use of P_i to measure pH may be inappropriate or even impossible. Several pH-dependent phosphomonoester peaks can function as alternatives including fructose-1-phosphate (51), methylenediphosphonic acid (MDP) (173), glucose-6-phosphate (127), 2-deoxyglucose-6-phosphate (12, 82, 86, 126), and IMP (106). Finally it should be noted that several ^{19}F probes of cytoplasmic pH have recently been introduced (64, 178) (see below).

The intracellular pH has been measured in a wide range of tissues. The first demonstration of the method was by Moon & Richards (123) in the red blood cell. Their results should be accepted with caution because significant errors in the pH measurement in red blood cells can occur if the pO_2 varies, since changes in the deoxyhemoglobin concentration lead to magnetic susceptibility shifts (73, 107). Recently, Adler et al have shown that in rabbit renal proximal tubule suspensions the pH determined from the P_i chemical shift gives the cytoplasmic value, while that calculated from the uptake of the weak acid 5,5-dimethyl-2,4-oxazolidine dione (DMO) is heavily weighted toward the mitochondrial value (3). This result is not unexpected since the weak acid would be concentrated into the alkaline mitochondrial compartment. Gillies et al (82) used the P_i shift at low extracellular pH and the 2-deoxyglucose-6-P shift at high extracellular pH (where the P_i peak was obscured by the extracellular P_i signal) to show that Ehrlich ascites tumor cells regulated their intracellular pH, maintaining pH_i at 7.1 at extracellular pH of 6.8–7.2.

Cohen et al (51) showed that in the presence of valinomycin the intracellular P_i signal from hepatocyte suspensions was split into two peaks that corresponded to the cytoplasmic and mitochondrial pH. They confirmed the cytoplasmic location of the upfield P_i peak by adding fructose and showed that the pH measured from the shift of cytoplasmic F-1-P had the same value (169). The pH gradient behaved as predicted by the Mitchell chemiosmotic theory (121). It is unclear why the mitochondrial P_i was visible in this experiment, but the reason may well be related to the preparative methods used. The intramitochondrial P_i signal has also been seen in mitochondrial suspensions (135–137). This enabled Ogawa et al to determine that the proton stoichiometry of the mitochondrial H^+-ATPase was between 2 and 3. Again it is unclear why the mitochondrial P_i and ATP were visible in these experiments, although the presence of EDTA in the suspension buffer may have depleted the Ca^{2+} stores, leading to larger amounts of free intramitochondrial phosphates. It would be interesting to compare the ^{31}P spectra of mitochondria prepared and maintained in EDTA-containing and EDTA-free buffers, or in buffers containing different calcium concentrations.

Tissue pH has now been measured in many organs, both perfused and intact. The pH of perfused rat heart was 7.04–7.10 (87, 108, 115–117) while the pH in vivo was 7.35 ± 0.05 (87, 129). The high value in vivo probably reflects some contribution from plasma. The myocardial pH in perfused rabbit heart was 7.2 ± 0.01 (74). In rat skeletal muscle tissue the pH was 7.01 ± 0.01 (1, 106), while in studies of normal human skeletal muscle the pH was 7.04–7.13 (66, 153, 177, 185). The pH in vivo of both rat and rabbit brain was 7.13–7.14 (141). In perfused kidney the pH was 7.1–7.2 (2, 151), while in vivo it was 7.2–7.3 (170). The pH of perfused rat liver was 7.2–7.3 (48, 95, 97). Stevens et al measured a pH_i of 7.4 ± 0.6 from the P_i shift in rat liver in vivo, but a lower value of 7.2 ± 0.05 from the ^{19}F shift of 5FUdR (175). While normal tissue pH ranged from 7.0 to 7.3, it can vary over a much larger range, particularly towards the acid end in the face of metabolic stress (see below).

Magnetization-Transfer Studies

The various techniques of magnetization transfer have offered biochemists the opportunity to measure the activity of a number of key enzymes in vivo. However, early optimism was replaced by skepticism in some quarters when studies of the creatine kinase reaction in vivo yielded apparently contradictory results.

Several reviews covering work up to 1983 have recently appeared (6, 105). We summarize their contents to place subsequent developments in context. Brown et al (33) first showed that saturation transfer could be used in intact cells when they measured the rate of ATP synthesis in *E. coli*. Several more detailed studies of H^+-ATPase in *E. coli* (6) and *Saccharomyces cerevisiae* (5, 41) followed. Magnetization transfer techniques have been used to measure the rate of ATP synthesis in perfused rat heart (114, 167) and rat brain in vivo (168). The rate of ATP synthesis in perfused rat heart was measured as 2.8 ± 0.3 μmol s^{-1}/g dry wt (114). In a later study it was found that while the rate was lower in hearts from hypothyroid rats than in those from euthyroid rats (167), the P:O ratio was higher in the hypothyroid group. This is possibly an indication of varying efficiency of mitochondrial oxidative phosphorylation. In perfused rat kidney, Freeman et al (75) measured a synthesis rate of 0.27 μmol s^{-1}/g wet wt, with a pseudo first order rate constant for $[P_i]$ of 0.35 ± 0.048 s^{-1}. They measured a P:O ratio of 2.45. More recently they have performed the same study in vivo, measuring a synthesis rate of 0.28 μmol s^{-1}/g wet wt, a rate constant of 0.46 s^{-1}, and a P:O ratio of 2.9 (158). In contrast, Koretsky et al (104) have measured the same rate constant in vivo as 0.12 ± 0.03 s^{-1}. The reason for the discrepancy between the two groups is unclear. Finally, Shoubridge et al (168) measured the ATP synthetase rate in the rat brain as 0.33 μmol s^{-1}/g wet wt.

In all the above cases the P:O ratio, where measured, was close to 3, indicating that the mitochondrial H^+-ATPase/ATP synthetase was far from equilibrium, with little or no back-flux (hydrolysis) in vivo. In their study of the same enzyme in yeast, Alger et al (5) suggested, based on a much higher P:O ratio of 87, that there was significant back-flux and that H^+-ATPase might be close to equilibrium in yeast. However, a later series of studies by Campbell et al (41) indicated that the P : O ratio was indeed close to 3. These studies also demonstrated that under conditions where the glycolytic pathway is active, a significant contribution to the saturation transfer effect may arise from the glyceraldehyde-3-P dehydrogenase/P-glycerokinase couple. This substrate-level phosphorylation of ADP is the probable explanation of the large P : O ratio obtained by Alger et al. It should be noted that direct measurement of the back-flux has so far proved impossible, due to low sensitivity (6, 105).

Many groups have studied the creatine kinase reaction using magnetization transfer methods (15, 59, 114–116, 168, 182). The underlying goal has been to determine whether the reaction is at equilibrium or in a nonequilibrium steady state. This information, it was hoped, would clarify the role of the PCr/CK system. Matthews et al (115), using saturation transfer, found that the forward (PCr + ADP → Cr + ATP) and reverse (Cr + ATP → PCr + ADP) fluxes were the same in KCl-arrested perfused rat heart; however with increasing perfusion pressure the apparent reverse rate decreased, while there was no significant change in the forward rate. The discrepancy between the forward and reverse rates was also found to depend on substrate, being smaller for a given perfusion pressure with acetate than with glucose (115, 182). Different forward and reverse rates have also been observed in brain; for instance, Shoubridge et al measured the forward and reverse rates in rat brain in vivo as 1.46 and 0.68 μmol s^{-1}/g wet wt respectively (168). In contrast, Balaban et al found that the forward and reverse rates in rat head were equal (2 μmol s^{-1}/g wet wt) when measured by the 2D-NOESY technique (15). In the same study the forward and reverse rates in rat leg muscle were both 13 μmol s^{-1}/g wet wt. While various interpretations of the different forward and back fluxes have been proposed, including a role for the PCr/CK system as an energy shuttle between the mitochondrial and cytoplasmic compartments (134) and exchange with NMR-invisible pools (105), Matthews et al have pointed out that the data might also be explained by the presence of other ATP-consuming reactions (115).

Recent work by Ugurbil (182) has elegantly resolved these problems. In a series of experiments using different substrates he showed that the imbalance in the forward and back fluxes depended on the substrate utilized. Further, he reasoned that if the discrepancy arose as a result of an increase in the fraction of ATP consumed by non-CK pathways, then simultaneous saturation of P_i should correct it. The proposed "multiple-saturation" experiment was performed in perfused rat heart, and the rates were shown to be equal under all conditions. It is hoped that this series of experiments will lead to better controlled and more interpretable studies in the future.

Studies of Metabolic Stress

The utility of ^{31}P NMR in the monitoring of high-energy phosphates and pH in tissue makes it a particularly sensitive probe of metabolic stress. Many groups have taken advantage of this feature to study the effects of stresses such as hypoxia (21, 103, 148), anoxia (19, 21, 91, 129), ischemia (60, 74, 103, 108, 133, 166, 170, 179), hypoglycemia (22, 148), and seizure (142, 143, 148) in various perfused and intact organs. The PCr/P$_i$ ratio has often been used as an index of the energetic status of the tissue, but we note here that under certain circumstances this may be an oversimplification (142).

Thulborn et al (179) reported that cerebral ischemia in gerbils caused a simultaneous fall in PCr and ATP, rise in P$_i$, and drop in pH. In studies of cerebral hypoxia in rabbits (148) and dogs (90), similar behavior was observed. Behar (21) has measured the rates of lactate production by ^{1}H NMR and cerebral acidification by ^{31}P NMR simultaneously in the anoxic rat brain. This enabled him to determine the buffering capacity of rat brain. In both rabbits (148) and rats (22) hypoglycemia induced by insulin caused a fall in PCr and ATP and a rise in P$_i$ and pH shortly after the EEG went silent. These changes were reversed by infusion of glucose. Petroff et al (142, 148) reported that bicuculline-induced status epilepticus caused a reciprocal fall in cerebral PCr and rise in P$_i$, together with a fall in cerebral pH. They attributed the changes in the PCr and P$_i$ levels not to "energy failure," but rather to a shift in the creatine kinase equilibrium driven by the acidification observed. In a subsequent study Petroff et al (143) combined ^{1}H and ^{31}P NMR to confirm that the acidification was a consequence of lactate production. They also found that in the case of mild seizures, the pH recovered far more rapidly than the lactate level. Decorps et al (58) reported a decline in cerebral high energy phosphates and a cerebral acidosis following intraperitoneal injection of small doses of KCN.

Since the first studies of perfused heart in 1977 (80, 100), ^{31}P NMR has been used many times to study various aspects of cardiac energy metabolism (for reviews see 98, 139). Thus global (13, 80, 99, 100) and partial (99) ischemia, anoxia (129), and the effects of various drugs (12, 117, 125, 133, 146) have been subjects of scrutiny. Cardiac ischemia leads to a decline in PCr and ATP, a rise in P$_i$, and a fall in pH (80, 98, 100, 139) to a degree that depends on the duration of the insult. Bailey et al (13) observed that perfused rat heart sustained no long-term impairment of contractility if the duration of ischemia was 14 min or less. On reperfusion they found that PCr levels rapidly exceeded preischemic values, while ATP rose more slowly. This was interpreted as evidence of the role of CK as an energy shuttle between mitochondria and myofibrils. Longer periods of ischemia led to multiple P$_i$ peaks on reperfusion, and ATP and PCr remained below control levels. Changes in high energy phosphates during transient ischemia were insignificant, and pH dropped only slightly. Contractility was nonetheless adversely affected, indicating that other factors must be involved in contractile failure. Perfusion of rat hearts with insulin and glucose prior to ischemia slowed the decline in ATP while causing a more profound acidification of the tissue (12). The effects of various drugs on pH and energy metabolism in the perfused heart have been studied. Propanolol (125, 146) and verapamil (133) both slowed the decline in PCr during ischemia, but only verapamil slowed the loss of ATP (133). Ouabain and R02-2985 (117) both caused a fall in PCr in the normoxic heart but left ATP unaffected, while epinephrine caused both PCr and ATP to drop (117). Effects of different substrate regimes on high energy phosphates have been documented in normal (115, 117, 182) and diabetic (144, 145) hearts.

Studies of metabolic insult in other organs have been fewer. Radda and his colleagues have investigated the effects of cold and warm ischemia on kidneys excised from rats (165, 166). They found that the lower temperature slowed the loss of ATP and tissue acidification, as well as decreasing the ultimate pH drop. More recent work (44, 170) has confirmed their findings in the intact animal. In studies of acute renal failure (ARF) in rats in vivo, Siegel et al documented similar changes following the onset of renal ischemia, and showed that functional recovery correlated with recovery of renal ATP levels, which were severely depleted for several hours following 45 min of ischemia (170). They further showed that postischemic infusion of ATP-$MgCl_2$ led to increased ATP levels at 2 hr, despite the lack of difference between control and ATP-$MgCl_2$ treated animals at the end of the infusion. This suggested that depletion of total adenine nucleotide pool size may be a causative factor in the onset of ischemic ARF. Siegel et al's suggestion has been supported by recent experiments in which AMP-$MgCl_2$ had the same effect as ATP-$MgCl_2$ (176). Glycerol infusions also produce ARF, and ^{31}P NMR studies have shown that such infusions severely deplete renal ATP levels (44). It should be noted, however, that fructose infusion also depletes renal ATP levels (44, 104), yet no ARF ensues.

Several groups have studied the effects of ischemia (85, 118), glycerol and fructose infusions (94, 95, 104), and hormones (49, 118) on perfused and in vivo liver. In the perfused starved rat liver, ischemia (94) caused a rapid drop in hepatic ATP; ATP became undetectable in 5–6 min. Also, pH fell to 6.8 in 15 min and remained at that level. Reflow after 30 min led to rapid resynthesis of ATP to about 50% of preischemic levels; this phenomenon is similar to that seen in the postischemic kidney (170). Fructose and glycerol infusions in perfused and in vivo rat liver elicited dose-dependent decreases in tissue ATP. ATP recovered slowly in the case of fructose (94, 95, 104) but did not recover in the case of glycerol (94). Cohen has recently reported increased levels of GPC following administration of glucagon to perfused rat liver (48a, 49).

Human Studies

The availability of horizontal magnets with fields of 1.5–2.0 T and bore sizes of 30 cm and up has enabled several groups to begin studies of metabolism in humans. These studies can be grouped as follows: (a) investigations of normal and aberrant muscle metabolism; (b) investigations of cerebral metabolism in neonates; and (c) investigations of metabolism in the adult brain and abdominal organs. In this section we summarize recent findings in this area.

The identification of errors in muscle metabolism by ^{31}P NMR requires that the characteristics of normal human muscle, both at steady state and following a defined perturbation, be fully established (153, 177). To this end Radda and his collaborators have studied the forearm muscles of a large number of normal subjects and found the following: At rest, pH_i was 7.03 ± 0.03, $PCr/(PCr + P_i)$ was 0.91 ± 0.2, PCr/ATP was 4.7 ± 0.7, P_i/ATP was 0.48 ± 0.13, [PCr] was 25.9 ± 3.9 mmol/kg wet wt and [P_i] was 2.6 ± 0.5 mmol/kg wet wt [these values are in agreement with those of Wilkie et al (185)]; during exercise, pH_i was a characteristic function of [PCr], and only began to decline once $\sim 60\%$ of the PCr had been used up; in ischemic muscle there was no metabolic recovery after exercise, indicating that glycolysis ceased when exercise was stopped; during recovery the rate of P_i decrease was faster than the rate of PCr synthesis, and during recovery the P_i signal often fell below its normal resting value (153, 177). This last feature has been suggested by Taylor et al (177) to be a consequence of the transport of P_i into mitochondria where it is no longer detectable by NMR. The slower resynthesis of PCr was taken to reflect mitochondrial oxidative capacity and to be an index of mitochondrial function.

A number of muscle disorders have been examined. In a patient with McArdle's syndrome, a phosphorylase deficiency, Ross et al found that rather than acidifying during aerobic or anaerobic exercise, the muscle pH rose (159). In four patients pH_i, which was a normal 7.05 at rest, rose to 7.2–7.3 (153). This alkalinization was a result of the breakdown of PCr, a H^+-consuming reaction, in the absence of any lactic acid production. In the case of patients deficient in the debranching enzyme amylo-1,6-glucosidase, a condition that also limits glycogen mobilization, a similar alkalinization was observed (153). In this group the resting pH_i was higher than normal (7.09–7.17). Two groups have described patients with phosphofructokinase deficiency (45, 66). The condition was characterized, not unexpectedly, by the appearance following exercise of a peak in the phosphomonoester region attributed to hexose monophosphate. Several cases of mitochondrial myopathy have also been examined. Gadian et al (79) studied a patient with myopathy, ophthalmoplegia, and raised basal metabolic rate, and found an elevated cytoplasmic P_i ($PCr/P_i = 3.8 \pm 0.7$; normal $= 10.0$ ± 2.7), despite normal PCr and resting pH. In a study of two patients with NADH-coenzyme Q reductase deficiency, Radda et al (152) found a lower rate of PCr resynthesis following exercise, consistent with a defect in the electron transport chain. These and most other muscle conditions so far studied have as their underlying cause a genetic defect. However, a patient with postviral exhaustion/fatigue syndrome was recently reported to exhibit anomalous muscle behavior (9), specifically an abnormally early and profound intracellular acidification at a lower degree of PCr depletion than in controls.

Two groups have reported ^{31}P NMR studies of the neonate brain. Cady et al (37) followed the cerebral metabolism and pH of infants born at 33–40 wk gestation, at ages 44 hr to 17 days. In addition to the ATP, PCr, phosphodiester, and P_i peaks seen in adult rat (21) and rabbit (148) brain at 1.9 T, they saw a large phosphomonoester peak, which they assigned to ribose-5-phosphate. More recent studies suggest that this phosphomonoester peak is actually phosphorylcholine and/or phosphorylethanolamine (103). The PCr/P_i ratio varied from 1.7 in normal brain to 0.2 in infants with severe birth asphyxia. Although Cady et al concluded that there was no pattern in the cerebral pH, Petroff & Prichard (140) have pointed out that the PCr/P_i ratio correlated well with the measured pH. They suggested that both may reflect the increasing proportion of cerebrospinal fluid in the region of brain that was examined, with increasing pH and decreasing PCr/P_i associated with a higher proportion of CSF. Younkin et al (189) reported similar spectra of neonate brain, and measured a cerebral pH of 7.1 ± 0.1. They too saw a large phosphomonoester peak, which they assigned to phosphorylcholine and/or phosphorylethanolamine. In a follow-up to the work of Cady et al, Hope et al (91) once again found high levels of PME, which they assigned this time as phosphorylethanolamine. Interestingly, they could detect no differences between the spectra of normal and birth-asphyxiated infants on the first day, but by the second to ninth day saw a significant decline in PCr/P_i. The availability of magnets of 1.5–2.3 T with 0.6–1.0 m bore has made it possible to examine adult brain (29, 30, 153) and heart (28). These spectra do not contain the high levels of phosphomonoesters seen in neonatal brain.

Finally, Griffiths et al (84) reported a study of the effects of chemotherapy on a human tumor in situ. The first ^{31}P NMR spectrum, collected when the patient was already on chemotherapy, was characterized by a large PME peak as well as the usual P_i, PDE, PCr, and ATP peaks, and a pH of 7.1. Later spectra showed a drop in PCr with no change in ATP, which Ng et al have found to be characteristic of continued tumor growth (131). This correlated with the clinically observed resistance of the tumor to chemotherapy.

^{13}C NMR

^{13}C NMR provides a potential window into wide areas of metabolism. Due to its low natural abundance (1.1%), ^{13}C NMR studies are restricted to investigations of highly concentrated compounds such as glycogen, fats, and triglycerides, or to studies utilizing ^{13}C-enriched substrates. This low natural abundance does, however, permit the use of labeling techniques, analogous to ^{14}C tracer methods, to determine the fluxes through various metabolic pathways. An advantage over the ^{14}C techniques is that the positions of the label within the metabolite can be easily determined, thereby allowing a wider variety of applications.

Several reviews detailing the application of ^{13}C NMR to studies of the metabolism of cells have appeared (96, 164). We therefore restrict our discussion to the application of ^{13}C NMR to studies of tissues and organs. Our discussion of the application of ^{13}C NMR to metabolism is divided into four areas: (a) perfused organ and in vivo studies of glycogen and fat metabolism, (b) the use of ^{13}C substrates in the elucidation of various metabolic fluxes, (c) a technique for improved spectral resolution, and (d) technical and instrumental considerations for ^{1}H decoupling of large biological samples.

It should be noted that ^{13}C NMR, although highly developed in the area of cell metabolism (20, 50, 52, 61, 63, 111–113, 180), has been less extensively applied to tissues and intact organs and has yet to gain the wide usage of ^{31}P NMR. As such, many reports of its utilization demonstrate its applicability to various metabolic areas rather than answering specific biochemical questions. The area of glycogen metabolism has gained the widest application, and is discussed here in detail. However, it should be noted that the great potential of ^{13}C NMR for elucidating in vivo flux rates of various pathways via the use of specifically labeled substrates should be an area of expanding utilization. Therefore we discuss as examples experiments utilizing ^{13}C-enriched substrates, showing the types of information that can be gained and the corresponding biochemical interpretations.

Glycogen and Fat Metabolism

Before discussing the study of glycogen via ^{13}C NMR it is necessary to discuss its visibility in the ^{13}C NMR spectrum since the extent and nature of its visibility is a critical factor (130).

Because of its relatively large molecular weight, $\sim 10^7$ d, and therefore its expected slow rotation, glycogen would be expected to contribute only very broad resonances to the ^{13}C NMR spectrum. However, relatively narrow resonances were detected from hepatic glycogen in perfused livers (53) and from the abdomen of an intact rat (7) after administration of labeled metabolites. These studies demonstrated that liver glycogen was at least partially visible, suggesting the existence of some internal motion. Subsequently Sillerud & Shulman (172) demonstrated that despite its large molecular weight glycogen was 100% visible in perfused rat liver. This determination was made by comparing the amount of glucose released after complete digestion of glycogen to glucose. Digestion was carried out in situ by addition of glucagon and in vitro, following a variety of glycogen extraction techniques, using amyloglucosidase. In all cases stoichiometric conversion of glycogen to glucose was observed, indicating that hepatic glycogen was 100% visible in the ^{13}C NMR spectrum. Similar results have been reported for hepatic (174) and cardiac glycogen (130). Cohen (48a, 49) has reported a visibility of 45% using the indirect means of comparing perfusate glucose to liver glycogen. Despite the indirectness of this measurement, it seems to be well established from the digestion comparisons that hepatic and cardiac glycogen are 100% visible, with the visibility attributed to rapid internal segmental motion ($\tau_r \approx 10^{-8}$ s) (172).

Due to its high concentration, 100% NMR visibility, and short T_1[~ 250 msec at 8.5 and 1.9 T (4, 172), allowing rapid pulsing techniques to be applied] natural abundance signals from the C-1 carbon of glycogen have been detected in livers at a variety of field strengths (4, 156, 172, 174). The ability to detect natural abundance levels of glycogen has allowed the investigation of various aspects of glycogen metabolism, including glucagon-stimulated glycogenolysis in perfused rat liver (172) and in vivo (156), without the need for costly labeled substrates. Stevens et al (174) used the natural abundance ^{13}C spectra of phosphorylase kinase–deficient rats and normal rats to demonstrate the accumulation of glycogen in the pathological rats. From these studies it is clear that the natural abundance ^{13}C NMR spectrum can be studied without the use of costly labels, making it biochemically applicable to larger animals. Canioni et al (42) and Reo et al (156) have noted that care must be taken in the surgical preparation of animals for these experiments; otherwise stress-related glycogenolysis may occur, thereby eliminating glycogen from the ^{13}C NMR spectrum. This stress-related glycogenolysis should not be a problem in noninvasive experiments. It is important to note that the C-1, C-6, and C-2–C-5 resonances of glycogen have been detected and resolved at 1.9 T (4), making these techniques applicable to field strengths that are currently available for studies of humans.

In addition to natural abundance levels of glycogen, triglycerides and phospholipids have been detected in a number of systems (16–18, 27, 42, 65, 132). To demonstrate the applicability of the natural abundance ^{13}C spectra of fatty acids to metabolism, Canioni et al (42) supplemented the diets of two groups of rats with carbohydrates (group A) and poly-unsaturated fats (group B). Comparison with rats fed a control diet revealed increases in monounsaturated fats and carbohydrates in group A and in polyunsaturated fats in group B.

Natural abundance spectra of small metabolites from human muscle and from frog muscle and brain (16–18, 65) have been reported. However, these spectra have been obtained on excised tissues and only after acquisition times of 10–15 hr, thereby limiting their applicability to tissue biopsies where metabolic degradation is not a limitation.

^{13}C Label Studies

Although information about highly concentrated storage compounds such as triglycerides and glycogen can be gained by investigations of natural abundance ^{13}C NMR spectra, studies of less concentrated metabolites are not possible with the natural abundance spectra. However, labeled compounds can be used to investigate the products of various metabolic pathways, yielding information about the relative fluxes of these pathways.

Typically, ^{13}C NMR studies involving enriched compounds monitor one or more of the following properties: (a) the rate of incorporation or degradation of a label, (b) the level of steady-state labeling, and (c) the position of the label in the carbon skeleton.

The infusion of 1-^{13}C glucose and its incorporation into hepatic glycogen in vivo was first demonstrated by Alger et al (7). The insulin-stimulated incorporation of glucose into myocardial glycogen in guinea pig and into hepatic glycogen in rat in vivo have been utilized by Neurohr et al (130) and Reo et al (156) to measure the rate of glycogen synthesis. In further experiments, Neurohr et al (130) observed that there was no significant exchange of glycogen after chasing with unlabeled glucose, while during subsequent anoxia ^{13}C-labeled glycogen was metabolized preferentially. Reo et al (156) and Sillerud & Shulman (172) have measured the rate of glycogen breakdown after infusion of ^{13}C-labeled glucose and observed an increased rate of degradation in the presence of glucagon. Using an infusion of ^{13}C acetate, Cross et al (54) have observed incorporation of ^{13}C label into acetoacetate and β-hydroxybutyrate in rat liver in vivo.

Cohen et al (49) used 2-^{13}C–labeled glycerol in perfused mouse liver to determine the extent of pentose phosphate cycling. 2-^{13}C glycerol will be directly incorporated into the C-2 and C-5 positions of glucose, while pentose phosphate cycling will place the label in the C-1 position. The labeling in the C-1 position indicated that $\sim 10\%$ of the hexoses had cycled through the pentose phosphate pathway. Bailey et al (11) and Neurohr et al (128), studying perfused rat heart and in vivo guinea pig heart, utilized an infusion of 2-^{13}C acetate to evaluate the activity of TCA cycling relative to glutamate incorporation. Both investigators observed incorporation of the label into the C-4 (directly) and the C-2 positions, indicating significant scrambling of the label by TCA cycling.

Resolution in ^{13}C NMR

Although ^{13}C NMR has the advantage of high spectral dispersion, ~ 200 ppm, studies at low fields (1.9 T) are still hampered by poor resolution. Bendall et al (26) have proposed two experiments to utilize the multiplicity of the ^{13}C-^{1}H coupling to obtain edited ^{13}C spectra. In the first, they proposed that overlap due to broad lipid resonances could be eliminated and spectra containing only CH, CH$_2$, or CH$_3$ carbons could be obtained. This experiment involves the transfer of polarization from the ^{1}H spins to the ^{13}C spins to which they are coupled. In the other, similar experiment, they proposed that separate spectra of CH and CH$_3$ groups and CH$_2$ groups could be obtained by using a ^{13}C spin echo sequence with gated ^{1}H decoupling during the ^{13}C evolution periods. Although this technique to date has only been applied to excised tissues by Bendall et al (26), it promises to be of utility for further improvement of spectral resolution in ^{13}C NMR experiments.

^1H Decoupling Considerations

Since nearly all ^{13}C NMR experiments require ^{1}H decoupling during acquisition, the ability to decouple the relevant ^{13}C spins is an area of primary interest. Although effective decoupling has been easily achieved for perfused systems, where volumes are small and required powers are low, large in vivo systems can present problems because of excessive heating due to ^{1}H RF. In this area, development has proceeded along two lines, the design of ^{1}H and ^{13}C coil systems, and the application of broad-band composite pulse decoupling techniques.

Due to the spatial constraints imposed by most organs, concentric ^{1}H and ^{13}C surface coils have often been used (2, 42, 155, 174). Significant improvements in the elimination of coupling between the coils have been described by Reo et al (155). However, these coils inherently provide a decoupled ^{1}H volume greater than that of the observed ^{13}C spins, thereby leading to extra RF absorption and heating. The design of a doubly-tuned coil, as described by Cross et al (55) and applied by den Hollander et al (62), allows for the matching of the ^{13}C and ^{1}H volumes, thereby minimizing heating. A second approach described by Cross (54) is the use of broad band composite pulse decoupling techniques (109). These pulse techniques allow decoupling of the ^{13}C NMR spectrum at one fifth the power required for CW decoupling. It is clear that with the application of ^{13}C NMR studies to large animals and humans at 1.9 T and higher field strengths, these broadband decoupling schemes will be of great importance.

^1H NMR

Unlike ^{13}C and ^{31}P NMR of tissues and organs, high resolution ^1H NMR spectra of metabolites cannot in general be obtained using simple pulse acquire techniques. Before high-resolution spectra can be obtained the large tissue H_2O resonance must be suppressed and broad resonances due to proteins and lipids must be reduced or eliminated. In addition, the small chemical shift range, ~ 10 ppm, and extensive multiplicity due to homonuclear j-coupling often result in severe spectral overlap, making quantitation and even observation of some metabolites difficult. Owing to these limitations, ^1H NMR has not been utilized to the same extent as ^{13}C and ^{31}P NMR in metabolic studies. However, the greater sensitivity (~ 60 and 16 times that of ^{13}C and ^{31}P NMR, respectively), the 100% natural abundance, and the ubiquity of protons in biological compounds make ^1H NMR an attractive tool for the biochemist, physiologist, and clinician studying the metabolism of tissues and intact organs.

In response to the advantages provided by ^1H NMR a flurry of technical development has occurred over the past few years, providing methods that have made the application of ^1H NMR to the study of in vivo metabolism possible. We discuss in detail the following recent technical developments: (a) methods of water suppression, (b) techniques to improve spectral resolution, and (c) sequences to observe ^1H-^{13}C coupled protons. We then summarize several studies that have taken advantage of these techniques. In beginning this discussion of ^1H techniques, we draw the reader's attention to the pioneering work of Campbell and co-workers (32, 39, 40) on cell suspensions. In several cases the spin echo techniques described by Campbell and co-workers have provided a basis for the spin echo techniques presently in use in in vivo studies.

Water Suppression

Before collection of ^1H NMR spectra, the water resonance, which is $\sim 10^4$–10^5 times larger than the resonances of most metabolites, must be suppressed to allow the signals from small metabolites to be digitized.

The form of water suppression most commonly used in in vivo studies is presaturation of the H_2O resonance using single-frequency irradiation of the H_2O resonance (150, 188). This method has been used in frog muscle (188) and brain (24) to suppress the H_2O resonance adequately. A variant of this technique utilizes a Dante sequence (124) to saturate the ^1H resonance of H_2O in perfused heart (181). However, this method of suppression commonly leaves a water resonance that is still much larger than the remainder of the ^1H spectrum. Although this degree of suppression may be adequate for digitization of small metabolite signals, the remaining H_2O signal may be large enough to require the use of baseline corrections in the course of spectral interpretation (24).

A second approach to water suppression takes advantage of the short T_2 of tissue water (40). Since the T_2s of most metabolites are significantly longer than the T_2 of tissue water, a spin echo sequence can be utilized to differentially suppress the water resonance. This technique has been applied at 1.9 T on rabbit brain by Behar et al (25) and at 8.5 T by Williams et al (186) while studying muscle using a Carr-Purcell Meiboom-Gill (CPMG) sequence (43). Additional suppression can be gained by combining the two techniques, presaturation and spin echo sequences, to suppress the water resonance in the in vivo rat brain at 8.5 T by a factor of 10^3 (162).

A third approach to water suppression is the use of semiselective and selective pulse trains that do not significantly perturb the H_2O resonance. As described by Plateau & Gueron (147) and Hore (92), these sequences derive their selectivity on the basis of chemical shift evolution during interpulse delay periods. The Plateau-Gueron sequence has been used by Arus et al (10) in a pulse acquire experiment using intact frog muscles at 11.1 T and 20° C. In addition to the excellent suppression obtained, a number of exchangeable resonances in the downfield spectrum were detected, which would not be visible if the water resonance were saturated. Among these, carnosine was used as a ^1H marker of pH. Additionally, resolution of creatine and phosphocreatine resonances was reported. However, many of these exchangeable resonances, due to their exchange characteristics, broaden significantly at temperatures above 30° C, making resolution and detection difficult.

The Hore 1331 sequences can be easily incorporated into spin echo sequences, making use of the short T_2 of tissue water, to provide in vivo water suppression by a factor of 10^4–10^5 (89). Gadian et al (77) have used a variant of this sequence to study histidinemic mice. Schnall et al (163) have used a 1331 pulse acquire sequence in the detection of ^1H metabolites in cat brain at 1.9 T. Similar suppression factors of 10^4–10^5 have been achieved by Jue et al (101) using a spin echo sequence composed of frequency-selective Dante pulse trains.

The spectral selectivity of these semiselective and selective sequences allows only certain regions of the ^1H spectrum to be observed during a single acquisition, thereby reducing the number of ^1H resonances and metabolites that can be detected. Additionally, anomalies in homonuclear j-modulation (31, 89) may occur owing to the spectral selectivity of the selective pulse. These anomalies may be difficult to interpret when a nonhomogeneous coil is used for excitation. However, at this time these semiselective spin echo sequences provide the greatest degree of suppression of the ^{1}H water resonance in vivo.

Spectral Resolution

Owing to the small chemical shift range, the extended multiplet structure due to j-coupling, and the ubiquity of protons, the natural ^{1}H spectrum generally contains extensive resonance overlap. Specifically, the broad resonances of proteins and fats often obscure and distort the intensities of most major metabolites (31). Before ^{1}H metabolites can be quantitatively interpreted these broad, intense resonances must be eliminated.

The broad resonances of proteins and some fats may be eliminated by exploiting their relatively short T_2 by use of a spin echo sequence (40). This technique has proved most successful in rat brain; virtually all brain lipids can be eliminated by a spin echo sequence with a 68 msec delay (162). However, for tissues other than brain significant fat resonances may continue to obscure the ^{1}H spectrum, even with longer delays. Rothman et al have described a double resonance technique (160, 162) that takes advantage of the effects of homonuclear j-modulation as described by Brown et al (32) and Campbell & Dobson (38). This technique uses gated selective CW decoupling of the j-coupled partner of the resonance of interest during a spin echo sequence. When the spectra collected in the absence and presence of decoupling are subtracted, all resonances not affected by the decoupling cancel. The resonance of interest, i.e. the resonance j-coupled to the spins that were irradiated, inverts its signal in the decoupled spectrum, thereby adding coherently in the difference spectrum. This technique has subsequently been used by Williams et al (186) to detect lactate in muscle. In addition to allowing for the observation of resonances in the absence of spectral overlap, the technique may achieve signal-to-noise enhancement through exponential multiplication rather than through Gaussian multiplication (67), which is required in the presence of multiple resonances.

Since Rothman's report (160, 162) analogous editing schemes have been reported that use semiselective and selective pulses in the spin echo sequence (89, 101). These sequences provide the advantage of additional water suppression. For these sequences, j-modulation can be induced or inhibited by the alternate application of a Dante pulse to the coupled spins. An additional benefit of these sequences is that the selectivity of the editing pulse or RF is not limited by the coupling constant. For CW decoupling–based editing experiments, the decoupling intensity in H_2 must be much larger than the effective coupling constant (8). The selectivity of the Dante editing pulse may be adjusted to eliminate unwanted subtraction artifacts due to off-resonance decoupling effects. A major limitation of the Dante-based editing techniques is the requirement of an accurate 180° pulse by the second RF source.

^1H-$[^{13}C]$ NMR

A major limitation to the utilization of ^{13}C NMR in in vivo systems is its low sensitivity. This sensitivity, even after the addition of ^{13}C-enriched substrates, is often insufficient to provide adequate time resolution of the initial kinetics of many metabolic pools. A technique describing the observation of ^{13}C-coupled protons by alternate decoupling of coupled ^{13}C nuclei during acquisition of ^1H spectra has been described (171). However, this technique is limited by the necessity of observing the edited intensity distributed among three resonances, of which two are inverted relative to the central resonance. Thus resolution may be impaired by the extended nature of the edited resonance and its overlap with other ^1H resonances that are ^{13}C-coupled.

A variant of the ^{13}C editing experiment described by Bendall et al (26), analogous to the ^1H editing techniques (160, 162), has been utilized by Rothman et al (161) in the measurement of the kinetics of lactate and glutamate in rat brain during infusion of 1-^{13}C glucose. This technique is to apply gated decoupling of the ^{13}C nuclei during a ^1H spin echo sequence to obtain inversion of the ^1H spins coupled to ^{13}C nuclei relative to those bonded to ^{12}C nuclei. As noted by the authors, this technique has the advantage of allowing the measurement of the fractional enrichment of the specific resonance, thereby allowing determination of the kinetics of metabolic pathways.

In Vivo ^1H NMR

Currently, ^1H NMR is most commonly employed to view changes in the C-3 lactate resonance under various physiological stresses (22, 24, 25, 163, 186). However, a wealth of information regarding many other metabolites may be gained. The variety and richness of the ^1H NMR spectrum can be seen in the application of ^1H NMR in conjunction with ^{31}P NMR for monitoring the effects of hypoglycemia upon rat brain in vivo (22). Correlations between the ^1H resonances of aspartate, glutamate, glutamine, and lactate and the ^{31}P resonances of inorganic phosphate, phosphocreatine, nucleotide triphosphates, and the sum of the nucleotide di- and monophosphates have been determined. As the techniques described in the previous sections become widely used, ^{1}H NMR will undoubtedly provide an effective tool for investigating metabolism in vivo.

^{19}F NMR

The ^{19}F nucleus has not been used as much as other nuclei in the study of metabolism in vivo, largely because under normal circumstances there are no fluorine-containing metabolites of sufficient concentration in animal tissues. However, a number of fluorinated drugs are currently in clinical use, and ^{19}F NMR offers a powerful method for monitoring their uptake and metabolism in vivo. Stevens et al (175) have used ^{19}F NMR to follow the metabolism of 5-fluorouracil (5FU) in rats, both in the kind of tumor against which this agent is normally directed and in liver. Following intravenous injection of 5FU Stevens et al were able to follow several intermediates in 5FU metabolism. In the implanted tumor 5FU was converted to 5FdUMP, the active antitumor metabolite. In contrast, no 5FdUMP was seen in liver, but the catabolites of 5FU, namely dihydro-fluorouracil and fluoro-β-alanine, were detected instead. The researchers noted that in the presence of thymidine a peak corresponding to 5FUdR was seen in the liver. The pK_a of 5FUdR was 7.5 and the pH measured from this probe in liver was 7.2 ± 0.05, compared with 7.4 ± 0.6 from the P_i peak in parallel ^{31}P NMR experiments.

There have been several studies of the uptake of fluorinated anesthetics into various tissues using ^{19}F NMR. Wyrwicz et al (187) followed the incorporation of halothane, methoxyflurane, and isoflurane into rabbit brain and found evidence that these agents partition into a hydrophilic as well as a lipophilic environment. Furthermore, the rate of clearance of the anesthetics was very low; ^{19}F signals were still detectable 98 hr after the anesthesia had ended. Burt et al (35, 36) have suggested that ^{19}F NMR spectroscopy of halothane and isoflurane can be used to differentiate normal and tumor tissue.

Finally we mention the recent introduction of mono-, di-, and trifluoro-methylamine for the measurement of cytoplasmic pH (64, 178). These three indicators have pK_a of 8.5, 7.3, and 5.9 respectively. Since their ^{19}F signals are easily resolvable they may be simultaneously introduced, permitting coverage of a wide pH range. They can be loaded into cells as methyl esters, which readily permeate the plasma membranes of many cells. The most generally useful indicator is dimethylfluoroalanine. Its proton-decoupled ^{19}F spectrum is AB in character, and the center-peak spacing between the two fluorine resonances is pH-dependent so no other internal standard is required. In the case of the mono- and trifluoroalanine a single resonance is observed with a pH-dependent chemical shift. Using these indicators

Taylor & Deutsch have found a regulation of cytoplasmic pH in lymphocytes (178) similar to that described by Gillies et al (82).

CONCLUSION

In this review we have tried to convey to the reader the range of information that is available from NMR studies of tissue metabolism. For many years NMR studies in vivo have been described as "promising;" it is our belief that in the past five years the field has fulfilled its promise. Furthermore, the noninvasive nature of the NMR measurement should make it the method of choice in many studies of whole organ physiology and pathophysiology. This noninvasive aspect of the technique, together with the advent of 0.6–1.0 m high–field strength magnets, leads us to expect rapid growth in NMR studies of normal and diseased states in humans in the next few years.

ACKNOWLEDGMENTS

The authors wish to thank Drs. K. L. Behar, D. L. Rothman, S. Campbell, J. R. Alger, and T. Ogino for helpful comments, and C. Jones for secretarial assistance. This work was supported by NSF grant PCM 8402670.

Literature Cited

1. Ackerman, J. J. H., Grove, T. H., Wong, G. G., Gadian, D. G., Radda, G. K. 1980. *Nature* 283:167–70
2. Ackerman, J. J. H., Lowry, M., Radda, G. K., Ross, B. D., Wong, G. G. 1981. *J. Physiol.* 319:65–79
3. Adler, S., Shoubridge, E. A., Radda, G. K. 1984. *Am. J. Physiol.* 247:C188–96
4. Alger, J. R., Behar, K. L., Rothman, D. L., Shulman, R. G. 1984. *J. Magn. Reson.* 56:334–37
5. Alger, J. R., den Hollander, J. A., Shulman, R. G. 1982. *Biochemistry* 21:2957–63
6. Alger, J. R., Shulman, R. G. 1984. *Quart. Rev. Biophys.* 17:83–124
7. Alger, J. R., Sillerud, L. O., Behar, K. L., Gillies, R. J., Shulman, R. G., et al. 1981. *Science* 214:660–62
8. Anderson, W. A., Freeman, R. 1962. *J. Chem. Phys.* 37:85–103
9. Arnold, D. L., Radda, G. K., Bore, P. J., Styles, P., Taylor, D. J. 1984. *Lancet* 1:1367–69
10. Arus, C., Barany, M., Westler, W. M., Markley, J. L. 1984. *FEBS Lett.* 165:231–37
11. Bailey, I. A., Gadian, D. G., Matthews, P. M., Radda, G. K., Seeley, P. J. 1981. *FEBS Lett.* 123:315–18
12. Bailey, I. A., Radda, G. K., Seymour, A. M. L., Williams, S. R. 1982. *Biochim. Biophys. Acta* 720:17–27
13. Bailey, I. A., Seymour, A. M. L., Radda, G. K. 1981. *Biochim. Biophys. Acta* 637:1–7
14. Balaban, R. S., Gadian, D. G., Radda, G. K. 1981. *Kidney Int.* 20:575–79
15. Balaban, R. S., Kantor, H. L., Ferretti, J. A. 1983. *J. Biol. Chem.* 258:12787–89
16. Barany, M., Arus, C., Chang, Y.-C. 1985. *Magn. Reson. Med.* 2:289–95
17. Barany, M., Doyle, D. D., Graff, G., Westler, W. M., Markley, J. L. 1982. *J. Biol. Chem.* 257:2741–43
18. Barany, M., Doyle, D. D., Graff, G., Westler, W. M., Markley, J. L. 1984. *Magn. Reson. Med.* 1:30–43
19. Barbour, R. L., Sotak, C. H., Levy, G. C., Chan, S. H. P. 1984. *Biochemistry* 23:6053–62
20. Barton, J. K., den Hollander, J. A., Hopfield, J. J., Shulman, R. G. 1982. *J. Bacteriol.* 151:177–85
21. Behar, K. L. 1985. *Nuclear magnetic resonance of the brain: Evaluation of ^{1}H, ^{31}P and ^{13}C spectra in normal and pathological states* in vivo. PhD thesis. Yale Univ. 275 pp.
22. Behar, K. L., den Hollander, J. A.,

Petroff, O. A. C., Hetherington, H. P., Prichard, J. W., Shulman, R. G. 1985. *J. Neurochem.* 44:1045–55

23. Deleted in proof
24. Behar, K. L., den Hollander, J. A., Stromski, M. E., Ogino, T., Shulman, R. G., et al. 1983. *Proc. Natl. Acad. Sci. USA* 80:4945–48
25. Behar, K. L., Rothman, D. L., Shulman, R. G., Petroff, O. A. C., Prichard, J. W. 1984. *Proc. Natl. Acad. Sci. USA* 81:2517–19
26. Bendall, M. R., den Hollander, J. A., Arias-Mendoza, F., Rothman, D. L., Behar, K. L., Shulman, R. G. 1985. *Magn. Reson. Med.* 2:56–64
27. Block, R. E. 1982. *Biochem. Biophys. Res. Commun.* 108:940–47
28. Bottomley, P. A. 1985. *Science* 229:769–72
29. Bottomley, P. A., Foster, T. B., Darrow, R. D. 1984. *J. Magn. Reson.* 59:338–42
30. Bottomley, P. A., Hart, H. R. Jr., Edelstein, W. A., Schenck, J. F., Smith, L. S., et al. 1984. *Radiology* 150:441–46
31. Brindle, K. M., Porteous, R., Campbell, I. 1984. *J. Magn. Reson.* 56:543–47
32. Brown, F. F., Campbell, I. D., Kuchel, P., Rabenstein, D. L. 1977. *FEBS Lett.* 82:12–16
33. Brown, T. R., Ugurbil, K., Shulman, R. G. 1977. *Proc. Natl. Acad. Sci. USA* 74:5551–53
34. Burt, C. T., Cohen, S. M., Barany, M. 1979. *Ann. Rev. Biophys. Bioeng.* 8:1–25
35. Burt, C. T., Moore, R. R., Roberts, M. F. 1983. *J. Magn. Reson.* 53:163–66
36. Burt, C. T., Moore, R. R., Roberts, M. F., Brady, T. J. 1984. *Biochim. Biophys. Acta* 805:375–81
37. Cady, E. B., Costello, A. M., Dawson, M. J., Delpy, D. T., Hope, P. L., et al. 1983. *Lancet* 1:1059–62
38. Campbell, I. D., Dobson, C. M. 1975. *J. Chem. Soc. Chem. Commun.* 18:750–51
39. Campbell, I. D., Dobson, C. M., Jeminet, G., Williams, R. J. P. 1974. *FEBS Lett.* 49:115–19
40. Campbell, I. D., Dobson, C. M., Williams, R. J. P., Wright, P. E. 1975. *FEBS Lett.* 57:96–99
41. Campbell, S. L., Jones, K. A., Shulman, R. G. 1985. *FEBS Lett.* In press
42. Canioni, P., Alger, J. R., Shulman, R. G. 1983. *Biochemistry* 22:4974–80
43. Carr, H. Y., Purcell, E. M. 1954. *Phys. Rev.* 94:630
44. Chan, L., Ledingham, G. G., Dixon, J. A., Thulborn, K. R., Waterton, J. C., et al. 1982. In *Acute Renal Failure*, ed. H. E. Eliahou, pp. 35–41. London: Libbey
45. Chance, B., Eleff, F., Bank, W., Leigh, J. S. Jr., Warnell, R. 1982. *Proc. Natl.*

Acad. Sci. USA 79:7714–18
46. Chance, B., Eleff, S., Leigh, J. S. Jr., Sokolow, D., Sapega, A. 1981. *Proc. Natl. Acad. Sci. USA* 78:6714–18
47. Chance, B., Radda, G. K., Seeley, P. J., Silver, I., Nakase, Y., et al. 1979. In *NMR and Biochemistry*, ed. S. J. Opella, P. Lu, pp. 269–81. New York: Dekker
48. Cohen, S. M. 1983. *Hepatology* 3:738–49
48a. Cohen, S. M. 1983. *J. Biol. Chem.* 258:14294–308
49. Cohen, S. M. 1984. *Fed. Proc.* 43:2657–62
50. Cohen, S. M., Glynn, P., Shulman, R. G. 1981. *Proc. Natl. Acad. Sci. USA* 78:60–64
51. Cohen, S. M., Ogawa, S., Rottenberg, H., Glynn, P., Yamane, T., et al. 1978. *Nature* 273:554–56
52. Cohen, S. M., Ogawa, S., Shulman, R. G. 1979. *Proc. Natl. Acad. Sci. USA* 76:1603–7
53. Cohen, S. M., Rongstad, R., Shulman, R. G., Katz, J. 1981. *J. Biol. Chem.* 256:3428–32
54. Cross, T. A., Pahl, C., Oberhansli, R., Aue, W. P., Keller, U., Seelig, J. 1984. *Biochemistry* 23:6398–402
55. Cross, V. R., Hester, R. K., Waugh, J. S. 1976. *Rev. Sci. Instrum.* 47:1486
56. Dawson, M. J., Gadian, D. G., Wilkie, D. R. 1978. *Nature* 274:861–66
57. Dawson, M. J., Gadian, D. G., Wilkie, D. R. 1980. *J. Physiol.* 299:465–84
58. Decorps, M., Lebas, J. F., Leviel, J. L., Confort, S., Remy, C., Benabid, A. L. 1984. *FEBS Lett.* 168:1–6
59. Degani, H., Laughlin, M., Campbell, S., Shulman, R. G. 1985. *Biochemistry* 24:5510–16
60. Delpy, D. T., Gordon, R. E., Hope, P. L., Parker, D., Reynolds, E. O. R., et al. 1982. *Pediatrics* 70:310–13
61. den Hollander, J. A., Behar, K. L., Shulman, R. G. 1981. *Proc. Natl. Acad. Sci. USA* 78:2693–97
62. den Hollander, J. A., Behar, K. L., Shulman, R. G. 1984. *J. Magn. Reson.* 57:311–13
63. den Hollander, J. A., Brown, T. R., Ugurbil, K., Shulman, R. G. 1979. *Proc. Natl. Acad. Sci. USA* 76:6096–100
64. Deutsch, C. J., Taylor, J. S. 1986. In *Nuclear Magnetic Resonance Spectroscopy of Cells and Organisms*, ed. R. K. Gupta. Boca Raton, Fla: CRC Press. In press
65. Doyle, D. D., Chalovich, J. M., Barany, M. 1981. *FEBS Lett.* 131:147–60
66. Edwards, R. H. T., Dawson, M. J., Wilkie, D. R., Gordon, R. E., Shaw, D. 1982. *Lancet* 1:725–31

67. Ernst, R. R. 1967. In *Advances in Magnetic Resonance*, Vol. 2, ed. J. S. Waugh, pp. 1–132. New York: Academic

68. Evanochko, W. T., Ng, T. C., Glickson, J. D. 1984. *Magn. Reson. Med.* 1:508–34

69. Evanochko, W. T., Ng, T. C., Lilly, M. B., Lawson, A. J., Corbett, T. H., et al. 1983. *Proc. Natl. Acad. Sci. USA* 80:334–38

70. Evanochko, W. T., Sakai, T. T., Ng, T. C., Krishna, N. R., Kim, H. D., et al. 1984. *Biochim. Biophys. Acta* 805:104–16

71. Evelhoch, J. L., Ewy, C. S., Siegfried, B. A., Ackerman, J. J. H. 1985. *Magn. Reson. Med.* 2:410–17

72. Deleted in proof

73. Fabry, M. E., SanGeorge, R. C. 1983. *Biochemistry* 22:4119–25

74. Flaherty, J. T., Weisfeldt, M. L., Bulkley, B. H., Gardner, T. J., Gott, V. L., Jacobus, W. E. 1982. *Circulation* 65:561–71

75. Freeman, D., Bartlett, S., Radda, G. K., Ross, B. D. 1983. *Biochim. Biophys. Acta* 762:325–36

76. Gadian, D. G. 1982. *Nuclear Magnetic Resonance and its Application to Living Systems.* Oxford: Oxford Univ. Press

77. Gadian, D. G., Proctor, E., Williams, S. R., Cady, E. B., Gardiner, R. M. 1986. *Magn. Reson. Med.* In press

78. Gadian, D. G., Radda, G. K. 1981. *Ann. Rev. Biochem.* 50:69–83

79. Gadian, D. G., Radda, G. K., Ross, B. D., Hockaday, J., Bore, P. J., et al. 1981. *Lancet* 2:774–75

80. Garlick, P. B., Radda, G. K., Seeley, P. J., Chance, B. 1977. *Biochem. Biophys. Res. Commun.* 74:1256–62

81. Gillies, R. J., Alger, J. R., den Hollander, J. A., Shulman, R. G. 1982. In *Intracellular pH: Its Measurement, Regulation, and Utilization in Cellular Functions*, ed. R. Nuccitelli, D. W. Deamer, pp. 79–104. New York: Liss

82. Gillies, R. J., Ogino, T., Shulman, R. G., Ward, D. C. 1982. *J. Cell Biol.* 95:24–28

83. Gonzalez-Mendez, R., Litt, L., Koretsky, A. P., VonColditz, J., Weiner, M. W., James, T. L. 1984. *J. Magn. Reson.* 57:526–33

84. Griffiths, J. R., Cady, E., Edwards, R. H. T., McCready, V. R., Wilkie, D. R., Wiltshaw, E. 1983. *Lancet* 1:1436–37

85. Griffiths, J. R., Iles, R. A. 1980. *Clin. Sci.* 59:225–30

86. Griffiths, J. R., Iles, R. A. 1982. *Biosci. Rep.* 2:719–25

87. Grove, T. H., Ackerman, J. J. H., Radda, G. K., Bore, P. J. 1980. *Proc. Natl. Acad. Sci. USA* 77:299–302

88. Gyulai, L., Bolinger, L., Leigh, J. S., Barlow, C., Chance, B. 1984. *FEBS Lett.* 178:137–42

89. Hetherington, H. P., Avison, M. J., Shulman, R. G. 1985. *Proc. Natl. Acad. Sci. USA* 82:3115–18

90. Hilberman, M., Subramanian, V. H., Haslegrove, J., Cone, J. B., Egan, J. W., et al. 1984. *J. Cereb. Blood Flow Metab.* 4:334–42

91. Hope, P. L., Cady, E. B., Tofts, P. S., Hamilton, P. A., Costello, A. M., et al. 1984. *Lancet* 2:366–70

92. Hore, P. J. 1983. *J. Magn. Reson.* 55:283–300

93. Hoult, D. I., Busby, S. J. W., Gadian, D. G., Radda, G. K., Richards, R. E., Seeley, P. J. 1974. *Nature* 252:285–87

94. Iles, R. A., Griffiths, J. R. 1982. *Biosci. Rep.* 2:735–42

95. Iles, R. A., Griffiths, J. R., Stevens, A. N., Gadian, D. G., Porteous, R. 1980. *Biochem. J.* 192:191–202

96. Iles, R. A., Stevens, A. N., Griffiths, J. R. 1982. *Prog. Nucl. Magn. Reson. Spectrosc.* 15:49–200

97. Iles, R. A., Stevens, A. N., Griffiths, J. R., Morris, P. G. 1985. *Biochem. J.* 229:141–45

98. Ingwall, J. S. 1982. *Am. J. Physiol.* 242:H729–44

99. Jacobus, W. E., Pores, I. H., Lucas, S. K., Weisfeldt, M. L., Flaherty, J. T. 1982. *J. Mol. Cell. Cardiol.* 14:13

100. Jacobus, W. E., Taylor, G. J., Hollis, D. P., Nunnally, R. L. 1977. *Nature* 265:756–58

101. Jue, T., Arias-Mendoza, F., Gonnella, N. C., Shulman, G. I., Shulman, R. G. 1985. *Proc. Natl. Acad. Sci. USA* 82:5246–49

102. Kantor, H. L., Briggs, R. W., Balaban, R. S. 1984. *Circ. Res.* 55:261–66

103. Kopp, S. J., Krieglstein, J., Freidank, A., Rachman, A., Siebert, A., Cohen, M. M. 1984. *J. Neurochem.* 43:1716–31

104. Koretsky, A. P., Wang, S., Murphy-Boesch, J., Klein, M. P., James, T. L., Weiner, M. W. 1983. *Proc. Natl. Acad. Sci. USA* 80:7491–95

105. Koretsky, A. P., Weiner, M. W. 1984. In *Biomedical Magnetic Resonance*, ed. T. L. James, A. R. Margulus, pp. 209–30. San Francisco: Radiol. Res. Educ. Found.

106. Kushmerick, M. J., Meyer, R. A. 1985. *Am. J. Physiol.* 248:C542–49

107. Labotka, R. J. 1984. *Biochemistry* 23:5549–55

108. Lavanchy, N., Martin, J., Rossi, A. 1984. *Cardiovasc. Res.* 18:573–82

109. Levitt, M. H., Freeman, R., Frenkiel, T. 1983. In *Advances in Magnetic Reson-*

ance, Vol. 11, ed. J. S. Waugh, pp. 47–109. New York: Academic

110. Lilly, M. B., Ng, T. C., Evanochko, W. T., Katholi, C. R., Kumar, N. G., et al. 1984. *Cancer Res.* 44: 633–38

111. London, R. E., Hildebrand, C. E., Olson, E. S., Matwiyoff, N. A. 1976. *Biochemistry* 15: 5480–86

112. London, R. E., Kollman, V. H., Matwiyoff, N. A. 1973. *J. Am. Chem. Soc.* 97: 3565–73

113. London, R. E., Matwiyoff, N. A., Kollman, V. H., Mueller, D. D. 1975. *J. Magn. Reson.* 18: 555–57

114. Matthews, P. M., Bland, J. L., Gadian, D. G., Radda, G. K. 1981. *Biochem. Biophys. Res. Commun.* 103: 1052–59

115. Matthews, P. M., Bland, J. L., Gadian, D. G., Radda, G. K. 1982. *Biochim. Biophys. Acta* 721: 312–20

116. Matthews, P. M., Bland, J. L., Radda, G. K. 1983. *Biochim. Biophys. Acta* 763: 140–46

117. Matthews, P. M., Williams, S. R., Seymour, A. M., Schwartz, A., Dube, G., et al. 1982. *Biochim. Biophys. Acta* 720: 163–71

118. McLaughlin, A. D., Takeda, H., Chance, B. 1979. *Proc. Natl. Acad. Sci. USA* 76: 5445–49

119. Melner, M. H., Sawyer, S. T., Evanochko, W. T., Ng, T. C., Glickson, J. D., Puett, D. 1983. *Biochemistry* 22: 2039–42

120. Meyer, R. A., Kushmerick, M. J., Brown, T. R. 1982. *Am. J. Physiol.* 242: C1–C11

121. Mitchell, P. 1961. *Nature* 191: 144–48

122. Mitsumori, F., Ito, O. 1984. *FEBS Lett.* 174: 248–52

123. Moon, R. B., Richards, J. H. 1973. *J. Biol. Chem.* 248: 7276–78

124. Morris, G. A., Freeman, R. 1978. *J. Magn. Reson.* 29: 433–62

125. Nakazawa, M., Katano, Y., Imai, S., Matsushita, K., Ohuchi, M. 1982. *J. Cardiovasc. Pharmacol.* 4: 700

126. Navon, G., Ogawa, S., Shulman, R. G., Yamane, T. 1977. *Proc. Natl. Acad. Sci. USA* 74: 87–91

127. Navon, G., Shulman, R. G., Yamane, T., Eccleshall, T. R., Lam, K. B., et al. 1979. *Biochemistry* 18: 4487–89

128. Neurohr, K. J., Barrett, E. J., Shulman, R. G. 1983. *Proc. Natl. Acad. Sci. USA* 80: 1603–7

129. Neurohr, K. J., Gollin, G., Barrett, E. J., Shulman, R. G. 1983. *FEBS Lett.* 159: 207–10

130. Neurohr, K. J., Gollin, G., Neurohr, J. M., Rothman, D. L., Shulman, R. G. 1984. *Biochemistry* 23: 5029–35

131. Ng, T. C., Evanochko, W. T., Hiramoto, R. N., Ghanta, V. K., Lilly, M. B., et al. 1982. *J. Magn. Reson.* 49: 271–86

132. Norton, R. S. 1982. *Bull. Magn. Reson.* 3: 29

133. Nunnally, R. L., Bottomley, P. A. 1981. *Science* 211: 177–80

134. Nunnally, R. L., Hollis, D. P. 1979. *Biochemistry* 18: 3642–46

135. Ogawa, S., Boens, C. C., Lee, T. M. 1981. *Arch. Biochem. Biophys.* 210: 740–47

136. Ogawa, S., Lee, T. M. 1984. *J. Biol. Chem.* 259: 10004–11

137. Ogawa, S., Rottenberg, H., Brown, T. R., Shulman, R. G., Castillo, C. L., Glynn, P. 1978. *Proc. Natl. Acad. Sci. USA* 75: 1796–800

138. Ogawa, S., Shulman, R. G., Glynn, P., Yamane, T., Navon, G. 1978. *Biochim. Biophys. Acta* 502: 45–50

139. Osbakken, M., Briggs, R. W. 1984. *Am. Heart J.* 108: 574–90

140. Petroff, O. A. C., Prichard, J. W. 1983. *Lancet* 2: 105–6

141. Petroff, O. A. C., Prichard, J. W., Behar, K. L., Alger, J. R., den Hollander, J. A., Shulman, R. G. 1985. *Neurology* 35: 781–88

142. Petroff, O. A. C., Prichard, J. W., Behar, K. L., Alger, J. R., Shulman, R. G. 1984. *Ann. Neurol.* 16: 169–77

143. Petroff, O. A. C., Prichard, J. W., Ogino, T., Avison, M. J., Alger, J. R., Shulman, R. G. 1986. *Ann. Neurol.* In press

144. Pieper, G. M., Murray, W. J., Salhany, J. M., Wu, S. T., Eliot, R. S. 1984. *Biochim. Biophys. Acta* 803: 241–49

145. Pieper, G. M., Salhany, J. M., Murray, W. J., Wu, S. T., Eliot, R. S. 1984. *Biochim. Biophys. Acta* 803: 229–40

146. Pieper, G. M., Todd, G. L., Wu, S. T., Salhany, J. M., Clayton, F. C., Eliot, R. S. 1980. *Cardiovasc. Res.* 14: 646

147. Plateau, P., Gueron, M. 1982. *J. Am. Chem. Soc.* 104: 7310

148. Prichard, J. W., Alger, J. R., Behar, K. L., Petroff, O. A. C., Shulman, R. G. 1983. *Proc. Natl. Acad. Sci. USA* 80: 2748–51

149. Prichard, J. W., Shulman, R. G. 1986. *Ann. Rev. Neurosci.* 9: 61–85

150. Rabenstein, D. L., Isab, A. A. 1979. *J. Magn. Reson.* 36: 281

151. Radda, G. K., Ackerman, J. J. H., Bore, P., Sehr, P., Wong, G. G. 1980. *J. Biochem.* 12: 277–81

152. Radda, G. K., Bore, P. J., Gadian, D. G., Ross, B. D., Styles, P., Morgan-Hughes, J. 1982. *Nature* 295: 608–9

153. Radda, G. K., Bore, P. J., Rajagopalan, B. 1983. *Br. Med. Bull.* 40: 155–59

154. Radda, G. K., Seeley, P. J. 1979. *Ann. Rev. Physiol.* 41: 749–69

155. Reo, N. V., Ewy, C. S., Siegfried, B. A., Ackerman, J. J. H. 1984. *J. Magn. Reson.* 58:76–84
156. Reo, N. V., Siegfried, B. A., Ackerman, J. J. H. 1984. *J. Biol. Chem.* 259:13664–67
157. Roberts, J. K. M., Wade-Jardetzky, N., Jardetzky, O. 1981. *Biochemistry* 20:3389–94
158. Ross, B. D., Freeman, D. M., Chan, L. 1984. *Adv. Exp. Med. Biol.* 178:455–64
159. Ross, B. D., Radda, G. K., Gadian, D. G., Rocker, G., Esiri, M., Smith, J. F. 1981. *N. Engl. J. Med.* 304:1338–43
160. Rothman, D. L., Arias-Mendoza, F., Shulman, G. I., Shulman, R. G. 1984. *J. Magn. Reson.* 60:430–36
161. Rothman, D. L., Behar, K. L., Hetherington, H. P., den Hollander, J. A., Bendall, M. R., et al. 1985. *Proc. Natl. Acad. Sci. USA* 82:1633–37
162. Rothman, D. L., Behar, K. L., Hetherington, H. P., Shulman, R. G. 1984. *Proc. Natl. Acad. Sci. USA* 81:6330–34
163. Schnall, M. D., Subramanian, V. H., Leigh, J. S., Gyulai, L., McLaughlin, A., Chance, B. 1985. *J. Magn. Reson.* 63:401–5
164. Scott, A. J., Baxter, R. L. 1981. *Ann. Rev. Biophys. Bioeng.* 10:151–74
165. Sehr, P. A., Bore, P. J., Papatheofanis, J., Radda, G. K. 1979. *Br. J. Exp. Pathol.* 60:632–41
166. Sehr, P. A., Radda, G. K., Bore, P. J., Sells, R. A. 1977. *Biochem. Biophys. Res. Commun.* 77:195–202
167. Seymour, A. L., Keough, J. M., Radda, G. K. 1983. *Biochem. Soc. Trans.* 11:376
168. Shoubridge, E. A., Briggs, R. W., Radda, G. K. 1982. *FEBS Lett.* 140:288–92
169. Shulman, R. G., Brown, T. R., Ugurbil, K., Ogawa, S., Cohen, S., den Hollander, J. A. 1979. *Science* 205:160–66
170. Siegel, N. J., Avison, M. J., Reilly, H. F., Alger, J. R., Shulman, R. G. 1983. *Am. J. Physiol.* 245:F530–34
171. Sillerud, L. O., Alger, J. R., Shulman, R. G. 1981. *J. Magn. Reson.* 45:142–50
172. Sillerud, L. O., Shulman, R. G. 1983. *Biochemistry* 22:1087–94
173. Slonczewski, J. L., Rosen, B. P., Alger, J. R., Macnab, R. M. 1981. *Proc. Natl. Acad. Sci. USA* 78:6271
174. Stevens, A. N., Iles, R. A., Morris, P. G., Griffiths, J. R. 1982. *FEBS Lett.* 150:489–93
175. Stevens, A. N., Morris, P. G., Iles, R. A., Sheldon, P. W., Griffiths, J. R. 1984. *Br. J. Cancer* 50:113–17
176. Stromski, M. E., Siegel, N. J., Thulin, G., Cooper, K., Avison, M. J., Shulman, R. G. 1986. *Am. J. Physiol.* In press
177. Taylor, D. J., Bore, P. J., Styles, P., Gadian, D. G., Radda, G. K. 1983. *Mol. Biol. Med.* 1:77–94
178. Taylor, J. S., Deutsch, C. 1983. *Biophys. J.* 43:261–67
179. Thulborn, K. R., DuBoulay, G. H., Duchen, L. W., Radda, G. K. 1982. *Cereb. Blood Flow Metab.* 2:299–306
180. Ugurbil, K., Brown, T. R., den Hollander, J. A., Glynn, P., Shulman, R. G. 1978. *Proc. Natl. Acad. Sci. USA* 75:3742–46
181. Ugurbil, K., Petein, M., Maidan, R., Michurski, S., Cohn, J. N., From, A. H. 1984. *FEBS Lett.* 167:73–78
182. Ugurbil, K., Petein, M., Maidan, R., Michurski, S., From, A. H. L. 1985. *Biochemistry* In press
183. Ugurbil, K., Rottenberg, H., Glynn, P., Shulman, R. G. 1978. *Proc. Natl. Acad. Sci. USA* 75:2244–48
184. Vink, R., Bendall, M. R., Simpson, S. J., Rogers, P. J. 1984. *Biochemistry* 23:3667–75
185. Wilkie, D. R., Dawson, M. J., Edwards, R. H. T., Gordon, R. E., Shaw, D. 1984. *Adv. Exp. Med. Biol.* 170:333–47
186. Williams, S. R., Gadian, D. G., Proctor, E., Sprague, D. B., Talbot, D. F., et al. 1985. *J. Magn. Reson.* 63:406–12
187. Wyrwicz, A. M., Pszenny, M. H., Schofield, J. C., Tillman, P. C., Gordon, R. E., Martin, P. A. 1983. *Science* 222:428–30
188. Yoshizaki, K., Seo, Y., Nishikawa, H. 1981. *Biochim. Biophys. Acta* 678:283–91
189. Younkin, D. P., Delivoria-Papadopoulos, M., Leonard, J. C., Subramanian, V. H., Eleff, S., et al. 1984. *Ann. Neurol.* 16:581–86

Ann. Rev. Biophys. Biophys. Chem. 1986. 15 : 403–56

RECOMBINANT LIPOPROTEINS: Implications for Structure and Assembly of Native Lipoproteins

David Atkinson and Donald M. Small

Biophysics Institute, Housman Medical Research Center, Departments of Medicine and Biochemistry, Boston University School of Medicine, Boston, Massachusetts 02118

CONTENTS

PERSPECTIVES AND OVERVIEW

Lipoprotein Structure and Metabolism

Lipoproteins are assemblies of lipids and specific apoproteins that transport lipids to and from tissues. From a structural standpoint, the

403

0883–9182/86/0610–0403$02.00

lipoproteins may be divided into two general categories, (*a*) lipoproteins that are emulsion or microemulsion particles, which contain a core of relatively nonpolar lipid stabilized by a surface of polar lipids and proteins, and (*b*) lipoproteins in which polar lipids are in a bilayer conformation, which do not contain a nonpolar core. The bilayer may be discs (bilayer segments) or even single bilayer vesicles. The major classes of lipoproteins in fasting human plasma are very low density lipoproteins (VLDL), low density lipoproteins (LDL), and high density lipoproteins (HDL) (128). During ingestion of a fatty meal, chylomicrons are also present in the plasma (Figure 1). Thus, the major classes of lipoproteins are chylomicrons (CM) and their core and surface remnants, VLDLs and their core and surface remnants, and a variety of HDL particles that may be microemulsions of different sizes or discoidal bilayer particles (Figure 1).

Chylomicrons are triglyceride-rich emulsion particles formed in the intestine during the absorption of fat. They largely contain a core of triacylglycerols and a little cholesterol ester derived mainly from dietary and bile cholesterol. These large particles (1000–5000 Å diameter) are an emulsion stabilized by a surface of phospholipids (137) and two specific classes of apoproteins: apoprotein B, of intestinal origin (also called B_i, Little B, B_{48}) and the apoA group of apoproteins (apoA-I, A-II, and A-IV) (56). These apoproteins are synthesized in the intestinal mucosal cells (56). They join the nascent chylomicron either during its synthesis in the endoplasmic reticulum or perhaps later, when the nascent chylomicron enters the Golgi apparatus (147, 213). ApoB appears necessary for the release of the chylomicron from the intestinal mucosal cell since, if apoB is absent as in abetalipoproteinemia, nascent chylomicrons accumulate within the cell and are not released into plasma (71).

The chylomicron, as secreted into lymph, appears to have extremely low free cholesterol such that the surface composition of free cholesterol is only about 5 wt% of the total surface lipid (121, 149). Thus, the

---

Figure 1 Structure and composition of the major lipoproteins in human plasma. Triglyceride-rich lipoproteins are secreted with intestinal apoB (B_i) and apoA (A). In plasma they are catabolized by lipoprotein lipase to a CM core B_i containing remnant and surface remnants that contain phospholipids, cholesterol, and some of the soluble apoproteins (A, C, E). VLDLs are also triglyceride-rich particles secreted from the liver with a large apoB (B_L) and with apoC. In plasma they are converted to VLDL remnants, IDL, and finally to LDL, a cholesterol ester-rich microemulsion-sized particle that contains almost exclusively $apoB_L$. HDL is a heterogeneous group of both spherical and discoidal particles. Only exchangeable apolipoproteins A, E, and C are present in HDL. The mean composition of spherical HDL shows that it is cholesterol ester, phospholipid-rich microemulsion with about 50 wt% apoprotein. Discoidal HDLs are 150–180 Å in diameter and are one bilayer thick (~ 50 Å) and contain almost no core lipids (TG and CE). The mean estimates of diameter and density range are given below the particles. For further discussion see text.

	Lymph CM	Plasma VLDL		IDL	LDL	mean	HDL
	TG 90	TG 65		TG 3	TG 3	TG 2	TG 0
	CE 2	CE 10		CE 50	CE 50	CE 20	CE 2
	PL 6	PL 15		PL 21	PL 21	PL 25	PL 50
	Chol 1	Chol 3		Chol 6	Chol 6	Chol 3	Chol 8
	Prot <1	Prot 7		Prot 20	Prot 20	Prot 50	Prot 40
Diameter (Å)	≈2000	≈1000	≈600	≈350	≈220	110	80	180
Density Range (gm/ml)	<1.006	<1.006	<1.006	1.006–1.063	1.006–1.063	1.063–1.21

chylomicron is an excellent sump for free cholesterol from both cellular and other lipoprotein sources. In the lymph, some free cholesterol and some apoproteins bind to the surface of the chylomicron but major physical changes occur once the chylomicron enters plasma (77, 121, 123, 149). In plasma, free cholesterol and apoproteins C and E bind avidly to the surface to increase the total protein and cholesterol on the nascent chylomicron two- or threefold. This causes an increase in surface components and a concomitant increase in lateral surface pressure. As a result, some phospholipid and all the apoA-IV are pushed off the surface to enter the HDL fraction. Thus, during the passage from lymph into plasma the chylomicron gains free cholesterol and apoproteins C and E, and loses apoprotein A-IV and phospholipid.

The chylomicron attaches to capillary endothelium where lipoprotein lipase, assisted by its cofactor apoprotein C-II, promotes triglyceride hydrolysis to fatty acids and partial glycerides (131). As triglyceride is lost, the core shrinks and finally gives rise to a 400–1000 Å core remnant particle that is largely depleted of triglyceride (121, 124, 146, 149). During the process of core shrinkage the surface becomes redundant, buckles to form bilayers, and dissociates from the chylomicron as a surface remnant (194, 195). These remnants contain phospholipid, some free cholesterol, and probably apoproteins A and C (149, 186). They apparently fuse with some elements of the HDL system to convert some of the smaller HDL particles (HDL_3) to larger phospholipid-rich HDL (HDL_{2a}) (4). Thus the action of lipoprotein lipase decreases the size of the core and produces an emulsion-to-bilayer transition (169) at the surface of the chylomicron (194, 195).

Partially ionized fatty acids and 2-monoacyl glycerols are probably initially present in the chylomicron surface (D. M. Small & J. A. Hamilton, unpublished work) but in the presence of large amounts of albumin they bind to albumin and are removed. The chylomicron remnant is much richer in free cholesterol in both the core and surface, and the total surface lipid contains approximately 25% free cholesterol (121). The remnant contains apoE, some apoCs and As, and $apoB_i$ (57, 58). This particle is recognized by both the B-E receptor (25) and the chylomicron receptor (apoE receptor in the liver) (104, 105), and it is rapidly removed by receptor-mediated endocytosis (93). It is important to note that lymph chylomicrons, which have a different lipid and apoprotein surface composition, are not readily recognized by the liver and thus are not taken up or metabolized by the liver (32, 146, 163). Changes in the composition of both lipids and apoproteins appear necessary for specific recognition of the remnant particle. There is little doubt that apoE is the ligand for both B-E and E receptors (138, 208).

VLDLs, so termed originally because of the flotation characteristics of these lipoproteins (128), are, by convention, particles secreted from the liver (3, 64, 72). Newly synthesized triacylglycerols appear 1 min after injection of ^{14}C palmitate in the membrane fraction of the endoplasmic reticulum, after 5 min in the endoplasmic reticulum contents, and only after 10 min in the Golgi fraction (72). The amount of triacylglycerol in the membranes is only 2.6 wt% that of the phospholipid, i.e. the membranes are at saturation levels of triacylglycerol for phospholipid bilayers (63, 120). These data are consistent with the budding model of chylomicron or VLDL formation (169). VLDLs may be as large as chylomicrons but the VLDL isolated from fasting plasma tends to be smaller and to fall in a density range greater than that of chylomicrons; thus came the original definition. We use "VLDL" to represent nascent particles from the liver and some of their plasma remnants.

VLDL particles are assembled in the endoplasmic reticulum and Golgi of the liver (3, 64, 72). ApoB is probably formed at the same time as the triacylglycerol (64) and its formation may be part of the process by which the particle separates from the endoplasmic reticulum or Golgi membrane (169). The particles are rich in triglyceride but may contain varying amounts of cholesterol depending upon the cholesterol input into the liver (120, 183). As formed in the Golgi, the particles have a core of triglyceride and a small amount of cholesterol ester, and a surface of phospholipid and specific apoproteins, probably one molecule of hepatic apoprotein B (synonyms: liver B, B-100, B_L) and probably some apoprotein C. Whether the apoC binds in the Golgi or upon secretion into plasma is not known.

Once in plasma the physical changes that occur in chylomicrons also occur in the nascent VLDL. Free cholesterol, apoproteins C and E, and possibly some apoA bind to the surface, some phospholipid is probably displaced, the VLDL binds to capillaries, and triglyceride hydrolysis ensues. It is probable that part of the VLDL in fasting plasma is a remnant or is certainly physically altered VLDL (121). This is illustrated by the chemical composition of the surface lipids of plasma VLDL and nascent VLDL. The surfaces of nascent VLDL are very low in cholesterol ($\sim 6\%$ total surface lipid) while cholesterol content for plasma VLDL is about 22% of the total surface lipid (121). On lipolysis of triglyceride a smaller core remnant (VLDL remnant, intermediate density lipoprotein) is formed (124). Some surface remnants of phospholipids and apoprotein C are lost (41) and enter the HDL fraction. The IDL particle is recognized by the liver, possibly through apoE. Then further lipolysis and remodeling occur, giving rise to a smaller remnant particle containing the remaining cholesterol ester and some phospholipid, very little triglyceride, and a copy of apoB. This particle is the LDL, a cholesterol ester-rich microemulsion.

HIGH DENSITY LIPOPROTEIN SYSTEM   The HDL system is composed of a heterogeneous class of small, more dense particles (density > 1.063 g/ml) (38, 98a, 103, 195). These particles are protein-rich, which accounts for their increased density; they also contain phospholipids, cholesterol esters, free cholesterol, and triacylglycerol. The metabolism of HDL is complicated but may be summarized briefly as shown in Figure 2.

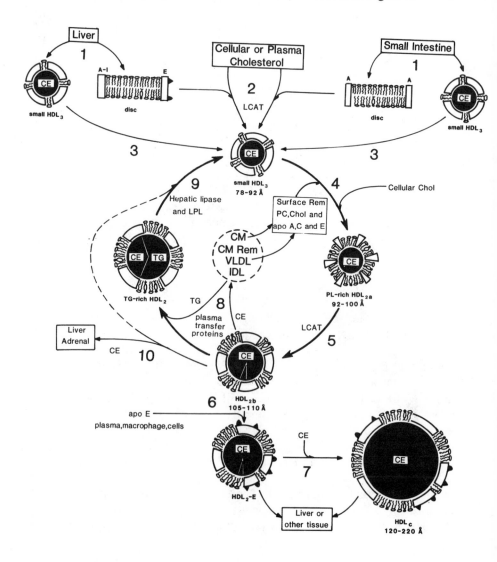

Nascent HDL particles may be secreted from either the liver (66) or the intestine (Figure 2, Step 1) (59). The liver is believed to secrete two kinds of particles: (a) Discoidal HDL consists of a bilayer of phospholipid with very little free cholesterol and an annulus of apoprotein protecting the hydrophobic phospholipid acyl chains from water. In the nascent liver disc some of the apoproteins are A apoproteins and some are probably E apoproteins (59, 112). Therefore mixed discs with different apoproteins are secreted. (b) A small (~ 80 Å diameter) spherical HDL_3 particle has apoA-I, phospholipid, and a very small amount of cholesterol ester in the core (155). The intestine also secretes discoidal and small spherical HDLs (56, 59). However, the discs and small nascent spherical HDLs appear to contain only A-I and A-II since the intestine does not synthesize significant quantities of apoE (36).

Discoidal HDLs with a very low concentration of free cholesterol that contain apoA-I may accept cholesterol from other tissues and cells and thus account for a net movement of cholesterol from tissues (Figure 2, Step 2) (195). These discoidal HDLs containing phospholipid and free cholesterol at the surface and apoA-I as the annulus (194, 195) provide an excellent substrate for the enzyme lecithin-cholesterol-acyltransferase (LCAT), (54, 55) which converts some of the phospholipid and free cholesterol to cholesterol ester (132). This reaction converts these discoidal particles into small HDL_3-like particles.

Figure 2   The HDL system. Particles are drawn to scale. Some of the following steps may occur simultaneously. *Step 1*: Newly synthesized nascent particles (discs and small spherical particles) are secreted from both the liver and the small intestine. *Step 2*: Discoidal apoproteins pick up net amount of free cholesterol and LCAT converts these discoidal nascent lipoproteins to small spherical HDL_3 particles. *Step 3*: The small nascent spherical HDL_3-like particles are presumed to enter the plasma and become part of the HDL_3 fraction. *Step 4*: During lipolysis of VLDL and chylomicrons, surface remnants of phospholipid, cholesterol, and exchangeable apolipoproteins fuse with small HDL_3 to produce a larger phospholipid-rich HDL_{2a}. This lipoprotein can take up cholesterol from cells or plasma. *Step 5*: LCAT reacts with HDL_{2a} to produce larger core HDL_{2b}. *Step 6*: ApoE may be transferred to HDL_{2b} to make an HDL_2-E particle that may be rapidly taken up by the liver or by the B-E receptor in tissues that require cholesterol. *Step 7*: Under special conditions (cholesterol feeding) increasing amounts of cholesterol ester may be incorporated into HDL_2 to make a larger cholesterol ester–rich HDL (HDL_c). This particle contains apoE and may be rapidly taken up by the liver or other tissues. *Step 8*: Cholesterol ester is continuously transferred out of the cholesterol ester–rich HDL_2 into cholesterol-poor chylomicrons, VLDL, and their remnants in exchange for triacylglycerols, potentially producing TG-rich HDL_2. *Step 9*: Triacylglycerol and some of the phospholipid are hydrolyzed by hepatic lipase and lipoprotein lipase, converting the particle back to a small HDL_3-like particle. *Step 10*: In specific tissues such as adrenal and liver, cholesterol esters appear to be extracted from HDL_2, leaving a cholesterol ester-depleted particle that probably reenters the cycle.

When most of the substrate (phosphatidylcholine and free cholesterol) on the surface is used up, the HDL_3 can accept a new source of phospholipid from the surface remnants produced during lipolysis of both chylomicrons and VLDL (186, 195). It is probable that surface remnants actually fuse with these small HDL_3s to produce a phospholipid-rich HDL_{2a} (Figure 2, Step 4) (186). Because there is an excess of surface components the apoproteins in the surface either may be pushed off or may have a looser configuration with a greater amount of apoprotein extending from the surface into the aqueous phase (195). These phospholipid-rich HDLs are also good sumps for free cholesterol and good substrates for LCAT. Thus, as free cholesterol enters from tissues and LCAT converts the excess phospholipid and free cholesterol into cholesterol ester, the particle core increases in volume and the redundancy in the particle surface tends to decrease to produce a new particle, HDL_{2b}, which has more cholesterol ester, i.e. a bigger core, than particle HDL_{2a} (Figure 2, Step 5) (4, 195).

A small fraction of HDL_{2b} particles can have apoE on the surface or can gain apoE from tissues that synthesize apoE (18, 36), for instance from macrophages (14), so that the HDL_2 particle becomes an HDL_2-E particle (Figure 2, Step 6). Such a particle may increase in size through further action of LCAT to produce a large HDL particle called HDL_c, which can approach the size of LDL (Figure 2, Step 7) (9, 108, 185). E-containing HDL particles are probably rapidly taken up by the liver (107) through either the B-E receptor or the chylomicron receptor (104), providing one mechanism for the return of cholesterol from tissues to the liver (139).

Since many animals have a variety of lipid transfer proteins in plasma, some modifications of core and surface can be carried out by transfer protein mechanisms. One mechanism that appears to operate in humans is that the cholesterol ester and triglyceride transfer proteins transfer cholesterol ester from HDLs to triglyceride-rich lipoproteins. In turn triglyceride is transferred from triglyceride-rich lipoproteins into HDLs (2, 13, 76, 111, 125, 126, 136, 188). As a consequence the lipoprotein particle core would become enriched in triglyceride and partially depleted in cholesterol ester (Figure 2, Step 8); however, the action of hepatic lipase and possibly lipoprotein lipase rapidly hydrolyzes the excess triglyceride and some of the phospholipid (34), probably for the most part the excess accumulated phosphatidylethanolamine (97), to reduce the size of the core and surface and regenerate a small HDL_3-like particle (Figure 2, Step 9). Thus the $HDL_3 \rightarrow HDL_2 \rightarrow HDL_3$ cycle is completed.

The rather discrete steps shown in Figure 2 of course occur more or less simultaneously and therefore would not be recognized if only fasting plasma were examined. However, when the HDL system is perturbed, for instance by fat-feeding, the shifts from HDL_3 to HDL_{2a} and HDL_{2b}-like particles become quite evident (see 4). Furthermore, during cholesterol feeding larger HDL particles containing apoE (HDL_c) are generated and become numerous (103). If LCAT is absent then nascent particles, primarily discs but also small spherical HDL_3-like particles, are present in large numbers and discs are not converted to spheres in plasma (132). Finally, in hepatic lipase deficiency (21) or in the presence of excess amounts of chylomicrons or VLDL triglyceride, HDL_2 tends to be triglyceride-rich.

Therefore the lipoproteins include particles that are large emulsions, chylomicron and VLDL (> 500 Å), microemulsions (80–500 Å) (IDL, LDL, HDL_c, HDL_2, HDL_3), vesicular lipoproteins (surface remnants), and discoidal bilayered lipoproteins (nascent HDL). In certain pathologic conditions such as cholestasis or liver disease, cholesterol-rich vesicular lipoproteins (lipoprotein X) are formed and circulate in the LDL density fraction. These do not contain a lipid core but the vesicles do contain an aqueous fluid core. Therefore when one considers models of lipoprotein recombinants one can direct the synthesis of recombinants toward large emulsions, small emulsions, microemulsions, bilayered discoidal lipoproteins, or bilayered vesicular lipoproteins. The purpose of this review is to discuss the different kinds of recombinants that have been made to model both the lipid structure of lipoproteins and the lipoprotein structures.

After a brief discussion of apoprotein structure and properties we discuss lipid recombinants, apoprotein surface lipid recombinants, and apoprotein emulsion and microemulsion recombinants. Recombinant lipoprotein systems in which the molecular complexity of the native particle has been simplified by the use of single species lipids and single classes of apoproteins provide well-defined model systems in which to study lipoprotein structure and molecular properties. These properties include details of the lipid-lipid interactions in the core of the particle, the lipid-lipid and lipid-protein interactions that determine surface organization and perhaps protein conformation, and the interactions between the core and surface components that help to stabilize the intact lipoprotein particle. In addition, recombinant lipoprotein complexes may serve as important models in which to study the effect of different composition and physical properties of lipoproteins and their role in the complexity of lipid metabolism.

Apoprotein Structure and Properties

The current knowledge of the primary structures of the lipoprotein-apoproteins is impressive. Traditional protein sequencing techniques and more recent nucleic acid sequencing methodology have provided the complete amino acid sequence of all (19, 22, 23, 74, 79, 92, 102, 145, 165) of the major exchangeable apoproteins (A-I, A-II, A-V, C-I, C-II, C-III, and E) of humans and/or other species. In addition, mutations or substitutions responsible for several of the apoprotein polymorphic forms or variants have now been determined and new mutants are rapidly being discovered. Since these details have been comprehensively reviewed by Mahley et al (106) and Zannis & Breslow (212) very recently we do not repeat them here.

The molecular weight of human apoB is large, $\sim 387,000$ (42). The apoB is probably a single peptide chain containing about 3500 amino acids. The messenger RNA has been estimated to be 18–20 kb (35a, 93a, 118a), consistent with that necessary to produce a very large peptide. Very recently molecular biological techniques have yielded several partial, largely unrelated sequences (35a, 93a, 118a, 206a). One is a carboxyl terminal of 1455 amino acids that may contain the liganding site for the LDL (B/E) receptor (93a). The structure of rat apoB is also being actively pursued, but no sequence has yet been derived (101a).

For reference purposes Table 1 presents the molecular weight, molecular volume, number of residues, and functions (if known) of the major apoprotein classes.

Table 1 The apoproteins

Protein	Number of amino acids	MW	Molecular volume (Å³) ($\bar{v} = 0.73$ cm³/g)	Function/cofactor[a]
A-I	243	28,100	133,808	LCAT activator
A-II	77	17,400	20,934	
A-IV	391	42,500	51,132	LCAT activator
B$_{intestine}$	—	220,000	264,685	(Mediates exocytosis of chylomicron from intestine)
B$_{liver}$	—	387,000	465,589	Ligand for LDL receptor
C-I	57	6,605	7,946	(Mediates exocytosis of VLDL from hepatocyte)
C-II	79	8,824	10,616	LPL activator
C-III	79	8,750	10,527	(Inhibits LPL)
D	—	$\sim 20,000$	24,061	(CE transfer)
E	299	34,200	41,145	Ligand for LDL receptor and E receptor on hepatocytes

[a] Tentative functions are indicated in parentheses.

Of major importance from a structural standpoint has been the recognition in the primary sequences of all of the exchangeable apoproteins of segments of amino acids predicted to form amphipathic helices. These sections of helical structure with a longitudinal distribution of apolar and polar amino acids between opposing sides of the helix have been strongly implicated in the lipid/protein interactions of all apoproteins (162). Apoprotein A-I, for example, has been predicted to contain a series of eight amphipathic helical regions, each 20–25 amino acids long (5, 11, 38, 118, 162).

The concept of the amphipathic α-helix and of its importance in plasma lipoprotein/apoprotein structure, first proposed by Segrest et al (162), has recently been strengthened by work describing the amino acid sequence of apoA-IV and comparing that sequence with that of apoprotein A-I (19). ApoA-IV is a 391–amino acid protein that contains a 22-residue sequence that is repeated at least 13 times within the protein (19). Although each repeating unit shows some sequence variability, this variability is conservative. Nine of the 22 amino acid units begin with proline and are demonstrated to have a high α-helical content and to form amphipathic α-helices. Such a repeating sequence had been predicted in apoA-I in earlier studies by McLachlan (118). A comparison of these sequences in apoA-IV and apoA-I shows that there is a striking similarity in the corresponding sequences in the two apoproteins, which suggests that the 22–amino acid segment of amphipathic α-helices is a common structural feature among the apoproteins.

Thus, the secondary structural features of the plasma apolipoproteins responsible for the apoproteins' structural role of interacting with lipid and stabilizing the lipoprotein surface are thought to be well understood. However, it is interesting that although amphipathic α-helices have been predicted in all the exchangeable apoprotein classes, as yet there has been no direct structural (crystallographic) demonstration of the existence of an amphipathic α-helix in the plasma lipoprotein-apoproteins. Details of the tertiary and quarternary organization of these proteins and the role of these levels of structure in the structure and stability of lipoproteins are completely uncharacterized.

One important consequence of the fact that the exchangeable apoproteins can exist in two environments (one associated with lipid at the lipoprotein surface and the other in free solution) is that the physical properties and stability of the apoproteins might be expected to depend on the environment of the apoprotein. Since the preferred environment of the apoprotein is in interaction with lipid at the lipoprotein surface, one might expect the free energy of the protein to be lower in this environment than in free solution (see section on apoA-I–surface lipid recombinants). Additionally, one might expect the apoprotein in free solution to self-associate as a function of concentration, which should occur when hydrophobic regions of the protein structure that would normally interact with lipid at the lipoprotein surface are exposed to the aqueous environment.

There have been many studies of the self-association properties of the different apoprotein classes. Early studies of Stone & Reynolds (177) suggested that apoA-I dimerizes above a concentration of 0.1 mg/ml, whereas later studies of Vitello & Scanu (200) suggested that the self-association may be more extensive and that a monomer-dimer-tetramer-octomer model may be appropriate for the association of apoA-I. This higher level of self-association is supported by studies of Formisano and co-workers (46) that examined the effect of pressure and ionic strength on the self-association of A-I. Additionally, cross-linking studies of Swaney & O'Brien (182) suggested a tetrameric and pentameric association of apoA-I at concentrations above 0.5 mg/ml. For apoA-II similar studies have produced differing results. Stone & Reynolds (177) suggested noncovalent dimers for apoA-II self-association between 0.4 and 1 mg/ml, whereas the studies of Vitello & Scanu (201) and Teng et al (197) suggest a monomer-dimer-trimer association for native human apoA-II.

Calorimetric and spectroscopic studies of plasma lipoprotein-apoproteins in solution have shown that the apoproteins in general have low free energy of stabilization when compared to the values normally determined for globular water-soluble proteins (192, 196). These low values for the free energy of stabilization presumably reflect the high degree of exposure of hydrophobic areas of the protein to the aqueous environment.

Tall et al (192, 196) studied human apoA-I by both differential scanning calorimetry and ultraviolet difference spectroscopy as a function of temperature, pH, concentration of the apoprotein, and urea concentration. The calorimetric studies clearly showed that apoA-I in solution undergoes reversible thermal denaturation between 42 and 71°C with a midpoint of 54°C and enthalpy of 64 Kcal/mol. Similar values have been estimated spectroscopically (39, 192). This is 10–30°C lower than the temperature for denaturation and particle disruption in native HDL (187, 190), which demonstrates that the protein is more stable with respect to temperature at the surface of a lipoprotein than in free solution.

The free energy of stabilization of the native conformation of A-I was estimated at 37 and 25°C, giving values of 2.4 and 2.7 Kcal/mol respectively. These values are small compared to those of typical globular proteins and demonstrate that A-I has a loosely folded tertiary structure (192). This loosely folded tertiary structure with a low free energy of stabilization may allow apoprotein A-I to adopt different tertiary conformations depending on the environment either at the surface of a lipoprotein or in free solution. In addition, this conformational flexibility may allow an apoprotein to adapt to changes in the composition, physical state, or structure of the lipoprotein surface. Recently others have made similar conclusions concerning apoC-II (109), A-II (151), and A-IV (37, 207) using primarily denaturation by quanidine hydrochloride and spectroscopic methods. Thus all the apoproteins have free energies of stabilization in the range of 0–4 Kcal/mol; these may be compared with typical values of 9–20 Kcal/mol for myoglobin, ribonuclease, or β-lactoglobulin.

Very few studies have been made of the overall morphology of apoproteins in solution that might give some insight into the tertiary organization of the apoprotein. A single study on apoprotein A-I in solution using ultracentrifugation, viscometry, and fluorescence polarization (12) has suggested that apoA-I in solution has an axial ratio of approximately 6 : 1, which indicates a fairly extended, elongated overall morphology for the protein. This morphology is best described by an elongated ellipsoid with a semi-major axis of 75.6 Å and a semi-minor axis of 12.6 Å.

In contrast to the exchangeable apoproteins, apoB, the major apoprotein of LDL, is poorly characterized. Understanding of the properties of apoB has not progressed because this protein is extremely insoluble in aqueous buffer and tends to form aggregates in the absence of high concentrations of denaturants or detergents. However, a variety of solubilization methods employing either chemical modification, denaturation, or detergents have now been developed in order to maintain apoB in solution. Most of these studies have been concerned with apoB$_{100}$, the form of apoB found on LDL and a ligand for the LDL receptor, and not with apoB$_{48}$, the form of apoB synthesized by the intestine and found in chylomicrons and chylomicron remnants. Methods for the solubilization of apoB have recently been reviewed by Walsh & Atkinson (204).

Estimates of the molecular weight of apoB have ranged from less than 100,000 to the order of 400,000. Additionally, apoB seems to be very sensitive to oxidation and cleavage by proteases and the molecular weight of apoB has been controversial for more than a decade. Steele & Reynolds (175) have provided evidence that even in the presence of SDS, apoB$_{100}$ is a stable dimer with a molecular weight on the order of 500,000. This stable dimer was thought to be the form of apoB present in the native LDL particle. However, most recently a very careful reassessment of the molecular weight of apoB has been undertaken by Elovson and co-workers (42) by a variety of methods including analytical ultracentrifugation in guanidine hydrochloride, column chromatography in guanidine hydro-chloride, and sodium dodecyl sulfate polacrylamide gel electrophoresis. These authors derive a value of 387,000 for the molecular weight of apoB.

Studies in our laboratory have concentrated on the use of sodium deoxycholate for the solubilization of apoB (202). Calorimetric and spectroscopic characterization of the protein have also been carried out on the mixed micellar complex of apoB and sodium deoxycholate (203). Differential scanning calorimetry of sodium deoxycholate–solubilized apoB shows a single thermal transition that occurs with a T_{max} of 52°C and an enthalpy of 0.22 cal/gm of apoB. Circular dichroism studies show that this thermal transition corresponds to thermal denaturation of the protein. Circular dichroism also indicates that apoB solubilized in sodium deoxycholate exhibits a secondary structure similar to that of apoB of native LDL over the temperature range 5–30°C. Some reversible structural changes occur between 30 and 50°C immediately prior to denaturation. Thermal denaturation with a T_{max} of 52°C occurs some 30°C lower than the denaturation/particle disruption of native LDL.

LIPID RECOMBINANTS RESEMBLING LIPOPROTEINS

Recombinants of Surface Lipids

Surface lipids are considered to be phospholipids and free cholesterol. Some core lipids (cholesterol esters and triacylglycerols) also partition slightly into the surface but they are usually ignored. Further products of triglyceride lipolysis (diglycerides, monoglycerides, and fatty acids) also partition strongly into the surface and may influence the properties of the surface. However, they also have been largely ignored as important constituents of the surface. Thus most of our discussion focuses on phospholipid and cholesterol. Surface lipids may form multilamellar bodies, unilamellar vesicles, or, with detergent, discoidal micelles.

LIPOSOMES   When mixed phospholipids or phosphatidylcholines (PC) are extracted from lipoproteins, dried from an organic solvent, and allowed to interact with water or buffers at physiologic pH and ionic strength, they swell to form multilamellar bodies (e.g. myelin figures, liposomes, and lamellar liquid crystals) (171). These particles can be observed under the microscope as myelin figures or anisotropic droplets with a positive sign of birefringence (171). In most naturally occurring phospha-tidylcholines, such as egg yolk PC (83, 166) and lipoprotein PC (137), the fatty acyl chains are a mixture of saturated and unsaturated chains. These chains are in the fluid state well below body temperature (171).

Thus, at body temperature in excess water these systems form concentric lamellar arrays consisting of 10–20 bilayers with melted chains encompassing a center void of aqueous fluid. The periodicity from bilayer to bilayer (d), established by X-ray diffraction, is about 62–64 Å for egg lecithin (83, 166). The surface area at the lipid-water interface is about 70–72 Å^2 per molecule. Because liposomes are large particles > 1 μM the surface is nearly flat.

The total thickness of the H_2O-lipid bilayer (~ 64 Å) is conventionally divided into a lipid layer, d_L (35.3 Å), and a water layer, d_w (29.7 Å). A more structural approach divides the bilayer into three regions, the hydrocarbon bilayer ~ 30 Å thick, two polar group–H_2O layers (about 11 Å each), and a water layer ~ 12 Å thick (166). If some charged species are included with phosphatidylcholine then the periodicity, d, may be increased due to an increased H_2O space. The outer surface of these particles has been assumed to be similar to the outer surface of a chylomicron or large VLDL.

Synthetic saturated lecithins with chains longer than 15 carbons have chain melting transitions above body temperature (171) and are in the "gel" state (hexagonally packed, frozen chains) at 37°C. Saturated lecithins with shorter chains [e.g. dimyristoyl PC (DMPC)] have chain melting transitions below body temperature and thus are in the melted chain lamellar liquid crystalline state (L_α) at 37°C.

The interactions of DMPC and dipalmitoyl PC (DPPC) with water are well understood (84, 85, 171). In a system with excess water, DMPC at equilibrium can form a stable crystalline phase with 2–4 water molecules at a low temperature that transforms at ~ 0°C to a bilayered gel ($L_{\beta'}$) phase. In $L_{\beta'}$ the chains are packed pseudohexagonally in a two-dimensional rectangular lattice and the bilayers incorporate about 29 wt% water. The lipid bilayer–H_2O periodicity is 62 Å and the tilt of the chains with respect to the plane of the bilayer is approximately 60°. The total calculated lipid bilayer thickness, d_L, is about 44 Å. When the temperature is increased through the "pretransition" at about 11°C, the lattice transforms into true hexagonal chain packing and the bilayer ripples to form the $P_{\beta'}$ phase. The ripple has a periodicity of 120 Å and the DMPC-H_2O bilayer is 64.8 Å thick. The molecules are tilted about 60° to the plane of the bilayer and the calculated bilayer thickness, d_L, is about 45 Å. At 23.9°C, the chains melt to form a lamellar liquid crystalline phase (L_α). Just above the transition L_α has a total lipid-H_2O periodicity of about 62 Å, contains 40% H_2O, and has a calculated lipid thickness of only 36 Å. The actual hydrocarbon thickness is only about 24 Å. Major heat absorptions and volume increases occur at the transition from true crystal to $L_{\beta'}$ and at the chain melting transition ($P_{\beta'} \rightarrow L_\alpha$) (171).

DPPC behaves in a similar fashion but its transition temperatures are higher,

$$\text{crystalline} \xrightarrow{25°C} L_{\beta'} \xrightarrow{35°C} P_{\beta'} \xrightarrow{42°C} L_{\alpha},$$

and the specific lattice parameters are different (171).

Cholesterol (C) may be added to the different phospholipid systems to mimic the cholesterol/phospholipid ratio of certain lipoproteins. Cholesterol appears to form a pseudo complex of about 2–3 PC per cholesterol in the gel state. When very little cholesterol is present, i.e. less than ~ 25 mol%, there are probably two regions in the bilayer, a 2–3 : 1 PC/C complex and pure PC. Above 25–30 mol% C the bilayer probably has a more uniform distribution of cholesterol. This complex probably dissipates in L_α although it is not at all certain that cholesterol is uniformly distributed throughout the bilayer in L_α. In L_α, cholesterol thickens the hydrocarbon region by about 5 Å (20, 98) and stiffens the bilayer (increases the viscosity) along the 12 Å length of the steroid ring system. The interior part of the bilayer where the isooctyl side chain is located is actually more fluid than the part of the bilayer without cholesterol. These regional differences in fluidity in a lecithin-cholesterol bilayer can be explained by the different cross-sectional areas of the sterol moiety, 38 Å^2/molecule, and the isooctyl side chain, 31 Å^2/molecule. In L_α, cholesterol can be incorporated up to a mole ratio of 1 : 1 (50 mol%, 33 wt% relative to egg yolk PC) in stable liposomes (20). Above 1 : 1 a metastable bilayer can be formed but it gradually crystallizes the excess cholesterol to cholesterol monohydrate (20, 98), thus returning the bilayer composition to the equilibrium 1 : 1 composition (30, 49).

Even though there is a fixed periodicity between layers the size distribution of liposomes is difficult to control. Therefore the surface area is variable. In general only about 5% or less of the total phospholipid in liposomes is on the surface, but this figure is quite variable. The utility of liposomes as lipoprotein models is clearly limited and it is unlikely that such multilamellar bodies actually appear in plasma except perhaps in severe cholestasis.

UNILAMELLAR VESICLES   Multilamellar liposomes may be subjected to ultrasonic radiation to produce small unilamellar liposomes or vesicles (75, 119). These are ~ 250 Å in diameter when made from pure phosphatidylcholine and somewhat larger when made with PC-cholesterol. Most of the vesicles consist of a single bilayer although there may be more than one bilayer in some vesicles. The inner and outer layers of the bilayer are not equivalent and this can be demonstrated by nuclear magnetic resonance (40, 160). Presumably the inner layer is more constrained in the polar region than the outer layer and the hydrocarbon chain packing must be somewhat different in the inner and outer layers as well.

Unilamellar vesicles may be easily made with molar cholesterol-to-phospholipid ratios of 0.5, but systems containing more than 1 mol cholesterol per mol phospholipid appear to be very metastable (119) and are not well characterized. With time they separate excess cholesterol as cholesterol monohydrate (30) in accordance with the phase diagram established for egg phosphatidylcholine and free cholesterol (20). Small unilamellar vesicles are not at an equilibrium state since their free energy in the gel state is probably considerably greater than that of the multilamellar vesicle in the gel state (171). If left at a low temperature vesicles gradually coalesce to form large unilamellar (135) and some multilamellar vesicles. The chain melting transition of vesicles has only about half the enthalpy expected. Further, the volume change in going from gel to liquid crystalline state is only about half the volume change that would be expected if all the chains were melted, but the volume above the transition is very similar to that in unsonicated systems (171). This suggests that the vesicles below the transition temperature (approximately 17°C) consist of patches of chains in the gel state and other patches that are partly melted. This state is unstable and promotes fusion.

Vesicles are thought to provide a surface that has similar curvature to low density lipoprotein ($D = \sim 250$ Å). This surface may be varied by changing the phospholipid composition or the phospholipid-to-cholesterol ratio. Cholesterol-containing PC vesicles are a structurally sound model for lipoprotein X, an abnormal vesicular lipoprotein in the plasma of patients with cholestasis or certain liver diseases (65). Because vesicles contain a trapped aqueous core, various aqueous-soluble materials including drugs can be placed in the vesicles. Further, liganding molecules for certain cell receptors can be attached to the surface to target the vesicles to cells. Such targeted vesicles have been used as delivery systems for drugs, toxins, etc.

Vesicles may be made by a variety of other techniques including French press (67), passage of mixed micelles over columns to remove the detergent (26), dialysis of mixed micelles (43), and solvent evaporation (15). Large unilamellar vesicles may also be made (135), but these have not been used as models for lipoproteins.

DISCOIDAL AGGREGATES (MICELLES)   The original description of discoidal micelles was made for the bile salt/lecithin system and for bile salt lecithin/cholesterol systems (167, 168, 174). When mixtures of bile salts and phosphatidylcholines in molar ratios of about 1:2 or greater are lyophilized and enough water is added (e.g. 70 g H_2O/20 g egg yolk phospholipid + 10 g sodium cholate) the bile salts dissolve the phospholipids into small discoidal micelles (172, 174). The size of the micelle depends on the bile salt–to-phospholipid ratio (28, 173); the smaller the ratio the larger the micelle. It is probable that these micelles contain a small amount of bile salts within the phospholipid bilayer (28), but most of the bile salts form an annulus around the exterior part of the bilayer protecting the hydrophobic aliphatic chains of the lecithin from water (168, 174). The hydrophobic side of the bile salts associates with the hydrophobic chains leaving the largely hydrophilic side containing hydroxyl groups and ionized carboxyls of the bile salts at the H_2O interface. The bile salts in these micelles are in rapid equilibrium with a low concentration of bile salt monomers in aqueous solution. Bile salts are quite effective in solubilizing lecithin.

If cholesterol is also mixed with the lecithin then solubilization of some cholesterol occurs, but the limits are quite stringent (29). Cholesterol does not appear to interact at the annulus with bile salts and therefore its solubility is relegated to the more central parts of the disc. A maximum of about 10 mol% cholesterol can be solubilized in such a mixed micelle (1, 29). When a concentrated mixed micelle solution is diluted, bile salts come off the disc to keep the molecularly dispersed aqueous phase concentration of bile salt constant. This leaves bilayers with naked hydrophilic edges that fuse to form large discs. Below a critical bile salt concentration the bilayers of lecithin fuse with one another to form vesicles. Such vesicles contain a small amount of associated bile salt and all the cholesterol that was in the original mixed micelle. The vesicles can also be formed by dialyzing the bile salts out of the system. The size of the vesicle appears to be determined by the amount of bile salts left in the vesicle and by the cholesterol-to-phospholipid ratio (161).

Recombinants of Surface and Core Lipids: Emulsions and Microemulsions

EMULSIONS: MODELS FOR LARGER TRIGLYCERIDE-RICH LIPOPROTEINS Among the earliest models for triglyceride-rich lipoproteins that were used extensively 20 years ago to measure the rates of triglyceride clearance were commercial preparations such as Intralipid (61). These emulsions are usually made of soybean or egg phosphatidylcholines (usually containing small amounts of other phospholipids) or mixed phospholipids (soy, corn, or safflower oil). The constituents are emulsified in an aqueous solution of glycerol that contains some antioxidants. An excess of multi- and unilamellar liposomes is added to "stabilize" the emulsion systems; thus the usual commercial mixtures contain not only glycerol but also a large excess of phospholipid that is not necessarily associated with emulsion particles. The standard Intralipid product, in which egg lecithin is used as the emulsifier, contains a small but varying amount of free cholesterol, and Intralipid from different distributors varies considerably (44). Although the commercial products are heterogeneous many investigators have subfractionated them to remove the glycerol and excess phospholipids and to prepare pure emulsions (44). Both large and small fractions have been obtained and have been used as models for triglyceride-rich lipoproteins both for apoprotein binding studies and for some metabolic studies. While these particles represent an improvement over multilamellar liposomes, use of them is limited by variable phospholipid composition, low and variable levels of cholesterol, and the potential presence of excess phospholipid in the form of multilamellar liposomes and vesicles.

In an attempt to understand the structure of a simple lipid model for triglyceride-rich lipoproteins we emulsified systems of egg lecithin/cholesterol/triolein in excess water by agitation or brief sonication. After equilibration the core and surface were separated by breaking the emulsion in the ultracentrifuge. The sinking surface and free-floating core fractions were collected, washed, and analyzed (120). Cholesterol and triolein partitioned between both the surface and core fractions while lecithin was found exclusively in the surface fractions (120). When the core contained 1% cholesterol or less (1 g C/100 g oil) the distribution coefficient of cholesterol between surface and core, $K_{Cs/o}$, was 28 at 24°C indicating that for every molecule present per gram of core oil, 28 surface molecules were present per gram of surface. Above 1% cholesterol in the core the distribution coefficient fell to 14 so that at higher cholesterol contents for every molecule in the core there were 14 in the surface. The halving of $K_{Cs/o}$ was ascribed to dimerization of the cholesterol in the oil phase at concentrations above 1%. The calculated values of $K_{Cs/o}$ at 37°C were 22 below 1% and 11 above.

Triolein also partitioned between core and surface. The surface composition of triolein was about 3.5 g/100 g phospholipid or 2.8 mol% (120). These results were corroborated by sonicating an excess of triolein in the presence of lecithin. The resulting vesicles also contained about 3% triolein oriented with all three ester groups in the interface and the chains parallel to the phospholipid chains. The β (2 position) carbonyl group of the triolein was less hydrated than the α (1, 3 positions) (63). Free cholesterol tended to diminish the amount of triolein in the surface. Thus, triglyceride-rich emulsions can be made in which the surface composition can be varied from 0% cholesterol and $\sim$3% triolein to up to 40% cholesterol and very little triolein. Such emulsions are simplified apoprotein-free chylomicron models (low surface cholesterol) or chylomicron remnant models (high surface cholesterol) (121).

To model triglyceride-rich lipoproteins more completely, increasing amounts of cholesterol ester in the form of cholesteryl oleate were added to triolein/cholesterol/lecithin emulsions and cores and surfaces isolated as previously described (122). Both cholesteryl oleate and triolein had very limited solubilities in the surface. In the absence of free cholesterol, triolein and cholesteryl oleate nearly reflected the core composition; if the core composition was 90% triolein and 10% cholesteryl oleate, the surface composition was quite similarly high in triolein (62). In the presence of free cholesterol the surface concentrations of both triolein and cholesteryl oleate were diminished.

Incorporation of cholesteryl oleate into the emulsions had another effect. Cholesterol incorporated into the system partitioned between core and surface, but the presence of cholesteryl oleate in the oil phase decreased the $K_{Cs/o}$ to about 9–11 at 37°C. This was a result of increased cholesterol partitioning into the core. It was assumed that association of free cholesterol with cholesterol ester in the oil phase increased the distribution into the oil phase. Even small amounts of cholesteryl oleate in the oil phase ($> 2\%$ the amount of triolein) decrease the distribution coefficient appreciably (122). When the oil is nearly pure cholesterol ester, $K_{Cs/o}$ approaches 6 (35, 99). Thus in a particle such as β-VLDL or LDL, for every molecule per gram of core lipid there are about six molecules of cholesterol per gram of surface phospholipid.

Although $K_{Cs/o}$ clearly favors the surface, when a particle is large its mass of core is very large compared to mass of surface and a large fraction of the total particle cholesterol can be partitioned into the core (121, 122). In large particles the overall particle cholesterol/phospholipid ratio obtained from the composition cannot be used to estimate the surface cholesterol/phospholipid ratio because at least 50% of the cholesterol may be partitioned into the core; therefore the true cholesterol/phospholipid ratio in the surface is much less than that of the total (121). However, by plotting particle composition on the appropriate phase diagrams and using the appropriate $K_{Cs/o}$ (121), both core and surface composition can be estimated.

In modeling the larger lipoproteins (chylomicrons, remnants, and VLDL) it is important to understand the distribution of cholesterol, cholesterol esters, and triacylglycerols between core and surface phases in the emulsions produced with phospholipid as the emulsifier. Such emulsions are now being used to study the binding of apoproteins and have been injected into animals to examine their metabolisms (110).

Similar emulsions can be made from extracted lipoprotein lipids (121). These of course contain a mixture of phospholipids, triacylglycerols, cholesterol esters, and some minor components such as partial glycerides and fatty acids. The distribution of free cholesterol in such particles is quite similar to that in the emulsions of chemically defined lipids (121, 122).

MICROEMULSIONS : MODELS FOR SMALLER CHOLESTEROL-RICH LIPO-PROTEINS As models for the lipid organization of LDL, protein-free homogeneous microemulsions have been prepared from specific phospholipids and cholesterol esters (51, 100, 164, 199), which constitute the major molecular components of LDL. These microemulsion systems can be formed with a single species of cholesterol ester and a single species of phospholipid. Fundamental details of the lipid-lipid interaction can be studied without the overwhelming complexity of the heterogeneous molecular composition found in native lipoprotein lipids. The microemulsion systems show many of the average properties of the LDL particle. They also demonstrate specific structural and physical properties that can only be observed in pure single-component lipid systems.

Typically microemulsions are formed by extensive sonication of both phospholipid and cholesterol ester components in an aqueous environment at a temperature above the main order-disorder transition temperature of both lipid components. Fractionation is then carried out by ultracentrifugation and agarose column chromatography (51). Stable, relatively homogeneous particles with a molar ratio of cholesterol ester to phospholipid of 0.8–0.9 can be formed. The particles have a Stokes radius of approximately 100 Å and have been shown by electron microscopy to have a spherical morphology and a diameter of about 200 Å. This is consistent with the surface-to-volume ratio of a particle with a cholesterol ester core, the surface of which is stabilized by a phospholipid monolayer. Microemulsion particles have been formed from either cholesteryl oleate (CO) or cholesteryl nervonate (CN) as the core cholesterol ester component and either egg yolk lecithin, dimyristoyl lecithin, or dipalmitoyl lecithin as the surface phospholipid monolayer. More recently the surface phospholipid egg yolk lecithin has been replaced with 1-palmitoyl-2-oleoyl lecithin (V. Ziemba & D. Atkinson, unpublished results).

A detailed study using differential scanning calorimetry, X-ray scattering and diffraction, and H_1 nuclear magnetic resonance spectroscopy has shown that in these microemulsion particles the cholesterol ester in the core undergoes a specific thermal transition; ΔH is about 0.7 cal per gram of cholesterol ester at a temperature that depends on the specific cholesterol ester in the core (51). Another thermal transition occurs owing to the fatty acyl chain order-disorder transition of the phospholipid that forms the surface monolayer (Figure 3). This transition typically has a ΔH of about 5 cal per gram of phospholipid.

Specifically in microemulsions formed with dimyristoyl lecithin as the phospholipid in the surface the order-disorder transition of the surface phospholipid occurs at 25°C, i.e. at a temperature 2° higher than in pure dimyristoyl lecithin multilamellar liposomes. The transition of the cholesterol esters in the core of the particle occurs at 46°C for the DMPC/CO system and at 54°C for the DMPC/CN system, once again at slightly higher temperatures than the analogous transition in the neat cholesterol esters (42°C and 52°C, respectively). These elevations in transition temperature for both the phospholipid surface component and the core cholesterol ester component suggest that the cholesterol esters are stabilized with respect to temperature by the surface phospholipid monolayer.

Most interestingly, in microemulsions formed with DPPC concomitant melting of the surface phospholipid and the core cholesterol esters was observed. This melting occurred at a T_m of 41°C for both the DPPC/CO and DPPC/CN systems, even though in the DPPC/CN system the corresponding transitions of the isolated lipids are at 42°C for DPPC and at 52°C for cholesteryl nervonate. This directly indicates the coupling of the order-disorder transitions of the core and surface components (51).

These systems formed with well-characterized, single species lipids show the general organization of LDL. The core cholesterol ester components undergo an order-disorder transition similar to that observed for cholesterol esters in the neat state (50a) and described for cholesterol esters in the core of the LDL particle (35). In addition, since these particles are formed with single species phospholipids, changes in physical state

---

Figure 3   Calorimetric transitions and schematic representation of the structural changes in model phospholipid/cholesterol ester microemulsion systems. *Egg PC systems*: Core cholesterol esters undergo an order-disorder transition at a temperature dependent on their molecular species (42° CO, 52° CN). *DMPC systems*: In addition to the thermal transition of the core cholesterol esters these microemulsions undergo a high enthalpy transition at ~25°C, which corresponds to the order-disorder transition in the phospholipid acyl chains of the surface. *DPPC systems*: Surface and core components undergo a thermal order-disorder transition simultaneously at ~42°C. In DMPC and DPPC monolayer systems the fatty acyl chains of the phospholipid are shown at low temperature tilted with respect to the surface of the particle as determined for the $L_{\beta'}$ and $P_{\beta'}$ phase of these lipids in bilayer systems. This tilt provides a greater projected surface area of the fatty acyl chains at the surface of the particle to compensate for the somewhat larger surface area of the phospholipid head group. In addition, a small number of molecules of cholesterol ester are shown partitioned into the surface of the particle.

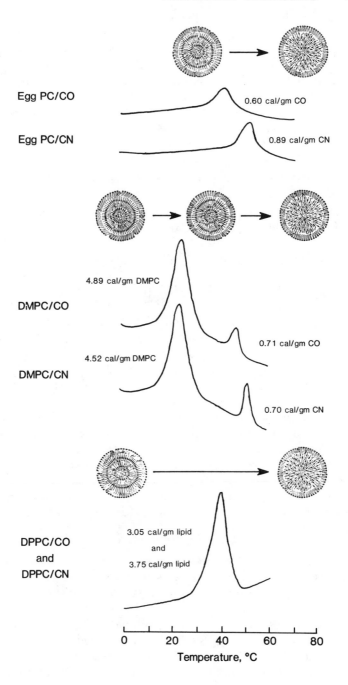

Egg PC/CO

0.60 cal/gm CO

Egg PC/CN

0.89 cal/gm CN

DMPC/CO

4.89 cal/gm DMPC

0.71 cal/gm CO

DMPC/CN

4.52 cal/gm DMPC

0.70 cal/gm CN

DPPC/CO
and
DPPC/CN

3.05 cal/gm lipid
and
3.75 cal/gm lipid

0   20   40   60   80
Temperature, °C

associated with the phospholipids at the surface of the particle can be observed. Changes occur in the interactions between the core and surface components which are dependent on the composition of the particle. These results may be extended to native lipoproteins and suggest that interactions between core and surface phases may change depending on lipid composition of the native LDL particle.

Via and co-workers (199) have described an alternative method of forming microemulsion particles. Mixtures of phospholipid, cholesterol esters, and small amounts of triglyceride solubilized in isopropanol are squirted into an aqueous environment. The isopropanol is then removed from the system by column chromatography and the resulting emulsion particles and microemulsion particles are fractionated. However, for these systems emulsions cannot be formed in the absence of small amounts of triglyceride, which appear necessary to stabilize the particle. Electron microscopy has demonstrated that the population of microemulsion particles is not as homogeneous as that of particles formed by sonication.

APOPROTEIN–SURFACE LIPID RECOMBINANTS

Shortly after the isolation of soluble apolipoproteins from HDL (10, 159), attempts were made to recombine the HDL lipids with the apoproteins to form reconstituted lipoproteins (158). Since those early studies many investigations have been carried out; these indicate that apoA-I and some of the other apoHDLs (apoC-I, apoC-III, apoE, and apoA-II) can recombine spontaneously with certain kinds of phospholipids to form either discoidal or vesicular complexes. The kinetics and thermodynamics of these reactions depend on (a) the apoprotein-phospholipid ratio, (b) the chemical nature of the phospholipid, (c) the physical state of the phospholipid, (d) the physical state of the protein, and (e) the method of mixing the components. The most widely studied and perhaps best understood of the phospholipid apoprotein A-I recombinants is the apoprotein-A-I/dimyristoyl phosphatidylcholine system (A-I–DMPC).

ApoA-I–Surface Lipid Recombinants

STRUCTURE AND THERMODYNAMICS OF SATURATED PHOSPHATIDYLCHOLINE–
A-I INTERACTIONS   In the early stages of reconstituted lipoprotein research the structure of the recombinant was unclear. However, by the early 1970s investigators using electron microscopy and other techniques noted that discoidal particles were present in some of these systems (5, 48, 69). A careful X-ray scattering study clearly indicated that a complex with 2.5:1 weight ratio of dimyristoyl phosphatidylcholine to A-I was a disc of 110 Å in diameter with a bilayer thickness of 55 Å (8). Furthermore,

the analysis indicated that the protein was on the outside perimeter of the disc covering the hydrocarbon chains; this finding was very similar to those of the earlier studies of bile salt/phospholipid micelles (167, 168, 174). The perimeter location of A-I on the discs was later clearly established by contrast variation neutron scattering experiments (6, 7).

The thermodynamics of these complexes were approached using a variety of techniques, especially precision differential scanning calorimetry as described by Privalov & Kechinashvili (144). Early studies (192, 196) indicated that apoA-I in aqueous solution without lipids had a very loosely folded conformation in water with a very small difference in free energy between the unfolded (denatured) state and native A-I in solution (i.e. a low free energy of stabilization). The study also indicated that binding to lipid greatly stabilized the apoprotein and increased its free energy of stabilization (196, 196a). ApoA-I in solution has a transition temperature in the range of 43–71°C with a peak at 54°C. It is most stable at about 30°C. The apoprotein was very easily denatured in fairly low concentrations of urea (192). When apoA-I was mixed with unsonicated liposomes of dimyristoyl lecithin at 24 or 37°C the cloudy liposomal suspensions were cleared if enough apoprotein was present. Clearing took approximately 6–12 hr. Similar solutions containing apoHDL (60% A-I, 30% A-II, 10% apoC) cleared within 1 hr, indicating that the reaction was more rapid with a mixture of apolipoproteins. When the A-I/DMPC recombinants were examined by column chromatography and electron microscopy (196a) discoidal particles were found that had greater diameter the larger the lipid-to-protein ratio. Between about 25% A-I/75% DMPC and 70% A-I/30% DMPC the discoidal diameter decreased from ∼150 to 60–70 Å. When the diameter was plotted against DMPC/apoprotein mole ratio (MW of apoHDL taken as 25,000) the correlation was linear and the slope was 1.37 Å per DMPC. This slope was very close to the 1.34 Å per DMPC calculated for a discoidal bilayer of DMPC with all the apoprotein on the perimeter. A similar slope was found for egg yolk lecithin/A-I recombinants (129).

The thermodynamic measurements (196a) are shown in Figure 4. The lipid transitions indicate that as protein is added to the system the chain transition at 23.5°C splits into two peaks, one at 23°C due to pure multilamellar lipid and the other at 26°C, which is the chain transition in the A-I/DMPC complex. At 20–30% apoA-I/70–80% DMPC, all the liposomes have been converted into large discoidal complexes and the chain transition is broad and centered at 26°C. The enthalpy of the DMPC chain transition falls as the protein concentration is raised until it reaches about half of the enthalpy of liposomes indicating that a boundary layer of lipid does not take part in the chain transition.

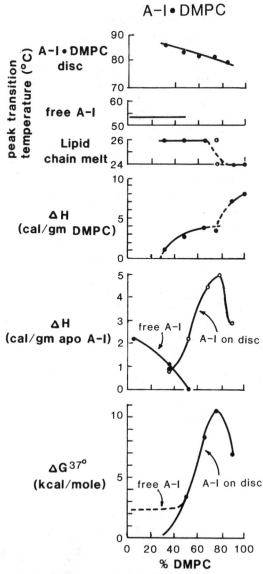

Figure 4 Thermodynamic data for apoA-I DMPC mixtures as a function of wt% of DMPC: temperature of the order-disorder transition of A-I in the discoidal complex falls but ΔH and ΔG$^{37°}$ increase as the % DMPC increases. Free A-I is found at low % DMPC and excess DMPC (not in discs) is found above 80% DMPC. (Plotted from data in References 196, 196a.)

The enthalpy of the protein unfolding transition is also shown in Figure 4. At high lipid ratios the transition temperature, enthalpy, and cooperativity of the protein transition are much higher than those of pure apoA-I. In fact the low free energy of stabilization reaches levels similar to those of other globular proteins. As more protein is added (i.e. as the disc becomes smaller) the transition broadens and decreases in enthalpy and cooperativity and the free energy of stabilization approaches that of free A-I. When approximately 60% protein is present a pure A-I peak is seen in addition to the A-I peak from the discoidal complex. This free A-I peak increases as the protein-to-lipid ratio increases beyond 6 : 4. Thus, free apoprotein is present in mixtures that contain >60% protein. Using the ΔH and the difference in specific heat above and below the transition, Tall et al calculated the free energy difference between the associated form and the unfolded free form of A-I (192, 196a). The free energy difference between the folded and unfolded non–lipid associated A-I is 2.4 Kcal/mol (Figure 4). The protein bound to the lipid discs has a considerably higher free energy of stabilization that increases as the discs enlarge (Figure 4). That is, as the perimeter of the disc becomes less curved the stabilization of the apoprotein increases. In the large discs this free energy is quite similar to the free energy of many globular proteins such as ribonuclease, and very different from that of the loosely folded free A-I in solution (192, 196, 196a).

Careful column chromatography, circular dichroism, and fluorescence polarization studies by Jonas and colleagues confirmed the size distribution of the discoidal lipoproteins (87–91). These studies further indicated that if sonicated dimyristoyl phosphatidylcholine vesicles (SUV) were incubated with apoA-I at very high lipid-to-protein ratios, then three or four apoprotein molecules could bind per vesicle without disruption of the vesicle (88). Thus, there appears to be a distinct difference between the interaction of apoA-I with a large excess of DMPC in liposomes and in SUVs. If the DMPC is present as multilamellar liposomes the small amount of A-I either binds to the surface of the multilamellar vesicle or perhaps forms a few large discs (196a), whereas if the DMPC is sonicated to form SUVs and a small amount of A-I is added, three or four molecules of A-I can bind to the surface without disrupting the vesicle. From size and stoichiometry it was calculated that the small discs (87, 89–91, 196a) contain two A-I molecules whereas larger discs contain three A-I molecules. The stoichiometry of A-I to DMPC did not appear to be precisely determined in a stepwise fashion in these experiments, and the free energy of A-I on the discs indicated that the free energy is lower the smaller the disc. As the disc approaches its lower limit of ~60–70 Å diameter (DMPC/A-I mole ratio ~12 : 1) A-I is present in an excess necessary to cover the perimeter and partly dissociates to self-associate with a much lower stability. Thus these small complexes with two apoA-I molecules and only 24–50 molecules of DMPC probably have only a few of the amphipathic helices associated with the DMPC chains.

The careful cross-linking studies of Swaney (178, 179) indicated that the stoichiometry of these mixtures was correct; in the small discs dimers of apoA-I were cross-linked whereas in the large discs trimers were cross-linked. Marked overlapping of dimers and trimers occurred in the intermediate size range. Swaney also showed that in mixed apolipoprotein discoidal aggregates A-I and A-II could be present on the same disc.

The calorimetry of the complexes indicates that the lipid transition temperature is higher by about 3°C but has only about half the enthalpy observed in DMPC liposomes. This could mean that the lipid in the complex is less ordered than DMPC in liposomes below the transition, but has the same order above the transition. Conversely, as has been found with brain proteolipid/DMPC interactions using Raman spectroscopy, the lipid could be similarly ordered below the transition but more ordered above the transition (33). However, contrary to the situation with proteolipid, acyl chains in the DMPC/A-I complex discs were more disordered below the transition than chains in DMPC liposomes, whereas at 30°C (above the transition) the number of gauche conformers and the degree of order were about the same (50). Therefore, it appears that there is some disordering of the chains in discs in the gel state. The disorder is probably in the boundary lipid as indicated in Figure 5 (50, 196a).

KINETICS OF A-I–DMPC INTERACTION

Effect of protein physical state   It is well known that apoA-I self-associates in lipid-free aqueous solutions (86, 176, 177, 200). The monomeric form is an elongated rod that self-associates into asymmetric oligimers (12).

Figure 5 Pseudo binary phase diagram of apoA-I DMPC mixtures. On the *right* pure DMPC undergoes a chain melting transition at 23°C. In presence of large amounts of DMPC, A-I and DMPC appear to separate into two phases, one made up of discs saturated with DMPC and the second the excess DMPC phase. At about 80% DMPC, 20% A-I, a single-phase zone begins, made up of large discs of DMPC with A-I of 150 Å diameter. Large discs tend to be thermodynamically the most stable in terms of the A-I protein denaturation (see Figure 4). As the ratio of protein to DMPC is increased, the disc becomes smaller and at a certain ratio of DMPC to A-I (70% A-I) the A-I is in excess. *Upper line*: Temperature of denaturation of A-I in the complex; the temperature of denaturation is much higher when A-I is present on the disc than when A-I is in solution. *Lower line*: Chain melting transition of DMPC; slightly higher in the disc than in the bulk phase of DMPC. *Far right*: Interaction of small amounts of A-I with sonicated DMPC vesicles. Small amounts of A-I can bind to the surface of DMPC vesicles without disrupting the vesicle to form a disc-like structure.

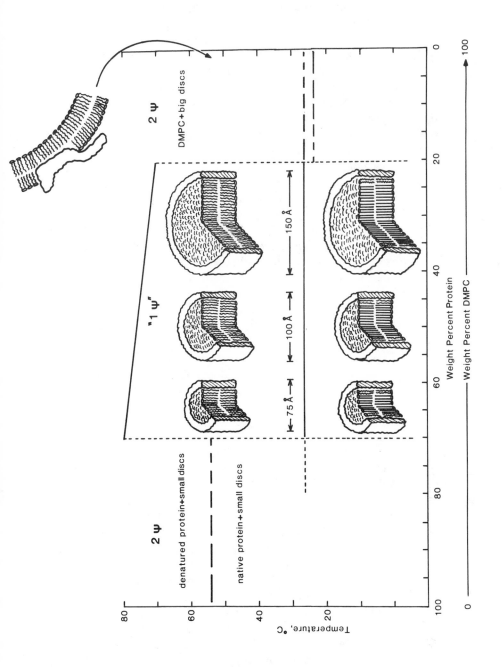

Since no fluorescent polarization changes were found as the concentration was raised it was argued that no major protein conformational change occurred with oligimerization (12). Studies on the rate of association of apoA-I with multilamellar DMPC liposomes were carried out with A-I in monomeric and oligimeric states (114). While the end recombinant products were identical when the starting lipid-to-protein ratios were the same for both cases, the rate of reaction was somewhat slower with the oligimeric protein. This suggests that the dissociation of oligimers might be a rate-limiting step in the formation of lipid protein recombinants. When aqueous A-I is unfolded, i.e. above 54°C, it does not readily interact with liposomes. For instance, it will not react with distearoyl phosphatidylcholine liposomes at 56°C. Thus, an intact secondary and tertiary conformation appears necessary for lipid interaction.

Effect of lipid physical state on kinetics of recombinant formation
Dimyristoyl phosphatidylcholine and other saturated phosphatidyl-cholines can exist in excess water in three basic states (see section on liposomes under Lipid Recombinants Resembling Lipoproteins). In the equilibrium state at low temperatures, phospholipids crystallize with a very small amount of water to form a true crystalline solid. Heating transforms this solid to a gel phase, $L_{\beta'}$, in which the chains are packed in a quasi-hexagonal lattice and tilted at about 60° to the plane of the bilayer. On passing through a small "pretransition" the layers ripple and the chains acquire a true hexagonal nonspecific packing. On further heating the chains melt, forming a lamellar liquid crystal (L_α). The major heat and volume changes occur at the crystal-to-gel transition and at the chain melting of the rippled gel phase (171). The volume of the lamellar liquid L_α phase is about 6–8% greater than the volume of the gel phase just below the transition (127). This volume change is seen in all of the acyl chain transitions of lipids (170). Since the bilayer is actually thinner in the liquid crystal than in the gel state, the transition would be accompanied by a major change in the surface area (as surface area times thickness is equal to volume). During the transition from liquid crystal to gel about 10% free volume is present in the surface due to the contraction and elongation of the chains. Thus, transient large cracks and spaces might occur at the phase transition (16).

The mean size of the group of small molecules that undergoes a transition can be calculated from the transition temperature and the width of the transition (144). Structural studies on DMPC (84, 85, 171) show that at the gel-to–liquid crystal transition the surface area per phospholipid in the gel phase is 47 Å^2. The actual cross-sectional area of the two chains is only 40.8 Å^2, but since the chains are tilted they occupy a somewhat larger area. The corresponding area of the hydrocarbon chain in the liquid crystalline state is 58 Å² (84). Thus, the surface area changes about 11 Å² per molecule at the transition in multilamellar liposomes. The cooperative unit can be as low as 100 if any impurities such as cholesterol are present. But even a cooperative unit of 100 would mean that a 1100-Å² space would be created at the border between the 100-molecule unit and the surrounding other phase. In a single area this would represent a spherical hole 37 Å in diameter. If the cooperative unit were as large as 500 molecules then the space created at the transition would be equivalent to about 5500 Å² or a hole 83.7 Å in diameter. If an annular space were created around the cooperative unit region it would be 9 Å wide, i.e. wide enough to allow an amphipathic helix to penetrate.

Pownall et al (141) studied the rates of recombinant formation of A-I with DMPC multilamellar vesicles at different temperatures. DMPC in the gel phase was unreactive whereas a maximum rate was reached at the phase transition. The rate decreased above the phase transition and was two orders of magnitude slower at 10°C above the transition. The researchers postulated that at the phase transition there were voids in the DMPC layers into which the apoA-I could insert; this would explain the increased rate of complex formation at the phase transition. The data also support the idea that small evanescent islands or domains of gel phase are present in the liquid crystal a few degrees above the transition since the rate is much faster 2°C above than 10°C above the transition.

INTERACTION OF A-I WITH OTHER PHOSPHOLIPIDS AND PHOSPHOLIPID-CHOLESTEROL LIPOSOMES AND VESICLES   Small amounts of cholesterol added to the system, up to 12 mol%, augmented the rate of interaction of A-I with PC liposomes even further than in the absence of cholesterol. However, larger amounts of cholesterol suppressed the rate (142), and above 33 mol% no interaction took place (189). Swaney noted further that in a variety of other pure phospholipids including phosphatidylcholines of different chain lengths (181), mixtures of phospholipids with different phase transition onsets (178, 181), and bovine brain sphingomyelin (180), the most rapid association and recombinant formation occurred just at the beginning of the phase transition when the deformations would have been the largest and the liquid crystal phase would have been the least important of the two phases.

Some modifications of the complex were noted with increases in the chain length of the phospholipid. For instance, dipalmitoyl phosphatidylcholine (chain transition in excess water at 42°C) was found to form complexes readily with apoHDL and much more slowly with apoA-I at around or slightly above the chain transition (196a). However, if unilamellar vesicles of DPPC are made then their reactions are relatively rapid and complexes of different stoichiometry may be formed. At high apoprotein/DPPC ratios the discoidal complexes are small and have two to three apoA-I molecules (1 A-I/100 DPPC molecules). Large discs with diameters of 208 Å and molecular weights of about 900,000 can be formed with three or four molecules of A-I (210). Larger unilamellar vesicles interacted more slowly with apoA-I than smaller vesicles to form discoidal particles (209). Thus, the radius of curvature affects the packing of phospholipids and the closer the structure is to a planar bilayer the less the interaction of A-I (209).

The original description of discs by methods such as electron microscopy and calorimetry indicated that the stoichiometry was two or three molecules of A-I per disc (196a). This was corroborated by protein cross-linking experiments (178, 179). However, it was clear that there was a great deal of overlap between particles with two A-Is and particles with three A-Is, which indicated that there may be some conformational differences in two such particles of the same size. While it was true that the larger the particle the more likely it was to have three A-Is, the electron microscopy studies could not discriminate the number of A-I molecules.

In a recent study using nondenaturing gradient gel electrophoresis (17), Brouillette et al (24) found there was a series of bands corresponding to fairly discrete particles; DMPC/A-I particles ranged from about 202 Å diameter (apparently a vesicle) through a series of diameter peaks of 168, 148, 126, 110, 98, 94, 92, and 89 Å for different mixtures of A-I and DML. Thus, there appears not only to be a distinction between particles containing two A-Is and particles containing three, i.e. between 126 Å and 145 Å diameter, but also between particles that have two or three A-Is but different amounts of phospholipid. These authors suggested that within a particle that contained just two A-Is, discrete semiquantitative heterogeneity was due to different lipid amounts that were caused by conformational heterogeneity in A-I. They suggested that as lipid is decreased, two or three amphipathic helices may dissociate from the smaller surface perimeter to interact with each other; the dissociation resembles the outfolding of an excessive surface. Thus, the most stable disc may be the largest disc that can be covered when all amphipathic helices are on the surface. As less lipid is present in the disc the helices pop off, giving rise to the semiquantitative heterogeneity of particles observed when this method is used (24). Such changes might give rise to changes in the free energy of the apoprotein.

The free energy state of the apoA-I allows one to make certain predictions about its interactions with lipids. There is no doubt that the protein is more stable on discs, particularly large discs, than in solution. The larger discs are extremely stable structures. They are probably at true equilibrium, unlike small unilamellar lecithin vesicles which are not at equilibrium. The kinetics of disc formation in mixtures of apoproteins and liposomes is clearly favored by the defects in the structure produced at the phase transition. However, even above the phase transition interaction takes place slowly.

In some instances with long chain lecithins a certain activation energy must be overcome. This can be accomplished by sonicating and producing metastable vesicles. It can also be accomplished by first allowing the bilayer to interact with a detergent such as sodium cholate or with other bile salts; this maintains a general structure of the lecithin bilayer in the micellar system, but chops it up into discoidal micelles (see section on discoidal aggregates). When apoproteins are added to such a system, even with long chain phospholipids that react extremely slowly in the form of liposomes, discoidal aggregates of apoA-I and long chain lecithin can be produced (116). Probably apoA-I and bile salt compete for hydrophobic lecithin surface in small mixed micelles. When the bile salt is dialyzed away or removed on a column, then discoidal aggregates of apoprotein and phospholipid remain.

The bile salt dialysis method (92, 116, 117, 129, 130) has been used widely to produce discoidal recombinants with phospholipids that are in the liquid crystalline state or that have particularly long chains or high cholesterol content and do not interact well with A-I in multilamellar or even in unilamellar states. Nichols et al (129, 130) used gradient gel electrophoresis to examine apoA-I/egg yolk phosphatidylcholine (EYPC) recombinants and noted, as was found for saturated lecithins (24), that a number of discrete discoidal particles were formed; the number depended on the apoA-I-to-EYPC ratio. Addition of cholesterol changed the distribution (129) and size, and reaction of lecithin-cholesterol-acyl-transferase (LCAT) with different-sized discs produced unique spherical products (130).

Using calorimetry to study the transitions of apoA on intact HDL, Tall et al found that both human HDL_2 and HDL_3 (187) and a series of bovine HDLs of different size (190) have apoA-I that has a higher transition than free A-I. However, this apoA-I has an estimated free energy of stabilization similar to but slightly higher than that of apoA-I in solution (1.7 vs 2.4 Kcal/mol apoA-I). Realizing that apoA-I was not very stable on intact HDL, Tall & Small (193) hypothesized and then showed that when intact HDL was mixed with DMPC liposomes A-I left the intact HDL and formed discoidal aggregates with the DMPC, leaving an HDL particle containing A-II and some naked hydrophobic surface.

These particles then fused to form large apoA-II–rich HDL particles. Thus, the free energy state or stability of the A-I can determine its movement between and interaction with other lipids. The unstable A-I conformation in spherical HDL could allow the A-I to dissociate from HDL under the proper conditions. A-I would then form new discoidal HDL with membrane phospholipid or remnants of phospholipid formed during the lipolysis of triglyceride-rich lipoproteins.

ApoA-II–Surface Lipid Recombinants

Similarly to the studies with A-I, many of the early studies on the interaction of A-II with lipids (particularly phospholipids) concentrated on conformational changes that occur in the protein. On interaction with lipid the conformational changes were investigated by circular dichroism or fluorescence spectroscopy. Few attempts were made to characterize the structure or morphology of the complex formed between the apoprotein and the phospholipids. The early studies of Jackson (80), for example, demonstrated that the α-helical content of apoA-II was reduced from approximately 46 to 37% on reduction and alkylation of the single cross-bridging disulfide bond in the apoA-II dimer. This reduction in α-helical content was reversed on interaction with sonicated phospholipid vesicles of egg yolk lecithin. The α-helical content of apoA-II increased from 46 to 61% for dimeric apoA-II and from 37 to 56% for reduced and alkylated apoA-II. Similar conclusions were derived on the interaction of A-II with lysolecithin or hexadethyl trimethyl ammonium bromide micelles (134). Additionally, these studies (134) showed that A-II was more resistant to quanidinium hydrochloride denaturation when in a complex with lysolecithin in micelles.

More complete studies have been carried out on the interaction of apoprotein A-II with the synthetic phospholipid dimyristoyl phosphatidylcholine (113, 115). These studies have demonstrated that the complex formed between apoA-II and DMPC depends both on the lipid-to-protein ratio of the incubation mixture and on the incubation temperature. The temperature dependence is related to the gel-to–liquid crystal transition temperature of the dimyristoyl lecithin (23–24°C).

Three complexes have been described with differing lipid-to-protein ratios. At an initial lipid-to-protein ratio of 45:1, a single complex of 2.3×10^5 d was quantitatively formed at all temperatures between 20 and 30°C (113). Under more lipid-rich conditions different complexes were formed depending on the temperature. At an initial molar ratio of 100–300:1 and at 24°C a 240:1 complex of 1.5×10^6 d was thought to be formed. This complex was formed from a precursor 75:1 molar ratio complex (3.43×10^5 d) under great excess of lipid. Later studies (115) showed that the formation of these complexes depended on the temperature of incubation with respect to the gel-to–liquid crystal transition temperature of dimyristoyl lecithin. At 20°C, below the liquid crystal transition, the 75:1 molar ratio complex was formed. This complex has a partial specific volume of 0.914 ml/g. At 24°C two different complexes were formed; one was similar to the 75:1 molar ratio complex formed at 20°C and the other resembled the 240:1 complex with a Stokes radius of 120 Å and partial specific volume of 0.948 ml/g. At 30°C, above the liquid crystal transition, the 45:1 complex was formed. This complex had a Stokes radius of 57 Å and partial specific volume of 0.892 ml/g. Electron microscopy showed that all of these complexes were discoidal structures with dimensions that depended on the lipid-to-protein ratio; the 45:1 complex had dimensions of 175 × 60 Å, the 75:1 complex had dimensions of 250 × 62 Å, and the 240:1 complex had dimensions of 500 × 55 Å.

Swaney (178) used binary mixtures of DMPC with distearoyl phosphatidylcholine (DSPC) or 1-palmitoyl, 2-oleyl phosphatidylcholine (POPC) to examine the effect of changes in liquid crystal transition temperature on the interaction of apoprotein A-II. His conclusions were similar to those described for apoA-I: Complex formation occurred most rapidly at the onset of the gel–liquid crystal transition.

The interaction of apoprotein A-II with a series of different phospholipid species has also been used to assess the free energy of association of A-II with different lipids (140). In this study reduced and carboxymethylated apoA-II was used. The free energy of association of A-II was measured for DMPC multilayers and single bilayer vesicles, bovine brain sphingomyelin vesicles, brominated egg lecithin in the absence and presence of 33 mol% cholesterol, and native human plasma HDL. These lipids showed free energy of association of approximately 6–8 Kcal/mol. The free energy of association was insensitive to temperature even close to the transition temperature of the lipid.

The kinetics of the association of A-II and dimyristoyl lecithin vesicles containing cholesterol has also been examined (198). A-II was found to associate more readily with vesicles than apoprotein A-I. As with A-I, optimal complex formation occurred at the phospholipid phase transition and at around 10 mol% cholesterol. However, in contrast to apoA-I, apoA-II still associated with vesicles containing up to 20 mol% cholesterol at temperatures up to 32°C. By electron microscopy these particles were viewed to be discoidal structures similar to those described in the absence of cholesterol.

Thus, in most respects (although in many instances the resulting complexes have not been characterized in such detail) the interaction of apoA-II with phospholipids seems to mirror that of apoA-I. Discoidal structures are usually formed. These are remnants of a phospholipid bilayer that is solubilized by apoprotein occupying a peripheral location at the edge of the bilayer remnant. As the composition and hence the stoichiometry of lipoprotein changes, complexes of different diameters are formed.

ApoC–Surface Lipid Recombinants

In all respects the interaction of the C apoproteins, including each of the individual C apoprotein classes, with lipids seems to mirror that described and catalogued for apoA-I and apoA-II.

The first demonstration that the individual C apolipoproteins, C-I, C-II, and C-III, could interact with lecithin to form discoidal structures in a similar fashion to the apoproteins A-I and A-II was carried out by Forte et al (47). Mixtures of the individual C apoproteins (C-I, C-II, C-III) with egg lecithin were sonicated and protein/lipid complexes were then isolated by ultracentrifugal flotation at different density ranges. In each case the bulk of the protein was recovered in the $d = 1.063$–1.21 g/ml fraction. Each of the C apolipoproteins was found to form a discoidal structure, as shown by electron microscopy, with a minor axis of approximately 40 Å and a major axis of 200 Å. At approximately the same time detailed studies on the interaction of apoC-I with egg lecithin by Jackson et al (81) demonstrated the formation of discoidal complexes and led to the development of the amphipathic helix concept. Subsequent studies have shown that the interaction with phospholipid is maximal at the phospholipid phase transition, and the phase transition temperature of the phospholipid is elevated approximately 2–3°C in complexes with the C apoproteins (27). The complex formed between dimyristoyl phosphatidylcholine and apoprotein C-III represents the only discoidal structure other than that formed between A-I and dimyristoyl lecithin that has been characterized structurally by means other than electron microscopy. Laggner et al (96) demonstrated by small-angle X-ray scattering methods that the structure of the complex could be described by ellipsoidal structure of 170 Å × 50 Å with a 10 Å–thick shell of high electron density containing the protein surrounding a core of low electron density within the particle.

ApoE–Surface Lipid Recombinants

Roth et al (154) have investigated the complex of apoprotein E with dimyristoyl lecithin, which has a discoidal structure very similar to those described for apoA-I complexes. ApoE/DMPC complexes have also been described by Innerarity et al (78), and have been used in studies of the apoB E receptor on human fibroblasts.

ApoB–Surface Lipid Recombinants

Few studies have examined the interaction of apoB with phosphatidylcholines, the major surface lipids in native LDLs. However, with the recent development of several methods (175, 202, 204) for the solubilization of apoB, studies on the interaction of the protein with different lipids are beginning to appear. Watt & Reynolds (206) have examined the interaction of apoB, solubilized by sodium dodecyl sulfate (SDS), with egg yolk lecithin, whereas the method developed in our laboratory uses sodium deoxycholate–solubilized apoB to form complexes with DMPC (202, 203).

In the method described by Watt & Reynolds (206) apoB solubilized with SDS was incubated with egg yolk lecithin at a 2000:1 molar excess of lipid to apoB. Excess SDS was removed from the mixture by dialysis and the complex was fractionated by ultracentrifugal flotation at a density of 1.1 g/ml, followed by a gel filtration chromatography on Sepharose CL 4B. A phospholipid/apoB complex containing 320 mol egg yolk lecithin per 5×10^5 g apoB (1:2 wt/wt) was isolated. Electron micrographs of negatively stained samples of the complex showed it to be approximately 140 Å in diameter, somewhat smaller than LDL. The micrographs were interpreted to suggest that the complex may be a flattened ellipsoid rather than having the spherical morphology of LDL. The apoB in this complex migrated as a single high-molecular-weight band on a 5% SDS polyacrylamide gel electrophoresis, showing that no degradation of the apoB had occurred during the formation of the complex. The apoB in the complex cross-reacted with rabbit antibodies to native LDL and circular dichroic spectroscopy indicated that the conformation of apoB in the complex was very similar to that observed in native LDL.

The complex between LDL apoB and dimyristoyl lecithin developed in this laboratory (202, 203) was formed from apoB and phospholipid both solubilized in sodium deoxycholate. Following incubation the bile salt was removed by extensive dialysis and a complex formed between apoB and the phospholipid. Studies over a wide range of DMPC-to-apoB incubation ratios from 10:1 to 1:1 (wt/wt) and characterization by ultracentrifugation and gel chromatography have shown that at all ratios a reasonably homogeneous DMPC/apoB complex with a weight ratio of 4:1 is formed. Density gradient fractionation of the complex has shown that it may be isolated with a median density of 1.095 g/ml, which is in agreement with the density calculated from the composition. Electron micrographs of negatively stained particles with the complex have shown the structural organization to resemble a single bilayer phospholipid vesicle with a diameter of approximately 210 Å into which apoB has been incorporated. This vesicular organization has been substantiated by trapping experiments employing tritiated dextran and carboxy-fluoroscein (203), and additionally by small-angle X-ray scattering experiments (M. T. Walsh & D. Atkinson, unpublished). 7.5% and 3% SDS polyacrylamide gel electrophoresis have shown that the apoB in the complex is stable and that no degradation occurred during the formation of the apoB/DMPC complex.

Differential scanning calorimetry and circular dichroic spectroscopy were used to study the physical properties of apoB in the apoB/DMPC complex and to compare the physical properties with those of sodium deoxycholate–solubilized apoB and apoB in native LDL (203). The DMPC/apoB complex exhibits a reversible thermal transition centered at 24°C (ΔH 3.34 Kcal/mol of DMPC), which is associated with the order-disorder transition of the hydrocarbon chains of the phospholipid. An irreversible transition that occurs over the range of 53–70°C ($T_m = 62°C$, $\Delta H = 2.09$ cal/g apoB) has been shown by circular dichroism to correspond to protein unfolding denaturation and to particle disruption. Circular dichroism has shown that apoB in the vesicular complex undergoes a series of conformational changes. The major alterations occur over the temperature of the order-disorder transition of the phospholipid; between 37 and 60°C the conformation is similar to that observed in native LDL. The irreversible transition, with a peak temperature of 62°C (which corresponds to protein unfolding, denaturation, and particle disruption) occurs at approximately 20°C lower than the corresponding transition in native LDL. This demonstrates that apoB is thermally less stable in the complex with phospholipid than in the native LDL particle.

APOPROTEIN EMULSION AND MICROEMULSION RECOMBINANTS

Apoprotein Emulsion Recombinants

INCUBATION AND REISOLATION EXPERIMENTS   Early work by Havel et al (70) indicated that when Intralipid was incubated with a $d > 1.006$ g/ml fraction from serum and reisolated it acquired a number of smaller exchangeable apoproteins very much like those associated with native VLDL. These apoproteins included apoC and perhaps some apoA and apoE. Such incubation and reisolation experiments have been used to study movements of exchangeable apoproteins from lipoprotein classes (e.g. HDL, LDL) to emulsions and to recover apoproteins that fell off during ultracentrifugation and were thus present in the $d > 1.21$ g/ml bottom. This method has also been used for a qualitative estimation of which apolipoproteins may be transferred from plasma to emulsions of different composition. For instance, when emulsion particles were made with egg lecithin and the nonpolar lipids methyl oleate, ethyl oleate, glycerol trioleate, and erythritol tetraoleate, differences were noted in the apoproteins bound to these emulsions after brief incubation in rat or human plasma. The triglyceride and erythritol tetraoleate emulsions bound A-I, A-IV, C-II, C-III, E, and albumin. Methyl oleate and ethyl oleate bound less protein but proportionately more albumin, less apoprotein, and almost no apoC. Isolated Intralipid particles bound more albumin and less apolipoproteins than the triolein emulsion (M. Badr, D. Kodali & T. R. Redgrave, unpublished). Thus, the absolute binding from any given plasma partly depends on the core lipids. Of course the distribution of core lipids in the surface (see 121) may be the key factor in determining which and how much apoprotein binds.

More specific binding data has been obtained with Intralipid and high density lipoproteins (44). There was some disruption of the particles by HDL apolipoproteins, which bound phospholipid and removed it from the surface. Connelly & Kuksis (31) showed that emulsions with cholesterol oleate cores bound fewer C apoproteins than emulsions with triolein cores. Their work showed that apoA-I bound more rapidly to large triacylglycerol-rich lipoproteins than to small particles. But Tajima et al (184) obtained opposite results.

QUANTITATIVE BINDING STUDIES   Quantitative binding studies using different apoprotein and/or substrate concentrations allow the calculation of an apoprotein dissociation constant (K_d) from the lipid substrate and a saturation number (N), the number of apolipoprotein molecules per unit of surface (i.e. per 1000 phospholipid molecules). Several of the apoproteins have been studied using well-characterized lipid particle surfaces as substrates. The data are summarized in Table 2. Some caveats are worth mentioning. First, the techniques vary from laboratory to laboratory. Second, there is a considerable variation between different methods within the same laboratory (184). Third, apoproteins were radiolabelled in some studies, and this could alter their binding properties (133). Nevertheless, the binding parameters of the exchangeable apo-proteins to surfaces in which egg lecithin was the primary emulsifier show some similarities (Table 2).

ApoA-I   In the studies in Table 2, the particles were prepared and suspended in an aqueous medium and mixed with an aqueous solution of the apolipoprotein. After a period of equilibration (~ 60 min) the particles were reisolated by flotation centrifugation or by small columns. Thus the protein had to penetrate from the aqueous phase and bind to the surface.

Table 2 Binding parameters for apoproteins

Apoprotein [V_{mol} (Å³)]	Surface	K_d (μM)	Mole peptide per 1000 PL	aa/PL	g peptide per 100 g PL	% Surface volume occupied by peptide	Reference
A-I [34,817]	EYL vesicle	0.90	1.74	0.42	6.29	4.77	211
	EYL vesicle with 20 mol% chol	0.30	3.70	0.90	13.37	10.14	211
	EYL/TO emulsion small (~279 Å)	0.20[a]	2.59	0.63	9.39	7.12	184
		0.16[b]	2.39	0.58	8.66	6.57	184
	EYL/TO emulsion large (~2250 Å)	3.17[a]	0.95	0.23	3.44	2.61	184
		4.24[b]	1.56	0.38	5.66	4.29	184
	CM model[c] (surface ~6% chol)	0.70	3.9	0.95	14.8	11.22	d
	CM model[c] (surface ~30% chol)	0.81	0.6	0.15	2.3	1.74	d
A-II [10,788]	EYL/TO emulsion small (~279 Å)	0.27[a]	8.61	1.24	19.35	14.67	184
		0.25[b]	10.00	1.44	22.47	17.04	184
	EYL/TO emulsion[e] large (~2290 Å)	0.63[a]	4.7	0.34	5.30	4.02	184
		0.30[b]	13.0	0.93	14.52	11.01	184

	pH						Ref.
A-IV [52,659]							
Isolated Intralipid TG-rich particles	6.0	5.3	0.962	0.36	5.28	4.00	207
	6.5	4.8	0.730	0.27	4.00	3.03	207
	7.0	9.3	0.762	0.28	4.18	3.17	207
	7.5	8.3	0.386	0.14	2.12	1.61	207
	8.0	9.3	0.435	0.16	2.39	1.81	207
	8.5	10.0	0.365	0.14	2.00	1.52	207
C-II [34,817]							
EYL/TO emulsion small (~279 Å)		1.07[a]	11.27	0.89	12.83	9.73	184
		0.45[b]	11.77	0.93	13.40	10.16	184
EYL/TO emulsion large (~2290 Å)		0.60[a]	8.99	0.71	10.24	7.76	184
		0.60[b]	8.35	0.66	9.51	7.21	184
C-III$_2$ [10,842]							
EYL/TO emulsion small (~279 Å)		0.72[a]	13.16	1.04	14.86	11.27	184
		0.53[b]	12.15	0.96	13.72	10.40	184
EYL/TO emulsion large (~2290 Å)		0.54[a]	8.23	0.65	9.29	7.04	184
		1.07[b]	10.63	0.84	12.00	9.10	184
E$_3$[f] [42,375]							
CM model[c] (surface ~6% chol)		1.17	3.8	1.14	17.5	13.27	[d]
CM rem model[c] (surface ~30% chol)		1.05	0.7	0.21	3.5	2.65	[d]
A-I Peptide [3252]							
EYL vesicles		1.92	15.1	0.33	5.13	3.89	211
EYL vesicles with 20 mol% chol		2.80	27.2	0.60	9.26	7.02	211

[a] Column method.
[b] Airfuge method.
[c] ApoA-1 and E$_3$ tested on the same day with the same emulsions. Emulsions CM model and CM remnant model made according to Miller & Small (122).
[d] A. Derksen & D. M. Small, unpublished.
[e] In absence of NaCl.
[f] Kindly provided by Dr. K. Weisgraber of the Gladstone Foundation Labs, San Francisco, California.

If the surface was egg lecithin vesicles with or without a small amount of cholesterol (20 mol%), the binding became stronger in the presence of a small amount of cholesterol. Furthermore, the number of peptides per 1000 phospholipids approximately doubled (211). Comparison of small and large egg yolk–lecithin/triolein emulsions without free cholesterol indicates that the K_d is much greater (i.e. binding is weaker) for the large particles even though the maximum number of molecules bound is similar (184). These data were taken to suggest that the packing of phospholipids is tighter on the large particles due to the more planar nature of the surface, and that this explains the higher binding constant.

Using a series of well-defined particles to model nascent chylomicrons and chylomicron remnants (122) we found that the binding constant was rather similar for the two particles, but that the number of molecules of A-I bound to the CM remnant model was much less (A. Derksen & D. M. Small, unpublished observations). Thus cholesterol appears to greatly decrease the surface available for A-I binding.

Other exchangeable apoproteins The dissociation constants (K_d) for apoA-II, apoC-II, and apoC-III (184) and for apoA-IV (207) are rather similar, as was the maximum binding when expressed as amino acids/PC. Size of the emulsion did not appear to influence the binding of A-II, C-II, or C-III. The rather constant amount of protein bound when expressed per phospholipid suggests that the surface of most of the particles that do not contain cholesterol can be compressed only approximately 8–12%, and that apoproteins that are able to penetrate the surface can do so only within the limits of this potential surface space. This space appeared to be increased by addition of a small amount of cholesterol (20 mol%) but decreased at higher surface concentration of cholesterol.

IN VIVO CORRELATIONS Using apoprotein-free lipid emulsions, chylomicron metabolism was simulated in the rat (110, 148). Physicochemically defined lipid emulsions that resemble chylomicrons and their remnants were made from egg lecithin, ^{14}C-triolein (TG), ^{3}H– cholesteryl oleate (CE), and free cholesterol (C). The models were constructed so that the surface composition of C was ~ 10 wt% for chylomicron emulsion models and 40 wt% for remnant models. The model particles were injected intra-arterially in nonanesthetized rats, and plasma samples were taken every 2 min. The livers were removed at 12 min and uptake was measured.

The plasma half-life of chylomicron models was TG = 2.8 min, CE = 5.4 min. Liver uptake was 11% TG and 48% CE. In constrast in the remnant model TG and CE were removed simultaneously (3.5 ± 3 vs 3.3 ± 3 min) and liver uptake was similar ($30\% \pm 5\%$, $35\% \pm 8\%$). Therefore the chylomicron model permitted lipolysis of TG to occur before the particle was removed by the liver whereas the remnant model went rapidly to the liver without TG hydrolysis in the plasma. In incubation and reisolation experiments both particles were able to adsorb pure apoE or apoC. However, when incubated with rat plasma the remnant model adsorbed much more apoE and less apoC than the chylomicron models. In recirculating liver perfusions, the remnant model was removed twice as rapidly as the chylomicron model.

Thus, in these apoprotein-free model emulsions the surface lipid composition determines the composition of the adsorbed apoproteins, which in turn determines the metabolic pathway. The high surface cholesterol in the remnant model totally inhibited lipolysis and directed the particles to the liver.

The addition of one exchangeable apoprotein can displace an apoprotein already bound to the surface of the particle (A. Derksen & D. M. Small, unpublished). The absolute affinities of the different apoproteins for specific particles have not been worked out, but A-II clearly replaces A-I quantitatively on HDL. In experiments in which Intralipid with or without apoA-I adsorbed to it was injected into human subjects, the clearance of the triglyceride was much more rapid with particles that contained apoA-I (45). When the particles were removed and repurified it was found that A-I had been replaced by apoC and the apoC content was higher than in the particles that had not originally contained A-I. Erkelens & Mocking suggested that the function of A-I was to provide a slot for apoC (45).

The possibility that one lipid-bound apoprotein may provide a binding site for another has not been carefully explored. However, large apoproteins such as apoB and apoE may have sites that can bind other apoproteins. Such binding would be protein-protein in nature but could be influenced by the conformation of the apoprotein bound to the lipid interface. Protein-protein interactions could change conformations, cover or alter binding sites, and perhaps account for some difference in lipoprotein metabolism.

Apoprotein Microemulsion Recombinants

MICROEMULSION–A-I RECOMBINANTS   The phospholipid, cholesterol ester, microemulsion particles described earlier (51) have been used as precursor particles for the formation of reassembled lipoprotein complexes that model LDL and large HDL particles. ApoA-I has been made to interact with preformed phospholipid cholesterol ester microemulsions and the resultant complexes have been characterized by gel filtration chromatography, differential scanning calorimetry, and circular dichroic spectroscopy (52). Microemulsions were used that contained a cholesteryl oleate core surface stabilized by either an egg or dimyristoyl lecithin or a dipalmitoyl lecithin surface monolayer.

Mixing of apoA-I with the microemulsions at 25°C resulted in complexes with a composition of the order of 10% protein, 45% phospholipid, and 45% cholesterol ester. The complexes were similar in size to the microemulsion precursor. The physical state of the phospholipid appeared to influence the nature of the apoprotein binding in a similar fashion to that described for apoA-I phospholipid interactions. Interaction of A-I with the microemulsion at the phase transition of the surface phospholipid resulted in at least two complexes of different size and stoichiometry, whereas the interaction of A-I with the phospholipid cholesterol micro-emulsion system above or below the phospholipid transition gave a single A-I microemulsion complex.

The apoA-I microemulsion complexes undergo at least two specific thermal transitions that depend on lipid composition similarly to their protein-free microemulsion precursor. The first transition arises from the core cholesterol esters and the second from the phospholipid that forms the surface monolayer. In addition, a third transition associated with particle disruption and protein unfolding occurs irreversibly at approximately 80°C. The free energy of this transition is similar to that observed in high density lipoproteins (187) but is different from that observed in discoidal phospholipid apoprotein A-I complexes (196a). This suggests that the association of A-I to the microemulsions is similar to that in HDL (187). The enthalpy of the surface and core transitions as well as the transition temperature of the core components were lower than those of the protein-free microemulsions (51), suggesting a direct apoprotein-core or surface-core interaction.

MICROEMULSION–ApoB RECOMBINANTS   Phospholipid cholesterol ester microemulsions also serve as precursor particles for the formation of a reassembled model LDL particle by the interaction of solubilized apoB with the preformed microemulsion particles (53). In our laboratory preformed microemulsion particles at room temperature are placed into an open-ended dialysis bag and then sodium deoxycholate-solubilized apoB in a small volume of buffer is slowly introduced. Under these conditions diffusion of bile salt across the dialysis membrane lowers the concentration in the dialysis bag well below the critical micellar concentration of the bile salt (168), allowing apoB to bind to the microemulsions. This method has been used to bind apoB with micro-emulsions formed from dimyristoyl lecithin, cholesteryl oleate, egg yolk lecithin, and cholesteryl oleate. Gel filtration chromatography was used to demonstrate that a stable complex was formed with a molar ratio of cholesterol esters to phospholipid of approximately 1 : 1, which was comparable to that observed in the initial microemulsion particle. The apoB-to-PC ratio was 0.32 (wt/wt) and was relatively constant. The model LDL complexes were shown by electron microscopy to have a circular morphology and a diameter of 220 Å, similar to native LDL. Complexes exhibited β migration on agarose gels and the apoB migrated as one high-molecular-weight band on 3% SDS polyacrylamide gel electrophoresis, demonstrating that no degradation of apoB$_{100}$ occurred during the preparation of the complexes.

Lundberg & Suominen published a variation in this methodology (101) in which apoB was also solubilized with sodium deoxycholate. However the sodium deoxycholate was removed from the apoB by dialysis prior to interacting with the preformed apoB microemulsion particles. The actual binding was accomplished with the aid of sonication of the microemulsion/apoB incubation mixture. The reassembled particles were isolated and characterized by density gradient ultracentrifugation and gel filtration chromatography on Sepharose 4B. Particles viewed by electron microscopy exhibited spherical morphology with a mean particle diameter of 210 Å. The cholesteryl oleate–to–egg lecithin ratio was 1.5 and the protein-to–egg lecithin ratio was 0.8.

The reassembled model LDL particles developed in our laboratory show many structural and physical properties similar to those of native LDL. In addition, the model LDL particles retain the thermodynamic characteristics unique to their precursor protein-free phospholipid cholesterol ester microemulsions. Figure 6 shows the calorimetric transitions observed in these model LDL particles. When egg lecithin forms the surface monolayer of the microemulsion the order-disorder transition of the cholesteryl oleate in the core of the particle occurs at 38°C with an enthalpy of 0.3 cal/g of cholesteryl oleate. The high-temperature transition at 85°C, which corresponds to particle disruption and protein unfolding, is also similar to that observed in native LDL. These two transitions are also observed in the complex formed with dimyristoyl lecithin forming the surface phospholipid. However, in addition a high-enthalpy transition occurs at 22°C, with an enthalpy of 2.6 cal/g of DMPC, which is associated with the order-disorder transition of the fatty acyl chains of the surface monolayer of phospholipid.

The transition temperatures and enthalpies of the core cholesterol esters and, in the case of the particle formed with dimyristoyl lecithin, of the surface phospholipid monolayer are, however, lower than those observed in the precursor protein-free microemulsion particles. Thus, apoB appears to either directly or indirectly influence the properties of both the core

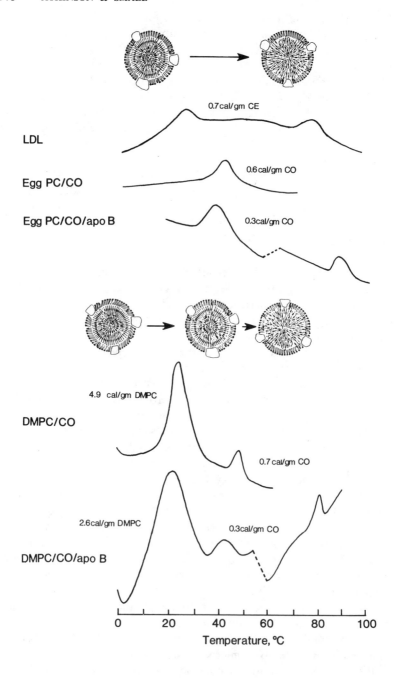

and the surface components. Also, in both particles the high-temperature (approximately 80°C) transition is similar to that described for native LDL. This may be contrasted with the denaturation transition that occurs some 20°C lower in complexes formed between apoB and phospholipid alone, suggesting that a complete microemulsion structural environment for the protein is necessary for the protein to obtain maximum thermal stability. At 4°C the secondary structure of apoB in the reassembled particle formed with microemulsions of egg lecithin and cholesteryl oleate is similar to that of native LDL. However, for complexes with dimyristoyl lecithin forming the surface monolayer the secondary structure shows less α-helix, suggesting that the protein conformation may depend at least in part on the composition and physical state of the surface phospholipid.

Reconstitution of Low Density Lipoproteins

Krieger and co-workers (94) have described an alternative route for the reassembly of LDL particles in which 99% of the core cholesterol esters and most of the free cholesterol are removed from LDL and subsequently replaced by exogenous neutral lipid. This method uses the extraction of the cholesterol esters from LDL bound to potato starch, a method first described by Gustafson (60). The neutral lipids are typically extracted by heptane at −10°C. The exogenous cholesterol esters, for example cholesteryl linoleate, are then reintroduced to the LDL particle in heptane. The heptane is evaporated and the reconstituted LDL is eluted from the potato starch. The chemical composition of these recombinant LDL particles resembles that of native LDL with the exception that cholesterol has been removed by the extraction, and that the cholesterol ester core is totally comprised of the exogenous single-species cholesterol ester.

The reassembled LDL particles show β migration on agarose gels and are precipitated by antibodies to LDL and by heparin manganese. The mean density of these particles is slightly greater than that of native LDL and slightly greater than the median density of 1.042 g/ml. However, there is some question as to whether all of the heptane or other organic solvent is removed from the core of these LDL particles.

Figure 6 Calorimetrically observed transitions and schematic representation of changes in structure of phospholipid/cholesterol ester/apoB complexes as models for low density lipoproteins. Shown are the differential scanning calorimetry traces of low density lipoprotein, microemulsion systems formed with egg PC and cholesteryl oleate or DMPC and cholesteryl oleate, and recombinant systems in which LDL apoprotein B has been recombined with the microemulsion systems. The transition of lipid components in the apoB systems mirrors that observed in protein-free precursor particles. In addition, a high temperature (∼80°C) transition corresponding to particle disruption/denaturation is observed in these systems as in native LDL.

Tall & Robinson (191) have reported that the reconstituted LDLs lack a thermal transition of the core cholesterol esters because heptane is retained within the core of the particle. A similar conclusion has been drawn in our laboratory (W. Horn, G. G. Shipley & D. M. Small, unpublished). However, Krieger et al (95), using electron paramagnetic resonance, has demonstrated a thermal transition of the core cholesterol esters in the reassembled particles. The biological activity of these reassembled particles has been reported in detail. The reconstituted LDL binds to the LDL receptor in cultured fibroblasts, has the same affinity as native LDL, and is taken up and hydrolyzed similarly to native LDL. Cholesterol released from lysosomal hydrolysis of the reassembled LDL modulates cholesterol metabolism intracellularly.

Reconstitution of High Density Lipoproteins

Very shortly after the first delipidation of high density lipoprotein and the isolation of total HDL apoproteins, attempts were made to reassemble an HDL complex using the HDL apoproteins and HDL lipids. These studies employed sonication of aqueous dispersions of the HDL lipids with the apoproteins (73, 156, 157) at 40°C, which is above the order-disorder transition of both the phospholipid and, most importantly, the cholesterol esters of native HDL. Indeed model HDL particles could be generated and isolated by ultracentrifugal flotation at density > 1.063 g/ml.

Later studies of Ritter & Scanu (152, 153) carefully defined the conditions and products formed in total reassembly of an HDL-like particle using the individual HDL apoprotein species apoA-I and apoA-II. These studies used a sonication technique similar to that originally described for the total HDL apoproteins. The resulting recombinant HDL particles were defined by density gradient ultracentrifugation and gel chromatography. In reassembled complexes with apoA-I only two types of HDL complexes were formed: (a) a small particle with a radius of 31 Å and (b) a large particle with a radius of ~ 39 Å. The particles contained two and three moles of apoA-I respectively. Additionally, these studies showed that the state of association of the apoA-I in the incubation mixture was important in the reassembly studies and that the interaction of apoA-I to form a reassembled HDL particle was maximal when apoA-I existed as a monomer in solution. A similar study on the interaction of A-II (153) to form recombinant HDL particles demonstrated that a reassembled complex could be formed that was much more heterogeneous in size, ranging from 90 to 200 Å, than that formed with apoA-I alone.

Recently in our laboratory (C. Glass, D. Atkinson & D. M. Small, unpublished) the original sonication methodology for HDL reassembly has been slightly modified to circumvent the demonstrated problem with the self-association of apoA-I in free solution. ApoA-I is solubilized in a small volume of 4 M urea and is slowly introduced into an aqueous dispersion of phospholipid and cholesterol ester during the sonication procedure. ApoA-I lipid complexes are then isolated by ultracentrifugal flotation between a density of 1.063 and 1.21 g/ml. By this method well-defined HDL complexes have been formed with a mean size of ~ 100 Å and with well-defined composition as shown by gel filtration chromatography.

ACKNOWLEDGMENTS

This work was supported by US Public Health Service grants HL-26335 and HL-07291.

Literature Cited

1. Admirand, W. H., Small, D. M. 1968. *J. Clin. Invest.* 47: 1045–52
2. Albers, J. J., Tollefson, J. H., Chen, C.-H., Steinmetz, A. 1984. *Arteriosclerosis* 4: 49–58
3. Alexander, C. A., Hamilton, R. L., Havel, R. J. 1976. *J. Cell Biol.* 69: 241–63
4. Anderson, D. W., Nichols, A. V., Forte, T. M., Lindgren, F. T. 1977. *Biochim. Biophys. Acta* 493: 55–68
5. Andrews, A. L., Atkinson, D., Barratt, M. D., Finer, E. G., Hauser, H., et al. 1976. *Eur. J. Biochem.* 64: 549–63
6. Atkinson, D., Shipley, G. G. 1984. In *Neutrons in Biology*, ed. B. P. Schoenborn, pp. 211–26. New York: Plenum
7. Atkinson, D., Small, D. M., Shipley, G. G. 1980. *Ann. NY Acad. Sci.* 348: 284–88
8. Atkinson, D., Smith, H. M., Dickson, J., Austin, J. P. 1976. *Eur. J. Biochem.* 64: 541–47
9. Atkinson, D., Tall, A. R., Small, D. M., Mahley, R. W. 1978. *Biochemistry* 78: 3930–33
10. Avignan, J. 1957. *J. Biol. Chem.* 226: 957–64
11. Baker, H. N., Gotto, A. M. Jr., Jackson, R. L. 1975. *J. Biol. Chem.* 260: 2725–38
12. Barbeau, D. L., Jonas, A., Teng, T., Scanu, A. M. 1979. *Biochemistry* 18: 367–69
13. Barter, P. J., Hopkins, G. J., Calvert, G. D. 1982. *Biochem. J.* 208: 1–7
14. Basu, S. K., Brown, M. S., Ho, Y. K., Havel, R. J., Goldstein, J. L. 1981.

Proc. Natl. Acad. Sci. USA 78: 7545–49
15. Batzri, S., Korn, E. D. 1973. *Biochim. Biophys. Acta* 298: 1015–19
16. Biltonin, R. L., Freire, E. 1978. *CRC Crit. Rev. Biochem.* 5: 85–124
17. Blanche, P. J., Gong, E. L., Forte, T. M., Nichols, A. V. 1981. *Biochim. Biophys. Acta* 665: 408–19
18. Blue, M.-L., Williams, D. L., Zucker, S., Khan, S. A., Blum, C. B. 1983. *Proc. Natl. Acad. Sci. USA* 80: 283–87
19. Boguski, M. S., Elshoubagy, N., Taylor, J. M., Gordon, J. I. 1984. *Proc. Natl. Acad. Sci. USA* 81: 5021–25
20. Bourges, M., Small, D. M., Dervichian, D. B. 1967. *Biochim. Biophys. Acta* 144: 189–202
21. Breckenridge, W. C., Little, J. A., Alaupovic, P., Wang, C. S., Kuksis, A., et al. 1982. *Atherosclerosis* 45: 161–74
22. Brewer, H. B. Jr., Fairwell, T., LaRue, A., Ronay, R., Hauser, H., Bronzert, T. J. 1978. *Biochem. Biophys. Res. Commun.* 80: 623–30
23. Brewer, H. B. Jr., Shulman, R., Herbert, P., Ronan, R., Wehrly, K. 1974. *J. Biol. Chem.* 249: 4975–84
24. Brouillette, C. G., Jones, J. L., Ng, T. C., Kercret, H., Chung, B. H., Segrest, J. P. 1984. *Biochemistry* 23: 359–67
25. Brown, M. S., Goldstein, J. 1983. *J. Clin. Invest.* 72: 743–47
26. Brunner, J., Skrabal, P., Hauser, H. 1976. *Biochim. Biophys. Acta* 455: 322–31
27. Cardin, A. D., Jackson, R. L., Johnson, J. D. 1982. *FEBS Lett.* 141: 193–97

28. Carey, M. C. 1984. *Hepatology* 4: 138–39S
29. Carey, M. C., Small, D. M. 1978. *J. Clin. Invest.* 61: 998–1026
30. Collins, J. J., Phillips, M. C. 1982. *J. Lipid Res.* 23: 291–98
31. Connelly, P. W., Kuksis, A. 1981. *Biochim. Biophys. Acta* 666: 80–89
32. Cooper, A. D. 1977. *Biochim. Biophys. Acta* 488: 464–74
33. Curatolo, W., Verma, S. P., Sakura, J. D., Small, D. M., Shipley, G. G., Wallach, D. F. H. 1978. *Biochemistry* 19: 1802–7
34. Deckelbaum, R., Eisenberg, S., Oschry, Y., Butbul, E., Sharon, I., Olivecrona, T. 1982. *J. Biol. Chem.* 257: 6509–17
35. Deckelbaum, R. J., Shipley, G. G., Small, D. M. 1977. *J. Biol. Chem.* 252: 744–54
35a. Deeb, S. S., Motulsky, A. G., Albers, J. J. 1985. *Proc. Natl. Acad. Sci. USA* 82: 4983–88
36. Driscoll, D. M., Getz, G. S. 1984. *J. Lipid Res.* 25: 1368–79
37. Dvorin, E., Mantulin, W. W., Rohde, M. F., Gotto, A. M. Jr., Pownall, H. J., Sherrill, B. C. 1985. *J. Lipid Res.* 26: 38–46
38. Edelstein, C., Kezdy, F. J., Scanu, A. M., Shen, B. W. 1979. *J. Lipid Res.* 20: 143–53
39. Edelstein, C., Scanu, A. M. 1980. *J. Biol. Chem.* 258: 5747–54
40. Eigenborg, K. E., Chan, S. I. 1980. *Biochim. Biophys. Acta* 599: 330–35
41. Eisenberg, S., Schurr, D. 1976. *J. Lipid Res.* 17: 578–87
42. Elovson, J., Jacobs, J. C., Schumaker, V. N., Puppione, D. L. 1985. *Biochemistry* 24: 1569–78
43. Enoch, H. G., Strittmatter, P. 1979. *Proc. Natl. Acad. Sci. USA* 75: 145–49
44. Erkelens, D. W., Chen, C., Mitchell, C. D., Glomset, J. A. 1981. *Biochim. Biophys. Acta* 663: 221–33
45. Erkelens, D. W., Mocking, J. A. J. 1985. *Metab. Clin. Exp.* 34: 222–26
46. Formisano, S., Brewer, H. B. Jr., Osborne, J. C. 1978. *J. Biol. Chem.* 253: 354–60
47. Forte, T., Gong, E., Nichols, A. V. 1974. *Biochim. Biophys. Acta* 334: 169–83
48. Forte, T. M., Nichols, A. V., Gong, E. L., Lux, S., Levy, R. I. 1971. *Biochim. Biophys. Acta* 248: 381–86
49. Freeman, R., Finean, J. B. 1975. *Chem. Phys. Lipids.* 14: 313–20
50. Gilman, T., Kauffman, J. W., Pownall,

H. J. 1961. *Biochemistry* 20: 656–61
50a. Ginsburg, G. S., Atkinson, D., Small, D. M. 1985. *Prog. Lipid Res.* 23: 135–67
51. Ginsburg, G. S., Small, D. M., Atkinson, D. 1982. *J. Biol. Chem.* 257: 8216–27
52. Ginsburg, G. S., Small, D. M., Atkinson, D. 1982. *Fed. Proc.* 41: 4392a/1021
53. Ginsburg, G. S., Walsh, M. T., Small, D. M., Atkinson, D. 1984. *J. Biol. Chem.* 259: 6667–73
54. Glomset, J. A., Norum, K. R. 1973. *Adv. Lipid Res.* 11: 1–65
55. Glomset, J. A., Norum, K. R., Nichols, A. V., et al. 1975. *Scand. J. Clin. Lab. Invest.* 35(142): 13–30
56. Green, P. H. R., Glickman, R. M. 1981. *J. Lipid Res.* 22: 1153–73
57. Green, P. H. R., Glickman, R. M., Riley, J. W., Quinet, E. 1980. *J. Clin. Invest.* 65: 911–19
58. Green, P. H. R., Glickman, R. M., Saudek, C. D., Blum, C. B., Tall, A. R. 1979. *J. Clin. Invest.* 64: 233–42
59. Green, P. H. R., Tall, A. R., Glickman, R. M. 1978. *J. Clin. Invest.* 61: 528–34
60. Gustafson, A. 1965. *J. Lipid Res.* 6: 512–17
61. Halberg, D. 1965. *Acta Physiol. Scand.* 64: 306–13
62. Hamilton, J. A., Miller, K. W., Small, D. M. 1983. *J. Biol. Chem.* 258: 12821–26
63. Hamilton, J. A., Small, D. M. 1981. *Proc. Natl. Acad. Sci. USA* 78(11): 6878–82
64. Hamilton, R. L. 1978. In *Disturbances in Lipid and Lipoprotein Metabolism*, ed. J. M. Dietschy, A. M. Gotto, J. A. Ontko, pp. 155–71. Baltimore: Waverly
65. Hamilton, R. L., Havel, R. J., Kane, J. P., Blaurock, H. E., Sata, T. C. 1971. *Science* 172: 475–78
66. Hamilton, R. L., Williams, M. C., Fielding, C. J., Havel, R. J. 1976. *J. Clin. Invest.* 58: 667–80
67. Hamilton, R. L. Jr., Goerke, J., Guo, L. S. S., Williams, M. C., Havel, R. J. 1980. *J. Lipid Res.* 21: 981–91
68. Deleted in proof
69. Hauser, H., Henry, R., Leslie, R. B., Stubbs, J. M. 1972. *Eur. J. Biochem.* 48: 583–94
70. Havel, R. J., Kane, J. P., Kashyap, M. I. 1973. *J. Clin. Invest.* 52: 32–38
71. Herbert, P. N., Assmann, G., Gotto, A. M. Jr., Fredrickson, D. S. 1983. In *The Metabolic Basis of Inherited Disease*, ed. J. B. Stanbury, J. B. Wyngarden, D. S. Fredrickson, J. L.

Goldstein, M. S. Brown, 29: 589–621. New York: McGraw-Hill. 5th ed.

72. Higgins, J. A., Hutson, J. L. 1984. *J. Lipid Res.* 25: 1295–1305

73. Hirz, R., Scanu, A. M. 1970. *Biochim. Biophys. Acta* 207: 364–67

74. Hospattankar, A. V., Farwell, T., Ronan, R., Brewer, H. B. Jr. 1984. *J. Biol. Chem.* 259: 318–22

75. Huang, C., Thompson, T. E. 1974. *Methods Enzymol.* 32: 485–89

76. Ihm, J. J., Harmony, A. K., Ellsworth, J., Jackson, R. L. 1980. *Biochem. Biophys. Res. Commun.* 93: 1114–20

77. Imaizumi, K., Fainaru, M., Havel, R. J. 1978. *J. Lipid Res.* 19: 721–22

78. Innerarity, T. L., Pitas, R. E., Mahley, R. W. 1979. *J. Biol. Chem.* 254: 4186–90

79. Jackson, R. L., Baker, H. N., Gilliam, E. B., Gotto, A. M. Jr. 1977. *Proc. Natl. Acad. Sci. USA* 74: 1942–45

80. Jackson, R. L., Morrisett, J. D., Pownall, H. J., Gotto, A. M. Jr. 1973. *J. Biol. Chem.* 248: 5218–24

81. Jackson, R. L., Morrisett, J. D., Sparrow, J. T., Segrest, J. P., Pownall, H. J., et al. 1974. *J. Biol. Chem.* 249: 5314–20

82. Jackson, R. L., Sparrow, J. T., Baker, H. N., Morrisett, J. D., Taunton, O. D., Gotto, A. M. Jr. 1974. *J. Biol. Chem.* 249: 5308–13

83. Janiak, M. J., Loomis, C. R., Shipley, G. G., Small, D. M. 1974. *J. Mol. Biol.* 86: 325–39

84. Janiak, M. J., Small, D. M., Shipley, G. G. 1976. *Biochemistry* 15: 4575–80

85. Janiak, M. J., Small, D. M., Shipley, G. G. 1980. *J. Biol. Chem.* 1255: 9753–59

86. Jonas, A. 1973. *Biochemistry* 12: 4503–7

87. Jonas, A. 1984. *Exp. Lung Res.* 6: 255–70

88. Jonas, A., Drengler, S. M., Patterson, B. W. 1980. *J. Biol. Chem.* 255: 2183–89

89. Jonas, A., Krajnovich, D. J. 1977. *J. Biol. Chem.* 252: 2194–99

90. Jonas, A., Krajnovich, D. J., Patterson, B. W. 1977. *J. Biol. Chem.* 252: 2200–5

91. Jonas, A., Mason, W. R. 1981. *Biochemistry* 20: 3801–5

92. Jonas, A., McHugh, H. T. 1983. *J. Biol. Chem.* 258: 10335–40

93. Jones, A. L., Hradek, G. T., Hornick, C., Renaud, G., Windler, E. T. T., Havel, R. J. 1984. *J. Lipid Res.* 25: 1151–58

93a. Knott, T. J., Rall, S. C. Jr., Innerarity,

T. L., et al. 1985. *Science* 230: 37–43

94. Krieger, M., Brown, M. S., Faust, J. R., Goldstein, J. L. 1978. *J. Biol. Chem.* 253: 4093–4101

95. Krieger, M., Peterson, J., Goldstein, J. L., Brown, M. S. 1980. *J. Biol. Chem.* 255: 3330–33

96. Laggner, P., Gotto, A. M. Jr., Morrisett, J. D. 1979. *Biochemistry* 18: 164–71

97. Landin, B., Nilsson, A., Two, J.-S., Schotz, M. C. 1984. *J. Lipid Res.* 25: 559–63

98. Lecuyer, H., Dervichian, D. G. 1969. *J. Mol. Biol.* 45: 39–57

98a. Lippel, K. L., ed. 1979. *Report of the High Density Lipoprotein Workshop, San Francisco, NIH Publ. No. 79-1661,* Washington, DC: Natl. Heart Lung Blood Inst.

99. Loomis, C. R. 1977. PhD thesis. Boston Univ.

100. Lundberg, B., Saarinen, E. R. 1975. *Chem. Phys. Lipids* 14: 260–62

101. Lundberg, B., Suominen, L. 1984. *J. Lipid Res.* 25: 550–58

101a. Lusis, A. J., West, R., Mehrabian, M., Reuben, M. A., Leboeuf, R. C., et al. 1985. *Proc. Natl. Acad. Sci. USA* 82: 4597–4601

102. Lux, S. E., John, K. M., Ronan, R., Brewer, H. B. Jr. 1972. *J. Biol. Chem.* 247: 7519–27

103. Mahley, R. W. 1982. *Med. Clin. North Am.* 66(2): 375–402

104. Mahley, R., Hui, D. Y., Innerarity, T. L., Weisgraber, K. H. 1981. *J. Clin. Invest.* 68: 1197–1206

105. Mahley, R. W., Innerarity, T. L. 1983. *Biochim. Biophys. Acta* 727: 197–222

106. Mahley, R. W., Innerarity, T. L., Rall, S. C. Jr., Weisgraber, K. H. 1984. *J. Lipid Res.* 25: 1277–94

107. Mahley, R. W., Innerarity, T. L., Weisgraber, K. H. 1980. *Ann. NY Acad. Sci.* 348: 265–77

108. Mahley, R. W., Weisgraber, K. H. 1979. See Ref. 98a, pp. 356–66

109. Mantulin, W. W., Rohde, M. F., Gotto, A. M. Jr., Pownall, H. J. 1980. *J. Biol. Chem.* 255: 8185–91

110. Maranhao, R. C., Lincoln, E. C., Brunengraber, H., Small, D. M., Redgrave, T. G. 1984. *Arteriosclerosis* 4(5): 566a

111. Marcel, Y. L., Vezina, C., Teng, B., Sniderman, A. 1980. *Atherosclerosis* 35: 127–33

112. Marsh, J. B. 1976. *J. Lipid Res.* 17: 85–90

113. Massey, J. B., Gotto, A. M., Pownall, H. J. 1980. *J. Biol. Chem.* 255: 10167–73

114. Massey, J. B., Gotto, A. M. Jr., Pownall, H. J. 1981. *Biochem. Biophys. Res. Commun.* 99: 466–74
115. Massey, J. B., Rohde, M. F., Van Winkle, W. B., Gotto, A. M. Jr., Pownall, H. J. 1981. *Biochemistry* 20: 1569–74
116. Matz, C. E., Jonas, A. 1982. *J. Biol. Chem.* 257: 4535–40
117. Matz, C. E., Jonas, A. 1982. *J. Biol. Chem.* 257: 4541–46
118. McLachlan, A. D. 1977. *Nature* 267: 465–66
118a. Mehrabian, M., Schumaker, V. N., Fareed, G. C., et al. 1985. *Nucleic Acids Res.* 13: 6937–53
119. Melchior, D. L., Scavitto, F. J., Steim, J. M. 1980. *Biochemistry* 19: 4828–34
120. Miller, K. W., Small, D. M. 1982. *J. Colloid Interface Sci.* 89(2): 466–78
121. Miller, K. W., Small, D. M. 1983. *J. Biol. Chem.* 258: 13772–84
122. Miller, K. W., Small, D. M. 1983. *Biochemistry* 22: 443–51
123. Minari, O., Zilversmit, D. B. 1963. *J. Lipid Res.* 4: 424–36
124. Mjos, O. D., Faergeman, O., Hamilton, R. L., Havel, R. J. 1975. *J. Clin. Invest.* 56: 603–15
125. Morton, R. E., Zilversmit, D. B. 1982. *J. Lipid Res.* 23: 1058–67
126. Morton, R. E., Zilversmit, D. B. 1983. *J. Biol. Chem.* 258: 11751–57
127. Nagle, J. F., Wilkinson, D. A. 1975. *Biophys. J.* 23: 159–75
128. Nelson, G. J. 1979. *Blood Lipids and Lipoproteins: Quantitation, Composition, and Metabolism*, p. 300. New York: Robert E. Krieger
129. Nichols, A. V., Gong, E. L., Blanche, P. J., Forte, T. M. 1983. *Biochim. Biophys. Acta* 750: 343–64
130. Nichols, A. V., Gong, E. L., Blanche, P. J., Forte, T. M., Shore, V. G. 1984. *Biochim. Biophys. Acta* 793: 325–37
131. Nikkila, E. A. 1983. See Ref. 71, 30: 622–42
132. Norum, K., Glomset, J. A., Nichols, A. V., et al. 1975. *Scand. J. Clin. Lab. Invest.* 35(142): 31–55
133. Osborne, J. C. Jr., Schaefer, E. J., Powell, G. M., Lee, N. S., Zech, L. A. 1984. *J. Biol. Chem.* 259: 347–53
134. Palumbo, G., Edelhoch, H. 1977. *J. Biol. Chem.* 252: 3684–85
135. Parente, R. A., Lentz, B. R. 1984. *Biochemistry* 23: 2353–62
136. Pattnaik, N. M., Montes, A., Hughes, L. B., Zilversmit, D. B. 1978. *Biochim. Biophys. Acta* 5390: 428–38
137. Patton, G., Bennett Clark, S., Fasulo, J., Robins, S. R. 1984. *J. Clin. Invest.* 73: 231–40
138. Pitas, R. E., Innerarity, T. L., Arnold, K. S., Mahley, R. W. 1979. *Proc. Natl. Acad. Sci. USA* 76: 2311–15
139. Pittman, R. C., Steinburg, D. 1984. *J. Lipid Res.* 25: 1557–85
140. Pownall, H. J., Hickson, D., Gotto, A. M. Jr. 1981. *J. Biol. Chem.* 256: 9849–54
141. Pownall, H. J., Massey, J. B., Kusserow, S. K., Gotto, A. M. Jr. 1978. *Biochemistry* 17: 1183–88
142. Pownall, H. J., Massey, J. B., Kusserow, S. K., Gotto, A. M. Jr. 1979. *Biochemistry* 18: 574–79
143. Pownall, H. J., Morrisett, J. D., Sparrow, J. T., Gotto, A. M. 1974. *Biochem. Biophys. Res. Commun.* 60: 779–86
144. Privalov, P. L., Kechinashvili, N. N. 1974. *J. Mol. Biol.* 86: 665–84
145. Rall, S. C. Jr., Weisgraber, K. H., Mahley, R. W. 1982. *J. Biol. Chem.* 257: 4171–78
146. Redgrave, T. G. 1970. *J. Clin. Invest.* 49: 465–71
147. Redgrave, T. G. 1983. In *International Review of Physiology*, Vol. 28, *Gastrointestinal Physiology IV*, ed. J. A. Young, 4: 103–30. Baltimore, Md: Univ. Park Press
148. Redgrave, T. G., Maranhao, R. C. 1985. *Biochim. Biophys. Acta* 835: 104–12
149. Redgrave, T. G., Small, D. M. 1979. *J. Clin. Invest.* 64: 162–71
150. Deleted in proof
151. Reynolds, J. A. 1976. *J. Biol. Chem.* 251: 6013–15
152. Ritter, M. C., Scanu, A. M. 1977. *J. Biol. Chem.* 252: 1200–16
153. Ritter, M. C., Scanu, A. M. 1979. *J. Biol. Chem.* 254: 2517–25
154. Roth, R. I., Jackson, R. L., Pownall, H. J., Gotto, A. M. Jr. 1977. *Biochemistry* 16: 5030–36
155. Sabesin, S. M., Hawkins, H. L., Kuiken, L., Ragland, J. B. 1977. *Gastroenterology* 72: 510–18
156. Scanu, A. M. 1972. *Biochim. Biophys. Acta* 265: 471–508
157. Scanu, A. M., Cump, E., Toth, J., Koga, S., Stiller, E., Albers, L. 1970. *Biochemistry* 9: 1327–35
158. Scanu, A., Hughes, W. L. 1974. *J. Biol. Chem.* 235: 2870–93
159. Scanu, A., Lewis, L. A., Bumpus, F. M. 1958. *Arch. Biochem. Biophys.* 74: 390–97
160. Schmidt, C. F., Barenholz, Y., Huang, C., Thompson, T. E. 1977. *Biochemistry* 16: 3948–54
161. Schurtenberger, P., Mazer, N. A., Kanzig, W. 1984. *Hepatology* 4: 143–47S

162. Segrest, J. P., Jackson, R. L., Morisett, J. D., Gotto, A. M. Jr. 1974. *FEBS Lett.* 38 : 247–58
163. Sherrill, B. C. 1978. In *Disturbances in Lipid and Lipoprotein Metabolism*, ed. J. M. Dietschy, A. M. Gotto Jr., J. A. Ontko, pp. 99–110. Bethesda, Md : Am. Physiol. Soc.
164. Shorr, L. D. 1978. PhD thesis. Boston Univ.
165. Shulman, R. S., Herbert, P. N., Wehrly, K., Fredrickson, D. S. 1975. *J. Biol. Chem.* 250 : 182–90
166. Small, D. M. 1967. *J. Lipid Res.* 8 : 551–57
167. Small, D. M. 1967. *Gastroenterology* 52 : 607–10
168. Small, D. M. 1971. In *The Bile Acids— Chemistry, Physiology and Metabolism*, ed. P. P. Nair, D. Kritchevsky, Vol. 1, 8 : 247–354. New York : Plenum
169. Small, D. M. 1981. In *Membranes, Molecules, Toxins and Cells*, ed. K. Bloch, L. Bolis, D. C. Tosteson, 2 : 11– 34. Boston : PSG
170. Small, D. M. 1986. In *Handbook of Lipid Research Series*, Vol. 4, *The Physical Chemistry of Lipids from Alkanes to Phospholipids*, ed. D. Hanahan, Chap. 2, pp. 21–42. New York : Plenum
171. Small, D. M. 1986. See Ref. 170, Chap. 12, pp. 475–522
172. Small, D. M., Bourges, M., Dervichian, D. G. 1966. *Biochim. Biophys. Acta* 125 : 563–80
173. Small, D. M., Bourges, M., Dervichian, D. G. 1966. *Nature* 211 : 816–18
174. Small, D. M., Penkett, S. A., Chapman, D. 1969. *Biochim. Biophys. Acta* 176 : 178–89
175. Steele, J. C. H., Reynolds, J. A. 1979. *J. Biol. Chem.* 254 : 1633–38
176. Stone, W. I., Reynolds, J. A. 1975. *J. Biol. Chem.* 250 : 3584–87
177. Stone, W. I., Reynolds, J. A. 1975. *J. Biol. Chem.* 250 : 8045–48
178. Swaney, J. B. 1980. *J. Biol. Chem.* 255 : 8791–97
179. Swaney, J. B. 1980. *J. Biol. Chem.* 255 : 8798–8803
180. Swaney, J. B. 1980. *J. Biol. Chem.* 258 : 1254–59
181. Swaney, J. B., Chang, B. C. 1980. *Biochemistry* 19 : 5637–44
182. Swaney, J. B., O'Brien, K. 1978. *J. Biol. Chem.* 253 : 7069–77
183. Swift, L. L., Manowitz, N. R., Dunn, G. D., LeQuire, V. S. 1980. *J. Clin. Invest.* 66 : 415–25
184. Tajima, S., Yokoyama, S., Yamamoto, A. 1983. *J. Biol. Chem.* 256 : 10073–82

185. Tall, A. R., Atkinson, D., Small, D. M., Mahley, R. W. 1977. *J. Biol. Chem.* 252 : 7288–93
186. Tall, A. R., Blum, C. B., Forester, G. P., Nelson, C. A. 1982. *J. Biol. Chem.* 257 : 198–207
187. Tall, A. R., Deckelbaum, R. J., Small, D. M., Shipley, G. G. 1977. *Biochim. Biophys. Acta* 487 : 145–53
188. Tall, A. R., Forester, L. R., Bongiovanni, G. L. 1983. *J. Lipid Res.* 24 : 277–89
189. Tall, A. R., Lange, Y. 1978. *Biochem. Biophys. Res. Commun.* 80(1) : 206–12
190. Tall, A. R., Puppione, D. L., Kunitake, S. T., Atkinson, D., Small, D. M., Waugh, D. 1981. *J. Biol. Chem.* 256 : 170–74
191. Tall, A. R., Robinson, L. A. 1979. *FEBS Lett.* 107 : 222–26
192. Tall, A. R., Shipley, G. G., Small, D. M. 1976. *J. Biol. Chem.* 251 : 3749–55
193. Tall, A. R., Small, D. M. 1977. *Nature* 265 : 163–64
194. Tall, A. R., Small, D. M. 1978. *N. Engl. J. Med.* 299 : 1232–36
195. Tall, A. R., Small, D. M. 1980. *Adv. Lipid Res.* 17 : 1–51
196. Tall, A. R., Small, D. M., Shipley, G. G., Lees, R. S. 1975. *Proc. Natl. Acad. Sci. USA* 72 : 4940–42
196a. Tall, A. R., Small, D. M., Deckelbaum, R. J., Shipley, G. G. 1977. *J. Biol. Chem.* 252 : 4701–17
197. Teng, T.-L., Barbeau, D. L., Scanu, A. M. 1978. *Biochemistry* 17 : 17–21
198. Van Tornout, P., Vercaemst, R., Lievens, M. J., Caster, H., Rosseneu, M., Assmann, G. 1980. *Biochim. Biophys. Acta* 601 : 509–23
199. Via, D. P., Craig, I. F., Jacobs, G. W., Van Winkle, W. B., Charlton, S. C., et al. 1982. *J. Lipid Res.* 23 : 570–76
200. Vitello, L. B., Scanu, A. M. 1976. *J. Biol. Chem.* 251 : 1131–36
201. Vitello, L. B., Scanu, A. M. 1976. *Biochemistry* 15 : 1161–65
202. Walsh, M. T., Atkinson, D. 1983. *Biochemistry* 22 : 3170–78
203. Walsh, M. T., Atkinson, D. 1985. *J. Lipid Res.* In press
204. Walsh, M. T., Atkinson, D. 1985. In *Methods in Enzymology Plasma Lipoproteins*, ed. J. P. Segrest, J. J. Albers. New York : Academic. In press
205. Walsh, M. T., Ginsburg, G., Atkinson, D. 1984. *57th Sci. Sess. Am. Heart Assoc. Meet., Miami Beach, Fla., Circulation* 70 (Suppl. II) : II-313 (Abstr.)
206. Watt, R. M., Reynolds, J. A. 1981. *Biochemistry* 20 : 3897–3901
206a. Wei, C.-F., Chen, S.-H., Yang, C.-Y.,

et al. 1985. *Proc. Natl. Acad. Sci. USA* 82: 7265–69
207. Weinberg, R. B., Spector, M. S. 1985. *J. Biol. Chem.* 260: 4914–21
208. Weisgraber, K. H., Innerarity, T. L., Harder, J. J., Mahley, R. W., Milne, R. W., et al. 1983. *J. Biol. Chem.* 258: 12348–54
209. Wetterau, J. R., Jonas, A. 1982. *J. Biol. Chem.* 257: 10961–66
210. Wetterau, J. R., Jonas, A. 1983. *J. Biol.*

Chem. 258: 2637–43
211. Yokoyama, S., Fukushima, D., Kupferberg, J. P., Kezdy, F. J., Kaiser, E. T. 1980. *J. Biol. Chem.* 255: 7333–39
212. Zannis, V. I., Breslow, J. L. 1985. In *Advances in Human Genetics*, ed. H. Harris, K. Hirschhorn, 3: 125–215. New York: Plenum
213. Zilversmit, D. B. 1978. See Ref. 163, pp. 69–82

Ann. Rev. Biophys. Biophys. Chem. 1986. 15 : 457–75
Copyright © 1986 by Annual Reviews Inc. All rights reserved

THE ROLE OF THE NUCLEAR MATRIX IN THE ORGANIZATION AND FUNCTION OF DNA

William G. Nelson, Kenneth J. Pienta,
Evelyn R. Barrack, and Donald S. Coffey

Departments of Urology, Oncology, and Pharmacology, The Johns
Hopkins University, School of Medicine, Baltimore, Maryland 21205

CONTENTS

PERSPECTIVES AND OVERVIEW

A mammalian nucleus contains a total length of almost 50 cm of DNA. The packing of this amount of DNA into a nucleus of 10 μm diameter presents mammalian cells with a formidable topological problem : The total length of cellular DNA must be reduced about 50,000-fold to fit within the confines of a single nucleus. Despite this tremendous DNA packing ratio, DNA contained within mammalian nuclei must have a dynamic conformation conducive to an active role in a variety of biologic processes. For example, the mammalian nucleus directs replication of the DNA in 30,000 to 90,000 small units termed replicons that are synthesized in a precise order and temporal sequence. During DNA synthesis, each of these DNA replicon units must pass through a very large multienzyme complex ($> 5 \times 10^6$ d) that represents the biochemical site for DNA replication; these enzyme

457

complexes have been termed replisomes (122, 166) or replitases (119, 133). Because each of these individual replicons passes through the site of replication within a 30-min period, the double helix of DNA must be unwound at a speed of over 100 rpm at each replicating site. These topological considerations and the precise ordering of DNA replication have been difficult to account for without considering the specific spatial organization and higher-order structure within the nucleus.

Such structural requirements are not unique to DNA replication; a variety of genetic processes appear to demand a spatial and temporal precision unattainable in the absence of structural order within the mammalian nucleus. What is the nature of this structural order? Despite the mechanistic diversity and complexity of DNA function in eukaryotes, three structural themes emerge as common features of genetic activity of these cells: (a) To permit diversity of function, DNA must be topographically and topologically partitioned into independent functional units or domains; (b) to carry out complex biosynthetic processes involving nucleic acids using multienzyme systems, eukaryotic cells appear to favor massive biosynthetic enzyme complexes, and these complexes must be strategically positioned with respect to DNA domains; and (c) to direct dramatic spatial and temporal rearrangements such as the segregation of genetic material during mitosis, nuclear structural elements must be dynamic and malleable (45). Clearly, this ordering of DNA must be accomplished by elements within the milieu of the eukaryotic nucleus other than chromatin (histones and DNA). We define the nuclear matrix as the dynamic structural subcomponent of the nucleus that directs the functional organization of DNA into domains and provides sites for the specific control of nucleic acids.

Insight into the problem of organization has been provided by the identification of a matrix structure within the nucleus (22, 24, 25). The isolated nuclear matrix is an insoluble, skeletal framework of the nucleus (22, 24, 25). This underlying structure is revealed following extraction from the nucleus of the nuclear membrane phospholipid and the chromatin. The nuclear matrix is generally isolated by a series of sequential extractions employing nonionic detergent (Triton X-100), brief digestion with DNase I, a hypotonic buffer containing very low concentrations of divalent cations (0.2 mM $MgCl_2$), and hypertonic salt buffer (e.g. 2 M NaCl). As shown in Figure 1, the resulting residual structure, the nuclear matrix, resembles the nucleus in size and shape and consists of a peripheral lamina that forms a continuous structure surrounding the periphery of the nuclear sphere and that represents a residual component of the nuclear envelope; residual nuclear pore complexes that are embedded in the lamina; an internal fibrogranular protein and RNA-containing network; and a residual nucleolus. The proteins comprising the matrix represent only about 10% of the total nuclear protein mass and comprise a subset of nonhistone proteins

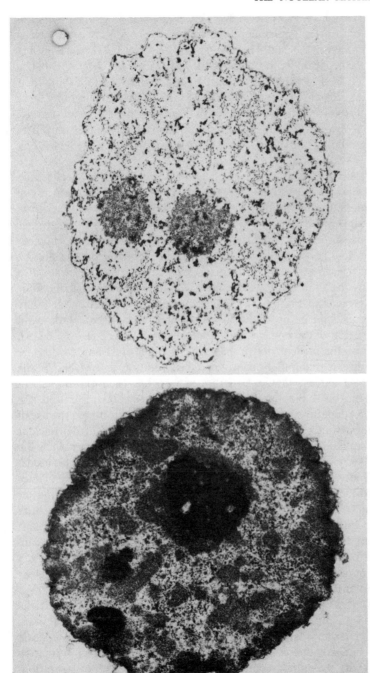

Figure 1 Comparison of the electron micrographs of an isolated rat liver nucleus (*left*) with a nuclear matrix preparation (*right*). Magnification 10,000 ×. The matrix structure on the *right* contains only 10% of the total nuclear proteins, 2% of the total phospholipid, and less than 1% of the total DNA that is present in the intact nuclear structure shown on the *left*. The nuclear matrix was prepared by sequentially extracting nuclei with low ionic strength buffer, 2 M NaCl, Triton X-100, and DNase treatment as described and discussed by Shaper et al (143).

(22, 24, 25; see 13, 19, 143 for reviews). A great deal of effort in a number of different laboratories is currently directed toward elucidating the nature of these proteins (26, 27, 88, 99, 116, 124–127, 132, 143, 154, 171), determining how they are organized and how they interact with each other (7, 30–32a, 38, 62, 67, 71, 81, 82, 91, 102, 115, 125, 131, 132a, 170), and establishing their specific intranuclear localization (40, 75, 157, 161, 164). Furthermore, there is much interest in ascertaining the fate of these proteins during mitosis (17, 28, 40, 74–76, 85, 96, 120, 135, 141, 155) and discovering whether they form the core scaffolding of the metaphase chromosomes (37, 53, 58, 59, 64, 103, 127–129).

The nuclear matrix is a universal feature of eukaryotic nuclei; such residual nuclear structures have now been isolated from a wide variety of mammalian and nonmammalian sources (see 13 for review). A variety of nomenclature has been used by different investigators to refer to the nuclear matrix: nuclear matrix, nuclear framework, nuclear skeleton, nuclear scaffold, nuclear ghost, nuclear cage, and chromatin-depleted nucleus. All these terms refer to very similar residual nuclear structures. There have also been numerous modifications of the original procedure described for isolating the nuclear matrix, and it is important to recognize that many factors can affect the nature of the final product. Details of the isolation and characterization of the nuclear matrix have been reviewed (13, 24, 66, 98, 143). Because the nuclear matrix has been isolated or visualized directly following drastic manipulations in vitro (e.g. addition of 2 M NaCl), the existence of a nuclear matrix structure in vivo has been questioned. In addition, the insoluble nature of the matrix proteins has raised the possibility of denaturation artifacts. These issues have prompted the use of alternative methods to confirm the nature and existence of the nuclear matrix in situ and to study specific facets of the nuclear matrix (24, 25, 33, 36–38, 40, 43, 65–67, 77, 78, 113). For example, Mirkovitch et al (113) used a lithium diiodosalicylate extraction procedure to study the attachment sites of DNA to the nuclear scaffold. Ciejek et al (43) isolated a nuclear matrix structure in 40% glycerol at −20°C to prevent RNA degradation in a study of the association of precursor RNA with the nuclear matrix. In an elegant series of studies, Fey et al (65; E. G. Fey, G. Krochmalnic, S. Penman, submitted for publication) employed a moderate salt extraction procedure using ammonium sulfate to best preserve the morphology of the nuclear matrix and hnRNP integrity.

Many studies have focused on determining whether the nuclear matrix is associated with important biological properties. Over the past several years evidence has accumulated that the nuclear matrix is not simply a static structure, but is rather a dynamic scaffolding system that is intimately associated with such fundamental nuclear processes as DNA organization, DNA replication, heterogeneous nuclear RNA (hnRNA) synthesis and processing, and hormone action (see for reviews 10, 13, 19, 20, 24, 30, 108, 143). These findings suggest that many important nuclear events occur not in solution but rather in association with relatively insoluble structural components that are firmly bound to the nuclear matrix.

The nuclear matrix appears to provide ordering and organization of complex processes heretofore thought of as soluble systems within the nucleoplasm. For example, it has been proposed that eukaryotic nuclear DNA is organized into supercoiled loops anchored at their bases to the nuclear matrix, and that during DNA replication these loops are reeled loop by loop through fixed replication sites on the nuclear matrix (4, 21, 23, 24, 49, 55, 90, 95, 110, 122, 152, 159, 160, 165, 166). This provides a plausible mechanism by which an enormous length of DNA can be ordered spatially during DNA replication such that the daughter strands remain untangled yet coupled in a precise fashion for separation during mitosis (see Figure 2). In addition, the demonstration that newly synthesized hnRNA and its

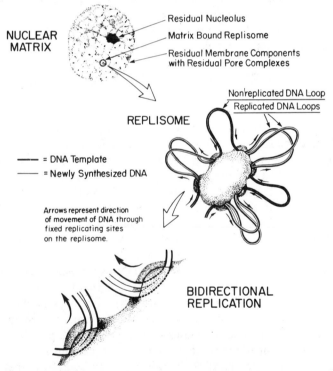

NUCLEAR
MATRIX
— Residual Nucleolus
— Matrix Bound Replisome
— Residual Membrane Components
  with Residual Pore Complexes

Nonreplicated DNA Loop
Replicated DNA Loops

REPLISOME

——— = DNA Template
———— = Newly Synthesized DNA

Arrows represent direction
of movement of DNA through
fixed replicating sites
on the replisome.

BIDIRECTIONAL
REPLICATION

Figure 2   Schematic of the mechanism of DNA replication on the nuclear matrix. *Top:* the nuclear matrix. *Middle:* one of the thousands of replisomes attached to the nuclear matrix, showing the manner in which loop domains move through fixed replicating sites. *Bottom:* the bidirectional movement of a strand through the replicating site. Modified from the study of Pardoll et al (122).

processing intermediates are associated with the insoluble nuclear matrix (18, 43, 87, 94, 95, 106, 109, 112, 138, 162) and the discovery that certain actively transcribed genes are preferentially associated with the nuclear matrix (44, 47, 50, 97, 100, 113, 117, 121, 136, 137, 146–148) have done much to further our understanding of the role of structural organization in cellular function. The nuclear matrix appears to be a major site of steroid hormone receptor binding in the nucleus (5, 6, 9–16, 35, 46, 54, 63, 79, 130, 134, 144, 156, 168, 169); this is consistent with the role of steroid hormones in stimulating hnRNA synthesis, a process that appears to occur in association with the nuclear matrix. Additionally, the nuclear matrix has been demonstrated to be an important cellular target for transforming proteins and oncogene products such as the myc protein (61, 114), large T antigen (34, 51, 52, 140, 153, 163), and the E1A protein (139).

The purpose of the present review is to provide the reader with a critical assessment of the role of the nuclear matrix in the organization of DNA within the nucleus. In addition, the current status of steroid hormone interactions with the nuclear matrix will be discussed. The reader may also consult other comprehensive reviews (10, 13, 19, 20, 24, 30, 108, 143) for additional details and more extensive references.

THE NUCLEAR MATRIX ORGANIZES DNA INTO LOOP-DOMAINS

The higher-order structure of DNA within the nucleus and within the metaphase chromosome remains one of the major unsolved problems in molecular biology. At least three higher-order levels of DNA organization have been identified in the past decade, including the nucleosome, the 30 nm solenoid, and the DNA loop-domain. In 1976 Cook et al (48) first proposed that loop structures are involved in the superhelical organization of eukaryotic DNA. New interest in DNA loop-domains arose from studies of DNA synthesis. Pardoll et al (122) found that the nuclear matrix contains fixed sites for the replication of DNA loops that were demonstrated to be equivalent to replicons, the basic lengths of DNA synthesized as continuous units (89). In 1980, Vogelstein et al (166) reported that DNA loop-domains are attached at their bases to the nuclear matrix. They observed that these DNA loop structures in the presence of ethidium bromide formed a large fluorescent halo surrounding and extending beyond the periphery of the nuclear matrix. Furthermore, they were able to demonstrate that these loops of DNA are topologically constrained at their bases by virtue of their attachment to the nuclear matrix. When the concentration of ethidium bromide was increased to high levels, they observed a rewinding of the DNA halo back to the periphery of the matrix (166). However, if the DNA was nicked, then the halo structure could not be rewound by elevating the concentration of ethidium bromide. These experiments suggest that the nuclear matrix organizes DNA such that it has topological properties equivalent to those of covalently closed circular DNA. Paulson & Laemmli (123) have provided evidence that these superhelical loop-domains are maintained through mitosis: They observed similar loop-domains attached to a residual chromosome core scaffold.

The DNA loop-domain defines a basic unit of higher-order DNA structure in eukaryotic cells. Vogelstein et al (166) directly measured the size of the fluorescent matrix-halo structures and determined from a halo radius of approximately 15 μm that a DNA loop would contain approximately 90,000 base pairs of DNA. Several investigators (60, 73, 84, 92) have estimated that the DNA loop-domain size ranges between 10,000 and 180,000 base pairs with an average of 63,000 base pairs for all of the studies combined. An average DNA loop-domain of 63,000 base pairs would contain 20 μm of DNA double helix. Each such loop would be large enough to contain 315 nucleosomes wound with 6 nucleosomes per turn into a solenoid such as that proposed by Finch & Klug (68). This 30 nm solenoid would form the filament of a loop; based on a total DNA content of 6 $\times$ 10^9 base pairs per diploid nucleus, there would be 95,000 such loops within a cell.

How these DNA loop-domains are organized within the interphase nucleus is not clear. Some studies suggest that the DNA loops hang like draperies suspended from their base on the nuclear lamina (39, 84, 118). In contrast, Pardoll et al (122) demonstrated that newly synthesized DNA was attached to an internal network of the nuclear matrix and not just located at the lamina. Recently, Smith & Berezney (149–152) have extended these observations to both in vivo and in vitro DNA synthesis and have demonstrated that over 70% of the replicating sites are contained within the nuclear matrix (152). A model featuring the attachment of DNA loop domains at their bases to an internal nuclear matrix structure that courses throughout the interior of an interphase nucleus seems to best reconcile all available experimental data. The fact that DNA loop-domains are preserved even in metaphase chromosomes (60, 73, 84, 92, 123), structures that are devoid of a nuclear envelope and lamina, supports the idea that DNA loop-domains are anchored to some structural element in the nuclear matrix, other than the nuclear lamina, in the interphase nucleus.

DNA LOOP-DOMAINS AND A MODEL FOR CHROMOSOME STRUCTURE

To better understand how DNA loops are packed into chromosomes for segregation at metaphase, Pienta & Coffey (128) have constructed scale models of a single chromatid of the #4 human chromosome, focusing on the organization of a 30 nm filament into loop-domains. They found that of four previously proposed models for chromosome structure (3, 8, 57, 101, 107, 142) they could accommodate the amount of DNA in the chromatid into the actual measured chromatid dimensions only by utilizing the radial loop model suggested by Marsden & Laemmli (107) and Adolph & Kreisman (3). The radial loop model features loops wrapped radially around the central axis of the chromatid, stacking to achieve the overall chromosome length. The analysis by Pienta & Coffey (128) revealed that the experimentally observed length and diameter of the chromatid is best approximated in a structure with 18 loops per radial turn, with each loop containing 60,000 base pairs (see Figure 3). This 18-loop unit defines a new higher-order structure of DNA organization, which they have termed the miniband. The miniband is equivalent to one full radial turn of 18 loops around the axis of the chromatid, and contains approximately 10^6 base pairs.

Pienta & Coffey (128) tested predictions of their chromosome model with several experimental observations. The predicted loop-domain size of 60,000 bases is comparable with loop-domains of 62,000 bases reported by Georgiev et al (73), 53,000 bases reported by Hancock and Boulikas (84), 54,000 bases reported by Igo-Kemenes et al (92), and $83,000 \pm 14,000$ bases reported by Earnshaw & Laemmli (60). In fact, the average value for all of the above studies, $63,000 \pm 14,000$ base pairs, agrees very well with the predictions of the model (128). In order to test their prediction of an 18-loop miniband, Pienta & Coffey (128) counted the number of loops per radial turn on chromatids pictured in scanning electron micrographs from four previous studies. They counted 17 ± 1.4 loops per radial turn in the studies of Utsumi (158), 17.8 ± 2.1 in those of Adolph (2), 16.8 ± 3.4 from Laemmli (101), and 16 ± 1.3 from DuPraw (57). The average observed number of loops per radial turn, 16.9 ± 1.9, closely approximates the prediction of 18 loops per miniband (128).

A schematic summary of the Pienta model, depicting the various hierarchies of chromosomal DNA organization, is presented in Figure 3. First, the 20 Å DNA double helix is wound twice around each histone octamer, forming the nucleosomes that each contain 160 base pairs. The nucleosomes comprise the 10 nm beads-on-a-string fiber. This fiber of nucleosomes is then wound in a solenoid fashion with 6 nucleosome particles per turn to form the 30 nm filament. A length of the 30 nm solenoid filament corresponding to 60,000 pairs of DNA forms a domain that loops out from attachment points in the nuclear matrix or chromosome scaffold. In the chromosome, these loops are radially disposed on the chromosome, 18 to each turn, forming the miniband. Finally, approximately 106 of these minibands are arranged along a central axis to form each chromatid of the complete #4 human chromosome.

A fascinating consequence of this model was the discovery and definition of the miniband subunit of chromosome structure. The width of a miniband, 30 nm, approaches that of the smallest bands seen in cytological Giemsa banding of metaphase chromosomes. Furthermore, the number of base pairs of DNA encompassed by a miniband approximates the size of the genetic unit referred to as a centimorgan. Thus the miniband is not only a useful concept for DNA packaging within chromosomes but may also provide a structural correlate for the centimorgan.

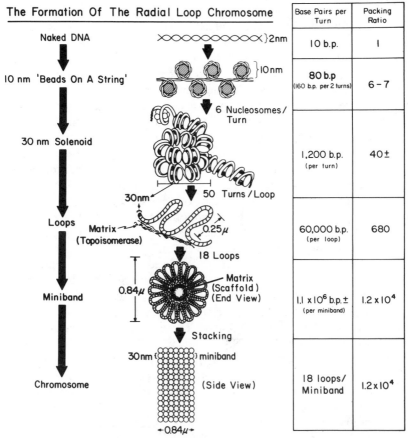

The Formation Of The Radial Loop Chromosome

	Base Pairs per Turn	Packing Ratio
Naked DNA	10 b.p.	1
10 nm 'Beads On A String'	80 b.p (160 b.p. per 2 turns)	6 – 7
30 nm Solenoid	1,200 b.p. (per turn)	40 ±
Loops	60,000 b.p. (per loop)	680
Miniband	1.1 x 10⁶ b.p.± (per miniband)	1.2 x 10⁴
Chromosome	18 loops/ Miniband	1.2 x 10⁴

Figure 3 Schematic of the organization of DNA within a chromosome. The double helix of DNA is wrapped twice around the histone octamer to form a nucleosome. Six nucleosomes form each turn of a solenoid and the 30 nm filament. The 30 nm filament forms the loop domain of DNA attached at the base to the nuclear matrix that contains topoisomerase II. Eighteen of the loops form a radial array around the core of the chromosome to produce a miniband. The continuous formation of these radial loops produces a stack of minibands to form a metaphase chromatid. Modified from the study of Pienta & Coffey (128).

DNA TOPOISOMERASES ASSOCIATED WITH THE NUCLEAR MATRIX

The organization of eukaryotic DNA into independent topological domains by structural subcomponents of both nuclei and chromosomes dominates the study of nuclear and chromosomal structure. Clearly, elements within the nuclear matrix and chromosome scaffold may have an important role in the dynamic control of DNA topology. Precisely how the loop-domains of DNA are anchored to these structures is not clear. Nevertheless, we do know that the interaction of DNA loop-domains with these structures must be tight enough to restrict or prevent the transfer of superhelical energy from loop to loop, i.e. the domains must be able to function independently. Additionally, these interactions must be preserved throughout mitosis. Lewis et al (103) have recently reported that both the interphase nuclear matrix and the metaphase chromosome scaffold contain a similar protein of 170 kd. Whatever the nature of the attachment of DNA loops to these structures, how their elements contribute to the control of DNA topology constitutes one of the paramount issues in the study of nuclear and chromosomal structure.

Control over DNA superhelicity is the province of the DNA topoisomerase enzymes. By regulating the superhelical structure of DNA, these enzymes are thought to have a major role in the control of a variety of biological processes including transcription and replication (for reviews see 41, 72, 104, 167). DNA topoisomerases can affect DNA superhelical structure by catalyzing transient single-strand (type I topoisomerases) and double-strand (type II topoisomerases) breaks in the phosphodiester backbone of DNA. The enzymes break and rejoin DNA by means of covalent enzyme-DNA substrate intermediates. In doing so, the enzymes can alter the linking number of a covalently closed circular DNA molecule or a constrained DNA loop-domain. The linking number refers to the number of times individual DNA strands in a DNA duplex are wrapped around each other. Reducing the linking number of a DNA molecule from that of a relaxed DNA double helix requires energy and is referred to as negative supercoiling. Most DNA isolated from living organisms is negatively supercoiled. In a negatively supercoiled DNA molecule, the DNA double helix can be thought of as underwound and torsionally strained. This torsional strain on the DNA helix facilitates the DNA strand separation required in biological processes such as replication and transcription.

In exciting recent work, Berrios et al (28) and Halligan et al (83) have identified a type II DNA topoisomerase associated with the nuclear matrix.

In another series of elegant studies, Earnshaw et al (58) and Earnshaw & Heck (59) have identified type II topoisomerase as a component of the chromosome scaffold. The presence of the type II DNA topoisomerase enzyme as a component of the nuclear matrix and chromosome scaffold endows these structures with a critical enzymatic role in the control as well as organization of DNA superhelical structure.

NUCLEAR MATRIX AND TRANSCRIPTION

The nuclear matrix has been implicated in RNA transcription in addition to DNA replication. Studies in this area have focused on the specific association of transcribed genes with the nuclear matrix (44, 47, 50, 86, 97, 100, 113, 117, 121, 136, 137, 146–148), the location of active RNA synthetic complexes on the nuclear matrix (1, 94, 95), and the fate of newly synthesized RNA with respect to the nuclear matrix (18, 43, 87, 94, 95, 105, 106, 112, 138, 162). Pardoll & Vogelstein (121) first found that ribosomal DNA is associated with the nuclear matrix isolated from rat liver, and numerous other investigators later determined that DNA associated with the nuclear matrix is enriched in transcribed sequences. In studies of the matrix association of a specific hormone-inducible gene, Robinson et al (136, 137) and Ciejek et al (44) found that sequences adjacent to and within the ovalbumin gene are associated with the nuclear matrix of hen oviducts but not with the nuclear matrix of other tissues, and that the nuclear matrix association of some of the sequences reverses upon hormone withdrawal (see also 97). Small et al (146, 148) have reported that heat shock genes are associated with nuclear matrix structures prepared from *Drosophila* nuclei independent of transcriptional rate. Mirkovitch et al (113) have also demonstrated a specific association of DNA sequences near transcribed genes with nuclear skeletal structures prepared from *Drosophila* cells, but have cautioned that the precise attachment point of the nuclear skeleton–DNA interaction might vary with the method of preparation of the structures.

As with DNA synthesis, fixed transcriptional complexes that synthesize RNA have been identified on the nuclear matrix (1, 94, 95). Some studies have suggested that over 95% of newly synthesized hnRNA is associated with the nuclear matrix (87, 94, 105, 106, 112, 162). Ciejek et al (43) found that precursor mRNAs are preferentially associated with the nuclear matrix compared to mature mRNAs. In addition, small nuclear RNA (snRNA) species are recovered along with the nuclear matrix (109, 111). These studies collectively suggest that the nuclear matrix may play a critical role in the synthesis and processing of RNA.

STEROID HORMONE RECEPTORS INTERACT WITH THE NUCLEAR MATRIX

Since the nuclear matrix may be involved in the replication and transcription of DNA, it is important to understand how the matrix might be regulated by biological factors. Insight into this type of control has been obtained by studies on sex steroid hormone action. Barrack and colleagues (6, 9–16, 54) have identified and characterized specific steroid receptors associated with the nuclear matrix of both estrogen- and androgen-responsive tissues: estrogen receptors on the nuclear matrix of rat uterus (13, 15), rat liver (6), and hen liver (12, 13, 16); and androgen receptors on the nuclear matrix of rat (12, 13, 54) and human (11) prostate. The nuclear matrix appears to be a significant intranuclear site of hormone receptor interactions, since approximately 50–75% of all the nuclear receptors are recovered in the isolated nuclear matrix. These receptors are steroid-specific and tissue-specific, and accumulate in the nuclear matrix only in response to an appropriate hormonal stimulus, not indiscriminately. For example, the liver of an egg-laying hen, in response to high blood levels of estrogen, synthesizes vitellogenin (the precursor of the major egg yolk proteins). Under these conditions, the liver nuclear matrix of the hen contains about 600 molecules of estrogen receptor per nuclear matrix sphere. In contrast, the liver nuclear matrix of an untreated rooster, which does not produce yolk proteins, contains only one eighth as many estrogen receptors as that of the egg-laying hen. However, the administration of pharmacological doses of estrogen to roosters or immature chicks results in a stimulation of vitellogenin mRNA synthesis and a 12-fold increase in the number of nuclear matrix–associated estrogen receptors (12, 13).

Similarly, in the rat ventral prostate, where growth and functions are androgen-dependent, the presence of androgen receptors on the nuclear matrix is specifically associated with androgen stimulation of the gland. The receptors are present in the prostate nuclear matrix of intact adult male rats; following withdrawal of androgen by castration a rapid loss (within 24 h) of these receptors from the nuclear matrix precedes the involution of the gland. Administration of androgen, but not of estrogen, restores these sites to normal receptor levels within 1 h (12, 13).

The association of steroid receptors with the nuclear matrix is also tissue-specific. For example, following treatment of an immature female rat with a physiological dose of estrogen (0.1 μg) that is sufficient to induce maximal uterine growth, estrogen receptors become associated with the uterine nuclear matrix but not with the liver nuclear matrix (15). This observation is consistent with the lack of responsiveness of the rat liver to these small doses of estrogen. In contrast, a higher dose of estrogen sufficient to induce a specific liver response does cause accumulation of estrogen receptors in the liver nuclear matrix (6, 10). Agutter & Birchall (5) have confirmed that the rat uterine nuclear matrix contains specific estradiol-binding sites and they have shown in addition that the rat lung nuclear matrix contains none.

For the studies described above, the identification of receptors on the nuclear matrix relied on the measurement of binding of specific ^{3}H-labeled steroids in a specific, saturable, and high affinity ($K_d \sim 1$ nM) manner. Using monoclonal antibodies raised against the estrogen receptor, we have recently obtained direct evidence that receptor proteins become associated with the nuclear matrix of estrogen target tissue (6). Using the high dose estrogen-treated female rat liver model, immunoblots of SDS-polyacrylamide gels of liver nuclear matrix proteins from estrogen-treated rats reveal the presence of a single immunoreactive band at 69 kd, indicating the presence of estrogen receptor. No immunoreactive proteins are detected by the receptor antibody in the liver nuclear matrix of unstimulated animals, corroborating the data obtained by ^{3}H-estradiol binding assays (6).

Available evidence indicates that the association of receptors with the nuclear matrix is a bona fide aspect of steroid hormone action, and is not a result of the high salt conditions used to isolate the nuclear matrix. As discussed above, receptors accumulate in the nuclear matrix in response to appropriate hormonal stimuli in vivo and this accumulation correlates with the stimulation of a biological response. In addition, as reported recently by Barrack (9, 10) cell-free reconstitution assays provide direct evidence that the nuclear matrix of animals from which hormones have been withdrawn (to deplete endogenous receptors) contains high affinity ($K_d = 10^{-10}$ M), saturable, and tissue-specific acceptor sites for steroid receptors (9, 10); moreover, these receptor-acceptor interactions occur in the absence of high-salt conditions. Thus, the binding of receptors to the nuclear matrix, which appears to contain the machinery for replication and transcription, is consistent with the known role of steroid receptor complexes in regulating many of these events. The accumulating evidence thus indicates a direct role for this receptor-acceptor-matrix interaction in regulating specific gene expression (see also discussion in previous reviews 10, 13, 14).

CHEMOMECHANICAL LINKAGES TO THE NUCLEAR MATRIX

The shape of a cell may be directly involved in determining cellular functions such as proliferation and differentiation. For example, Folkman & Moscona (69) could precisely control the shape of normal cells in vitro by varying the substratum adhesiveness of the culture plates to which the cells were attached. With appropriate control experiments, the authors concluded that cell shape was coupled to DNA synthesis and cell growth. They discussed their findings in relation to such phenomena as density-dependent inhibition of cell growth and anchorage dependence; these authors also discussed the possible relationship of cell shape to response to serum growth factors. Gospodarowicz et al (80) also observed that the mitogenic response of a given cell was determined by the cell shape. They observed that corneal epithelial cells adopted a flattened shape when maintained in vitro on plastic and were sensitive to fibroblast growth factor but not to epidermal growth factor. In contrast, when these cells were maintained on a layer of collagen they assumed a tall columnar form and responded to epidermal growth factor. With additional experiments, the authors concluded that the extracellular matrix upon which the cells rested determined the cell shape that in turn regulated the cell's proliferation properties (80). Simons et al (145) have found dramatic structural changes in nuclei from target lymphocytes treated with cyclosporin A or calmodulin inhibitors. Cause and effect have not been resolved in final detail; nevertheless, these types of experiments are generating what may become compelling evidence to support such structure-function relationships.

If cell shape does directly control cell replication and differentiation it is important to resolve the factors that determine and modulate cell shape and the mechanisms by which this information might be transmitted within the cell to control nuclear functions.

Cells contain extensive and elaborate three-dimensional skeletal networks that form integral structural components of the plasma membrane, the cytoplasm, and the nucleus (see Figure 4). If these matrix systems were interconnected and could undergo dynamic phase shifts in structure and conformation, e.g. by polymerization, depolymerization, crosslinking, biochemical modifications, or contractile movements, then one might visualize how a change in one matrix component could transmit and couple these changes to the rest of the system. Perhaps by such a mechanism externally applied signals might be transmitted along this communications network from one part of the cell to another. Alternatively, these interconnecting networks might help transport informational molecules to their effector sites. A third function of these matrix systems might be to act as "solid-state catalysts" or organizational support structures that could facilitate the interactions of molecules. Since cellular skeleton networks are composed of a number of different types of macromolecules, one could imagine a system in which all three types of interactions might be feasible. The intrinsic appeal of such an organization is the capacity for vectorial chemomechanical coupling as a means of signal transduction (see Figure 4).

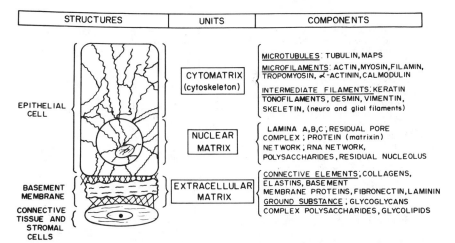

STRUCTURES	UNITS	COMPONENTS

Figure 4 Schematic of the overall tissue matrix system with a list of some of the major components. Reprinted with permission from Isaacs et al (93).

For further discussion of these concepts, see additional reviews (14, 29, 65, 70, 93, 124, 143).

The nuclear matrix is not a rigid static skeleton but is rather a dynamic and interactive component of the nucleus that orders the DNA and facilitates the formation of large enzymatic complexes and regulatory units that might control the topology and function of DNA. The dynamic state of the nucleus is very evident at the time of mitosis when visual phase shifts such as prophase and telophase occur in the nucleus. The interphase nucleus may also have such ordered arrangements that vary throughout the cell cycle. Studies using monoclonal antibodies to nuclear matrix proteins clearly indicate dramatic and dynamic changes occurring throughout the cell cycle (40). Much is left to be resolved about this complex nuclear structure, but it is probably time to leave behind the earlier concepts of soluble enzymes and to focus our attention on the more difficult problem of resolving the mechanisms by which the nuclear matrix controls DNA function.

ACKNOWLEDGMENTS

Work performed in the authors' laboratories was supported by NIH Grant AM 22000. WGN was supported by MSTP 5T32 GM 07309. We thank Mary Buedel and Ruth Middleton for their assistance in preparing this manuscript.

Literature Cited

1. Abulafia, R., Ben-Ze'ev, A., Hay, N., Aloni, Y. 1984. *J. Mol. Biol.* 172:467
2. Adolph, K. W. 1980. *Chromosoma* 76:23
3. Adolph, K. W., Kreisman, L. R. 1983. *Exp. Cell Res.* 147:155
4. Aelen, J. M., Opstelten, R. J., Wanka, F. 1983. *Nucleic Acids Res.* 11:1181
5. Agutter, P. S., Birchall, K. 1979. *Exp. Cell Res.* 124:453
6. Alexander, R. B., Greene, G. L., Barrack, E. R. 1986. *Endocrinology.* In press
7. Allen, S. L., Berezney, R., Coffey, D. S. 1977. *Biochem. Biophys. Res. Commun.* 75:111
8. Bak, A. L., Zeuthen, J., Crick, F. H. C. 1977. *Proc. Natl. Acad. Sci. USA* 74:1595
9. Barrack, E. R. 1983. *Endocrinology* 113:430
10. Barrack, E. R. 1986. In *The Nuclear Matrix*, ed. R. Berezney, New York: Plenum. In press
11. Barrack, E. R., Bujnovszky, P., Walsh, P. C. 1983. *Cancer Res.* 43:1107
12. Barrack, E. R., Coffey, D. S. 1980. *J. Biol. Chem.* 255:7265
13. Barrack, E. R., Coffey, D. S. 1982. *Recent Prog. Horm. Res.* 38:133
14. Barrack, E. R., Coffey, D. S. 1983. In *Gene Regulation by Steroid Hormones II*, ed. A. K. Roy, J. H. Clark, p. 239. New York: Springer-Verlag
15. Barrack, E. R., Hawkins, E. F., Allen, S. L., Hicks, L. L., Coffey, D. S. 1977. *Biochem. Biophys. Res. Commun.* 79:829
16. Barrack, E. R., Hawkins, E. F., Coffey, D. S. 1979. *Adv. Exp. Med. Biol.* 117:47
17. Bekers, A. G., Gijzen, H. J., Taalman, R. D. F. M., Wanka, F. 1981. *J. Ultrastruct. Res.* 75:352
18. Ben-Ze'ev, A., Aloni, Y. 1983. *Virology* 125:475
19. Berezney, R. 1984. In *Chromosomal Nonhistone Proteins*, Vol. 4, ed. L. S. Hnilica, p. 119. Boca Raton: CRC
20. Berezney, R., ed. 1986. *The Nuclear Matrix*. New York: Plenum. In press
21. Berezney, R., Buchholtz, L. A. 1981. *Exp. Cell Res.* 132:1
22. Berezney, R., Coffey, D. S. 1974. *Biochem. Biophys. Res. Commun.* 60:1410
23. Berezney, R., Coffey, D. S. 1975. *Science* 189:291
24. Berezney, R., Coffey, D. S. 1976. *Adv. Enzyme Regul.* 14:63
25. Berezney, R., Coffey, D. S. 1977. *J. Cell Biol.* 73:616
26. Berrios, M., Blobel, G., Fisher, P. A.

1983. *J. Biol. Chem.* 258:4548
27. Berrios, M., Filson, A. J., Blobel, G., Fisher, P. A. 1983. *J. Biol. Chem.* 258:13384
28. Berrios, M., Osheroff, N., Fisher, P. A. 1985. *Proc. Natl. Acad. Sci. USA* 82:4142
29. Bissell, M. J., Hall, H. G., Parry, G. 1982. *J. Theor. Biol.* 99:31
30. Bouteille, M., Bouvier, D., Seve, A. P. 1983. *Int. Rev. Cytol.* 83:135
31. Bouvier, D., Hubert, J., Seve, A.-P., Bouteille, M. 1985. *Eur. J. Cell Biol.* 36:323
32. Bouvier, D., Hubert, J., Seve, A.-P., Bouteille, M., Moens, P. B. 1984. *J. Ultrastruct. Res.* 87:112
32a. Brasch, K. 1982. *Exp. Cell Res.* 140:161
33. Brasch, K., Sinclair, G. D. 1978. *Virchows Arch. B* 27:193
34. Buckler-White, A. J., Humphrey, G. W., Pigiet, V. 1980. *Cell* 22:37
35. Buttyan, R., Olsson, C. A., Sheard, B., Kallos, J. 1983. *J. Biol. Chem.* 258:14366
36. Capco, D. G., Krochmalnic, G., Penman, S. 1984. *J. Cell Biol.* 98:1878
37. Capco, D. G., Penman, S. 1983. *J. Cell Biol.* 96:896
38. Capco, D. G., Wan, K. M., Penman, S. 1982. *Cell* 29:847
39. Cavazza, B., Irefiletti, V., Piolo, F., Ricci, E., Patrone, E. 1983. *J. Cell Sci.* 62:81
40. Chaly, N., Bladon, R., Setterfield, G., Little, J. E., Kaplan, J. G., Brown, D. L. 1984. *J. Cell Biol.* 99:661
41. Champoux, J. J. 1978. *Ann. Rev. Biochem.* 47:447
42. Deleted in proof
43. Ciejek, E. M., Nordstrom, J. L., Tsai, M. J., O'Malley, B. W. 1982. *Biochemistry* 21:4945
44. Ciejek, E. M., Tsai, M. J., O'Malley, B. W. 1983. *Nature* 306:607
45. Coffey, D. S. 1983. In *Genes and Proteins in Oncogenesis*, ed. I. B. Weinstein, H. J. Vogel, p. 339. New York: Academic
46. Colvard, D. S., Wilson, E. M. 1984. *Biochemistry* 23:3479
47. Cook, P. R., Brazell, I. A. 1980. *Nucleic Acids Res.* 8:2895
48. Cook, P. R., Brazell, I. A., Jost, E. 1976. *J. Cell Sci.* 22:303
49. Cook, P. R., Lang, J. 1984. *Nucleic Acids Res.* 12:1069
50. Cook, P. R., Lang, J., Hayday, A., Lania, L., Fried, M., et al. 1982. *EMBO J.* 1:447

51. Covey, L., Choi, Y., Prives, C. 1984. *Mol. Cell. Biol.* 4:1384
52. Deppert, W. 1978. *J. Virol.* 26:165
53. Detke, S., Keller, J. M. 1982. *J. Biol. Chem.* 257:3905
54. Diamond, D. A., Barrack, E. R. 1984. *J. Urol.* 132:821
55. Dijkwel, P. A., Mullenders, L. H. F., Wanka, F. 1979. *Nucleic Acids Res.* 6:219
56. Donnelly, B. J., Lakey, W. H., McBlain, W. A. 1984. *J. Urol.* 131:806
57. DuPraw, E. J. 1970. *DNA and Chromosomes.* New York: Holt, Rinehart & Winston
58. Earnshaw, W. C., Halligan, B., Cooke, C. A., Heck, M. M. S., Liu, L. F. 1985. *J. Cell Biol.* 100:1706
59. Earnshaw, W. C., Heck, M. M. S. 1985. *J. Cell Biol.* 100:1716
60. Earnshaw, W. C., Laemmli, U. K. 1983. *J. Cell Biol.* 96:84
61. Eisenman, R. N., Tachibana, C. Y., Abrams, H. D., Hann, S. R. 1985. *Mol. Cell. Biol.* 5:114
62. Engelhardt, P., Plagens, U., Zbarsky, I. B., Filatova, L. S. 1982. *Proc. Natl. Acad. Sci. USA* 79:6937
63. Epperly, M., Donofrio, J., Barham, S. S., Veneziale, C. M. 1984. *J. Steroid Biochem.* 20:691
64. Feinberg, A. P., Coffey, D. S. 1982. In *The Nuclear Envelope and the Nuclear Matrix*, ed. G. G. Maul, p. 293. New York: Liss
65. Fey, E. G., Capco, D. G., Krochmalnic, G., Penman, S. 1984. *J. Cell Biol.* 99:203s
66. Deleted in proof
67. Fey, E. G., Wan, K. M., Penman, S. 1984. *J. Cell Biol.* 98:1973
68. Finch, J. T., Klug, A. 1976. *Proc. Natl. Acad. Sci. USA* 73:1897
69. Folkman, J., Moscona, A. 1978. *Nature* 273:345
70. Fulton, A. B. 1984. *The Cytoskeleton: Cellular Architecture and Choreography.* New York: Chapman & Hall
71. Gallinaro, H., Puvion, E., Kister, L., Jacob, M. 1983. *EMBO J.* 2:953
72. Gellert, M. 1981. *Ann. Rev. Biochem.* 50:879
73. Georgiev, G. P., Nedspasov, S. A., Bakayev, V. U. 1978. In *The Cell Nucleus*, Vol. 6, ed. H. Busch, p. 3. New York: Academic
74. Gerace, L., Blobel, G. 1980. *Cell* 19:277
75. Gerace, L., Blum, A., Blobel, G. 1978. *J. Cell Biol.* 79:546
76. Gerace, L., Comeau, C., Benson, M. 1984. *J. Cell Sci.* 1:137 (Suppl.)
77. Ghosh, S., Paweletz, N., Ghosh, I. 1978. *Exp. Cell Res.* 111:363
78. Goldfischer, S., Kress, Y., Coltoff-Schiller, B., Berman, J. 1981. *J. Histochem. Cytochem.* 29:1105
79. Gonor, S. E., Lakey, W. H., McBlain, W. A. 1984. *J. Urol.* 131:1196
80. Gospodarowicz, D., Greenburg, G., Birdwell, C. R. 1978. *Cancer Res.* 38:4155
81. Goueli, S. A., Ahmed, K. 1984. *Arch. Biochem. Biophys.* 234:646
82. Haggis, G. H., Schweitzer, I., Hall, R., Bladon, T. 1983. *J. Microsc.* 132:185
83. Halligan, B. D., Small, D., Vogelstein, B., Hsieh, T.-S., Liu, L. F. 1984. *J. Cell Biol.* 99:128a
84. Hancock, R., Boulikas, T. 1982. *Int. Rev. Cytol.* 79:165
85. Henry, S. M., Hodge, L. D. 1983. *Eur. J. Biochem.* 133:23
86. Hentzen, P. C., Rho, J. H., Bekhor, I. 1984. *Proc. Natl. Acad. Sci. USA* 81:304
87. Herman, R., Weymouth, L., Penman, S. 1978. *J. Cell Biol.* 78:663
88. Hodge, L. D., Mancini, P., Davis, F. M., Heywood, P. 1977. *J. Cell Biol.* 72:194
89. Huberman, J. A., Riggs, A. D. 1968. *J. Mol. Biol.* 32:327
90. Hunt, B. F., Vogelstein, B. 1981. *Nucleic Acids Res.* 9:349
91. Ierardi, L. A., Moss, S. B., Bellve, A. R. 1983. *J. Cell Biol.* 96:1717
92. Igo-Kemenes, T., Horz, W., Zachau, M. G. 1982. *Ann. Rev. Biochem.* 51:89
93. Isaacs, J. T., Barrack, E. R., Isaacs, W. B., Coffey, D. S. 1981. In *The Prostatic Cell: Structure and Function*, Part A, ed. G. P. Murphy, A. A. Sandberg, J. P. Karr, p. 1. New York: Liss
94. Jackson, D. A., McCready, S. J., Cook, P. R. 1981. *Nature* 292:552
95. Jackson, D. A., McCready, S. J., Cook, P. R. 1984. *J. Cell Sci.* 1:59 (Suppl.)
96. Jost, E., Johnson, R. T. 1981. *J. Cell Sci.* 47:25
97. Jost, J. P., Seldran, M. 1984. *EMBO J.* 3:2005
98. Kaufmann, S. H., Coffey, D. S., Shaper, J. H. 1981. *Exp. Cell Res.* 132:105
99. Kaufmann, S. H., Shaper, J. H. 1984. *Exp. Cell Res.* 155:477
100. Kirov, N., Djondjurov, L., Tsanev, R. 1984. *J. Mol. Biol.* 180:601
101. Laemmli, U. K. 1979. *Pharmacol. Rev.* 30:469
102. Lafond, R. E., Woodcock, H., Woodcock, C. L., Kundahl, E. R., Lucas, J. J. 1983. *J. Cell Biol.* 96:1815
103. Lewis, C. D., Lebkowski, J. S., Daly, A. K., Laemmli, U. K. 1984. *J. Cell Sci.* 1:103 (Suppl.)
104. Liu, L. F. 1983. *CRC Crit. Rev. Biochem.* 15:1

105. Long, B. H., Huang, C.-Y., Pogo, A. O. 1979. *Cell* 18:1079
106. Mariman, E. C., van Eekelen, C. A. G., Reinders, R. J., Berns, A. J. M., van Venrooij, W. J. 1982. *J. Mol. Biol.* 154:103
107. Marsden, M. P. F., Laemmli, U. K. 1979. *Cell* 17:849
108. Maul, G. G., ed. 1982. *The Nuclear Envelope and the Nuclear Matrix*. New York: Liss
109. Maundrell, K., Maxwell, E. S., Puvion, E., Scherrer, K. 1981. *Exp. Cell Res.* 136:435
110. McCready, S. J., Godwin, J., Mason, D. W., Brazell, I. A., Cook, P. R. 1980. *J. Cell Sci.* 46:365
111. Miller, T. E., Huang, C.-Y., Pogo, A. O. 1978. *J. Cell Biol.* 76:692
112. Miller, T. E., Huang, C.-Y., Pogo, A. O. 1978. *J. Cell Biol.* 76:675
113. Mirkovitch, J., Mirault, M. E., Laemmli, U. K. 1984. *Cell* 39:223
114. Moelling, K., Benter, T., Bunte, T., Pfaff, E., Deppert, W., et al. 1984. *Curr. Top. Microbiol. Immunol.* 113:198
115. Muller, M., Spiess, E., Werner, D. 1983. *Eur. J. Cell Biol.* 31:158
116. Nakayasu, H., Ueda, K. 1983. *Exp. Cell Res.* 143:55
117. Nelkin, B. D., Pardoll, D. M., Vogelstein, B. 1980. *Nucleic Acids Res.* 8:5623
118. Nicolini, C. 1983. *Anticancer Res.* 3:63
119. Noguchi, H., Reddy, G. P. V., Pardee, A. B. 1983. *Cell* 32:443
120. Ottaviano, Y., Gerace, L. 1985. *J. Biol. Chem.* 260:624
121. Pardoll, D. M., Vogelstein, B. 1980. *Exp. Cell Res.* 128:466
122. Pardoll, D. M., Vogelstein, B., Coffey, D. S. 1980. *Cell* 19:527
123. Paulson, J. R., Laemmli, U. K. 1977. *Cell* 12:817
124. Penman, S., Fulton, A., Capco, D., Ben-Ze'ev, A., Wittelsberger, S., Tse, C. F. 1982. *Cold Spring Harbor Symp. Quant. Biol.* 46:1013
125. Peters, K. E., Comings, D. E. 1980. *J. Cell Biol.* 86:135
126. Peters, K. E., Okada, T. A., Comings, D. E. 1982. *Eur. J. Biochem.* 129:221
127. Pieck, A. C., van der Velden, H. M., Rijken, A. A., Neis, J. M., Wanka, F. 1985. *Chromosoma* 91:137
128. Pienta, K. J., Coffey, D. S. 1984. *J. Cell Sci.* 1:123 (Suppl.)
129. Pienta, K. J., Nelson, W. G., Coffey, D. S. 1986. See Ref. 20
130. Pietras, R. J., Szego, C. M. 1984. *Biochem. Biophys. Res. Commun.* 123:84
131. Potashkin, J. A., Zeigel, R. F., Huberman, J. A. 1984. *Exp. Cell Res.* 153:374

132. Pouchelet, M., St.-Pierre, E., Bibor-Hardy, V., Simard, R. 1983. *Exp. Cell Res.* 149:451
132a. Raveh, D., Ben-Ze'ev, A. 1984. *Exp. Cell Res.* 153:99
133. Reddy, G. P. V., Pardee, A. B. 1983. *Nature* 304:86
134. Rennie, P. S., Bruchovsky, N., Cheng, H. 1983. *J. Biol. Chem.* 258:7623
135. Riley, D. E., Keller, J. M. 1978. *J. Cell Sci.* 29:129
136. Robinson, S. I., Nelkin, B. D., Vogelstein, B. 1982. *Cell* 28:99
137. Robinson, S. I., Small, D., Idzerda, R., McKnight, G. S., Vogelstein, B. 1983. *Nucleic Acids Res.* 11:5113
138. Ross, D. A., Yen, R. W., Chae, C. B. 1982. *Biochemistry* 21:764
139. Sarnow, P., Hearing, O., Anderson, C. W., Reich, N., Levine, A. J. 1982. *J. Mol. Biol.* 162:565
140. Sato, C., Nishizawa, K., Yamaguchi, N. 1984. *Cell Struct. Funct.* 9:305
141. Schatten, G., Maul, G. G., Schatten, H., Chaly, N., Simerly, C., et al. 1985. *Proc. Natl. Acad. Sci. USA* 82:4727
142. Sedat, J., Manuelidis, L. 1977. *Cold Spring Harbor Symp. Quant. Biol.* 42:331
143. Shaper, J. H., Pardoll, D. M., Kaufmann, S. H., Barrack, E. R., Vogelstein, B., Coffey, D. S. 1979. *Adv. Enzyme Regul.* 17:213
144. Simmen, R. C., Means, A. R., Clark, J. H. 1984. *Endocrinology* 115:1197
145. Simons, J. W., Noga, S. J., Colombani, P. M., Beschorner, W. E., Coffey, D. S., Hess, A. D. 1985. *J. Cell Biol.* In press
146. Small, D. 1985. *The organization of DNA with respect to supercoiled loops and the nuclear matrix.* PhD dissertation. Johns Hopkins Univ., Baltimore, Md. 155 pp.
147. Small, D., Nelkin, B., Vogelstein, B. 1982. *Proc. Natl. Acad. Sci. USA* 79:5911
148. Small, D., Nelkin, B., Vogelstein, B. 1985. *Nucleic Acids Res.* 13:2413
149. Smith, H. C., Berezney, R. 1980. *Biochem. Biophys. Res. Commun.* 97:1541
150. Smith, H. C., Berezney, R. 1982. *Biochemistry* 21:6751
151. Smith, H. C., Berezney, R. 1983. *Biochemistry* 22:3042
152. Smith, H. C., Puvion, E., Buchholtz, L. A., Berezney, R. 1984. *J. Cell Biol.* 99:1794
153. Staufenbiel, M., Deppert, W. 1983. *Cell* 33:173
154. Staufenbiel, M., Deppert, W. 1983. *Eur. J. Cell Biol.* 31:341
155. Stick, R., Schwarz, H. 1983. *Cell* 33:949
156. Swaneck, G. E., Alvarez, J. M. 1985.

Biochem. Biophys. Res. Commun. 128:
   1381
157. Tsutsui, Y., Saga, S., Hoshino, M. 1984.
   Cell Struct. Funct. 9:345
158. Utsumi, K. R. 1981. Cell Struct. Funct.
   6:395
159. Valenzuela, M. S., Mueller, G. C., Das-
   gupta, S. 1983. Nucleic Acids Res. 11:
   2155
160. van der Velden, H. M. W., van Willigen,
   G., Wetzels, R. H. W., Wanka, F. 1984.
   FEBS Lett. 171:13
161. van Eekelen, C. A. G., Salden, M. H. L.,
   Habets, W. J. A., van de Putte, L. B. A.,
   van Venrooij, W. J. 1982. Exp. Cell Res.
   141:181
162. van Eekelen, C. A. G., van Venrooij, W.
   J. 1981. J. Cell Biol. 88:554

163. Verderame, M. F., Kohtz, D. S., Pollack, R. E. 1983. J. Virol. 46:575
164. Vogelstein, B., Hunt, B. F. 1982. Biochem. Biophys. Res. Commun. 105:1224
165. Vogelstein, B., Nelkin, B., Pardoll, D.,
   Hunt, B. F. 1982. See Ref. 64, p. 169
166. Vogelstein, B., Pardoll, D. M., Coffey,
   D. S. 1980. Cell 22:79
167. Wang, J. C. 1985. Ann. Rev. Biochem.
   54:665
168. Wilson, B. D., Albrecht, C. F., Wium, C.
   A. 1982. S. Afr. Med. J. 61:44
169. Wilson, E. M., Colvard, D. S. 1984. Ann.
   NY Acad. Sci. 438:85
170. Wolfe, J. 1980. J. Cell Biol. 84:160
171. Zbarsky, I. B. 1981. Mol. Biol. Rep. 7:
   139

SUBJECT INDEX

477

CUMULATIVE INDEXES

CONTRIBUTING AUTHORS, VOLUMES 11–15

CHAPTER TITLES, VOLUMES 11–15
INDEXED BY KEYWORD

491

Annual Reviews Inc.

A NONPROFIT SCIENTIFIC PUBLISHER

| ORDER FORM |

4139 El Camino Way, Palo Alto, CA 94306-9981, USA • (415) 493-4400

nnual Reviews Inc. publications are available directly from our office by mail or telephone (paid by credit card or
rchase order), through booksellers and subscription agents, worldwide, and through participating professional
cieties. Prices subject to change without notice.

Individuals: Prepayment required on new accounts by check or money order (in U.S. dollars, check drawn on
U.S. bank) or charge to credit card — American Express, VISA, MasterCard.
Institutional buyers: Please include purchase order number.
Students: $10.00 discount from retail price, per volume. Prepayment required. Proof of student status must be
provided (photocopy of student I.D. or signature of department secretary is acceptable). Students must send
orders direct to Annual Reviews. Orders received through bookstores and institutions requesting student rates
will be returned.
Professional Society Members: Members of professional societies that have a contractual arrangement with
Annual Reviews may order books through their society at a reduced rate. Check with your society for information.

egular orders: Please list the volumes you wish to order by volume number.
tanding orders: New volume in the series will be sent to you automatically each year upon publication. Cancel-
ion may be made at any time. Please indicate volume number to begin standing order.
republication orders: Volumes not yet published will be shipped in month and year indicated.
alifornia orders: Add applicable sales tax.
ostage paid (4th class bookrate/surface mail) **by Annual Reviews Inc.** Airmail postage extra.

ANNUAL REVIEWS SERIES		Prices Postpaid per volume USA/elsewhere	Regular Order Please send:	Standing Order Begin with:
			Vol. number	Vol. number

nnual Review of ANTHROPOLOGY (Prices of Volumes in brackets effective until 12/31/85)

[Vols. 1-10	(1972-1981).	$20.00/$21.00]		
[Vol. 11	(1982).	$22.00/$25.00]		
[Vols. 12-14	(1983-1985).	$27.00/$30.00]		
Vols. 1-14	(1972-1985).	$27.00/$30.00		
Vol. 15	(avail. Oct. 1986).	$31.00/$34.00	Vol(s). _____	Vol. _____

nnual Review of ASTRONOMY AND ASTROPHYSICS (Prices of Volumes in brackets effective until 12/31/85)

[Vols. 1-2, 4-19	(1963-1964; 1966-1981).	$20.00/$21.00]		
[Vol. 20	(1982).	$22.00/$25.00]		
[Vols. 21-23	(1983-1985).	$44.00/$47.00]		
Vols. 1-2, 4-20	(1963-1964; 1966-1982).	$27.00/$30.00		
Vols. 21-23	(1983-1985).	$44.00/$47.00		
Vol. 24	(avail. Sept. 1986).	$44.00/$47.00	Vol(s). _____	Vol. _____

nnual Review of BIOCHEMISTRY (Prices of Volumes in brackets effective until 12/31/85)

[Vols. 30-34, 36-50	(1961-1965; 1967-1981).	$21.00/$22.00]		
[Vol. 51	(1982).	$23.00/$26.00]		
[Vols. 52-54	(1983-1985).	$29.00/$32.00]		
Vols. 30-34, 36-54	(1961-1965; 1967-1985).	$29.00/$32.00		
Vol. 55	(avail. July 1986).	$33.00/$36.00	Vol(s). _____	Vol. _____

nnual Review of BIOPHYSICS AND BIOPHYSICAL CHEMISTRY (Prices of Vols. in brackets effective until 12/31/85)
(Formerly Annual Review of Biophysics and Bioengineering)

[Vols. 1-10	(1972-1981).	$20.00/$21.00]		
[Vol. 11	(1982).	$22.00/$25.00]		
[Vols. 12-14	(1983-1985).	$47.00/$50.00]		
Vols. 1-11	(1972-1982).	$27.00/$30.00		
Vols. 12-14	(1983-1985).	$47.00/$50.00		
Vol. 15	(avail. June 1986).	$47.00/$50.00	Vol(s). _____	Vol. _____

nnual Review of CELL BIOLOGY

Vol. 1	(1985).	$27.00/$30.00		
Vol. 2	(avail. Nov. 1986).	$31.00/$34.00	Vol(s). _____	Vol. _____

nnual Review of COMPUTER SCIENCE

Vol. 1	(avail. late 1986).	**Price not yet established**	Vol. _____	Vol. _____

nnual Review of EARTH AND PLANETARY SCIENCES (Prices of Volumes in brackets effective until 12/31/85)

[Vols. 1-9	(1973-1981).	$20.00/$21.00]		
[Vol. 10	(1982).	$22.00/$25.00]		
[Vols. 11-13	(1983-1985).	$44.00/$47.00]		
Vols. 1-10	(1973-1982).	$27.00/$30.00		
Vols. 11-13	(1983-1985).	$44.00/$47.00		
Vol. 14	(avail. May 1986).	$44.00/$47.00	Vol(s). _____	Vol. _____

ANNUAL REVIEWS SERIES	Prices Postpaid per volume USA/elsewhere	Regular Order Please send:	Standing Order Begin with:

Annual Review of **ECOLOGY AND SYSTEMATICS** (Prices of Volumes in brackets effective until 12/31/85)

[Vols. 1-12	(1970-1981) . . . . $20.00/$21.00]	
[Vol. 13	(1982) . . . . $22.00/$25.00]	
[Vols. 14-16	(1983-1985) . . . . $27.00/$30.00]	
Vols. 1-16	(1970-1985) . . . . $27.00/$30.00	
Vol. 17	(avail. Nov. 1986) . . . . $31.00/$34.00	Vol(s). _____ Vol. _____

Annual Review of **ENERGY** (Prices of Volumes in brackets effective until 12/31/85)

[Vols. 1-6	(1976-1981) . . . . $20.00/$21.00]	
[Vol. 7	(1982) . . . . $22.00/$25.00]	
[Vols. 8-10	(1983-1985) . . . . $56.00/$59.00]	
Vols. 1-7	(1976-1982) . . . . $27.00/$30.00	
Vols. 8-10	(1983-1985) . . . . $56.00/$59.00	
Vol. 11	(avail. Oct. 1986) . . . . $56.00/$59.00	Vol(s). _____ Vol. _____

Annual Review of **ENTOMOLOGY** (Prices of Volumes in brackets effective until 12/31/85)

[Vols. 9-16, 18-26	(1964-1971; 1973-1981) . . . . $20.00/$21.00]	
[Vol. 27	(1982) . . . . $22.00/$25.00]	
[Vols. 28-30	(1983-1985) . . . . $27.00/$30.00]	
Vols. 9-16, 18-30	(1964-1971; 1973-1985) . . . . $27.00/$30.00	
Vol. 31	(avail. Jan. 1986) . . . . $31.00/$34.00	Vol(s). _____ Vol. _____

Annual Review of **FLUID MECHANICS** (Prices of Volumes in brackets effective until 12/31/85)

[Vols. 1-5, 7-13	(1969-1973; 1975-1981) . . . . $20.00/$21.00]	
[Vol. 14	(1982) . . . . $22.00/$25.00]	
[Vols. 15-17	(1983-1985) . . . . $28.00/$31.00]	
Vols. 1-5, 7-17	(1969-1973; 1975-1985) . . . . $28.00/$31.00	
Vol. 18	(avail. Jan. 1986) . . . . $32.00/$35.00	Vol(s). _____ Vol. _____

Annual Review of **GENETICS** (Prices of Volumes in brackets effective until 12/31/85)

[Vols. 1-15	(1967-1981) . . . . $20.00/$21.00]	
[Vol. 16	(1982) . . . . $22.00/$25.00]	
[Vols. 17-19	(1983-1985) . . . . $27.00/$30.00]	
Vols. 1-19	(1967-1985) . . . . $27.00/$30.00	
Vol. 20	(avail. Dec. 1986) . . . . $31.00/$34.00	Vol(s). _____ Vol. _____

Annual Review of **IMMUNOLOGY**

Vols. 1-3	(1983-1985) . . . . $27.00/$30.00	
Vol. 4	(avail. April 1986) . . . . $31.00/$34.00	Vol(s). _____ Vol. _____

Annual Review of **MATERIALS SCIENCE** (Prices of Volumes in brackets effective until 12/31/85)

[Vols. 1-11	(1971-1981) . . . . $20.00/$21.00]	
[Vol. 12	(1982) . . . . $22.00/$25.00]	
[Vols. 13-15	(1983-1985) . . . . $64.00/$67.00]	
Vols. 1-12	(1971-1982) . . . . $27.00/$30.00	
Vols. 13-15	(1983-1985) . . . . $64.00/$67.00	
Vol. 16	(avail. August 1986) . . . . $64.00/$67.00	Vol(s). _____ Vol. _____

Annual Review of **MEDICINE** (Prices of Volumes in brackets effective until 12/31/85)

[Vols. 1-3, 5-15, 17-32	(1950-52; 1954-64; 1966-81) . . . . . . $20.00/$21.00]	
[Vol. 33	(1982) . . . . $22.00/$25.00]	
[Vols. 34-36	(1983-1985) . . . . $27.00/$30.00]	
Vols. 1-3, 5-15, 17-36	(1950-52; 1954-64; 1966-85) . . . . . . $27.00/$30.00	
Vol. 37	(avail. April 1986) . . . . $31.00/$34.00	Vol(s). _____ Vol. _____

Annual Review of **MICROBIOLOGY** (Prices of Volumes in brackets effective until 12/31/85)

[Vols. 18-35	(1964-1981) . . . . $20.00/$21.00]	
[Vol. 36	(1982) . . . . $22.00/$25.00]	
[Vols. 37-39	(1983-1985) . . . . $27.00/$30.00]	
Vols. 18-39	(1964-1985) . . . . $27.00/$30.00	
Vol. 40	(avail. Oct. 1986) . . . . $31.00/$34.00	Vol(s). _____ Vol. _____

Annual Review of **NEUROSCIENCE** (Prices of Volumes in brackets effective until 12/31/85)

[Vols. 1-4	(1978-1981) . . . . $20.00/$21.00]	
[Vol. 5	(1982) . . . . $22.00/$25.00]	
[Vols. 6-8	(1983-1985) . . . . $27.00/$30.00]	
Vols. 1-8	(1978-1985) . . . . $27.00/$30.00	
Vol. 9	(avail. March 1986) . . . . $31.00/$34.00	Vol(s). _____ Vol. _____